Multiscale, Nonlinear and Adaptive Approximation II

Ronald DeVore • Angela Kunoth
Editors

Multiscale, Nonlinear and Adaptive Approximation II

Springer

Editors
Ronald DeVore
Department of Mathematics
Texas A&M University
College Station, TX, USA

Angela Kunoth
Department of Mathematics and Computer Science, Division of Mathematics
University of Cologne
Cologne, Germany

ISBN 978-3-031-75801-0 ISBN 978-3-031-75802-7 (eBook)
https://doi.org/10.1007/978-3-031-75802-7

Mathematics Subject Classification (2020): 65, 41

This Springer imprint is published by the registered company Springer Nature Switzerland AG
The registered company address is: Gewerbestrasse 11, 6330 Cham, Switzerland

Dedicated to Wolfgang Dahmen on the Occasion of his 75th Birthday

Preface

On the occasion of his 75th birthday in October 2024, friends, collaborators, and admirers of Wolfgang Dahmen have organized this volume which touches on various of his research interests. As its predecessor "Multiscale, Nonlinear and Adaptive Approximation", published in 2009, this volume will provide an excursion into many important topics in applied and computational mathematics. These include nonlinear, sparse or adaptive approximation, multiscale, wavelet and reduced basis methods, numerical schemes for partial differential integrations including parameters or randomness, learning theory, high-dimensional integrals and tensor networks, and electron microscopy.

College Station, Texas, USA
Cologne, Germany
September 2024

Ronald A. DeVore
Angela Kunoth

Acknowledgements

We are deeply grateful to Francesca Ferrari, Dr. Remi Lodh and Casey Russell from Springer for realizing this book project, and Tolunay Yilmaz for proofreading.

Contents

List of Contributors

Markus Bachmayr
Institut für Geometrie und Praktische Mathematik, RWTH Aachen University, Templergraben 55, 52062 Aachen, Germany, e-mail: bachmayr@igpm.rwth-aachen.de

Hassan Ballout
Cemosis, IRMA UMR 7501, University of Strasbourg, CNRS, e-mail: hassan.ballout@etu.unistra.fr

Ido Ben Shaul
School of Mathematical Sciences, Tel Aviv University, e-mail: ido.benshaul@gmail.com

Benjamin Berkels
Aachen Institute for Advanced Study in Computational Engineering Science (AICES), RWTH Aachen University, Germany, e-mail: berkels@aices.rwth-aachen.de

Peter Binev
University of South Carolina, Columbia, SC 29208, USA, e-mail: binev@math.sc.edu

Andrea Bonito
Department of Mathematics, Texas A&M University, College Station, TX 77845, USA, e-mail: bonito@tamu.edu

Dietrich Braess
Mathematisches Institut, Ruhr-Universität Bochum, 44780 Bochum, Germany, e-mail: Dietrich.Braess@rub.de

Albert Cohen
Laboratoire Jacques-Louis Lions, Sorbonne Université, 4 place Jussieu, 75005 Paris, France, e-mail: albert.cohen@sorbonne-universite.fr

Jan-Christopher Cohrs
Aachen Institute for Advanced Study in Computational Engineering Science (AICES), RWTH Aachen University, Germany, e-mail: cohrs@aices.rwth-aachen.de

Shai Dekel
School of Mathematical Sciences, Tel Aviv University, e-mail: shaidekel6@gmail.com

Ronald DeVore
Department of Mathematics, Texas A&M University, College Station, TX 77840, USA, e-mail: rdevore@math.tamu.edu

Lars Grasedyck
Institut für Geometrie und Praktische Mathematik, RWTH Aachen University, Templergraben 55, 52056 Aachen, Germany, e-mail: lgr@igpm.rwth-aachen.de

Diane Guignard
Department of Mathematics and Statistics, University of Ottawa, Ottawa, ON K1N 6N5, Canada, e-mail: dguignar@uottawa.ca

Wolfgang Hackbusch
Max-Planck-Institut *Mathematik in den Naturwissenschaften*, Inselstr. 22, 04103 Leipzig, Germany, e-mail: wh@mis.mpg.de

Helmut Harbrecht
Departement Mathematik und Informatik, Universität Basel, Spiegelgasse 1, 4051 Basel, Schweiz, e-mail: helmut.harbrecht@unibas.ch

Michael Herty
Institut für Geometrie und Praktische Mathematik, RWTH Aachen University, Templergraben 55, D-52056 Aachen, Germany, e-mail: herty@igpm.rwth-aachen.de

Kamen G. Ivanov
Institute of Mathematics and Informatics, Bulgarian Academy of Sciences, Sofia, Bulgaria, e-mail: kamen@math.bas.bg

Gerard Kerkyacharian
University Paris Diderot-Paris 7, LPMA, Paris, France, e-mail: kerk@math.univ-paris-diderot.fr

Adrian Kolb
Institut für Geometrie und Praktische Mathematik, RWTH Aachen University, Templergraben 55, D-52056 Aachen, Germany, e-mail: kolb@eddy.rwth-aachen.de

Sebastian Krämer
Institut für Geometrie und Praktische Mathematik, RWTH Aachen University, Templergraben 55, 52056 Aachen, Germany, e-mail: kraemer@igpm.rwth-aachen.de

Angela Kunoth
Department of Mathematics and Computer Science, Division of Mathematics, University of Cologne, Weyertal 86–90, 50931 Cologne, Germany, e-mail: akunoth@uni-koeln.de

George Kyriazis
Department of Mathematics and Statistics, University of Cyprus, 1678 Nicosia, Cyprus, e-mail: kyriazis@ucy.ac.cy

Yvon Maday
Sorbonne Université, CNRS, Université Paris Cité, Laboratoire Jacques-Louis Lions (LJLL), F-75005 Paris, France e-mail: yvon.maday@sorbonne-universite.fr

Dieter Moser
Institut für Geometrie und Praktische Mathematik, RWTH Aachen University, Templergraben 55, 52056 Aachen, Germany, e-mail: moser@igpm.rwth-aachen.de

Olga Mula
Eindhoven University of Technology, Den Dolech 2, P.O. Box 513, 5600 Eindhoven, NL, e-mail: o.mula@tue.nl

Siegfried Müller
Institut für Geometrie und Praktische Mathematik, RWTH Aachen University, Templergraben 55, D-52056 Aachen, Germany, e-mail: mueller@igpm.rwth-aachen.de

Michael Multerer
Istituto Eulero, Università della Svizzera italiana, Via la Santa 1, 6962 Lugano, Svizzera, e-mail: michael.multerer@usi.ch

Mathias Oster
RWTH Aachen, Templergraben 55, 52062 Aachen, Germany, e-mail: oster@igpm.rwth-aachen.de

Guergana Petrova
Department of Mathematics, Texas A& M University, College Station, TX 77843, USA, e-mail: gpetrova@tamu.edu

Pencho Petrushev
Department of Mathematics, University of South Carolina, Columbia, SC 29208, USA, e-mail: pencho@math.sc.edu

Dominique Picard
Université Paris-Cité, Laboratoire de Probabilités, Statistique et Modélisation (LPSM, UMR 8001) Bâtiment Sophie Germain, 8 place Aurélie Nemours 75013 Paris, e-mail: dominiquepicard@gmail.com

Christophe Prud'homme
Cemosis, IRMA UMR 7501, Université de Strasbourg, CNRS, e-mail: prudhomme@unistra.fr

Reinhold Schneider
Department of Mathematics, TU Berlin, Str. des 17. Junis 135, 10623 Berlin, Germany, e-mail: schneidr@math.tu-berlin.de

Rob Stevenson
Korteweg-de Vries (KdV) Institute for Mathematics, University of Amsterdam, P.O. Box 94248, 1090 GE Amsterdam, The Netherlands. e-mail: rob.p.stevenson@gmail.com

Eitan Tadmor
Department of Mathematics, Institute for Physical sciences and Technology (IPST), University of Maryland, MD 20742-4015, USA, e-mail: tadmor@umd.edu

Przemysław Wojtaszczyk
Institut of Mathematics, Polish Academy of Sciences, ul. Sniadeckich 8, 00-656, Warsawa, Poland, e-mail: Wojtaszczyk@impan.pl

Prologue to Multiscale, Nonlinear and Adaptive Approximation II

Ronald A. DeVore and Angela Kunoth

This volume is dedicated to our dear friend and colleague Wolfgang Dahmen on the occasion of his 75th birthday. To a great extent, it is a continuation of the publication **Multiscale, Nonlinear and Adaptive Approximation** which was published by Springer Verlag fifteen years ago in honor of Wolfgang's 60th birthday. As such, we found it appropriate to include (below) the introduction to that volume which provided, in some detail, the enormous contributions of Wolfgang to the subjects of Approximation Theory, Numerical Computation, and Engineering Applications.

The pointed contributions of Wolfgang to these areas have not slowed at all in the interim. He has continued to make major advances to numerical methods and their application to the pressing numerical challenges of this century which include learning from data, the construction of efficient platforms for numerical methods, and targeted applications such as his extensive analysis of electron microscopy. We briefly comment on the impact of Wolfgang's research in these areas.

Note: The updated **references** are at the end of that introduction in the next chapter.

1 Learning from data

Certainly one of the dominant themes in numerical analysis during the last decade has been the problem of approximating an unknown function f from data observations. This is the core of artificial intelligence and the numerical process of deep learning.

Ronald A. DeVore
Department of Mathematics, Texas A&M University, College Station, TX 77843, USA, e-mail: rdevore@tamu.edu

Angela Kunoth
Department of Mathematics and Computer Science, Division of Mathematics, University of Cologne, Weyertal 86–90, 50931 Cologne, Germany, e-mail: akunoth@uni-koeln.de

R. DeVore, A. Kunoth (eds.), *Multiscale, Nonlinear and Adaptive Approximation II*,
https://doi.org/10.1007/978-3-031-75802-7_1

Wolfgang already had a footprint in learning with the papers [160, 166, 168] which analyzed the performance of linear and nonlinear methods. This included adaptive piecewise polynomial approximation. The goal of that work was not only to describe how adaptivity could be employed in numerical algorithms for learning but also to obtain rigorous a priori error bounds for the performance of adaptive learning algorithms. This was accomplished for methods described by low dimensional polynomial approximation on adaptively generated partitions.

Quantitative bounds for learning rates (accuracy vs. the number of observations) are obtained under model class assumptions on the function f to be learned. The novelty in [202, 213] was to employ model class assumptions that stipulate how well f can be approximated by known linear spaces V_n of dimension n. This paper uncovered the importance of the geometry between V_n and the space spanned by the measurement functionals. Optimal algorithmic performance is described in terms of the angle between these two spaces.

One of the primary applications of learning is to enhance the efficiency of computing solutions to PDEs by using data observations of the solution u. These methods are commonly referred to as Data Assimilation in the numerical PDE community. The papers [202, 211, 215] give quantitative bounds on the optimal error that can be achieved in terms of the number m of observations. Wolfgang's research in this arena would later impact his work on reduced models and electron microscopy.

2 Novel numerical platforms

Given the task to recover an unknown function f, one builds a numerical algorithm through some form of approximation such as finite elements, wavelets, or more general linear or nonlinear spaces. Here the challenge is to have the form of approximation match the the stipulated application setting. Much of Wolfgang's work centers on quantifying the optimal performance that a given platform of approximation yields in specified applications. His work on finite element and wavelet methods in this area is well known and highly cited. There remains a significant challenge to uncover the optimal approximation method to apply in several important application areas. These include linear and nonlinear transport, parametric PDEs, and PDEs involving many variables. Wolfgang's research has been at the heart of understanding novel approximation schemes to address these challenges.

Following the pioneering work of Maday, Patera and their collaborators, the problem arose as to which approximation method would be best suited for recovering the solutions to a parametric family of PDEs. Wolfgang and his collaborators investigated this problem from various vantage points. The highly cited papers [181, 208, 213] studied how to construct an optimal linear space for resolving parametric PDEs. It is shown that a certain greedy algorithm selects an optimal linear space for building the numerical recovery algorithm. It is known that nonlinear methods of approximation typically beat linear methods in performance and hence should be used in building numerical algorithms. Because of the complexity of the solution manifold, it remains a challenge to understand how nonlinear methods should be built for recovering the solution to parametric PDEs. The publication [214] proposes and analyzes a first nonlinear approach based on approximation from a library of linear spaces.

A promising approach to numerically resolving high-dimensional PDEs has been through tensor constructions. From the viewpoint of approximation theory, the main challenge is to understand which functions can be well approximated by low-rank tensors, and how to realize such highly nonlinear approximations computationally. The paper [201] shows that solutions to certain high-dimensional elliptic PDEs have rapidly convergent approximations by sums of elementary tensors that can be employed in the numerical recovery of the solution. In a related line of work, solvers for more general elliptic problems based on hierarchical tensor representations producing near-optimal tensor ranks were proposed and analyzed in the series of papers [195, 199, 200, 208].

A recurring theme in Wolfgang's work has been to understand how best to numerically capture the solution to transport-dominated PDEs. Together with his collaborators, he has constantly returned to this research area. A main issue to be resolved is the design of stable variational formulations for problems with solutions that can exhibit shocks or boundary layers. He made progress on variational formulations of first-order transport equations in the papers [183, 205] and for convection-diffusion equations in [182]. This work led to new results on adaptive methods. The paper [206] is based on anisotropic approximation schemes, whereas in [223], error reduction for an adaptive scheme for a linear transport equation is established based on a discontinuous Petrov-Galerkin (DPG) discretization. Wolfgang also used a related DPG approach in new methods for the radiative transfer equation [210, 221].

3 Electron microscopy

Wolfgang has always maintained an eye towards applications in his research. A large portion of his research on applications centered on electron microscopy. He was exposed to the numerical challenges in this area during his time at the University of South Carolina. His approach has been a synergetic synthesis of "model-based" and "data-driven" methodologies. He was able to show how his novel research in compressed sensing and learning from data successfully improve existing numerical methods in electron microscopy. The papers [175, 176, 177, 178, 194] provided theoretical analysis which helped to shape new generations of numerical methods in this field including sparse recovery and denoising. The developed approach of non-rigid registration of series of low-dose aberration-corrected Z-contrast scanning transmission electron microscopy images [184] resulted in unprecedented sub-picometre precision measurements of atom positions [185, 193] that allow for the identification of previously undetectable atomic shifts in materials. This has an impact on the advancement of the studies of new materials by better estimating the atomic compositions in special configurations like interfaces between two different materials or defects in crystalline structures.

Acknowledgements We thank Markus Bachmayr and Peter Binev for several valuable additions to this prologue.

Introduction: Wolfgang Dahmen's mathematical work (as of 2009)

Ronald A. DeVore and Angela Kunoth

Abstract This volume is testimony to the rich and amazingly diverse mathematical life of Wolfgang Dahmen. The cornerstones of Wolfgang's research are deep theoretical analysis and extensive interdisciplinary projects in high-performance scientific computing. This article touches on some of the highlights of his work and its impact in application domains.

1 Introduction

Wolfgang Dahmen is at the peak of his mathematical career — witness his recent forays into learning theory, compressed sensing, and high-dimensional problems. His accomplishments to date have unquestionable diversity and depth. Perhaps the two characteristics that best identify his mathematics are the constant exploration of new and emerging fields as well as the quest for the relevancy of his work to application domains. The subsequent contributions to this volume will certainly validate this view. We therefore take the opportunity in this introductory article to give some overall feeling on how this all came about.

2 The early years: Classical approximation theory

Wolfgang grew up in a small village called Linnich some forty kilometers from Aachen tucked into the northwest corner of Germany near the Dutch border. So it was natural for him to pursue his university studies at the RWTH in Aachen. He was quickly identified as a mathematical talent and offered a Research Assistantship in the Paul Leo Butzer program in Harmonic Analysis and Approximation — a group notable for its breadth and talent. He began investigating problems in classical approximation theory and set out to settle the conjecture of Golumb and Korovkin

R. DeVore, A. Kunoth (eds.), *Multiscale, Nonlinear and Adaptive Approximation II*,
https://doi.org/10.1007/978-3-031-75802-7_2

on whether convolution operators could simultaneously realize best approximation orders for the entire spectrum of smoothness classes. He disproved this conjecture in [1, 2] and these results became the core of his dissertation thesis. He went on to give general connections between the behaviour of norms of operators, asymptotically best approximation for a particular range, and spectral convergence [3, 4, 5, 6, 7]. He followed this by using these techniques to prove in [8] a slightly corrected conjecture of Stechkin. These results would already establish him as a prominent young analyst.

3 Bonn, Bielefeld, Berlin, and multivariate splines

Wolfgang took an assistant position with Karl Scherer at Universität Bonn in 1976. Bonn was at this time a hub of activity in approximation theory among the faculty and assistants as well as the steady stream of visitors who frequented the Institut für Angewandte Mathematik. It was because of some of this new exposure that Wolfgang Dahmen switched to the newly developing field of multivariate splines promoted by Carl de Boor. At that time there was not much known about such smooth piecewise polynomials in more than one spatial dimension, apart from classical finite element approaches. Although many engineering sciences built their approximations on splines, theoretical results for genuine multivariate analogues were not known.

The starting point for Wolfgang's work was a suggestion by Carl de Boor to define multivariate splines through a generalization of a geometrical interpretation of univariate B-Splines, going back to Iso Schoenberg. This eventually led to the development of a cohesive and deep theory that is well documented in the article of Carl de Boor in this volume.

There were several obvious gains that emanated from Wolfgang's entry into multivariate splines. One was the exposure to the interconnectivity of various branches of mathematics. In this case, multivariate splines intersect with commutative algebra, combinatorics, number theory, and geometry, in addition to the obvious connections to finite element methods. Another big plus for Wolfgang was his collaboration with Charles A. Micchelli. Wolfgang received an IBM PostDoctoral Fellowship for 1979/1980 which was the starting point for the very fruitful and long-term Dahmen–Micchelli collaboration.

Wolfgang returned to Germany and took his first professor position in Bielefeld in 1981. His work on multivariate splines was now at its peak. Multivariate B-splines are generated from volumes of intersections of hyperplanes with simplices. While they have many beautiful properties, they are not so easy to work with numerically. So they were soon replaced by box splines which are defined by replacing the simplex by a cube. Although box splines result in uniform grid structures, the theory for box splines needed very different techniques [28, 29, 31, 32, 38].

This box spline theory involved a number of interesting algebraic and combinatoric problems [33, 34, 37, 38]. Indeed, box-spline techniques had very surprising applications to questions of combinatorial number theory like reciprocal relations, Bell's theorem and linear systems of diophantic equations [41, 43, 65, 66, 68], culminating with far-reaching results in commutative algebra by Dahmen–Micchelli and by Rong-Qing Jia. Finally, in the long paper [67] together with Andreas Dress and Charles Micchelli, Wolfgang investigated the application of concepts from homological algebra for the treatment of the central problem of the determination of the exact dimension of the intersection of the null space of a family of endomorphisms which are characterized by a certain combinatorial structure.

3.1 Computer aided geometric design

The Dahmen–Micchelli collaboration also had lasting impact in Computer Aided Geometric Design (CAGD) [42]. This work began while in Bielefeld but continued well during his stay at the Freie Universität in Berlin (1987-1992). A number of their papers in this area treat data fitting and interpolation problems [49, 50, 51].

Univariate splines have important properties like being variation-diminishing which is strongly connected to the concept of total positivity. In its full strength, the latter is inherently one-dimensional. However, there exists a strong coupling to Polya frequency functions which can in turn be interpreted in the multivariate case as well. This viewpoint resulted in a number of partly function-theoretic-based investigations [10, 20] as well as results concerning biinfinite matrices [18, 44, 76]. Roughly speaking, a central result in this context is that totally positive matrices (which appear, for instance, for splines, Tchebycheff systems, or in the theory of small oscillations) can be completely factorized into totally positive 2-band matrices. This in turn has a number of important consequences, like corner cutting in CAGD, knot insertion by repeated convex combinations of knots, and variation-diminishing properties of B-splines.

A relevant property for data fitting is shape preservation, where the goal is to preserve properties like monotonicity or convexity [53, 56, 59, 60]. Moreover, a completely new approach for modelling surfaces with smooth interfaces based on piecewise algebraic surfaces was proposed in [54, 57, 59].

The motivation for several of Wolfgang's investigations was the premise that parametric representations of surfaces are very unsuitable for graphics of high quality. While in Berlin, Wolfgang had a close collaboration with the nearby company Mental Images founded in 1986. During this collaboration, the intent was to develop a modelling library which was able to reduce all CAD formats used at that time to a unified format and to which after polygonalization a highly efficient ray tracer could be applied. Since algebraic representations would also support the ray tracer, one could circumvent polygonalization and the handling of different data formats. The prototype of such a library has been developed and implemented together with Mental Images and has in the aftermath triggered a number of related projects. Nowadays, their website states that "Mental Images is the recognized international leader in providing rendering and 3D modeling technology to the entertainment, computer-aided design, scientific visualization, architecture, and other industries that require sophisticated images".

3.2 Subdivision and wavelets

An integral part of CAGD is played by subdivision algorithms for quickly generating, displaying and controlling geometrical surfaces. Here Wolfgang made major contributions [37, 39, 45, 46, 48] culminating in the monograph [62], written together with Alfred Cavaretta and Charles Micchelli and published in the Memoirs of the American Mathematical Society.

CAGD and subdivision in particular were a forerunner of the development of wavelets which dominated the applied harmonic analysis and image processing communities in the 80's and 90's. So it was natural that the emphasis of Wolfgang Dahmen's work at the beginning of the 90's became the theory of multiscale analysis with a particular eye to the application of these new sophisticated tools to the numerics of partial differential and integral equations. From 1992 on, this was fueled by the fact that his new position at RWTH Aachen brought ample research opportunities with engineers.

Another important factor in this new direction was his exposure to nonlinear approximation during the Bonn years. Nonlinear spline approximation was a favorite topic of Karl Scherer. Two of the main proponents of nonlinear methods were Dietrich Braess and Ronald A. DeVore who were frequent visitors to Bonn during the 70's. In fact, Dahmen and DeVore had several collaborations through the years but this became particularly intense with the wavelet evolution when they teamed up with the young star Albert Cohen. The impact of their work would be felt in many applied disciplines including image compression, compressed sensing, learning theory, and numerical partial differential equations (PDEs).

4 Wavelet and multiscale methods for operator equations

4.1 Multilevel preconditioning

Wolfgang's work in PDEs took on great momentum at the end of the 80's. His first major contributions centered around preconditioning elliptic operators. At that time, multigrid methods were known to provide fast numerical solvers for the system of linear equations arising from finite element discretization. A much-discussed question from the point of finite elements at the end of the 80's was under which conditions multilevel preconditioners such as the hierarchical and the BPX preconditioner (proposed by James H. Bramble, Joseph E. Pasciak, and Jinchao Xu) provided uniformly bounded spectral condition numbers for the system matrix. This is an essential point to guarantee fast iterative solvers. Corresponding conditions were derived in [75] and, specifically, the uniform boundedness of the BPX preconditioner was established; a result independently obtained by Peter Oswald. From the angle of wavelets, the corresponding ingredients for optimal preconditioning became much more transparent [91, 92]. In particular, in [75] uniform boundedness was established for the first time also for adaptively refined grids in a classical finite element framework. Two articles on preconditioning systems of PDEs by multilevel ingredients are collected in this volume: Angela Kunoth's on optimized preconditioning with wavelets, and the survey by Long Chen, Ricardo Nochetto, and Jinchao Xu on BPX and multigrid preconditioning.

Wavelet-based preconditioners directly give norm equivalences for functions in Sobolev spaces and are therefore not restricted to operators of positive order. This led Wolfgang and his collaborators to a number of investigations for pseudo-differential and boundary integral operators for whose compression of operators is an even more relevant issue, see Section 4.2.

Within the last few years, discontinuous Galerkin methods for PDEs have become increasingly popular, due to the fact that one can relatively easily increase the polynomial degree and therefore ensure higher order convergence where solutions are smooth. Very recently, multilevel preconditioners for such interior discontinuous Galerkin methods have been presented for the first time in [171, 172].

These studies led to two well-received extensive surveys on wavelet and multiscale methods for the numerical solution of partial differential equations in an Acta Numerica article [108] and, a few years later, [126].

4.2 Compression of operators

The goal of the multiscale approach to operator equations is to extract efficient representations for both the operator and the solution. For boundary integral formulations, conventional discretizations yield fully populated matrices and were therefore not feasible for 3D problems in reasonable time. Multipole expansions and panel clustering methods became very popular as a way of thinning out the system matrix and making it more amenable to computation.

Multiscale wavelet methods arrived on the scene a little later spurred on by the observations of Gregory Beylkin, Ronald Coifman, and Vladimir Rokhlin that certain (one-dimensional) operators are almost sparse in a wavelet representation. These authors showed that for a fixed tolerance ε a matrix vector multiplication can be realized in $O(N \log N)$ arithmetic operations with accuracy ε. Wolfgang and his collaborators, mainly Siegfried Prößdorf und Reinhold Schneider, were set on developing a rigorous theory that quantified the gain of multiscale methods [79, 80, 84, 86, 89, 90, 97, 101]. The results in [146] were the first to show for a large class of elliptic operators, including those of negative order, that the system matrices can be compressed up to optimal complexity $O(N)$ while at the same time admitting optimal preconditioning of the system matrix and the solution of the resulting problems with asymptotically optimal convergence order. It also made the relevant point to bring preconditioning into play. (This topic was not addressed by Beylkin et al. since they had chosen an operator of order zero for which no conditioning issue arises.) Surveys about the different stages of the investigations for pseudodifferential and boundary integral operators were provided in [97, 103, 108].

So far, these results referred to discretizations on uniform grids. Another milestone for complexity reduction was achieved by the introduction of adaptively refined a–posteriori discretizations. Optimal complexity estimates for adaptive methods based on wavelets for integral equations were proved in [162].

A survey of the main results and the current state of the art of wavelet methods for integral equations is provided by the article by Helmut Harbrecht and Reinhold Schneider in this volume.

5 Adaptive solvers

Multiscale decompositions have long played an important role in image processing which leads to efficient compression of images. It is natural to try to bring these ideas into the realm of numerical PDEs. But of course, new problems arise since the solution is not known to us but can only be seen through computation.

Wolfgang had already collaborated with Albert Cohen and Ron DeVore on various aspects of image compression and multiscale systems but in the latter part of the 1990s, they begin to turn their attention to PDE solvers with the goal of providing a rigorous foundation for both theory and algorithms for adaptive methods. Of course, adaptive finite element methods had been around for some time. However, theoretical results which established their advantages over non-adaptive methods were still lacking.

It seemed natural to first tackle adaptivity in the context of wavelet methods since these were already known to have the advantages mentioned above of yielding highly compressible matrix representations. The result was a penetrating theory and algorithms much beyond the state of the art known for partial differential and integral operators. It all began with the work [118] for linear elliptic partial differential equations which not only developed convergent algorithms but also algorithms with asymptotically optimal rates of convergence when compared to the wavelet-best N term approximation. This seminal work was followed by extensions to more general settings including nonlinear problems [133, 147]. Other collaborations of Wolfgang extended and substantially refined the notion of adaptive methods of optimal complexity to more general problems including saddle point problems [139, 144], adaptivity steered by goals different than the energy norm [165], and problems in optimal control constrained by an elliptic partial differential equation [150].

Interestingly, these results also applied to operators of negative order. Specifically for ensuring optimal complexity, a number of techniques known from optimal coding and compression [121] had to be intertwined in a sophisticated way, yielding a novel approximate matrix-vector multiplication of optimal complexity. The derivation of adaptive methods with optimal complexity estimates for nonlinear stationary partial differential equations posed yet another difficulty which was attacked in [147] based on tree approximations. Recently, these techniques have been extended for the first time to derive convergent adaptive schemes for elliptic eigenvalue problems in [173].

A survey of the basic principles and the main results is provided in the Encyclopedia Article [151] or the longer survey article [153]. Rob Stevenson gives a very nice overview of the state of the art of adaptive wavelet methods for operator equations in this volume.

The principles for proving convergence of adaptive methods for linear elliptic PDEs based on finite elements have been known since Willy Dörfler's article in '96. However, it was a much-discussed question in practical finite element codes whether a heuristically used derefinement/coarsening step was really needed for 'good complexity'. Establishing a sound theory was a difficult issue since a finite element discretization does not characterize in a natural way the underlying Sobolev space or the solution like in wavelet theory. Again tree approximations turned out to be the key to success, resulting in [145] for the first time in optimal complexity estimates for adaptive finite element methods. This seminal paper triggered numerous followers and extended into a rich theory, for which an account is given in the article of Ricardo Nochetto, Kunibert G. Siebert and Andreas Veeser in this volume.

6 Construction and implementation

The numerical implementation of wavelet algorithms is not without challenges. The main obstacle is to design multiscale systems for domains and manifolds which arise in applications. The classical development of wavelets is tied to $\mathbb{R}^n$ or, via periodization, to the torus. Wolfgang was at the heart of multiscale constructions for practical settings which took place largely in the 1990s. The initial studies [76, 78, 81] were devoted to systematically developing Fourier-free concepts and clarifying the basic relations of stability, Riesz bases, and norm equivalences. These investigations led to Fourier-free stability criteria [81, 92, 95] which include biorthogonality as well as regularity and approximation properties. This results in the preservation of the relevant functional analytic properties of function spaces on domains and manifolds.

Flexible constructions of biorthogonal wavelets for a whole range of Sobolev spaces on domains and manifolds were provided in [88, 93, 100, 104, 105, 107, 112, 113, 115]. These tools also led to new evaluation algorithms [117, 158]. For instance, the exact computation of integrals of products of derivatives of scaling functions can be reduced to solving linear systems whose size is independent of the discretization [71]. Based on biorthogonal wavelets one can systematically construct for any spatial discretization stable pairs of approximation spaces for saddle point problems like for the Stokes problem [94] or for treating essential boundary conditions by Lagrange multipliers [122]. In addition, they admit optimal preconditioners. Starting from wavelets on the interval [104], tensorizing these, and finally using parametric mappings, one can construct wavelets in particular for boundary integral equations on manifolds in 3D [100, 107, 115]. The construction in [107] is intrinsically tied to characterizations of function spaces on manifolds through manifold decompositions developed earlier by Zbigniew Ciesielski and Tadeusz Figiel using orthogonalized B-splines. This approach also leads naturally to domain decompositions for boundary integral equations. Finally, in [112] local wavelet bases with the desired stability and compression properties were constructed for standard finite element decompositions in up to 3D.

A discussion of implementation and numerical experiments for adaptive wavelet schemes for linear elliptic partial differential equations, exhibiting the theoretically predicted convergence rates in an exemplary manner was provided in [125].

The foundation for many efficient algorithms are norm equivalences for multiscale expansions which can be used for preconditioning. So far these concepts have been derived for discretizations, initially on uniform grids. On the other hand, in view of problems in dimensions higher than three, in the past years also partition-of-unity methods which can be applied to essentially nonuniform grids have become fashionable. However, for these methods, there was up to [169] no efficient and analytically proven optimal preconditioner available. Decompositions of spaces on nonuniform or anisotropic grids are the subject of an ongoing collaboration of Wolfgang Dahmen with Shai Dekel and Pencho Pestrushev, see their contribution in this volume.

7 Hyperbolic partial differential equations and conservation laws

Adaptive methods are also frequently applied to non–elliptic problems, primarily non-stationary hyperbolic problems. Wolfgang and his collaborators became interested in this area around the time that Ami Harten developed his compression approach for conservation laws. Harten's approach was improved in an essential way in [119] both with respect to practicability (2D problems on curvilinear grids) as well as conceptually with respect to the amount of expensive flux evaluations in an adaptive grid refinement, see the contribution by Siegfried Müller in this volume. For the compressible Navier-Stokes equations, these ideas were elaborated in [141].

Another non-elliptic example centers around adaptive multigrid methods for convection dominated problems in [123]. Here adaptivity not only reduces the complexity but also stabilizes a standard Galerkin discretization so that no modification by artificial introduction of viscosity is necessary.

8 Engineering collaborations

Wolfgang was not immune to collaborating with Engineers and he had an excellent environment for such collaborations in Aachen. For example, he worked with mechanical engineers at RWTH Aachen, specifically Josef Ballmann, to efficiently simulate transport phenomena, grid generation, hypersonic flow problems, and the interaction of aerodynamics and structure. This also resulted in theoretical results on Riemann solvers for non-convex flux functions including phase transitions for hyperbolic conservation laws [154] or the well-posedness of modeling problems for nonlinear elasticity [161].

Wolfgang's long-term collaboration with the chemical engineer Wolfgang Marquardt at RWTH Aachen is one of the prominent examples of bringing novel mathematical concepts into practical applications to substantially improve simulation results. Here real-time optimization of dynamical chemical processes requires compression of systems of ordinary differential equations [102, 111, 127, 128, 129, 135]–[138].

9 The present

Wolfgang's research program continues to find interesting new avenues. In the past few years, Wolfgang Dahmen's work has primarily been driven by problems in learning theory and compressed sensing. These investigations have taken place largely in collaboration with Albert Cohen and Ron DeVore. Their goal, quite similar to the adaptive PDE program, is to understand in what sense various algorithms in these areas are optimal. This program has led to a series of results [160, 166, 167, 168, 170, 174] which clarify the gains of sparsity and nonlinearity. These remarkable accomplishments are well documented in the article by Ron DeVore in this volume.

10 Final remarks

Wolfgang Dahmen has had an illustrious mathematical career full of high points. Perhaps his most significant recognition was the awarding of the Gottfried Wilhelm Leibniz-Award in 2002, the highest scientific award in Germany given to him by the Deutsche Forschungsgemeinschaft (German Science Foundation). The authors of this introduction and his many collaborators thank him for his friendship and years of stimulating mathematics. We look forward to more in the future.

Publications by Wolfgang Dahmen (as of Summer 2024)

1. (with E. Görlich) A conjecture of Golomb on optimal and nearly-optimal approximation, Bull. Amer. Math. Soc., **80** (1974), 1199-1202.
2. (with E. Görlich) Asymptotically optimal linear approximation processes and a conjecture of Golomb, In: Linear Operators and Approximation II, ed. by P.L. Butzer and B. Sz.-Nagy, ISNM 25, Birkhäuser, Basel, 1974.
3. (with E. Görlich) Best approximation with exponential error orders and intermediate spaces, Math. Z., **148** (1976), 7-21.
4. (with E. Görlich) The characterization problem for best approximation with exponential error orders and evaluation of entropy, Math. Nachr., **76** (1977), 163-179.
5. Trigonometric approximation with exponential error orders. I. Construction of asymptotically optimal processes; generalized de la Vallée Poussin sums, Math. Ann., **230** (1977), 57-74.
6. Trigonometric approximation with exponential error orders. II. Properties of asymptotically optimal processes; impossibility of arbitrarily good error estimates, Journal of Mathematical Analysis and Applications, **68** (1979), 118-129.
7. Trigonometric approximation with exponential error orders. III. Criteria for asymptotically optimal processes, Proc. of the International Conference on Approximation and Function Spaces, ed. by Z. Ciesielski, Gdansk, 1979.
8. On best approximation and de la Vallée sums, Mat. Zametkii, **5** (1978), 671-683.
9. (with K. Scherer) Best approximation with piecewise polynomials with variable knots and degrees, J. Approx. Theory, **26** (1979), 1-13.
10. Multivariate B-splines - recurrence relations and linear combinations of truncated powers, in: Multivariate Approximation Theory, ed. by W. Schempp, K. Zeller, Birkhäuser, Basel, (1979), 64-82.
11. Polynomials as linear combinations of multivariate B-splines, Math. Z., **169** (1979), 93-98.
12. Konstruktion mehrdimensionaler B-Splines and ihre Anwendungen auf Approximationsprobleme, in: Numerische Methoden der Approximationstheorie, vol. **5**, ed. by L. Collatz, G. Meinardus, H. Werner, Birkhäuser, Basel, (1980), 84-110.
13. On multivariate B-splines, SIAM J. Numer. Anal., **17** (1980), 179-191.
14. (with R. DeVore and K. Scherer) Multidimensional spline approximation, SIAM J. Numer. Anal., **17** (1980), 380-402.
15. Approximation by smooth multivariate splines on non-uniform grids, in: Quantitative Approximation, ed. by R. DeVore and K. Scherer, Academic Press, 1980, 99-114.
16. (with C.A. Micchelli) On limits of multivariate B-splines, J. d'Analyse Math., **39** (1981), 256-278.
17. (with A.S. Cavaretta, C.A. Micchelli and P.W. Smith) On the solvability of certain systems of linear difference equations, SIAM J. Math. Anal., **12** (1981), 833-841.
18. (with A.S. Cavaretta, C.A. Micchelli and P.W. Smith) A factorization theorem for banded matrices, Linear Algebra and its Applications, **39** (1981), 229-245.
19. (with C.A. Micchelli) On entire functions of affine lineage, Proc. Amer. Math. Soc., **84** (1982), 344-346.
20. Approximation by linear combinations of multivariate B-splines, J. Approx. Theory, **31** (1981), 299-324.
21. (with C.A. Micchelli) Computation of integrals and inner products of multivariate B-splines, Numer. Funct. Anal. and Optimiz. **3** (1981), 367-375.
22. (with C.A. Micchelli) On the linear independence of multivariate B-splines. I. Triangulations of simploids, SIAM J. Numer. Anal., **19** (1982), 993-1012.
23. Adaptive approximation by smooth multivariate splines, J. Approx. Theory, **36** (1982), 119-140.
24. (with C.A. Micchelli) Numerical algorithms for least squares approximation by multivariate B-splines, in: Numerische Methoden der Approximationstheorie, ed. by L. Collatz, G. Meinardus and H. Werner, Birkhäuser, Basel, 1981, 85-114.

25. (with C.A. Micchelli) On the linear independence of multivariate B-splines. II. Complete configurations, Math. Comp., **41** (1982), 143-163.
26. (with C.A. Micchelli) Some remarks on multivariate B-splines, in: Muiltivariate Approximation Theory, ed. by W. Schempp, and K. Zeller, ISNM 61, Birkhäuser, Basel, 1982, 81-87.
27. (with C.A. Micchelli) Multivariate splines - a new constructive approach, in: Surfaces in Computer Aided Geometric Design, ed. by R.E. Barnhill and W. Boehm, North-Holland, 1982, 191-215.
28. (with C.A. Micchelli) Translates of multivariate splines, Linear Algebra and its Applications, **52/53** (1983), 217-234.
29. (with C.A. Micchelli) On the approximation order of certain multivariate spline spaces, Journal of the Australian Mathematical Society, Ser. **B 26** (1984), 233-246.
30. (with C.A. Micchelli) Recent Progress in multivariate splines, in: Approximation Theory IV, ed. by C.K. Chui, L.L. Schumaker and J. Ward, Academic Press, 1983, 27-121.
31. (with C.A. Micchelli) On the approximation order of criss- cross finite element spaces, Journal of Computational and Applied Mathematics, **10** (1984), 255-273.
32. (with C.A. Micchelli) Some results on box splines, IBM RC 10094, July 1983, Bull. AMS, **1** (1984), 147-150.
33. (with C.A. Micchelli) On the local linear independence of translates of a box spline, Studia Mathematica **82** (1985), 243-263.
34. (with C.A. Micchelli) On the solution of certain systems of linear partial difference equations and linear dependence of translates of box splines, Trans. Amer. Math. Soc., **292** (1985), 305-320.
35. (with C.A. Micchelli) Subdivision algorithms for the generation of box spline surfaces, Computer Aided Geometric Design **1** (1984), 115-129.
36. (with C.A. Micchelli) On the multivariate Euler-Frobenius polynomials, in: Constructive Theory of Functions, ed. by B. Sendov, P. Petrushev, R. Maleev, S. Tashev, Publishing House of the Bulgarian Academy of Sciences, 1984, 237-243.
37. (with N. Dyn and D. Levin) On the convergence rates of subdivision algorithms for box spline surfaces, Constr. Approx., **1** (1985), 305-322.
38. (with C.A. Micchelli and P.W. Smith) Asymptotically optimal sampling schemes for periodic functions, Math. Proc. Camb. Phil. Soc., **99** (1986), 171-177.
39. Subdivision algorithms converge quadratically, Journal Computational and Applied Mathematics, **16** (1986), 145-158.
40. (with C.A. Micchelli) Line average algorithm: A method for the computer generation of smooth surfaces, Computer Aided Geometric Design **2** (1985), 77-85.
41. (with C.A. Micchelli) On the number of solutions to systems of linear diophantine equations and multivariate splines, Trans. Amer. Math. Soc., **308** (1988), 509-532.
42. (with C.A. Micchelli) Convexity of multivariate Bernstein polynomials and box spline surfaces, Studia Scientiarum Mathematicarum Hungarica, **23** (1988), 265-287.
43. (with C.A. Micchelli) Combinatorial aspects of multivariate splines, in: "Multivariate Approximations III", ed. by W. Schempp, K. Zeller, Birkhäuser, Basel, 1985, 120-137.
44. (with C.A. Micchelli, P.W. Smith) On factorization of bi-infinite totally positive block Toeplitz matrices, Rocky Mountain J. of Math., **16** (1986), 335-364.
45. Subdivision algorithms - recent results, some extensions and further developments, in Algorithms for Approximation, ed. by J.C. Mason, M.G. Cox, Clarendon Press, Oxford, 1987, 21-49.
46. (with C.A. Micchelli) On the piecewise structure of discrete box splines, Computer Aided Geometric Design, **3** (1986), 185-191.
47. (with C.A. Micchelli) Statistical encounters with B-splines, AMS contemporary Mathematics, **59** (1986), 17-48.
48. (with C.A. Micchelli) Algebraic properties of discrete box splines, Constr. Approx., **3** (1987), 209-221.
49. (with C.A. Micchelli) Some remarks on ridge functions, Approximation Theory and its Applications, **3** (1987), 139-143.

50. (with C.A. Micchelli) Theory and Applications of exponential splines, in: Proceedings of the International. Workshop on Multivariate Approximation, Santiago de Chile, Dec. 15-19, 86, 37-46, 1987.
51. (with C.A. Micchelli) On multivariate E-splines, Advances in Mathematics, **76** (1989), 33-93.
52. (with C.A. Micchelli, T.N.T. Goodman) Compactly supported fundamental functions for spline interpolation, Numerische Mathematik, **52** (1988), 639-664.
53. (with C.A. Micchelli, T.N.T. Goodman) Local spline interpolation schemes in one and several variables, in: Approximation and Optimization, Proceedings Havana, 1987, A. Gomez, F. Guerra, M.A. Jimenez, G. Lopez (Eds.) Springer Lecture Notes in Mathematics **1354**, 11-24, 1988.
54. (with R. Gmelig Meyling, J. Ursem) Scattered data interpolation by bivariate C^1-piecewise polynomials, Approximation Theory and its Applications, **6** (1990), 6-29.
55. (with L. Elsner) Algebraic multigrid methods and the Schur complement, in: Robust Multi-Grid Methods, Proceedings of the Fourth GAMM-Seminar, Kiel, January 22-24, 1988, Notes on Numerical Fluid Mechanics, **23** (1988), 57-68.
56. (with C.A. Micchelli) Convexity and Bernstein Polynomials on k-Simploids, Acta Mathematicae Applicatae Sinica, **1** (1990), 50-66.
57. (with A.S. Cavaretta, C.A. Micchelli) The volume of restricted moment spaces, Rendiconti del Circolo Matematico di Palermo, Serie II, **38** (1989), 419-429.
58. (with C.A. Micchelli) Local dimension of piecewise polynomial spaces, syzygies and solutions of systems of partial differential equations, Mathematische Nachrichten, **148** (1990), 117-136.
59. Smooth piecewise quadric surfaces, in: Mathematical Methods in Computer Aided Geometric Design, T. Lyche, L.L. Schumaker, eds. Academic Press, 1989, 181-193.
60. A basis for certain spaces of multivariate polynomials and exponentials, Algorithms for Approximation II, J.C. Mason, M.G. Cox (Herausg.), Chapman and Hall, 1990, 80-98.
61. (with R.Q. Jia, C.A. Micchelli) On linear dependence relations for integer translates of compactly supported distributions, Mathematische Nachrichten, **151** (1991), 303-310.
62. (with A.S. Cavaretta, C.A. Michelli) Stationary subdivision, Memoirs of the Amer. Math. Soc., No. **453**, 1991.
63. (with R.Q. Jia, C.A. Micchelli) Linear dependence of cube splines revisited, in "Approximation Theory VI", C.K. Chui, L.L. Schumaker, J.D. Ward eds., Academic Press, Volume I, 1989, 161-164.
64. (with C.A. Micchelli) Stationary subdivision, fractals and wavelets, FU Preprint A-89-21, in: the Proceedings of the NATO ADVANCED STUDY INSTITUTE series, Computation of curves and Surfaces, Hersg. M. Gasca, W. Dahmen, C.A. Micchelli, Kluwer Academic Publishers, 3-26, 1990.
65. (with C.A. Micchelli) On stationary subdivision and the construction of compactly supported orthonormal wavelets, in: Multivariate Approximation and Interpolation, W. Haussmann, K. Jetter (Herausg.), Birkhäuser, 1990, 60-90.
66. (with T. M. Thamm) Cubicoids: Modeling and visualization, Computer Aided Geometric Design, **10** (1993), 89-108.
67. (with A. Dress, C.A. Micchelli) On multivariate splines, matroids and the ext-functor, Advances in Applied Mathematics, **17** (1996), 251-307.
68. Convexity and Bernstein-Bézier polynomials, in: Curves and Surfaces, P.J. Laurent, A. Le Méhauté, L.L. Schumaker eds., Academic Press, 1991, 107-134.
69. (with H.P. Seidel, C.A. Micchelli) Blossoming begets B-spline bases built better by B-patches, Mathematics of Computation, **59** (1992), 97-115.
70. (with R. DeVore, C.A. Micchelli) On monotone extensions of boundary data, Numer. Math., **60** (1992), 477-492.
71. (with C.A. Micchelli) Using the refinement equation for evaluating integrals of wavelets, SIAM J. Numer. Anal., **30** (1993), 507-537.
72. (with J.M. Carnicer) Characterization of local strict convexity preserving interpolation methods by C^1 functions, Journal of Approximation Theory, **77** (1994), 2-30.

73. (with J.M. Carnicer) Convexity preserving interpolation and Powell-Sabin elements, Computer Aided Geometric Design, **9** (1992), 279-289.
74. (with P. Oswald, X.Q. Shi) C^1-conforming hierarchical bases, Journal of Computational and Applied Mathematics, 9(1993), 263-281.
75. (with A. Kunoth) Multilevel preconditioning, Numer. Math., **63** (1992), 315-344.
76. (with C.A. Micchelli) Banded matrices with banded inverses II: Locally finite decompositions of spline spaces, Constr. Approx., **9** (1993), 263-281.
77. (with C.A. Micchelli) Continuous refinement equations and subdivision, Advances in Computational Mathematics, **1** (1993), 1-37.
78. Decomposition of refinable spaces and applications to operator equations, Numerical Algorithms, **5** (1993), 229-245.
79. (with S. Prössdorf, R. Schneider) Wavelet approximation methods for pseudodifferential equations I: Stability and convergence, Mathematische Zeitschrift, **215** (1994), 583-620.
80. (with S. Prössdorf, R. Schneider) Wavelet approximation methods for pseudodifferential equations II: Matrix compression and fast solution, Advances in Computational Mathematics, **1** (1993), 259-335.
81. (with J.M. Carnicer, J.M. Peña) Local decomposition of refinable spaces and wavelets, Applied and Computational Harmonic Analysis, **3** (1996), 127-153.
82. (with M Butzlaff, S. Diekmann, A. Dress, E. Schmitt, E. V. Kitzing) A hierarchical approach to force field calculations through spline approximations, Journal of Mathematical Chemistry, **15** (1994), 77-92.
83. (mit B. Raabe, T.-M. Thamm) Optimal geometry representation for rendering, in: Visualisierung in Mathematik, Technik und Kunst, Grundlagen und Anwendungen, A. Dress, G. Jäger (eds.), Vieweg-Verlag, Braunschweig, 1999, 23–49.
84. (with S. Prössdorf, R. Schneider) Multiscale methods for pseudodifferential equations, in: 'Recent Advances in Wavelet Analysis' L.L. Schumaker, G. Webb eds., Academic Press, 1993, 191-235.
85. (with S. Dahlke, V. Latour) Smooth refinable functions and wavelets obtained by convolution products, Applied and Computational Harmonic Analysis, **2** (1995), 68-84.
86. (with B. Kleemann, S. Prössdorf, R. Schneider) A multiscale method for the double layer potential equation on a polyhedron, in: Advances in Computational Mathematics, H.P. Dikshit, C.A. Micchelli, eds., World Scientific, 1994, 15-57.
87. Some remarks on multiscale transformations, stability and biorthogonality, in: Wavelets, Images and Surface Fitting, P.J. Laurent, A. Le Méhauté, L.L. Schumaker eds., Academic Press, 157-188, 1994.
88. (with A. Cohen, R. DeVore) Multiscale decompositions on bounded domains, Trans. Amer. Math. Soc., No. 8, **352** (2000), 3651–3685.
89. (with S. Prössdorf, R. Schneider) Multiscale methods for pseudo-differential equations on smooth manifolds, in: Proceedings of the International Conference on Wavelets: Theory, Algorithms, and Applications, C.K. Chui, L. Montefusco, L. Puccio (eds.), Academic Press, 385-424, 1994.
90. (with S. Prössdorf, R. Schneider) Wavelets zur schnellen Lösung von Randintegralgleichungen und angewandte harmonische Analyse, Z. Angew. Math. Mech. 74, No 6 (1994), 505-507.
91. Multiscale Techniques - Some Concepts and Perspective, in: Proceedings of the International Congress of Mathematicians, Zürich 94, Birkhäuser, Basel, 1995, 1429-1439.
92. Stability of multiscale transformations, Journal of Fourier Analysis and Applications, **2** (1996), 341-361.
93. (with S. Dahlke, E. Schmitt, I. Weinreich) Multiresolution analysis and wavelets on S^2 and S^3, Numerical Functional Analysis and Optimization, **16** (1995), 19-41.
94. (with A. Kunoth, K. Urban) Wavelet-Galerkin method for the Stokes equations, Computing, **56** (1996), 259-301.
95. (with C.A. Micchelli) Biorthogonal wavelet expansions, Constr. Approx., **13** (1997), 293–328.

96. Multiskalen-Methoden und Wavelets – Konzepte und Anwendungen, DMV–Jahresbericht, **97** (1995), 97-114.
97. (with B. Kleemann, S. Prössdorf, R. Schneider) Multiscale methods for pseudodifferential equations, *Proceedings des ICIAM /GAMM Hamburg 1995 - Kongresses*, (1996). Z. Angew. Math. Mech. 76 Suppl.1 7–10.
98. Multiscale analysis, approximation, and interpolation spaces, in: Approximation Theory VIII, Vol 2., Wavelets and Multilevel Approximation, C.K. Chui, L.L. Schumaker (eds.), World Scientific Publishing Co, 1995, 47-88.
99. (with S. Dahlke, R. Hochmuth, R. Schneider) Stable multiscale bases and local error estimation for elliptic problems, Applied Numerical Mathematics, **23** (1997), 21-48.
100. (with R. Schneider) Composite wavelet bases for operator equations, Math. Comp., **68** (1999), 1533-1567.
101. (with B. Kleemann, S. Prössdorf, R. Schneider) Multiscale methods for the solution of Helmholtz and Laplace equations, in: Boundary Element Methods, Reports from the Final Conference of the Priority Research Programme 1989-1995 of the German Research Foundation, Oct. 2-4, 1995 in Stuttgart, W. Wendland, ed., Springer-Verlag, 1996.
102. (with R. V. Watzdorf, K. Urban, W. Marquardt) A wavelet-Galerkin method applied to separation processes, in: *Scientific Computing in Chemical Engineering*, S. Keil, W. Mackens, H. Voß, J. Werther (Hersg.), Springer-Verlag, Berlin, 246-252, 1996.
103. (with A. Kunoth, R. Schneider) Operator equations, multiscale concepts and complexity, in: Proceedings, *1995 AMS-SIAM Summer Seminar, Mathematics of numerical Analysis: Real Number Algorithms* J. Renegar, M. Shub, S. Smale (eds.), Park City, 1995, Lectures in Applied Mathematics, **32** (1996), 225-261.
104. (with A. Kunoth, K. Urban) Biorthogonal spline-wavelets on the interval – Stability and moment conditions, Applied and Computational Harmonic Analysis, **6** (1999), 132-196.
105. (with A. Kunoth, K. Urban) Wavelets in numerical analysis and their quantitative properties, in: Surface Fitting and Multiresolution Methods, A. Le Méhauté, C. Rabut, L.L. Schumaker (eds.), Vanderbilt University Press, 1997, 93-130.
106. (with S. Dahlke, R. DeVore) Nonlinear approximation and adaptive techniques for solving elliptic operator equations, in: Wavelet Multiscale Methods for PDEs, Academic Press, W. Dahmen, A. Kurdila, P. Oswald eds., London, 1997, 237-283, 1997.
107. (with R. Schneider) Wavelets on manifolds I. Construction and domain decomposition, SIAM Journal on Mathematical Analysis, **31** (1999), 184-230.
108. Wavelet and Multiscale Methods for Operator Equations (invited contribution), Acta Numerica, Cambridge University Press, **6** (1997), 55–228.
109. (with D. Braess) A cascade algorithm for the Stokes equations, Numer. Math., **82** (1999), 179-191.
110. (with S. Müller, T. Schlinkmann) Multigrid and multiscale decompositions, in: Large-Scale Scientific Computations of Engineering and Environmental Problems, M. Griebel, O.P. Iliev, S.D. Margenov, P.S. Vassilevski, eds., Notes on Numerical Fluid Mechanics, Vol. 62, Vieweg, Braunschweig/Wiesbaden, 18–41, 1998.
111. (with T. Binder, L. Blank, W. Marquardt) Towards multiscale dynamic data reconciliation, in: *Nonlinear Model based Process Control* NATO ASI, (R. Berber and C. Kravaris, eds.), Kluwer Academic Publishers, 623–665, 1998.
112. (with R. Stevenson) Element-by element construction of wavelets satisfying stability and moment conditions, SIAM J. Numer. Anal., **37** (No. 1) (1999), 319–325.
113. (with B. Han, R.-Q. Jia, A. Kunoth) Biorthogonal multiwavelets on the interval: Cubic Hermite splines, Constr. Approx. **16** (2000), 221–259.
114. B-splines in analysis, algebra and applications, Traveaux Mathématiques, 1999, 15–76.
115. (with R. Schneider) Wavelets with complementary boundary conditions – Function spaces on the cube, Results in Mathematics, **34** (1998), 255–293.
116. (with D. Braess, C. Wieners) A multigrid algorithm for the mortar finite element method, SIAM J. Numer. Anal., **37** (No. 1) (1999), 48–69.
117. (with R. Schneider, Y. Xu) Nonlinear functions of wavelet expansions – Adaptive reconstruction and fast evaluation, Numer. Math., **86** (2000), 49-101.

118. (with A. Cohen, R. DeVore) Adaptive wavelet methods for elliptic operator equations – Convergence rates, Math. Comp. **70** (2001), 27–75.
119. (with B. Gottschlich-Müller, S. Müller) Multiresolution schemes for conservation laws, Numer. Math., **88** (2001), 399-443.
120. (with D. Braess) Stability estimates of the mortar finite element method for 3-dimensional problems, East West J. Numer. Math., **6** (4) (1998), 249–264.
121. (with A. Cohen, I. Daubechies, R. DeVore) Tree approximation and optimal encoding, Applied and Computational Harmonic Analysis, **11** (2001), 192–226.
122. (with A. Kunoth) Appending boundary conditions by Lagrange multipliers: Analysis of the LBB condition, Numer. Math., **88** (2001), 9–42.
123. (with S. Müller, T. Schlinkmann) On an adaptive multigrid solver for convection-dominated problems, SIAM J. Scient. Comp., **23** (No 3)(2001), 781–804.
124. (with A. Kunoth, R. Schneider) Wavelet least squares methods for boundary value problems, Siam J. Numer. Anal. **39**, No. 6, (2002), 1985–2013.
125. (with A. Barinka, T. Barsch, P. Charton, A. Cohen, S. Dahlke, K. Urban) Adaptive wavelet schemes for elliptic problems – implementation and numerical experiments, SIAM J. Sci. Comp., **23**, No. 3 (2001), 910–939.
126. Wavelet methods for PDEs – Some recent developments, J. Comp. Appl. Math., **128** (2001), 133–185.
127. (with T. Binder, L. Blank, W. Marquardt) Iterative algorithms for multiscale state estimation, Part I, Concepts, J. Opt. Theo. Appl. **111**(3)(2001), 529–551.
128. (with T. Binder, L. Blank, W. Marquardt) An adaptive multiscale method for real time moving horizon optimization, Proc. American Control Conference 2000, Chicago, Illinois June 2000, pp 4234 - 4238, Omnipress.
129. (with T. Binder, L. Blank, W. Marquardt) Grid refinement in multiscale dynamic optimization, Proc. European Symposium on Computer Aided Process Engineering 10, Florence, Italy, Mai 2000, (Ed. S. Pierucci), Elsevier, Amsterdam, pp 31-37.
130. (with I. Graham, B. Faehrmann, W. Hackbusch, S. Sauter) Inverse inequalities on non-quasiuniform meshes and application to the mortar element method, Math. Comp. **73** (2004), 1107-1138.
131. (with A. Cohen, I. Daubechies, R. DeVore) Harmonic analysis of the space BV, Revista Matematica Iberoamericana **19** (2003), 1-29.
132. (with Th. Binder, L. Blank, W. Marquardt) Regularization of dynamic data reconciliation problems by projection, Proc. IFAC Symposium on Advanced Control of Chemical Processes, ADCHEM 2000, Pisa, Italy, June 2000. Ed. L. T. Biegler, A. Brambilla, C. Scali, pp 689 - 694, Vol. 2
133. (with A. Cohen, R. DeVore) Adaptive wavelet methods II - Beyond the elliptic case, Foundations of Computational Mathematics, **2** (2002), 203–245.
134. (with H.P. Dikshit, A. Ojha) On Wachspress quadrilateral patches, Computer Aided Geometric Design, **17** (2000), 879–890.
135. (with T. Binder, L. Blank, W. Marquardt) Iterative algorithms for multiscale state estimation, Part II: Numerical investigations, J. Opt. Theo. Appl. **111**(3)(2001), 529–551.
136. (with T. Binder, L. Blank, W. Marquardt) Multiscale concepts for moving horizon optimization, in: *Online Optimization for Large Scale Systems* (M. Grötschel, S.O. Krumke, and J.Rambau, eds.), Springer, Berlin, 341–362, 2001.
137. (with T. Binder, L. Blank, W. Marquardt) On the regularization of dynamic data reconciliation problems, Journal of Process Control, **12** (2002), 557-567.
138. (with T. Binder, L. Blank, W. Marquardt) Iterative multiscale methods for process monitoring, in: Proc. of 'Fast Solutions of Discretized Optimization Problems', WIAS Berlin, May 8–12, Birkhäuser Verlag, 2000.
139. (with S. Dahlke, K. Urban) Adaptive wavelet methods for saddle point problems – Convergence rates, SIAM J. Numer. Anal., **40** (No. 4) (2002), 1230–1262.
140. (with D. Braess) The mortar element method revisited – What are the right norms ? in: Domain Decomposition Methods in Science and Engineering, N. Debit, M. Garbey, R. Hoppe, D. Keyes, Y. Kutznetzov, J. Périaux, eds., International Center for Numerical Methods in Engineering (CIMNE), Barcelona, 2002, 27–40.

141. (with F. Bramkamp, B. Gottschlich-Müller, Ph. Lamby, M. Hesse, S. Müller, J. Ballmann, K.-H. Brakhage) H-adaptive multiscale schemes for the compressible Navier-Stokes equations – Polyhedral discretization, data compression and mesh generation, in: Notes on Numerical Fluid Mechanics, Flow Modulation and Fluid-Structure-Interaction at Airplane Wings (ed. by J. Ballmann), Springer-Verlag, Vol. 84, 125–204, 2003.
142. Wavelets als mathematisches Mikroskop, Festschrift, 100 Jahre Berliner Mathematische Gesellschaft, WB-Druck, Rieden im Allgäu, Berlin, 2001.
143. (with T. Klint, K. Urban) On fictitious domain formulations for Maxwell's equations, J. Foundation of Computational Mathematics, **3** (2)(2003), 135–160.
144. (with K. Urban, J. Vorloeper) Adaptive wavelet methods – basic concepts and applications, in: Wavelet Analysis – Twenty Years Developments, Ding–Xuan Zhou ed., World Scientific, New Jersey, 2002, 39–80.
145. (with P. Binev, R. DeVore) Adaptive Finite Element Methods with Convergence Rates, Numer. Math., **97** (2004), 219–268.
146. (with H. Harbrecht, R. Schneider) Compression techniques for boundary integral equations – Asymptotically optimal complexity estimates, SIAM J. Numer. Anal., **43**(6) (2006), 2251–2271.
147. (with A. Cohen, R. DeVore) Adaptive wavelet schemes for nonlinear variational problems, SIAM J. Numer. Anal., **41**(5) (2003), 1785–1823.
148. (with A. Cohen, R. DeVore) Sparse evaluation of compositions of functions using multiscale expansions, SIAM J. Math. Anal., 35 (2003), 279–303.
149. (with P. Binev, R. DeVore, P. Petruchev) Approximation classes for adaptive methods, Serdica Math. J., **28** (No 4) (2002), 391–416.
150. (with A. Kunoth) Adaptive wavelet methods for linear–quadratic elliptic control problems: Convergence rates, SIAM J. Contr. Optim. **43** (5) (2005), 1640–1675.
151. (with A. Cohen, R. DeVore) Adaptive wavelet techniques in Numerical Simulation, in: *Encyclopedia of Computational Mechanics*, (R. De Borste, T. Hughes, E. Stein, eds.), Wiley-Interscience, 2004, 157–197.
152. (with A. Barinka, S. Dahlke) Adaptive application of operators in standard representation, Advances in Computational Mathematics, **24** (2006), 5–34.
153. Multiscale and Wavelet Methods for Operator Equations, C.I.M.E. Lecture Notes in Mathematics, *Multiscale Problems and Methods in Numerical Simulation*, Springer Lecture Notes in Mathematics, vol 1825, Springer-Verlag, Heidelberg, 2003, 31–96.
154. (with S. Müller, A. Voß) Riemann problem for the Euler equations with non-convex equation of state including phase transitions, in: *Analysis and Numerics for Conservation Laws*, G. Warnecke ed., Springer-Verlag, Berlin-Heidelberg, 2005, 137–162.
155. (with M. Campos-Pinto, A. Cohen, R. DeVore) On the stability of nonlinear conservation laws in the Hausdorff metric, Journal of Hyperbolic Differential Equations, **2**(1) (2005), 1-14.
156. (with P. Petrushev) "Push the error" algorithm for nonlinear n-term approximation, Constructive Approximation, **23**(3)(2006), 261–304.
157. (with P. Binev, R. DeVore, N. Dyn) Adaptive approximation of curves, in: Approximation Theory: a volume dedicated to Borislav Bojanov, 43-57, Prof. M. Drinov Acad. Publ. House, Sofia, 2004.
158. (with A. Barinka, R. Schneider) Fast computation of adaptive wavelet expansions, Numer. Math., **110**(4) (2007), 549–589.
159. (with A. Kurdila, M. Nechyba, R. Prazencia, P. Binev, R. DeVore, R. Sharpley) Vision-based control of micro-air-vehicles — Progress and problems in estimation, 43rd IEEE Conference on Decision and Control, Dec. 14–17, 2004, Atlantis, Paradise Island, Bahamas.
160. (with P. Binev, A. Cohen, R. DeVore, V. Temlyakov) Universal algorithms for learning theory - Part I : piecewise constant functions, Journal of Machine Learning Research, **6** (2005), 1297-1321.
161. (with R. Massjung, J. Hurka, J. Ballmann) On well-posedness and modeling for nonlinear aeroelasticity, Notes on Numerical Fluid Mechanics, Springer Verlag, vol. 84, 2003.
162. (with H. Harbrecht, R. Schneider) Adaptive methods for boundary integral equations — Complexity estimates, Math. Comp., **76** (259) (2007), 1243–1274.

163. (with M. Jürgens) Error controlled regularization by projection, ETNA, **25** (2006), 67–100.
164. (with P. Binev, A. Cohen, R. DeVore) Universal algorithms for learning theory - Part II : piecewise polynomial functions, Constr. Approx. **26**(2) (2007), 127-152.
165. (with A. Kunoth, J. Vorloeper) Convergence of adaptive wavelet methods for goal-oriented error estimation, in: Numerical Mathematics and Advanced Applications, Proceedings of ENUMATH 2005, A. Bermúdez de Castro, D. Gómez, P. Quintela, P. Salgado, eds., Springer-Verlag, Berlin-Heidelberg, 2006.
166. (with A. Barron, A. Cohen, R. DeVore) Approximation and learning by greedy algorithms, Annals of Statistics, **3** (No 1)(2008), 64–94.
167. (with A. Cohen, R. DeVore) Compressed Sensing and best k-term approximation, J. Amer. Math. Soc. 22 (2009), no. 1, 211–231.
168. (with P. Binev, A. Cohen, R. DeVore) Universal Piecewise Polynomial Estimators for Machine Learning, in: Curve and Surface Design: Avignon 2006, P. Chenin, T. Lyche, L.L. Schumaker, eds., Nashboro Press, Brentwood, TN, 2007, 48–77.
169. (with S. Dekel, P. Petrushev) Multilevel preconditioning for partition of unity methods — Some analytic concepts, Numer. Math., **107** (2007), 503–532.
170. (with A. Cohen, R. DeVore) A taste of compressed sensing, SPIE Conference Proceedings: Independent Component Analyses, Wavelets, Unsupervised Nano-Biometric Sensors, and Neural Networks V, Harold H. Szu, Jack Agee, editors, Vol. 6576, 9 April, 2007, ISBN: 9780819466983.
171. (with K. Brix, M. Campos Pinto) A multilevel preconditioner for the interior penalty discontinuous Galerkin method, SIAM J. Numer. Anal., **46**(5) (2008), 2742–2768.
172. (with K. Brix, M. Campos Pinto, R. Massjung) A multilevel preconditioner for the interior penalty discontinuous Galerkin method II - Quantitative studies, Commun. Comput. Phys., **5** (2009), 296-325.
173. (with T. Rohwedder, R. Schneider, A. Zeiser) Adaptive eigenvalue computation - complexity estimates, Numer. Math. 110 (2008), no. 3, 277–312.
174. (with A. Cohen, R. DeVore) Instance optimal decoding by thresholding in compressed sensing, Contemp. Math., 505 American Mathematical Society, Providence, RI, 2010, 1–28.
175. (with P. Binev, F. Blanco-Silva, D. Blom, R. Sharpley, T. Vogt) Super-Resolution Image Reconstruction by Nonlocal Means applied to High-Angle Annular Darkfield Scanning Transmission Electron Microscopy (HAADF–STEM), IGPM Report, RWTH-Aachen, Sep. 2009.
176. (with P. Binev, R. DeVore, P. Lamby, D. Savu, R. Sharpley) Compressed Sensing and Electron Microscopy. In: Modeling Nanoscale Imaging in Electron Microscopy, Series: Nanostructure Science and Technology, Springer-Verlag, ISBN 978-1-4614-2190-0, 2012, T. Vogt, W. Dahmen, P. Binev (eds.)
177. (with P. Binev, F. Blanco-Silva, D. Blom, P. Lamby, R. Sharpley, T. Vogt) High Qual- ity Image Formation by Nonlocal Means Applied to High-Angle Annular Darkfield Scanning Transmission Electron Microscopy (HAADF-STEM). In: Modeling Nanoscale Imaging in Electron Microscopy, Series: Nanostructure Science and Technology, Springer-Verlag, ISBN 978-1-4614-2190-0, 2012, T. Vogt, W. Dahmen, P. Binev (eds.)
178. (with B. Berkels, A. B. Yankovich, F. Shi, P. M. Voyles, R. Sharpley, P. Binev) High Precision STEM Imaging by Non-Rigid Alignment and Averaging of a Series of Short Exposures, in: Microscopy & Microanalysis 2012, July 29 – Aug. 2, 2012, proceedings.
179. (with S. Dekel, P. Petrushev) Two-level-split decomposition of anisotropic Besov spaces, Constr. Approx. 31 (2010), no. 2, 149–194.
180. (with P. Binev, Ph. Lamby) Fast high-dimensional approximation with sparse occupancy trees, J. Comput. Appl. Math. 235 (2011), no. 8, 2063–2076.
181. (with P. Binev, A. Cohen, R. DeVore, G. Petrova, P. Wojtaszczyk) Convergence rates for greedy algorithms in reduced basis methods, SIAM J. Math. Anal. 43 (2011), no. 3, 1457–1472.
182. (with A. Cohen, G. Welper) Adaptivity and variational stabilization for convection-diffusion equations, ESAIM Math. Model. Numer. Anal. 46 (2012), no. 5, 1247–1273.

183. (with C. Huang, Chr. Schwab, G. Welper) Adaptive Petrov-Galerkin methods for first order transport equations, SIAM J. Numer. Anal. 50 (2012), no. 5, 2420–2445.
184. (with B. Berkels, P. Binev, D. Blom, R. Sharpley, T. Vogt) Optimized Imaging Using Non-Rigid Registration, Ultramicroscopy, 25-Nov-2013, Vol. 138, 46–56.
185. (with A. B. Yankovich, B. Berkels, P. Binev, S.I. Sanchez, S.A. Bradley, and P.M. Voyles) Picometer precision STEM imaging of Pt Nanocatalysts, Nature Communications 5, Article number: 4155 (2014) doi:10.1038/ncomms5155
186. (with M. Bachmayr, R. DeVore, L. Grasedyck) Approximation of high-dimensional rank one tensors, Constr. Approx. 39 (2014), no. 2, 385–395.
187. (with Chr. Plesken, G. Welper) Double greedy algorithms: reduced basis methods for transport dominated problems, ESAIM Math. Model. Numer. Anal.48 (2014), no.3, 623–663.
188. (with T. Gotzen, S. Müller, M. Rom) Numerical simulation of transpiration cooling through porous material, Internat. J. Numer. Methods Fluids 76 (2014), no. 6, 331–365.
189. (with P. Binev, A. Cohen, R. DeVore) Classification algorithms using adaptive partitioning, Ann. Statist. 42 (2014), no. 6, 2141–2163.
190. (with Ch. Huang, G. Kutyniok, W-Q Lim, Chr, Schwab, G. Welper) Efficient resolution of anisotropic structures, Lect. Notes Comput. Sci. Eng., 102, Springer, Cham, 2014, 25–51.
191. (with K. Brix, C. Canuto) Robust preconditioners for DG-discretizations with arbitrary polynomial degrees, Lect. Notes Comput. Sci. Eng., 98 Springer, Cham, 2014, 537–545.
192. (with G. Deolmi, S. Müller) Effective boundary conditions for compressible flows over rough boundaries, Math. Models Methods Appl. Sci. 25 (2015), no. 7, 1257–1297.
193. (with A.B. Yankovich, B. Berkels, P. Binev, P.M. Voyles) High-precision scanning transmission electron microscopy at coarse pixel sampling for reduced electron dose, Ad- vanced Structural and Chemical Imaging, (2015) 1:2, DOI 10.1186/s40679-015-0003-9
194. (with N. Mevenkamp, P. Binev, P.M. Voyles, A.B. Yankovich, B. Berkels) Poisson Noise Removal from High-Resolution STEM images based on periodic Block-matching, Advanced Structural and Chemical Imaging. DOI 10.1186/s40679- 015-0004-8, December 2015, 1:3, first on-line, 25 March 2015
195. (with M. Bachmayr) Adaptive near-optimal rank tensor approximation for high-dimensional operator equations, Found. Comput. Math. 15 (2015), no. 4, 839–898.
196. (with K. Brix, C. Canuto) Nested dyadic grids associated with Legendre-Gauss-Lobatto grids, Numer. Math. 131 (2015), no. 2, 205–239.
197. (with K. Brix, M. Campos Pinto, C. Canuto) Multilevel preconditioning of discontinuous Galerkin spectral element methods. Part I: geometrically conforming meshes, IMA J. Numer. Anal. 35 (2015), no. 4, 1487–1532.
198. How to best sample a solution manifold?, Appl. Numer. Harmon. Anal. Birkhäuser/Springer, Cham, 2015, 403–435.
199. (with M. Bachmayr) Adaptive low-rank methods: problems on Sobolev spaces, SIAM J. Numer. Anal. 54 (2016), no. 2, 744–796.
200. (with M. Bachmayr) Adaptive low-rank methods for problems on Sobolev spaces with error control in L_2, ESAIM Math. Model. Numer. Anal. 50 (2016), no. 4, 1107–1136.
201. (with R. DeVore, L. Grasedyck, E. Süli) Tensor-sparsity of solutions to high-dimensional elliptic partial differential equations, Found. Comput. Math. 16 (2016), no. 4, 813–874.
202. (with P. Binev, A. Cohen, R. DeVore, G. Petrova, P. Wojtaszczyk) Data assimilation in reduced modeling, SIAM/ASA J. Uncertain. Quantif. 5 (2017), no. 1, 1–29.
203. (with A. Cohen, R. DeVore) Orthogonal matching pursuit under the restricted isometry property, Constr. Approx. 45 (2017), no. 1, 113–127.
204. (with G. Deolmi, S. Müller) Effective boundary conditions: a general strategy and application to compressible flows over rough boundaries, Commun. Comput. Phys. 21 (2017), no. 2, 358–400.
205. (with D. Broersen, R. Stevenson) On the stability of DPG formulations of transport equations, Math. Comp. 87 (2018), no. 311, 1051–1082.
206. (with G. Kutyniok, W-Q Lim, Chr. Schwab, G. Welper) Adaptive anisotropic Petrov-Galerkin methods for first order transport equations, J. Comput. Appl. Math. 340 (2018), 191–220.

207. (with G. Deolmi, S. Müller, M. Albers, P.S. Meysonnat, W. Schröder) Effective boundary conditions for turbulent compressible flows over a riblet surface, Springer Proc. Math. Stat., 236, Springer, Cham, 2018, 473–485.
208. (with M. Bachmayr, A. Cohen) Parametric PDEs: sparse or low-rank approximations?, IMA J. Numer. Anal. 38 (2018), no. 4, 1661–1708.
209. (with R. Stevenson) Adaptive strategies for transport equations, Comput. Methods Appl. Math. 19 (2019), no. 3, 431–464.
210. (with F. Gruber, O. Mula) An adaptive nested source term iteration for radiative transfer equations, Math. Comp. 89 (2020), no. 324, 1605–1646.
211. (with A. Cohen, R. DeVore, J. Nichols) Reduced basis greedy selection using random training sets, ESAIM Math. Model. Numer. Anal. 54 (2020), no. 5, 1509–1524.
212. (with M. Bachmayr) Adaptive low-rank approximations for operator equations: accuracy control and computational complexity, Contemp. Math., 754 American Mathematical Society, [Providence], RI, 2020, 1–44.
213. (with A. Cohen, R. DeVore, J. Fadili, O. Mula, J. Nichols) Optimal reduced model algorithms for data-based state estimation, SIAM J. Numer. Anal. 58 (2020), no. 6, 3355–3381.
214. (with A. Cohen, O. Mula, J. Nichols) Nonlinear reduced models for state and parameter estimation, SIAM/ASA J. Uncertain. Quantif. 10 (2022), no. 1, 227–267.
215. (with R. Stevenson, J. Westerdiep) Accuracy controlled data assimilation for parabolic problems, Math. Comp. 91 (2022), no. 334, 557–595.
216. (with M. Wang, Z. Wang) Nonlinear reduced DNN models for state estimation, Commun. Comput. Phys. 32 (2022), no. 1, 1–40.
217. (with A. Cohen, R. DeVore) State estimation—the role of reduced models, ICIAM 2019 SEMA SIMAI Springer Ser., 1 Springer, Cham, 2022, 57–77.
218. (with H. Monsuur, R. Stevenson) Least squares solvers for ill-posed PDEs that are conditionally stable, ESAIM Math. Model. Numer. Anal. 57 (2023), no. 4, 2227–2255.
219. (with A. Cohen, M. Dolbeault, A. Somacal) Reduced order modeling for elliptic problems with high contrast diffusion, ESAIM Math. Model. Numer. Anal. 57 (2023), no. 5, 2775–2802.
220. (with P. Binev, A. Bonito, A. Cohen, R. DeVore, G. Petrova) Solving PDEs with incomplete information, SIAM J. Numer. Anal. 62 (2024), no. 3, 1278–1312.
221. (with O. Mula) Accuracy controlled schemes for the eigenvalue problem of the radiative transfer equation, Preprint, July 2023, http://arxiv.org/abs/2307.07780.
222. (with A. Cohen, M. Dolbeault, A. Somacal) Reduced order modeling for elliptic problems with high contrast diffusion coefficients, Jan. 2023, to appear in: ESAIM: M2A.

Multilevel Representations of Random Fields and Sparse Approximations of Solutions to Random PDEs

Markus Bachmayr and Albert Cohen

Abstract In the analysis of random fields as well as in applications, such as in uncertainty quantification, representations of random fields as function series play an important role. While the most usual such expansions are based on Karhunen-Loève decompositions, in many cases of interest, other expansions with independent coefficients exist and may be more relevant. In particular, for certain classes of Gaussian random fields, one can obtain such expansions in terms of function systems with wavelet-type multilevel structure. In this article, we review constructions of such representations that are suitable for numerical methods, as well as their impact on sparse approximations of solutions to random partial differential equations.

1 Introduction

For a given set D, such as a real interval, a domain of $\mathbb{R}^d$ or a manifold, a random field on D is a family $\big(b(x)\big)_{x\in D}$ where $b(x)$ is a real-valued random variable on a joint probability space $(\Omega, \mathcal{A}, \mu)$ for each $x \in D$. We can identify this family with a function $b\colon D \times \Omega \to \mathbb{R}$ via $b(x, \omega) = b(x)(\omega)$ for $x \in D$ and $\omega \in \Omega$. When D is a real interval (as in the classical example of Brownian motion), b is called a stochastic process, and the function $b(\cdot, \omega)$ for fixed ω is called a sample path; in the case of general random fields, $b(\cdot, \omega)$ is called a realization.

If $b(x) \in L_2(\Omega, \mu)$ for each $x \in D$, the random field b has a mean function on D given by $\mathbb{E}b(x) = \int_\Omega b(x)\,\mathrm{d}\mu$, $x \in D$, and a covariance function

Markus Bachmayr
Institut für Geometrie und Praktische Mathematik, RWTH Aachen University, Templergraben 55, 52062 Aachen, Germany, e-mail: bachmayr@igpm.rwth-aachen.de

Albert Cohen
Laboratoire Jacques-Louis Lions, Sorbonne Université, 4 place Jussieu, 75005 Paris, France, e-mail: albert.cohen@sorbonne-universite.fr

R. DeVore, A. Kunoth (eds.), *Multiscale, Nonlinear and Adaptive Approximation II*,
https://doi.org/10.1007/978-3-031-75802-7_3

$$K(x,x') = \mathbb{E}\big(b(x) - \mathbb{E}b(x)\big)\big(b(x') - \mathbb{E}b(x')\big), \quad x,x' \in D.$$

If $\mathbb{E}b = 0$, the random field is called centered. A particular case are Gaussian random fields, where for each choice of $x_1, \dots, x_N \in D$, one has that $(b(x_1), \dots, b(x_N))$ is a Gaussian random vector. In this case, the mean and covariance functions – the latter always positive semidefinite – uniquely determine the random field, in the sense that they determine its probability law.

The covariance operator

$$T : v \mapsto \int_D K(\cdot, x) v(x) dx, \tag{1}$$

is positive, self-adjoint, compact and of trace class in $L_2(D)$, under the assumption that $\mathbb{E}\|b\|^2_{L_2(D)} < \infty$. This leads to the so-called Karhunen-Loève (KL) representation, which for a centered process takes the form

$$b = \sum_{j=1}^{\infty} y_j \phi_j, \tag{2}$$

where the ϕ_j are the eigenfunctions of T and the $y_j = \langle b, \phi_j \rangle_{L_2(D)}$ are centered scalar random variables. The KL representation is characterized by the double orthogonality property

$$\int_D \phi_j(x)\phi_k(x)dx = 0 \quad \text{and} \quad \mathbb{E}(y_j y_k) = 0 \quad \text{when } j \neq k.$$

In the case of a centered Gaussian process, the variables y_j are standard normal Gaussian (after convenient normalization of the ϕ_j) and pairwise independent.

In many relevant practical cases of random fields, the KL basis functions ϕ_j are globally supported over D. This occurs in particular when the field is *stationary*, so that the covariance function has the form

$$K(x,x') = k(x - x')$$

and T is a convolution operator. In the case of periodic covariance functions corresponding to random fields on a torus, the corresponding operator T is diagonalized by the Fourier complex exponentials.

One could, however, be interested in describing of the random field through other expansions

$$b = \sum_{j=1}^{\infty} y_j \psi_j, \tag{3}$$

where the basis functions ψ_j are better localized. One main theoretical motivation is the study the fine smoothness properties of b, which are best described by multilevel expansions into wavelet bases.

More recently, it was also observed that local expansions can be beneficial in the context of uncertainty quantification for PDEs that are parametrized by the random field b. Indeed, for numerical methods based on sparse polynomial expansions in the variables y_j, stronger convergence results can be obtained when these variables are associated to local bases.

The goal of this paper is to deliver a short survey on the development of multilevel expansions of random fields, with a particular focus on their recent applications to these numerical approximation methods. The construction of such multilevel expansions is adressed in §2. Note that one may either (i) insist on using a standard wavelet basis but give up the decorrelation property $\mathbb{E}(y_j y_k) = 0$, or (ii) maintain this property, and in particular use i.i.d. standard variables in the case of a Gaussian field, but use wavelets adapted to the given random field.

We follow the second approach and study the resulting multilevel expansions for relevant random fields on general domains and on the sphere. These may be seen as a generalization of the well-known Levy-Ciesielski decomposition of the Brownian motion into the Schauder hat function basis.

We then discuss in §3 results on the convergence of sparse polynomial approximations of solutions of PDEs parametrized by the random field b. In §4, we turn to numerical methods that can exploit expansions of random fields in terms of localized functions.

2 Multiscale series representations of random fields

Random fields can be specified in different ways. One option would be to option would be to define b in (3) directly by some choice of functions $(\psi_j)_{j\geqslant 1}$ and random variables $(y_j)_{j\geqslant 1}$. However, in applications such as spatial statistics, one is typically interested in defining random fields using interpretable statistical information, such as moments or cumulants. In this regard, Gaussian random fields play a special role, in that they are uniquely defined by their mean and covariance functions, that is, by their first and second cumulants.

In this section, we thus focus on centered Gaussian random fields and consider ways of obtaining expansion (2) or (3) for given covariance functions. We use the Brownian motion as a guiding example, paving the way towards multileval expansions of more general random fields.

2.1 Gaussian random fields

Let $b = (b(x))_{x\in D}$ be a Gaussian random field. Then if the realizations of b are almost surely in a separable Banach space $\mathcal{B}$, the random field also defines a Gaussian measure Γ_b on $\mathcal{B}$. One possible instance is $\mathcal{B} = L_2(D)$.

The regularity of realizations of Gaussian random fields is closely connected to the regularity of their covariance functions. For any D such that the Kolmogorov-Chentsov theorem applies (see, for example, [64, 1]), such as domains $D \subset \mathbb{R}^d$ or manifolds, one obtains as a corollary that if $K \in C^{2\alpha}(\overline{D} \times \overline{D})$, then b has a modification with realizations that are almost surely in $C^\beta(\overline{D})$ for any $\beta < \alpha$. As a consequence, since the random field then also defines a Gaussian measure on the separable Banach space $C(\overline{D})$ with $\Gamma_b(C^\beta(\overline{D})) = 1$, one may also define Γ_b on $\mathcal{B} = C(\overline{D})$. Accordingly, the covariance operator in (1) is then treated as an operator from $C(\overline{D})'$ to $C(\overline{D})$ with

$$T\colon \mathcal{B}' \to \mathcal{B},\ \varphi \mapsto T\varphi = \mathbb{E}\big(\varphi(b)b\big).$$

A covariance function K gives rise to a *reproducing kernel Hilbert space* [2] associated to the kernel K, which we denote by $\mathcal{H}_K$. Its inner product is defined first for functions $K(\cdot,x)$, $x \in D$, by $\langle K(\cdot,x), K(\cdot,x')\rangle_{\mathcal{H}_K} = K(x,x')$ for $x,x' \in D$, and $\mathcal{H}_K$ is obtained as the closure of finite linear combinations of the form $\sum_{i=1}^n \alpha_i K(\cdot,x_i)$ with $\alpha_i \in \mathbb{R}$, $x_i \in D$ for $i = 1,\dots,n$, in this norm. One then has the reproduction property

$$\langle h, K(\cdot,x)\rangle_{\mathcal{H}_K} = h(x) \quad \text{for all } h \in \mathcal{H}_K, x \in D. \tag{4}$$

The space $\mathcal{H}_K$ is usually referred to as the *Cameron-Martin space* of the Gaussian measure Γ_b associated to the random field (see, for example, [17]). Note that almost surely, realizations of b are *not* in $\mathcal{H}_K$.

2.2 Expansions of Brownian motion

The Wiener process W that describes Brownian motion on $[0,\infty)$ can be characterized as the centered Gaussian process with covariance function $\mathbb{E}\big(W(t)W(s)\big) = \min\{s,t\}$. The resulting process has the property that $W(0) = 0$ and $W(t) - W(s) \sim \mathcal{N}(0, t-s)$ for $0 \leqslant s < t$ with independent increments.

Restricting to the interval $D = [0,1]$ and conditioning on the event $W(1) = 0$, one obtains the Brownian bridge B with covariance function

$$K(s,t) = \mathbb{E}\big(B(t)B(s)\big) = \min\{s,t\} - st, \quad s,t \in [0,1]. \tag{5}$$

The KL representation of this process takes the form (2), that is, $B = \sum_{j=1}^\infty y_j \phi_j$, with $y_j \sim \mathcal{N}(0,1)$ i.i.d. and

$$\phi_j(x) = \frac{\sqrt{2}}{\pi j}\sin(\pi j x), \quad j \in \mathbb{N}. \tag{6}$$

One instance of an alternative representation with i.i.d. $y_j \sim \mathcal{N}(0,1)$ as in (3) is the *Lévy-Ciesielski representation:* decomposing each $j \in \mathbb{N}$ uniquely as $j = 2^\ell + k$ with $\ell \in \mathbb{N}_0$ and $k \in \{0,\dots,2^\ell - 1\}$ and defining

$$\psi_j(x) = \psi_{\ell,k}(x) = 2^{-\ell/2}\sigma(2^\ell x - k), \tag{7}$$

where $\sigma(x) = \max\{1 - |2x - 1|, 0\}$, one also has

$$B = \sum_{j=1}^{\infty} y_j \psi_j. \tag{8}$$

The family $(\psi_j)_{j\in\mathbb{N}}$ is also known as the Faber-Schauder basis. Since B is associated to a Gaussian measure on $C(\overline{D})$, a canonical definition of the limit in (8) is that the sum on the right converges almost surely in $C(\overline{D})$ and the limit has the same law as B.

Note the different support and decay properties of the functions in the two expansions: whereas $\|\phi_j\|_{L_\infty} \eqsim \|\phi_j\|_{L_2} \eqsim j^{-1}$, we have $\|\psi_j\|_{L_\infty} \eqsim j^{-1/2}$, but $|\operatorname{supp}\psi_j| \eqsim j^{-1}$ and $\|\psi_j\|_{L_2} \eqsim j^{-1}$. A particular type of decay property satisfied by (8) that will be of importance to us later is the scale-wise decay

$$\sup_{x\in[0,1]} \sum_{k=0}^{2^j-1} |\psi_{\ell,k}(x)| \leqslant 2^{-\ell/2}. \tag{9}$$

The local convergence properties of the expansion in terms of the functions ψ_j in (7) make it particularly suitable for studying the Hölder regularity of realizations, for which they were initially used by Ciesielski [22]. It was later also used for investigating the Besov regularity of Brownian motion [65] and the Besov-Orlicz regularity of certain fractional Brownian motions [23].

2.3 Expansions of general Gaussian random fields

For a Hilbert space $\mathcal{H}$, a family of functions $\Psi = (\psi_j)_{j\in\mathbb{N}}$ with $\psi_j \in \mathcal{H}$, $j \in \mathbb{N}$, is a *tight frame* of $\mathcal{H}$ if the span of Ψ is dense in $\mathcal{H}$ and if for some $a > 0$,

$$\sum_{j=1}^{\infty} \left|\langle h, \psi_j\rangle_{\mathcal{H}}\right|^2 = a\|h\|_{\mathcal{H}}^2 \quad \text{for all } h \in \mathcal{H}.$$

If in addition the ψ_j, $j \in \mathbb{N}$, are normalized such that $a = 1$, then Ψ is called a *Parseval frame* of $\mathcal{H}$. If Ψ is a Parseval frame, then

$$h = \sum_{j=1}^{\infty} \langle h, \psi_j\rangle_{\mathcal{H}} \psi_j \quad \text{for all } h \in \mathcal{H}. \tag{10}$$

Note in particular that any orthonormal basis of $\mathcal{H}$ is a Parseval frame of $\mathcal{H}$.

The following characterization of expansions of Gaussian random fields in terms of i.i.d. standard Gaussian coefficients was obtained in [58].

Theorem 2.1 *Let b be a Gaussian random field on D with covariance function K and with realizations in the separable Banach space $\mathcal{B}$, and let $\mathcal{H}_K$ be the corresponding Cameron-Martin space. Then $b = \sum_{j=1}^{\infty} y_j \psi_j$ with i.i.d. $y_j \sim \mathcal{N}(0,1)$, with convergence almost surely in $\mathcal{B}$, if and only if $(\psi_j)_{j\in\mathbb{N}}$ is a Parseval frame of the Cameron-Martin space $\mathcal{H}_K$.*

The "if" part of this theorem is not difficult to see: for given K, let $(\psi_j)_{j\in\mathbb{N}}$ be a Parseval frame of $\mathcal{H}_K$ and let $\tilde{b} = \sum_{j=1}^{\infty} y_j \psi_j$. Then with the reproduction property (4), we obtain

$$\begin{aligned}
\mathbb{E}\big(\tilde{b}(x)\tilde{b}(x')\big) &= \sum_{j,j'\in\mathbb{N}} \psi_j(x)\psi_{j'}(x')\mathbb{E}(y_j y_{j'}) \\
&= \sum_{j\in\mathbb{N}} \psi_j(x)\psi_j(x') \\
&= \sum_{j\in\mathbb{N}} \langle K(\cdot,x),\psi_j\rangle_{\mathcal{H}_K}\psi_j(x') \\
&= K(x,x'),
\end{aligned}$$

and thus the centered Gaussian random fields b and $\tilde{b}$ have the same law.

The following result that provides a construction of Parseval frames of Cameron-Martin spaces was also obtained in [58].

Theorem 2.2 *Under the assumptions of Theorem 2.1, for the covariance operator $T\colon \mathcal{B}' \to \mathcal{B}$ associated to b, let $T = SS'$ with $S\colon \mathcal{G} \to \mathcal{B}$, where $\mathcal{G}$ is a real separable Hilbert space. Let $(\zeta_j)_{j\in\mathbb{N}}$ be a Parseval frame of some closed subspace of $\mathcal{G}$ containing $(\ker S)^\perp$. Then $(S\zeta_j)_{j\in\mathbb{N}}$ is a Parseval frame of $\mathcal{H}_K$.*

Remark 2.1 The factorization in Theorem 2.2 can be viewed as a generalization of a standard approach to sampling Gaussian random vectors in $\mathbb{R}^n$: let the centered random vector under consideration has covariance $C \in R^n$. Then if $C = SS^\top$ with a $S \in \mathbb{R}^{n\times m}$, and if $z \sim \mathcal{N}(0, I)$, then $Sz \sim \mathcal{N}(0, C)$.

In the case of the KL expansion, one takes $\mathcal{B} = \mathcal{G} = L_2(D)$ and $S = T^{1/2}$ in Theorem 2.2, and applies S to an orthonormal basis of eigenfunctions of the (compact) operator $T\colon L_2(D) \to L_2(D)$ to obtain an orthonormal basis of $\mathcal{H}_K$ (see, for example, [58] or [44]). However, as we previously illustrated in the case of the Brownian bridge, many other choices are possible based on Theorems 2.1 and 2.2, either by choosing a different orthonormal basis of $L_2(D)$ or a more general Parseval frame.

One instance of a choice of orthonormal basis of $L_2(D)$ that leads to expansions in terms of localized functions is considered in the case of the full space $D = \mathbb{R}^d$ in [14]. This is done under the assumption that the operator T^{-1} defines a pseudodifferential operator of positive order on $\mathbb{R}^d$, with certain additional regularity assumptions on its symbol. The starting point is then a suitable wavelet basis $(\psi_\lambda)_{\lambda\in\vee}$ of $L_2(\mathbb{R}^d)$, such as Meyer wavelets [61] or Daubechies wavelets [33]; $\vee$ is a countable set of scale-space indices, where we denote by $|\lambda|$ the wavelet level of $\lambda \in \vee$. The square root of T is then expressed as

$$T^{1/2} = \frac{2}{\pi} \int_0^\infty \left(T^{-1} + t^2 I\right)^{-1} dt,$$

and wavelet compressibility estimates for the operators $\left(T^{-1} + t^2 I\right)^{-1}$ that are uniform in t are then used to obtain localization estimates for the functions $T^{1/2}\psi_\lambda$. In particular, for each λ, with $x_\lambda \in \mathbb{R}^d$ denoting the center of $\operatorname{supp}\psi_\lambda$, one obtains estimates of the form

$$|T^{1/2}\psi_\lambda(x)| \lesssim \frac{2^{|\lambda|(d-m)/2}}{(1 + 2^{|\lambda|}\|x - x_\lambda\|_2)^{d+\gamma}}. \tag{11}$$

Here $m > 0$ is the order of T^{-1} and $\gamma > 0$ depends on properties of its symbol.

While the above approach from [14] covers a fairly large class of covariance operators, it is rather difficult to construct the resulting expansion computationally. In the direction of more practical strategies, direct orthogonalization in the Cameron-Martin space has been studied in [46]. However, this leads to an expansion in terms of localized functions only in special cases. We next consider two instances where additional structure of the covariance function K is used to obtain a practically realizable expansion with localization from Theorem 2.2.

2.4 Stationary Gaussian random fields on domains

Stationary random fields on $\mathbb{R}^d$ are characterized by invariance of their probability law under translations, which in the case of centered Gaussian random fields is equivalent to the existence of an even function $k : \mathbb{R}^d \to \mathbb{R}$ such that for all $x, x' \in \mathbb{R}^d$,

$$K(x, x') = k(x - x'). \tag{12}$$

In applications, such random fields are often considered on bounded domains $D \subset \mathbb{R}^d$. A particular instance is the family of Matérn covariances [60] with two parameters $\lambda, \nu > 0$ given by

$$k(x) = \frac{2^{1-\nu}}{\Gamma(\nu)} \left(\frac{\sqrt{2\nu}\|x\|_2}{\lambda}\right)^\nu K_\nu\left(\frac{\sqrt{2\nu}\|x\|_2}{\lambda}\right).$$

Here K_ν is the modified Bessel function of the second kind. The Fourier transform of k takes the simple form

$$\hat{k}(\omega) = \int_{\mathbb{R}^d} k(x) e^{-ix\cdot\omega}\, dx = c_{\nu,\lambda} \left(\frac{2\nu}{\lambda^2} + \|\omega\|_2^2\right)^{-(\nu+d/2)},$$

where $c_{\nu,\lambda} = 2^d \pi^{d/2}\Gamma(\nu + d/2)(2\nu)^\nu/(\Gamma(\nu)\lambda^{2\nu})$. From this, it is readily seen that the RKHS $\mathcal{H}_K$ coincides with the Hilbertian Sobolev space $H^{d/2+\nu}(D)$ defined as the restriction to D of $H^{d/2+\nu}(\mathbb{R}^d)$.

Note that the associated operator T^{-1} defines a pseudodifferential operator of order $d + \nu/2$ on $\mathbb{R}^d$, and thus the results of [14] apply. However, since the functions in (11) are not compactly supported, this approach does not lead to expansions of stationary random fields on bounded domains D.

Expansions on bounded domains have been obtained in [8] by periodization: the domain D is embedded into a torus $\mathbb{T}$, and the given stationary random field b is extended to a periodic random field b_p on $\mathbb{T}$. To this end, we assume that $D \subset [-\delta/2, \delta/2]^d$ with a $\delta > 0$, and take $\mathbb{T} = [-\tau, \tau]^d$ with suitable $\tau > \delta$, so that $\mathbb{T}$ contains all possible differences of elements of D. With a sufficiently smooth and even cutoff function ϕ such that $\phi = 1$ on $[-\delta, \delta]^d$ and $\phi = 0$ outside $[-(2\tau - \delta), (2\tau - \delta)]^d$, we define the truncated covariance $k_\mathrm{t}(z) = k(z)\phi(z)$ and obtain its periodization as

$$k_\mathrm{p}(x) = \sum_{n \in Z^d} k_\mathrm{t}(x + 2\tau n). \tag{13}$$

The even function k_p defines a periodic random field on $\mathbb{T}$ provided that it is positive semidefinite, which holds precisely when $\hat{k}_\mathrm{t} = (2\pi)^{-d}\hat{k} * \hat{\phi} \geqslant 0$ on $\mathbb{R}^d$. Since the conditions on ϕ imply that $\hat{\phi}$ cannot be uniformly nonnegative, this condition is not necessarily satisfied. Under additional conditions on k, however, it can be ensured by choosing τ sufficiently large.

This kind of periodization and associated conditions for positive semidefiniteness arise in similar form in efficient procedures for sampling Gaussian random fields by Fourier methods, in particular circulant embedding [67], which can be interpreted as a grid-dependent periodization with non-smooth cutoff function. As considered in [11], a periodization with smooth cutoff as in (13) also leads to a grid-independent variant of circulant embedding.

As a consequence of periodicity, the covariance operator associated to k_p on $\mathbb{T}$ is diagonalized by Fourier exponentials. Taking again $(\psi_{\mathrm{p},\lambda})_{\lambda \in \mathcal{V}}$ to be Meyer wavelets, now in periodized form as an orthonormal basis of $L_2(\mathbb{T})$, the explicit Fourier representation with finitely many terms of these wavelets leads to an equally explicit representation of $\Psi_{\mathrm{p},\lambda} = T^{1/2}\psi_{\mathrm{p},\lambda}$. It then remains to verify the decay properties of these modified wavelets. This is done in [8] under the assumptions

$$\hat{k}(\omega) \eqsim \left(1 + \|\omega\|_2^2\right)^{-m/2}$$

with an $m > d$, that $k \in W^{\lceil m \rceil, 1}(\mathbb{R}^d)$, and

$$|\partial^\alpha \hat{k}(\omega)| \lesssim \left(1 + \|\omega\|_2^2\right)^{-(m+|\alpha|)/2}, \quad |\alpha| \leqslant M, \tag{14}$$

for an $M \geqslant d + 1$. This includes Matérn covariances as a particular case. One then obtains a decay estimate of the same form as the one from [14] in (11),

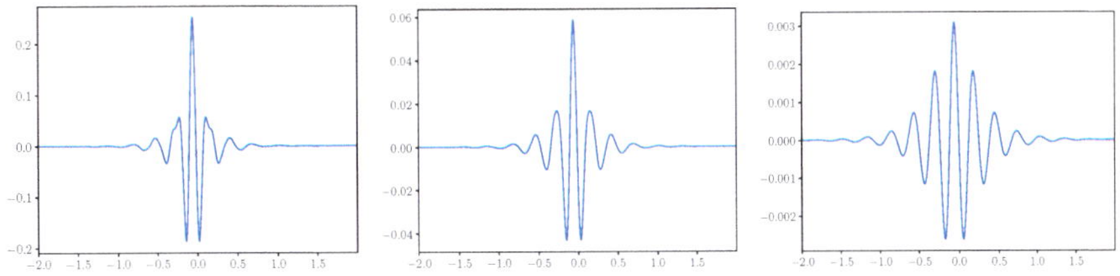

(a) $\nu = \frac{1}{2}$ **(b)** $\nu = 2$ **(c)** $\nu = 8$

Fig. 1: Plots of functions $\Psi_{\mathrm{p},\lambda}$ in (16) for $|\lambda| = 5$, and for a Matérn covariance with correlation length $\frac{1}{4}$ and different values of the smoothness parameter ν.

$$|\Psi_{\mathrm{p},\lambda}(x)| \lesssim \frac{2^{|\lambda|(d-m)/2}}{(1 + 2^{|\lambda|}\|x - x_\lambda\|_2)^{\tilde{M}}}, \tag{15}$$

where $\tilde{M} \leqslant M$ depends also on the smoothness of the cutoff function ϕ; if both are smooth and (14) holds for any M, then (15) yields decay faster than any polynomial order. For the original random field b, we then obtain on D the representation

$$b = \sum_{\lambda \in \vee} y_\lambda \Psi_{\mathrm{p},\lambda} \tag{16}$$

with i.i.d. $y_\lambda \sim \mathcal{N}(0, 1)$. The functions $\Psi_{\mathrm{p},\lambda}$ can be evaluated efficiently using the fast Fourier transform, see [8, §5]. Examples for Matérn covariances are shown in Figure 1.

2.5 Isotropic random fields on the sphere

Random fields on the sphere $\mathbb{S}^2$ arise naturally in applications where measurement data defined on a sphere play a role, such as in geophysics and in astrophysics. The analogue of stationary random fields on $\mathbb{R}^d$ are *isotropic* random fields on $\mathbb{S}^2$, whose probability law is invariant under spherical rotations. Assuming now b to be such an isotropic random field, the covariance function $K(s, s')$ for $s, s' \in \mathbb{S}^2$ of b can be written in the form $K(s, s') = \rho(\langle s, s' \rangle)$ with a function $\rho\colon [-1, 1] \to \mathbb{R}$, that is, K depends only on the angle between s and s'. The associated covariance operator T is diagonalized by the *spherical harmonics* $(Y_{\ell,m})_{\ell\in\mathbb{N}_0, m=-\ell,\ldots,\ell}$, which are an orthonormal basis of $L_2(\mathbb{S}^2)$. Its sequence of eigenvalues $(A_\ell)_{\ell\in\mathbb{N}_0}$, each of multiplicity $2\ell + 1$, is called the *power spectrum* of b. For detailed treatments of basic properties of such random fields and of spherical harmonics, see, for example, [59, 56].

A Parseval frame of $L_2(\mathbb{S}^2)$ with multilevel structure is provided by *spherical needlets* introduced in [63]. These are defined, for $j \in \mathbb{N}_0$ and $k = 1, \ldots, n_j$, as

$$\psi_{jk}(s) = \sqrt{\lambda_{jk}} \sum_{\ell=0}^{\infty} \theta\big(2^{-j}(2\ell+1)\big) \sum_{m=-\ell}^{\ell} Y_{\ell m}(\xi_{jk}) Y_{\ell m}(s), \tag{17}$$

where for each j, $\lambda_{jk} > 0$ and $\xi_{jk} \in \mathbb{S}^2$ for $k = 1, \dots, n_j$ with $n_j = O(2^{2j})$ are weights and points of a quadrature of sufficiently high polynomial exactness on $\mathbb{S}^2$, and $\theta \in C^r(\mathbb{R})$, $r > 2$, is a cutoff function with $\operatorname{supp} \theta = [\frac{1}{2}, 2]$ and $|\theta(t)|^2 + |\theta(2t)|^2 = 1$ for $t \in [\frac{1}{2}, 2]$. In addition to being a Parseval frame, these functions also satisfy a localization property shown in [63] (see also [59]),

$$|\psi_{jk}(s)| \lesssim 2^j (1 + 2^j \arccos\langle s, \xi_{jk}\rangle)^{-r}, \tag{18}$$

that is, for larger j, each ψ_{jk} is localized near ξ_{jk}. The needlets can be used for the analysis of local properties of random fields: as shown in [13], one has asymptotic decay of correlations between needlet coefficients in the sense that

$$\operatorname{Corr}\big(\langle b, \psi_{jk}\rangle_{L_2}, \langle b, \psi_{jk'}\rangle_{L_2}\big) \lesssim \big(1 + 2^j \arccos\langle \xi_{jk}, \xi_{jk'}\rangle\big)^{-M}$$

with an $M > 0$. The standard needlets (17), however, do not lead to expansions of such random fields with i.i.d. coefficients.

To obtain such representations, by Theorem 2.2, as observed in [9], it suffices to apply $T^{1/2}$ to the needlets as elements of a Parseval frame of $L_2(\mathbb{S}^2)$, which yields

$$T^{1/2}\psi_{jk}(s) = \sqrt{\lambda_{jk}} \sum_{\ell=0}^{\infty} \sqrt{A_\ell}\, \theta\big(2^{-j}(2\ell+1)\big) \sum_{m=-\ell}^{\ell} Y_{\ell m}(\xi_{jk}) Y_{\ell m}(s) \tag{19}$$

That the functions $T^{1/2}\psi_{jk}$ are a Parseval frame of $\mathcal{H}_K$ and thus yield the desired expansion of b is again immediate, but it remains to establish their decay properties.

This is done in [9, Thm. 12] under an additional condition on the higher-order forward differences of the sequence $(\sqrt{A_\ell})_{\ell\in\mathbb{N}_0}$, defined recursively by $\Delta_\ell^0 = \sqrt{A_\ell}$ and $\Delta_\ell^i = \Delta_{\ell+1}^{i-1} - \Delta_\ell^{i-1}$ for $i \in \mathbb{N}$, $\ell \in \mathbb{N}_0$. Then provided that for some $\tilde{r} > 2$ and $\beta > 0$,

$$|\Delta_\ell^r| \lesssim (1+\ell)^{-(1+\beta+\tilde{r})}, \quad \ell \in \mathbb{N}_0, \tag{20}$$

we have the localization estimate

$$|T^{1/2}\psi_{jk}(s)| \lesssim 2^{-\beta j}(1 + 2^j \arccos\langle s, \xi_{jk}\rangle)^{-\min\{r,\tilde{r}\}}.$$

Note that the condition (20) plays the role of a discrete analog of (14). The decay of $T^{1/2}\psi_{jk}$ depends both on the smoothness r of the cutoff function θ and on the power spectrum of the random field. Here $\beta > 0$ is related to the Hölder regularity of b: as one can see from the criterion in [56, Thm. 4.6], if (20) holds, then realizations of b are almost surely in $C^\gamma(\mathbb{S}^2)$ for any $\gamma < \beta$. An example of decay as in (20) are Matérn-like random fields on $\mathbb{S}^2$ with power spectrum

$$A_\ell = (1+\ell)^{-2(1+\beta)}, \quad \ell \in \mathbb{N}_0, \tag{21}$$

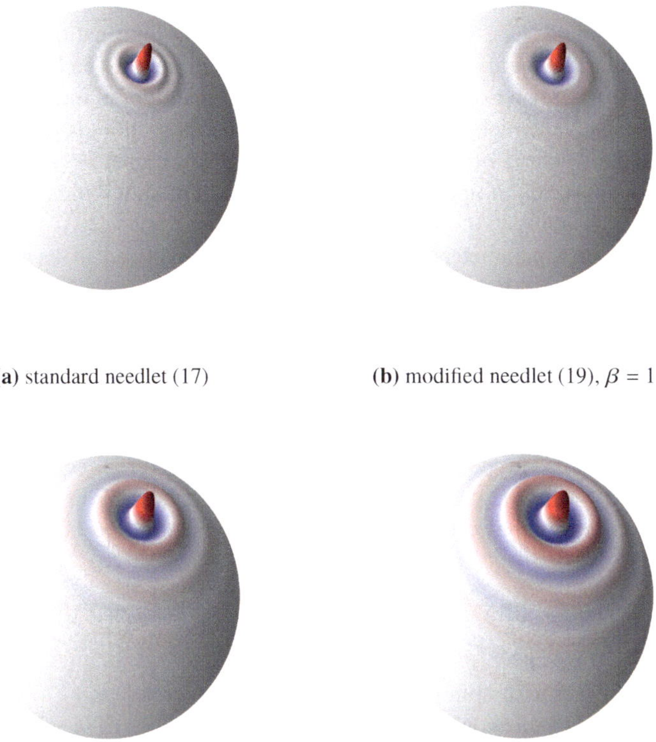

(a) standard needlet (17) **(b)** modified needlet (19), $\beta = 1$

(c) modified needlet (19), $\beta = 5$ **(d)** modified needlet (19), $\beta = 10$

Fig. 2: Comparison of standard and modified needlets: **(a)** standard needlet as in (17); modified needlets for $j = 6$ with power spectrum as (21) with **(b)** $\beta = 1$, **(c)** $\beta = 1$, **(d)** $\beta = 10$.

where (20) holds with any $\tilde{r}$. Resulting modified needlets (19) corresponding to power spectra of this form are shown in Figure 2.

3 Sparse approximations of solutions of random PDEs

One important application of series representations of random fields is in method for the efficient approximation of solutions of PDEs with random coefficients. A particularly well-studied model case are second-order diffusion problems of the form

$$-\nabla \cdot \big(a \nabla u\big) = f \text{ in } D, \qquad u = 0 \text{ on } \partial D, \tag{22}$$

where the uncertainty in the diffusion coefficient a is modelled by assuming a to be a random field. One basic approach to extracting information on the random solution u of (22) would be to use Monte Carlo sampling, by which one can estimate expectations and higher-order moments of u (or of quantities of interest derived from u). An alternative, which we consider in detail in this section, is to use expansions of the random field a in terms of countably many parameters $y_1, y_2, \ldots$ and try to directly construct an approximation of the random solution as a mapping $y = (y_1, y_2, \ldots) \mapsto u(y)$.

For each such parameter sequence y, (22) is considered in its standard weak formulation: find $u \in V = H^1_0(D)$ such that

$$\int_D a(y) \nabla u(y) \cdot \nabla v \, dx = \int_D f v \, dx \quad \text{for all } v \in V, \tag{23}$$

where $f \in V'$ and uniformly positive $a(y) \in L_\infty(D)$ for each y are given.

A frequently used type of random coefficients in (22) are *lognormal* coefficients of the form $a = \exp(b)$, where b is a Gaussian random field. With a series expansion of b as discussed in the previous section, such coefficients thus take the form

$$a(y) = \exp\big(b(y)\big), \quad b(y) = \sum_{j=1}^{\infty} y_j \psi_j \quad \text{where } y_j \sim \mathcal{N}(0,1) \text{ i.i.d.} \tag{24}$$

Here one imposes assumptions ensuring that almost surely, $\|b\|_{L_\infty} < \infty$, a sufficient condition being Hölder continuity of the covariance function of b, as noted in Section 2.1. Then almost surely, $\sup_D a < \infty$ and $\inf_D a > 0$. However, realizations of a can take values that are arbitrarily large or close to zero.

A simpler model are coefficients with affine parameterization

$$a(y) = \bar{a} + \sum_{j=1}^{\infty} y_j \psi_j, \quad y_j \sim \mathcal{U}(-1,1) \text{ i.i.d.} \tag{25}$$

with a fixed, uniformly positive $\bar{a} \in L_\infty(D)$. Such coefficients are treated under a *uniform ellipticity assumption* [26]

$$\alpha_0 = \inf_D \Big(a_0 - \sum_{j=1}^{\infty} |\psi_j| \Big) > 0, \tag{26}$$

which ensures that $a \geqslant \alpha_0 > 0$ and a is uniformly bounded for all realizations.

A series expansion of a given random field as in (24) or (25) determines a coordinate system in terms of which $y \mapsto u(y)$ can be approximated and thus reduces the problem to approximating a family of PDE solutions depending on parameters y. For elliptic problems as in (22), a suitable class of approximations are sparse product polynomial approximations in the parametric variables $y_1, y_2, \ldots$, which in this context are also known as *polynomial chaos* approximations.

In the most general setting, we obtain such approximations by choosing a sequence of polynomials $(P_n^{(j)})_{n\in\mathbb{N}_0}$ with $P_0^{(j)} = 1$ and $\deg P_n^{(j)} = n$ for each parametric variable y_j, $j \in \mathbb{N}$. We then introduce corresponding product polynomials

$$P_\nu(y) = \prod_{j=1}^{\infty} P_{\nu_j}^{(j)}(y_j), \quad \nu \in \mathcal{F}, \tag{27}$$

with the countable set of polynomial degree sequences

$$\mathcal{F} = \left\{\nu \in \mathbb{N}_0^{\mathbb{N}} : \nu_i \neq 0 \text{ for finitely many } i \in \mathbb{N}\right\}.$$

With a finite subset of degrees $\Lambda \subset \mathcal{F}$, the solution is then approximated in the form

$$u(y) \approx \sum_{\nu\in\Lambda} c_\nu P_\nu(y) \tag{28}$$

with $c_\nu \in V$ for each $\nu \in \Lambda$. Such approximations were pioneered in the computational engineering literature in the early 1990s, initially with truncation to a fixed finite number of parametric variables; see, for example, [42, 43]. An analysis of a direct treatment of the case (25) with countable parametric dimensionality was given in [25, 26].

In problems as in (24) and (25), where a certain probability law is assumed for the coefficient a, one is often interested in approximations in the mean-squared sense with respect to the underlying probability measure. In terms of expansions (24) or (25) with i.i.d. coefficients, this is a product measure $\mu = \bigotimes_{j\in\mathbb{N}} \mu_j$ supported on $Y \subseteq \mathbb{R}^{\mathbb{N}}$; if the underlying random field is Gaussian as in (24), we have $\mu_j = \mathcal{N}(0,1)$ for all j with $Y = \mathbb{R}^{\mathbb{N}}$, whereas in the case (25), one has the compactly supported measures $\mu_j = \mathcal{U}(-1,1)$ for all j with $Y = [-1,1]^{\mathbb{N}}$.

Under natural conditions, in both cases, the parameter-dependent solution u satisfies $u \in L_2(Y, V; \mu)$. Choosing $(P_n^{(j)})_{n\in\mathbb{N}}$ as the orthonormal polynomials with respect to μ_j, the associated product polynomials (27) are an orthonormal basis of $L_2(Y, V; \mu)$; for a proof, see, for example, [66]. We thus have the expansion

$$u(y) = \sum_{\nu\in\mathcal{F}} u_\nu P_\nu(y), \quad u_\nu = \int_Y u(y) P_\nu(y)\, d\mu(y) \in V, \tag{29}$$

with convergence in $L_2(Y, V; \mu)$. By retaining the summands with indices in $\Lambda \subset \mathcal{F}$, we obtain an approximation with error in $L_2(Y, V; \mu)$ satisfying

$$\left\| u - \sum_{\nu\in\Lambda} u_\nu P_\nu \right\|_{L_2(Y,V;\mu)} = \left(\sum_{\nu\notin\Lambda} \|u_\nu\|_V^2 \right)^{1/2}. \tag{30}$$

Best approximations with Λ of a given cardinality thus correspond to best n-term approximations of the sequence $(\|u_\nu\|_V)_{\nu\mathcal{F}}$. Their algebraic convergence rates can be characterized by the following classical result usually attributed to Stechkin, see [24, Lemma 3.6] for a proof; see also [32, Sec. 7.4] for a bibliographical discussion.

Lemma 3.1 *Let $0 < p < q < \infty$ and let $(c_\nu)_{\nu\in\mathcal{F}} \in \ell_p(\mathcal{F})$ be a sequence of positive numbers. Let $\Lambda_n \subset \mathcal{F}$ be a set of indices with n largest c_ν. Then*

$$\Big(\sum_{\nu\notin\Lambda_n} c_\nu^q\Big)^{1/q} \leqslant C(n+1)^{-s}, \quad \text{where } C = \|(c_\nu)_{\nu\in\mathcal{F}}\|_{\ell_p},\ s = \frac{1}{p} - \frac{1}{q}.$$

In order to apply this result to the right-hand side of (30) with $q = 2$, we thus need to establish $(\|u_\nu\|_V)_{\nu\mathcal{F}} \in \ell_p(\mathcal{F})$ for a $p < 2$, with smaller p leading to better convergence rates. This leads to semidiscrete approximations, since the coefficients in the approximation in (30) are still elements of the infinite-dimensional function space V. Error estimates for an additional spatial discretization replacing V by finite-dimensional subspaces can be derived from summability of stronger norms, such as higher-order Sobolev norms.

3.1 Affine-parametric coefficients

Let us first consider the diffusion problem in weak form (23) with affinely parametrized coefficient (25) such that the uniform ellipticity condition (26) holds.

In this case, by holomorphy properties of the extension of $y \mapsto u(y)$ from $Y = [-1,1]^{\mathbb{N}}$ to sequences of complex arguments, one can consider a Taylor expansion of u. With $t_\nu = \frac{1}{\nu!}\partial_y^\nu u(0)$, it takes the form

$$u(y) = \sum_{\nu\in\mathcal{F}} t_\nu y^\nu, \quad y \in Y. \tag{31}$$

Here the notion of unconditional convergence requires some care, see [24].

Concerning expansions in terms of orthonormal polynomials as in (29), as a basic example we consider the uniform measure on $Y = [-1,1]^{\mathbb{N}}$, denoted by σ. The corresponding orthonormal polynomials are the product Legendre polynomials $L_\nu(y) = \prod_{j=1}^\infty L_{\nu_j}(y_j)$, with coefficients

$$u_\nu = \int_Y u(y)\, L_\nu(y)\, d\sigma(y) \in V. \tag{32}$$

Initial results on the convergence of sparse polynomial approximations in [25, 26] were based on complex-variable arguments, under assumptions on the summability of the sequence $(\|\psi_j\|_{L_\infty})_{j\in\mathbb{N}}$ for the functions ψ_j, $j \in \mathbb{N}$, in (25). The main result of this type obtained in [26] reads as follows.

Theorem 3.1 *For* (23) *with affinely parametrized coefficient* (25) *satisfying* (26), *assume that* $(\|\psi_j\|_{L_\infty})_{j\in\mathbb{N}} \in \ell_p(\mathbb{N})$ *for some* $p \in (0,1)$. *Then* $(\|t_\nu\|_V)_{\nu\in\mathcal{F}}$ *and* $(\|u_\nu\|_V)_{\nu\in\mathcal{F}}$ *are in* $\ell_p(\mathcal{F})$.

For an overview of such complex-variable techniques, including their application to different classes of problems, we refer to [24]. A different strategy using real-variable arguments is used in [7], based on assumptions that are suitable for taking advantage of localization in the expansion (25). For the case of the Taylor expansion (31), the following is shown in [7].

Theorem 3.2 *For* (23) *with affinely parametrized coefficient* (25), *assume that there exist* $\rho_j > 0$, $j \in \mathbb{N}$, *such that*

$$\sup_D \frac{1}{\bar{a}} \sum_{j=1}^{\infty} \rho_j |\psi_j| < 1. \tag{33}$$

Then

$$\sum_{\nu\in\mathcal{F}} (\rho^\nu \|t_\nu\|_V)^2 < \infty. \tag{34}$$

Note that when $\rho_j = 1$ for all j, since $\bar{a}$ is uniformly positive, the condition (33) is easily seen to be equivalent to (26). For general ρ_j, it can thus be interpreted as a *weighted* uniform ellipticity condition. A weighted summability estimate analogous to Theorem 3.2 is also shown in [7] for the Legendre coefficients (32).

Theorem 3.3 *Under the assumptions of Theorem 3.2,*

$$\sum_{\nu\in\mathcal{F}} (a_\nu^{-1} \rho^\nu \|u_\nu\|_V)^2 < \infty, \quad \textit{where } a_\nu = \prod_{j=1}^{\infty} \sqrt{2\nu_j + 1}. \tag{35}$$

The proof is based on the identity

$$u_\nu = a_\nu \int_Y \frac{1}{\nu!} \partial_y^\nu u(y) \prod_{j=1}^{\infty} \frac{(1-y_j^2)^{\nu_j}}{2^{\nu_j}} \, d\sigma(y)$$

for the Legendre coefficients (32) that follows from the Rodrigues' formula for the Legendre polynomials, and on applying the bounds for Taylor coefficients to appropriately transformed problems to estimate $\frac{1}{\nu!}\partial_y^\nu u(y)$. The argument thus also applies to other orthogonal polynomials for which such an expression is available, such as Jacobi polynomials associated to general beta distributions.

The weighted estimates (34) and (35) immediately yield p-summability estimates that can be used to obtain rates of best n-term approximation with Lemma 3.1.

Corollary 3.1 *Let* (33) *hold with* $\rho_j > 1$, $j \in \mathbb{N}$, *such that* $(\rho_j^{-1})_{j\in\mathbb{N}} \in \ell_q(\mathbb{N})$ *with* $q \in (0,\infty)$ *satisfying* $\frac{1}{q} = \frac{1}{p} - \frac{1}{2}$ *for a* $p \in (0,2)$. *Then both* $(\|t_\nu\|_V)_{\nu\in\mathcal{F}}$ *and* $(\|u_\nu\|_V)_{\nu\in\mathcal{F}}$ *are in* $\ell_p(\mathcal{F})$.

With p and q as above, this can be seen as follows: for the Taylor coefficients, from (34), by Hölder's inequality we obtain

$$\sum_{\nu\in\mathcal{F}} \|t_\nu\|_V^p \leqslant \left(\sum_{\nu\in\mathcal{F}} (\rho^\nu \|t_\nu\|_V)^2\right)^{p/2} \left(\sum_{\nu\in\mathcal{F}} (\rho^\nu)^{-q}\right)^{1-p/2}, \tag{36}$$

where

$$\sum_{\nu\in\mathcal{F}} (\rho^\nu)^{-q} = \prod_{j=1}^\infty \left(\sum_{k=0}^\infty \rho_j^{-qk}\right) = \prod_{j=1}^\infty (1-\rho_j^{-q})^{-1}$$

with convergence of the product on the right precisely when $(\rho_j^{-1})_{j\in\mathbb{N}} \in \ell_q(\mathbb{N})$. The statement for the Legendre coefficients follows similarly using (35).

With the condition on weighted pointwise summability (33) that is used in Theorems 3.2 and 3.3, we can exploit the localization properties of multilevel expansions for a as considered in §2. Assume that we have an affine parameterization (25) in the form

$$a(y) = \bar{a} + \sum_{\lambda\in\vee} y_\lambda \psi_\lambda, \tag{37}$$

where $\vee$ is again a countable set of scale-space indices for the wavelet-type family of functions $(\psi_\lambda)_{\lambda\in\vee}$, where for some $\alpha > 0$,

$$\sup_D \sum_{|\lambda|=\ell} |\psi_\lambda| \leqslant C_1 2^{-\alpha\ell} \tag{38}$$

for all $\ell \geqslant 0$, with $\{\lambda \in \vee : |\lambda| = \ell\} \leqslant C_2 2^{d\ell}$, and where d is the dimension of D. The scale-wise decay property (38) is shared by the function systems arising in the different expansions in §2, where α is related to the Hölder smoothness of realizations.

Under the condition (38), if $a(y)$ in (37) satisfies the uniform ellipticity condition (26), then we can choose $\rho_\lambda = 1 + c2^{\beta|\lambda|}$ for any $\beta < \alpha$ and sufficiently small $c > 0$ to obtain

$$\sup_D \sum_{\lambda\in\vee} \rho_\lambda |\psi_\lambda| < 1.$$

Choosing an ordering $(\lambda_j)_{j\in\mathbb{N}}$ of $\vee$ by increasing level and setting $\psi_j = \psi_{\lambda_j}$, (38) is thus satisfied with $\rho_j = \rho_{\lambda_j}$ such that

$$\rho_j^{-1} \lesssim j^{-\beta/d}.$$

Thus $(\rho_j^{-1})_{j\in\mathbb{N}} \in \ell_q(\mathbb{N})$ for any $q > d/\alpha$. By Corollary 3.1 and Lemma 3.1, we arrive at the following result: For each $n \in \mathbb{N}$, let $\Lambda_n \subset \mathcal{F}$ be a set of n largest norms $\|u_\nu\|_V$, $\nu \in \mathcal{F}$. Then

$$\left\| u - \sum_{\nu\in\Lambda_n} u_\nu L_\nu \right\|_{L_2(Y,V;\sigma)} \leqslant C_{u,s} n^{-s} \quad \text{for any } s < \frac{\alpha}{d}. \tag{39}$$

The above described techniques have also been adapted to give similar estimates for other problem classes, for instance for saddle point problems [20] and for problems on affinely parameterized domains [50].

Remark 3.1 The result of Theorem 3.1 is in general sharp if a is given in terms of ψ_j, $j \in \mathbb{N}$, that have global supports in D, as in the case of KL eigenfunctions, but due to the summability condition on $\|\psi_j\|_{L_\infty}$ is not sharp for expansions of a with multilevel structure. In the latter case, the estimate (39) based on Theorem 3.2 and 3.3 is sharp, but these two theorems in turn do not give a sharp result for globally supported ψ_j.

So far, we have only discussed polynomial approximability in the stochastic variables. Numerically realizable approximations, however, require additional discretization in space, and to this end we thus need to consider approximations from subspaces of the form

$$\tilde{\mathcal{V}}_N = \left\{ \sum_{\nu \in \Lambda} v_\nu L_\nu \colon v_\nu \in V_\nu, \nu \in \Lambda \subset \mathcal{F} \right\} \subset L_2(Y, V; \sigma) \tag{40}$$

with finite-dimensional $V_\nu \subset V$ for each $\nu \in \Lambda$, with total dimension

$$N = \dim \tilde{\mathcal{V}}_n = \sum_{\nu \in \Lambda} \dim V_\nu .$$

Rates for approximation with fixed hierarchies of uniformly refined spatial discretizations can be obtained from summability of higher-order Sobolev norms as obtained in [4] assuming appropriate regularity of the problem data, and in particular under conditions similar to (33) on the derivatives of ψ_j, $j \in \mathbb{N}$. In this case, each space V_ν is selected from a hierarchy $V_1 \subset V_2 \subset \ldots \subset V$. However, in particular in the case of multilevel expansions of a, this type of spatial approximation is not very effective: as noted in [4], the rate $\frac{\alpha}{d}$ in (39) with respect to the number of polynomial terms n for the semidiscrete approximation reduces to a rate $\frac{\alpha}{2d}$ with respect to the total number of degrees of freedom N. In other words, for a given accuracy, fully discrete approximations from a fixed spatial hierarchy have squared costs compared to the semidiscrete ones as in (39).

One can do substantially better by allowing an independent adaptive choice of each space $V_\nu \subset V$ in (40), corresponding to independent adaptive grid refinements or different selections of wavelet basis elements for each given ν. Convergence rates for such fully adaptive approximations are obtained in [4] via summability of the norms of t_ν and u_ν in interpolation spaces $[V, W^{2,\tau}(D)]_\theta$ for $\theta \in (0,1)$ and $1 \leqslant \tau < 2$. Here we assume sufficiently regular f and D and expansions of a of the form (37) satisfying (26) as well as

$$\sup_D \sum_{|\lambda|=\ell} |\psi_\lambda| \lesssim 2^{-\alpha\ell}, \quad \sup_D \sum_{|\lambda|=\ell} |\nabla\psi_\lambda| \lesssim 2^{-(\alpha-1)\ell}$$

for $\alpha \in (0,1]$. As a consequence of [4, Cor. 8.8], with $\tilde{\mathcal{V}}_N$ as in (40) with adaptively chosen V_ν for each $\nu \in \mathcal{F}$, we obtain

$$\min_{v \in \tilde{\mathcal{V}}_N} \|u - v\|_{L_2(Y,V;\sigma)} \leqslant C_s N^{-s} \quad \text{for any } s < \begin{cases} \frac{\alpha}{d}, & d \geqslant 2, \\ \frac{2}{3}\alpha, & d = 1. \end{cases} \tag{41}$$

Thus, remarkably, when making full use of the sparsity of spatial coefficients in the polynomial approximation, the additional spatial discretization leaves the rate unchanged.

3.2 Lognormal coefficients

The case of lognormal coefficients as in (24) poses some additional challenges, both due to the lack of uniform ellipticity and to the nonlinear parameterization in terms of y. In this case, with γ denoting the product Gaussian measure on $Y = \mathbb{R}^{\mathbb{N}}$, we consider approximations in $L_2(Y,V;\gamma)$ by the product Hermite orthonormal polynomials $(H_\nu)_{\nu\in\mathcal{F}}$, and we are interested in the summability of norms of the Hermite coefficients

$$u_\nu = \int_Y u(y)\, H_\nu(y)\, d\gamma(y). \tag{42}$$

A first result on approximability of u in the lognormal case by Hermite polynomials was obtained in [54]: if $(j\|\psi_j\|_{L_\infty}) \in \ell_p(\mathbb{N})$ with $0 < p \leqslant 1$, then $(\|u_\nu\|_V)_{\nu\in\mathcal{F}} \in \ell_p(\mathcal{F})$.

Under assumptions that parallel those in the affine case in [7], summability results for $(\|u_\nu\|_V)_{\nu\in\mathcal{F}}$ that can account for localization in the expansion (24) (and improve on the estimates in [54] in the general case) are shown in [6]. Here we assume that for the functions $(\psi_j)_{j\in\mathbb{N}}$ in (24), there exist $\rho_j > 0$, $j \in \mathbb{N}$, such that with a $q \in (0,\infty)$,

$$\sup_D \sum_{j=1}^{\infty} \rho_j |\psi_j| < \infty \quad \text{and} \quad (\rho_j^{-1})_{j\in\mathbb{N}} \in \ell_q(\mathbb{N}). \tag{43}$$

This assumption already ensures that, despite the lack of uniform ellipticity in this problem, for any $k \in \mathbb{N}_0$ and $\mu \in \mathcal{F}$,

$$\int_Y \|\partial_y^\mu u(y)\|_V^k \, d\gamma(y) < \infty.$$

A central observation in [6] is that under these conditions, for any $r \in \mathbb{N}_0$,

$$\sum_{\nu\in\mathcal{F}} b_\nu \|u_\nu\|_V^2 = \sum_{\|\mu\|_{\ell_\infty} \leqslant r} \frac{\rho^{2\mu}}{\mu!} \int_Y \|\partial_y^\mu u(y)\|_V^2 \, d\gamma(y), \tag{44}$$

where

$$b_\nu = \prod_{j=1}^{\infty}\left(\sum_{k=0}^{r}\binom{\nu_j}{k}\rho_j^{2k}\right).$$

The identity (44) follows from the Rodrigues' formula and Parseval's identity for the Hermite polynomials. Energy estimates that make use of the specific structure of (23) yield boundedness of the right-hand side in (44), and from the resulting weighted summability one obtains the following by Hölder's inequality.

Theorem 3.4 *For* (23) *with lognormal coefficient* (24), *let* (43) *hold with* $q \in (0, \infty)$. *Then* $(\|u_\nu\|_V)_{\nu\in\mathcal{F}} \in \ell_p(\mathcal{F})$ *with* $p \in (0, 2)$ *such that* $\frac{1}{q} = \frac{1}{p} - \frac{1}{2}$.

The conclusions from this theorem are analogous to those of Corollary 3.1: if for $(\psi_\lambda)_{\lambda\in\vee}$ such that

$$a(y) = \exp\left(\sum_{\lambda\in\vee} y_\lambda\psi_\lambda\right), \quad y_\lambda \sim \mathcal{N}(0,1) \text{ i.i.d.}$$

there exist $C_1, C_2 > 0$ such that with an $\alpha > 0$,

$$\sup_D \sum_{|\lambda|=\ell} |\psi_\lambda| \leqslant C_1 2^{-\alpha\ell}, \quad \{\lambda \in \vee : |\lambda| = \ell\} \leqslant C_2 2^{d\ell} \tag{45}$$

for all $\ell \geqslant 0$, we can again choose $\rho_j > 0$ such that $(\rho_j^{-1})_{j\geqslant 1} \in \ell_q(\mathbb{N})$ for all $q > d/\alpha$, and for best n-term Hermite approximations supported on $\Lambda_n \subset \mathcal{F}$, analogously to we obtain

$$\left\| u - \sum_{\nu\in\Lambda_n} u_\nu H_\nu \right\|_{L_2(Y,V;\gamma)} \leqslant C_{u,s} n^{-s} \quad \text{for any } s < \frac{\alpha}{d}. \tag{46}$$

Summability of higher-order Hilbertian Sobolev norms has been shown along similar lines in [4]. This corresponds to spatial discretizations chosen from a fixed hierarchy of spaces, with the same limitations as discussed above in the case of affine parameterizations. In the case of lognormal coefficients, no result comparable to (41) is known on individual adaptive approximations of Hermite coefficients. One obstacle is that for the required non-Hilbertian Bochner spaces, some crucial techniques used in (44), such as Parseval's identity, are not available.

The strategy outlined above has also been adapted to different settings, for instance to certain optimal control problems in [19], and to lognormal elliptic problems on the sphere $\mathbb{S}^2$, for which a foundational analysis was given in [51]. As shown in [9], the expansions discussed in §2.5 are applicable in this case, satisfying a condition analogous to (45) on $\mathbb{S}^2$, provided that the underlying random field is isotropic with suitably decaying power spectrum. One thus obtains convergence results as in (46) also for lognormal elliptic problems on $\mathbb{S}^2$.

Further results on Hermite approximations for lognormal problems have been obtained in [31] using a different technique. While also making use of (44), it is based on estimating the right-hand side using holomorphy with respect to y, rather than using energy estimates. The obtained convergence rates are slightly lower than those obtained in [6, 4]; for example, based on the results in [31], one arrives at (46) with the limiting rate $\frac{\alpha}{d}$ replaced by $\frac{\alpha}{d} - \frac{1}{2}$, assuming that $\alpha > \frac{d}{2}$ rather than $\alpha > 0$. However, due to the use of holomorphy in place of problem-dependent techniques, one gains flexibility in the application of these results.

4 Numerical methods

Expansions of random fields with wavelet-like localization can be of use in various different computational tasks that typically arise in uncertainty quantification. For example, they can be used for drawing samples of random fields. This is computationally nontrivial, and standard methods for doing this efficiently, such as circulant embedding [67, 11], are typically based on discretizations of the entire domain D (or, in case of random fields on $\mathbb{S}^2$, on discretizations covering the entire surface of the sphere; see, for example, [56]). In particular when values of realizations in highly irregularly spaced or locally concentrated points are required, multilevel expansions of random fields that can efficiently be evaluated numerically, as in §2.4 and §2.5, can provide an alternative. In the case of $\mathbb{S}^2$, this is discussed in detail in [9, §6.2].

Here, we now focus on the numerical realization of sparse polynomial approximations of solutions to random PDEs, in particular on how the results of §3 on the achievable convergence rates can be matched algorithmically. In the general setting of approximation by orthogonal polynomials as in (29), our task is thus to find, for a given computational budget, a finite subset $\Lambda \subset \mathcal{F}$, spatial discretization subspaces $V_\nu \subset V$ for $\nu \in \Lambda$, as well as coefficient functions $v_\nu \in V_\nu$ that together yield an approximation

$$u(y) = \sum_{\nu\in\mathcal{F}} u_\nu P_\nu(y) \approx \sum_{\nu\in\Lambda} v_\nu P_\nu(y). \tag{47}$$

At least concerning the choice of Λ, the results on weighted summability of §3 are in principle constructive. As pointed out in [7, Rem. 3.4] and [6, Rem. 5.2], if the exact polynomial cofficients u_ν, $\nu \in \mathcal{F}$, of u in (47) satisfy

$$\sum_{\nu\in\mathcal{F}} w_\nu^2 \|u_\nu\|_V^2 < \infty \tag{48}$$

with some sequence of positive weights $(w_\nu)_{\nu\in\mathcal{F}}$, one can proceed as follows: let $\nu_1^*, \nu_2^*, \ldots \in \mathcal{F}$ be chosen such that $w_{\nu_1^*} \leqslant w_{\nu_2^*} \leqslant \ldots$, and let $\Lambda_n = \{\nu_i^* : 1 \leqslant i \leqslant n\}$, then by (30) we have

$$\Big\| u - \sum_{\nu\in\Lambda_n} u_\nu P_\nu \Big\|_{L_2(Y,V;\mu)} \leqslant w_{\nu_{n+1}^*}^{-1} \Big(\sum_{\nu\notin\Lambda_n} w_\nu^2 \|u_\nu\|_V^2 \Big)^{1/2} \lesssim w_{\nu_{n+1}^*}^{-1} .$$

In both cases (39) and (46), one can show that for the respective weights, one has $w_{\nu_n^*}^{-1} \leqslant C_s n^{-s}$ for any $s < \frac{\alpha}{d}$. At least concerning the selection of polynomial terms, the problem is thus reduced to a sorting of weight sequences.

However, although realizing the expected convergence rate, this construction may in general yield index sets Λ that are far larger than required for a given error. Moreover, it does not give information on how to choose the spaces V_ν, $\nu \in \Lambda$, in particular when these are chosen independently for each ν as in (41). Beyond such practical restrictions, it will also be of interest to treat problems where estimates of the form (48) are not available. This leads us to consider algorithms that, for each given total approximation error, can identify near-optimal $\Lambda \subset \mathcal{F}$ and approximations $v_\nu \in V_\nu \subset V$ for each $\nu \in \Lambda$ as in (47) using only *a posteriori* criteria, that is, using computable information derived from approximate solutions.

4.1 Stochastic Galerkin methods in the affine case

Established adaptive methods for PDEs can be adapted to problems with random coefficients, where discretization spaces (40) are generated by iterative refinement. This requires a suitable variational formulation of the full problem on $D \times Y$.

Let us consider this first in the case of affine parameterizations (25) with the uniform measure σ on $Y = [-1, 1]^{\mathbb{N}}$. Based on the weak formulation (23) of our elliptic model problem, on the space $\mathcal{V} = L_2(Y, V; \sigma)$, we define the operator $\mathcal{B}\colon \mathcal{V} \to \mathcal{V}'$ by

$$\langle \mathcal{B}v, w\rangle = \int_Y \int_D \Bigl(\bar{a} + \sum_{\lambda \in \vee} y_\lambda \psi_\lambda\Bigr) \nabla v(y) \cdot \nabla w(y)\, dx\, d\sigma(y) \tag{49}$$

for $v, w \in \mathcal{V}$ and the linear functional $F \in \mathcal{V}'$ by

$$F(v) = \int_Y \int_D f\, v\, dx\, d\sigma(y), \quad v \in \mathcal{V}. \tag{50}$$

Then a suitable variational formulation on $\mathcal{V}$ reads: find $u \in \mathcal{V}$ such that

$$\langle \mathcal{B}u, v\rangle = F(v) \quad \text{for all } v \in \mathcal{V}. \tag{51}$$

Under the uniform ellipticity condition (26), the operator $\mathcal{B}$ is bounded and elliptic on $\mathcal{V}$. Thus well-posedness of this spatial-parametric variational formulation follows by standard Lax-Milgram theory. The same holds, for any choice of $\tilde{\mathcal{V}}_N$ as in (40), for the so-called stochastic Galerkin discretization of finding $u_N \in \tilde{\mathcal{V}}_N$ such that $\langle \mathcal{B}u_N, v\rangle = F(v)$ for all $v \in \tilde{\mathcal{V}}_N$, and by Cea's lemma, this Galerkin approximation has the quasi-optimality property

$$\|u - u_N\|_{\mathcal{V}} \lesssim \min_{w \in \tilde{\mathcal{V}}_N} \|u - w\|_{\mathcal{V}}\,.$$

As $\mathcal{B}$ is an isomorphism, one also has $\|\mathcal{B}u_N - F\|_{\mathcal{V}'} \eqsim \|u_N - u\|_{\mathcal{V}}$ with $\mathcal{V}' = L_2(Y, V'; \sigma)$, and as usual for adaptive schemes, one can do systematic refinement of discretizations based on the residual $\mathcal{B}u_N - F$. Compared to standard adaptive finite element or wavelet methods, however, we are facing the added difficulty of identifying indices from $\mathcal{F}$ for resolving the infinite-dimensional parametric dependence.

A first aim is to to generate sequences of approximations that converge at the rates obtained in §3.1, which can be seen to be sharp by simple examples; that is, the method should yield discretizations of optimal cardinality. The second, more ambitious aim is to compute such approximations using a number of operations that scales linearly with the number of degrees of freedom in these approximations.

For $\mathcal{B}$ with affine coefficient as in (49), the residual has a particular structure: if $\tilde{u} \in \mathcal{V}$ has nonzero polynomial coefficients contained in $\Lambda \subset \mathcal{F}$, then the nonzero polynomial coefficients of $\mathcal{B}\tilde{u}$ (and hence of the residual $\mathcal{B}\tilde{u} - F$) are contained in $\Lambda \cup \mathcal{M}_\Lambda$, where $\mathcal{M}_\Lambda = \{\nu \in \mathcal{F} \colon \; \exists k \in \mathbb{N} \colon (\nu + e_k \in \mathcal{F}) \vee (\nu - e_k \in \mathcal{F})\}$ is also called the *margin* of Λ.

Note that the margin is an infinite set: for example, when $0 \in \Lambda$, one has $\{e_k \in \mathcal{F} \colon k \in \mathbb{N}\} \subset \mathcal{M}_\Lambda$, and thus some selection needs to be made in approximating the residual using finitely many polynomial coefficients. This can be done by evaluating all elements in the margin up to some maximum parametric dimension larger than the ones present in Λ, as for example in [35, 36, 30, 15], or with a further subselection based on operator compression as in [47, 48, 5, 12, 10]. The underlying approximations of the operator $\mathcal{B}$ are based on a truncation of the summation in (25). In case of a KL-type expansion of the random coefficient with $\|\psi_j\|_{L_\infty} \lesssim j^{-t}$ with $t > 1$, the error in $\mathcal{B}$ resulting from approximating $a(y)$ by $\bar{a} + \sum_{j=1}^{J} y_j \psi_j$ can be bounded up to a constant by

$$\sum_{j>J} \|\psi_j\|_{L_\infty} \lesssim J^{-(t-1)}.$$

Note that in this case one has a loss in the rate of decay from t to $t-1$, a discrepancy that is a major obstruction for adaptive methods of optimal complexity.

This issue can be circumvented when a is given by a multilevel expansion. In this case, $\mathcal{B}$ can be approximated with finitely many terms by $\mathcal{B}_\ell \colon \mathcal{V} \to \mathcal{V}'$, defined for $\ell \in \mathbb{N}_0$ by

$$\langle \mathcal{B}_\ell v, w \rangle = \int_Y \int_D \Bigl(\bar{a} + \sum_{|\lambda| \leqslant \ell} y_\lambda \psi_\lambda \Bigr) \nabla v(y) \cdot \nabla w(y) \, dx \, d\sigma(y), \qquad v, w \in \mathcal{V}, \tag{52}$$

where we keep the first ℓ levels in the expansion of a. As observed in [5], under assumption (38), we have

$$\Bigl\| \sum_{|\lambda| > \ell} |\psi_\lambda| \Bigr\|_{L_\infty} \leqslant \frac{C_1}{1 - 2^{-\alpha}} 2^{-\alpha \ell}. \tag{53}$$

In other words, due to the localization of the functions ψ_λ, the tail of the expansion of a decays at the same rate as the individual norms $\|\psi_\lambda\|_{L_\infty}$. From (53), we immediately get the following estimate.

Lemma 4.1 *Let* $(\psi_\lambda)_{\lambda \in \vee}$ *satisfy* (38)*, then* $\|\mathcal{B} - \mathcal{B}_\ell\|_{\mathcal{V} \to \mathcal{V}'} \lesssim 2^{-\alpha \ell}$.

This observation was used in [5, 12] for adaptive compression of $\mathcal{B}$ to steer the refinement in $\mathcal{F}$ such that optimal computational complexity of the method is possible. To actually achieve this, however, the spatial refinement – that is, the selection of the spaces V_ν for the chosen $\nu \in \mathcal{F}$ – needs to be of matching efficiency. So far, an analysis of optimality is available for two methods using wavelet discretizations in space.

The scheme analyzed in [5] is based on a direct wavelet compression of the individual second-order operators in the summation in (52). The resulting scheme yields discretizations of optimal cardinality by a coarsening step, but at total computational costs that are not optimal (and approach optimality only under very strong smoothness requirements on the spatial basis functions). A different wavelet-based strategy is used in [12], making use of the additional structure of spline wavelets in a hierarchical tree-based spatial refinement. In this case, optimal cardinality is achieved without coarsening, and with a computational complexity that is linear up to logarithmic factors with respect to the number of coefficients. This means in particular that under the conditions of the approximation result (41), for $d \geqslant 2$, an error $\varepsilon > 0$ in $L_2(Y, V; \sigma)$-norm is achieved using $O(\varepsilon^{-1/s})$ coefficients and using $O(\varepsilon^{-1/s})$ elementary operations for any $s < \frac{\alpha}{d}$.

An adaptive method using multiscale expansions of a in a similar way, but based on finite element discretizations on an independent adaptive mesh for each Legendre coefficient of u, is given in [10]. The method is shown to be convergent with a fixed error reduction in each refinement step; unlike previous approaches with independent adaptive meshes as in [30, 15], convergence is guaranteed without a saturation assumption. Although optimality of this method remains open, the numerical tests in [10] indicate that the computational costs in relation to (41) also have the above optimality property. Meshes generated by the method for $d = 2$, with D the L-shaped domain, are shown in Figure 3.

4.2 Stochastic Galerkin methods for lognormal problems

The case of problems with lognormal coefficients (24) with $Y = \mathbb{R}^{\mathbb{N}}$ is more difficult, since the most obvious stochastic Galerkin formulation is not uniformly well-posed: it is not difficult to see that the bilinear form defined in analogy to (49) by

$$\int_Y \int_D \exp\Big(\sum_{\lambda \in \vee} y_\lambda \psi_\lambda\Big) \nabla v(y) \cdot \nabla w(y)\, dx\, d\gamma(y), \qquad v, w \in L_2(Y, V; \gamma), \tag{54}$$

is neither bounded nor elliptic on $L_2(Y, V; \gamma)$. Although each discretization resulting from a restriction to a finite set of product Hermite polynomials is uniquely solvable, the condition numbers of such discretizations deteriorate under refinement, which is particularly problematic for adaptive methods.

Stabilized stochastic Galerkin for lognormal elliptic problems have been considered in [40, 45, 62]. They are based on introducing weights in (54) that depend only

Fig. 3 Legendre coefficients u_ν and meshes for (23) with $d = 2$ and D the L-shaped domain, generated by the adaptive stochastic Galerkin finite element method in [10]. The diffusion coefficient a is represented in terms of hierarchical piecewise linear hat functions that are a rescaled two-dimensional analogue of the Schauder basis (7), here with $\alpha = 1$ in (38). The top coefficient u_0 is equal to $\mathbb{E}u$.

on the stochastic parameters y, which leads to Petrov-Galerkin formulations or to modified trial spaces. The construction of rigorous adaptive methods based on the practical formulations of this type is still difficult, since the spaces on which one obtains boundedness and inf-sup-stability for the corresponding operators do not match. In our present context, a major issue is also that these formulations require too strong summability conditions on $(\|\psi_j\|_{L_\infty})_{j\in\mathbb{N}}$, which are appropriate for KL-type expansions $\sum_{j=1}^{\infty} y_j\psi_j$ of the Gaussian random field in (24), but make such stabilizations unsuitable for taking advantage of multilevel expansions of random fields. The construction of adaptive methods with similar optimality properties as in the case of affine parameterizations with multilevel structure thus remains open.

4.3 Stochastic collocation and discrete least squares

Rather than from a spatial-parametric variational formulation, polynomial approximations can also be determined from solution samples by sparse interpolation or by discrete least-squares formulations. Here, reconstructions of u are based on a certain number N of (approximate) point evaluations $u(y^i)$ for well-chosen $y^i \in Y$, $i = 1, \ldots, N$. Such methods are also called *non-intrusive*, since they can rely on existing codes for approximating solutions for fixed parameter values.

Especially in the case of lognormal coefficients (24), on the one hand, such methods are attractive because they can be constructed based on the well-posedness of (23) for almost all y, rather than on the well-posedness of a stochastic Galerkin formulation. On the other hand, in the absence of a stochastic Galerkin residual, less information is available for steering the refinements in adaptive schemes.

In sparse polynomial interpolation, also called *stochastic collocation* [3], polynomial terms are added in increments by sequentially adding interpolation points from a product grid. The construction of such methods is based on sequences of subsets $\Lambda \subset \mathcal{F}$ that are *downward-closed:* if $\nu \in \Lambda$, then also $\mu \in \Lambda$ for every μ such that $\mu \leqslant \nu$ holds componentwise. In [28], the results on weighted summability of §3.1 and §3.2 are discussed under this additional constraint.

Stochastic collocation for elliptic problems with lognormal coefficients has been studied in many works; for an overview, we refer to [3, 24]. In the case of infinitely many parameters, under general assumptions typically used for KL expansions, convergence is analyzed for affine coefficients in [21, 34] and for lognormal coefficients in [37, 38]. The numerical experiments in [38] include a direct comparison of KL and multiscale random field expansions. It indicates, on the one hand, that the available approximability results for lognormal problems with KL expansion are not sharp (unlike those for the multilevel case as in (46)); and that in such a setting of generically applicable schemes, one does not profit from multilevel expansions of the random coefficient a. The construction of methods that can take advantage of such structures appears to be open. In particular, although there exist stochastic collocation methods that adaptively choose both polynomial degrees in the parameters and spatial meshes (see, for example, [49, 39, 34, 16]), these methods are not yet designed to take locality in the expansion of a into account.

Another approach to constructing approximations from parametric point evaluations are *weighted discrete least-squares methods* [27]. In this case, for a finite $\Lambda \subset \mathcal{F}$, an approximation of u in terms of the appropriate orthonormal polynomials $(P_\nu)_{\nu\in\Lambda}$ is found as follows. With i.i.d. samples $y^i \in Y$, $i = 1, \dots, m$, drawn from a particular measure adapted to $\operatorname{span}(P_\nu)_{\nu\in\Lambda}$, with accordingly adapted weights $w_\Lambda(y^i)$, the approximation $\tilde u$ is found as the solution of the minimization problem

$$\min_{\tilde u \in \tilde{\mathcal{V}}_\Lambda} \sum_{i=1}^m w_\Lambda(y^i) \|\tilde u(y^i) - u(y^i)\|_V^2, \quad \tilde{\mathcal{V}}_\Lambda = \Bigl\{ \sum_{\nu\in\Lambda} v_\nu P_\nu : v_\nu \in \tilde V \Bigr\},$$

with a finite-dimensional subspace $\tilde V \subset V$. In the case of lognormal coefficients, in [29] this method is shown to achieve quasi-best approximations from $\tilde{\mathcal{V}}_\Lambda$ in expectation, with the near-optimal oversampling $m \gtrsim n \log n$. In this analysis, the choice of a single spatial discretization subspace $\tilde V$ is essential; due to the use of random samples, in this setting it is also difficult to profit from adaptive spatial approximations.

4.4 Quadrature

Rather than in the full solutions $y \mapsto u(y)$ of a random PDE, one may also rather be interested in expectations $\mathbb{E}(Q(u))$ of some *quantity of interest* Q derived from u, where in the simplest case, one considers linear functionals $Q \in V'$.

Following closely the analysis of stochastic collocation as in [37], sparse quadratures for problems with Gaussian random fields, and in particular for lognormal problems as in (24), are constructed and analyzed in [18]. Another well-studied alternative are *quasi-Monte Carlo* methods, in particular randomly shifted lattice rules. The approximation of integrals of quantities of interest by such rules using localized expansions of a has been studied in the affine case in [41], and in the lognormal case both in [53, 52] and independently in [55]. Here the use of multilevel expansions of a leads to improved efficiency of numerical generation of the lattice rules, since certain weights required in their construction can be chosen to have product form. While a hierarchy of spatial meshes is used in a multilevel quadrature scheme in [52] and adaptive spatial refinement has been considered with KL-type expansions of a in [57], the question of exploiting further sparsity in spatial discretizations for a in multilevel expansion appears to be open also for such methods.

5 Conclusions and open problems

While expansions of random fields with wavelet-type multilevel structure have originated from the analysis of certain classes of random fields, they can also be used for efficient numerical methods for computational tasks in uncertainty quantification. In particular, they lead to convergence estimates for sparse polynomial approximations for random PDEs that are stronger than those known for standard KL-type expansions of random coefficients. Such approximations can be realized by numerical methods, in particular by stochastic Galerkin methods with adaptive refinement both in the spatial discretization and in the stochastic variables. Let us now summarize some of the remaining open problems.

Several questions remain concering the results on sparse polynomial approximability discussed in §3. The convergence rates of fully adaptive approximations in the affine case as in (41) are known only for $\alpha \in (0, 1]$, corresponding to rough random fields. The numerical results in [12, 10] give a strong indication that such a result also holds for $\alpha > 1$, but the technique of proof used in [4] is not applicable in this case. Such a result including spatial adaptivity is also not available for lognormal problems. Note also that while for KL-type expansions in the affine case, Theorem 3.1 from [26] yields a result that can be seen to be sharp, the numerical tests in [38] indicate that the best available result for KL expansions in the lognormal case, derived from Theorem 3.4, is not yet sharp, so that the comparison between the different expansions in the lognormal case is not yet clear.

Concerning numerical methods, for affine coefficients, stochastic Galerkin methods that realize the strongest rates (41) for fully adaptive approximations are available; in the case of spatial discretizations by wavelets, these methods have also been shown to be of optimal computational complexity. For lognormal coefficients, no such schemes are available, and here the first question in fact concerns stable stochastic Galerkin variational formulations that are suitable for taking advantage of multilevel expansions of the underlying Gaussian random fields. For other classes of methods such as stochastic collocation, a main open question is what can be achieved for such expansions by independent spatial adaptivity for individual coefficients in the polynomial approximation.

Finally, practical expansions of random fields with multilevel localization have been constructed for certain important special cases such as stationary random fields, but it will be also be of interest to obtain efficiently computable wavelet-type expansions for nonstationary random fields on domains and for random fields on more general manifolds.

References

1. Andreev, R., Lang, A.: Kolmogorov-Chentsov theorem and differentiability of random fields on manifolds. Potential Anal. **41**(3), 761–769 (2014)
2. Aronszajn, N.: Theory of reproducing kernels. Trans. Amer. Math. Soc. **68**, 337–404 (1950)

3. Babuška, I., Nobile, F., Tempone, R.: A stochastic collocation method for elliptic partial differential equations with random input data. SIAM Rev. **52**(2), 317–355 (2010)
4. Bachmayr, M., Cohen, A., Dũng, D., Schwab, C.: Fully discrete approximation of parametric and stochastic elliptic PDEs. SIAM J. Numer. Anal. **55**(5), 2151–2186 (2017)
5. Bachmayr, M., Cohen, A., Dahmen, W.: Parametric PDEs: sparse or low-rank approximations? IMA J. Numer. Anal. **38**(4), 1661–1708 (2018)
6. Bachmayr, M., Cohen, A., DeVore, R., Migliorati, G.: Sparse polynomial approximation of parametric elliptic PDEs. Part II: Lognormal coefficients. ESAIM Math. Model. Numer. Anal. **51**(1), 341–363 (2017)
7. Bachmayr, M., Cohen, A., Migliorati, G.: Sparse polynomial approximation of parametric elliptic PDEs. Part I: Affine coefficients. ESAIM Math. Model. Numer. Anal. **51**(1), 321–339 (2017)
8. Bachmayr, M., Cohen, A., Migliorati, G.: Representations of Gaussian random fields and approximation of elliptic PDEs with lognormal coefficients. J. Fourier Anal. Appl. **24**(3), 621–649 (2018)
9. Bachmayr, M., Djurdjevac, A.: Multilevel representations of isotropic Gaussian random fields on the sphere. IMA J. Numer. Anal. **43**(4), 1970–2000 (2023)
10. Bachmayr, M., Eigel, M., Eisenmann, H., Voulis, I.: A convergent adaptive finite element stochastic galerkin method based on multilevel expansions of random fields. arXiv Preprint arXiv:2403.13770 (2024)
11. Bachmayr, M., Graham, I.G., Nguyen, V.K., Scheichl, R.: Unified analysis of periodization-based sampling methods for Matérn covariances. SIAM J. Numer. Anal. **58**(5), 2953–2980 (2020)
12. Bachmayr, M., Voulis, I.: An adaptive stochastic Galerkin method based on multilevel expansions of random fields: convergence and optimality. ESAIM Math. Model. Numer. Anal. **56**(6), 1955–1992 (2022)
13. Baldi, P., Kerkyacharian, G., Marinucci, D., Picard, D.: Asymptotics for spherical needlets. Ann. Statist. **37**(3), 1150–1171 (2009)
14. Benassi, A., Jaffard, S., Roux, D.: Elliptic Gaussian random processes. Rev. Mat. Iberoamericana **13**(1), 19–90 (1997)
15. Bespalov, A., Praetorius, D., Rocchi, L., Ruggeri, M.: Convergence of adaptive stochastic Galerkin FEM. SIAM Journal on Numerical Analysis **57**(5), 2359–2382 (2019)
16. Bespalov, A., Silvester, D.: Error estimation and adaptivity for stochastic collocation finite elements Part II: Multilevel approximation. SIAM J. Sci. Comput. **45**(2), A784–A800 (2023)
17. Bogachev, V.I.: Gaussian measures, *Mathematical Surveys and Monographs*, vol. 62. American Mathematical Society, Providence, RI (1998)
18. Chen, P.: Sparse quadrature for high-dimensional integration with Gaussian measure. ESAIM Math. Model. Numer. Anal. **52**(2), 631–657 (2018)
19. Chen, P., Ghattas, O.: Sparse polynomial approximation for optimal control problems constrained by elliptic pdes with lognormal random coefficients. arXiv Preprint arXiv:1903.05547 (2019)
20. Chen, P., Ghattas, O.: Sparse polynomial approximations for affine parametric saddle point problems. Vietnam J. Math. **51**(1), 151–175 (2023)
21. Chkifa, A., Cohen, A., Schwab, C.: Breaking the curse of dimensionality in sparse polynomial approximation of parametric PDEs. J. Math. Pures Appl. (9) **103**(2), 400–428 (2015)
22. Ciesielski, Z.: Hölder conditions for realizations of Gaussian processes. Trans. Amer. Math. Soc. **99**, 403–413 (1961)
23. Ciesielski, Z., Kerkyacharian, G., Roynette, B.: Quelques espaces fonctionnels associés à des processus gaussiens. Studia Math. **107**(2), 171–204 (1993)
24. Cohen, A., DeVore, R.: Approximation of high-dimensional parametric PDEs. Acta Numer. **24**, 1–159 (2015)
25. Cohen, A., DeVore, R., Schwab, C.: Convergence rates of best N-term Galerkin approximations for a class of elliptic sPDEs. Found. Comput. Math. **10**(6), 615–646 (2010)
26. Cohen, A., DeVore, R., Schwab, C.: Analytic regularity and polynomial approximation of parametric and stochastic elliptic PDE's. Anal. Appl. (Singap.) **9**(1), 11–47 (2011)

27. Cohen, A., Migliorati, G.: Optimal weighted least-squares methods. The SMAI Journal of Computational Mathematics **3**, 181–203 (2017)
28. Cohen, A., Migliorati, G.: Multivariate approximation in downward closed polynomial spaces. In: Contemporary computational mathematics—a celebration of the 80th birthday of Ian Sloan. Vol. 1, 2, pp. 233–282. Springer, Cham (2018)
29. Cohen, A., Migliorati, G.: Near-optimal approximation methods for elliptic PDEs with lognormal coefficients. Math. Comp. **92**(342), 1665–1691 (2023)
30. Crowder, A.J., Powell, C.E., Bespalov, A.: Efficient adaptive multilevel stochastic Galerkin approximation using implicit a posteriori error estimation. SIAM Journal on Scientific Computing **41**(3), A1681–A1705 (2019)
31. Dũng, D., Nguyen, V.K., Schwab, C., Zech, J.: Analyticity and Sparsity in Uncertainty Quantification for PDEs with Gaussian Random Field Inputs, *Lecture Notes in Mathematics*, vol. 2334. Springer, Cham (2023)
32. Dũng, D., Temlyakov, V., Ullrich, T.: Hyperbolic cross approximation. Advanced Courses in Mathematics. CRM Barcelona. Birkhäuser/Springer, Cham (2018). Edited and with a foreword by Sergey Tikhonov
33. Daubechies, I.: Orthonormal bases of compactly supported wavelets. Comm. Pure Appl. Math. **41**(7), 909–996 (1988)
34. Eigel, M., Ernst, O.G., Sprungk, B., Tamellini, L.: On the convergence of adaptive stochastic collocation for elliptic partial differential equations with affine diffusion. SIAM J. Numer. Anal. **60**(2), 659–687 (2022)
35. Eigel, M., Gittelson, C.J., Schwab, C., Zander, E.: Adaptive stochastic Galerkin FEM. Comput. Methods Appl. Mech. Engrg. **270**, 247–269 (2014)
36. Eigel, M., Gittelson, C.J., Schwab, C., Zander, E.: A convergent adaptive stochastic Galerkin finite element method with quasi-optimal spatial meshes. ESAIM Math. Model. Numer. Anal. **49**(5), 1367–1398 (2015)
37. Ernst, O.G., Sprungk, B., Tamellini, L.: Convergence of sparse collocation for functions of countably many Gaussian random variables (with application to elliptic PDEs). SIAM J. Numer. Anal. **56**(2), 877–905 (2018)
38. Ernst, O.G., Sprungk, B., Tamellini, L.: On expansions and nodes for sparse grid collocation of lognormal elliptic PDEs. In: Sparse grids and applications—Munich 2018, *Lect. Notes Comput. Sci. Eng.*, vol. 144, pp. 1–31. Springer, Cham (2021)
39. Feischl, M., Scaglioni, A.: Convergence of adaptive stochastic collocation with finite elements. Comput. Math. Appl. **98**, 139–156 (2021)
40. Galvis, J., Sarkis, M.: Approximating infinity-dimensional stochastic Darcy's equations without uniform ellipticity. SIAM J. Numer. Anal. **47**(5), 3624–3651 (2009)
41. Gantner, R.N., Herrmann, L., Schwab, C.: Quasi-Monte Carlo integration for affine-parametric, elliptic PDEs: local supports and product weights. SIAM J. Numer. Anal. **56**(1), 111–135 (2018)
42. Ghanem, R.G., Spanos, P.D.: Stochastic finite elements: a spectral approach. Springer-Verlag, New York (1991)
43. Ghanem, R.G., Spanos, P.D.: Spectral techniques for stochastic finite elements. Arch. Comput. Methods Engrg. **4**(1), 63–100 (1997)
44. Giné, E., Nickl, R.: Mathematical foundations of infinite-dimensional statistical models, *Cambridge Series in Statistical and Probabilistic Mathematics*, vol. [40]. Cambridge University Press, New York (2016)
45. Gittelson, C.J.: Stochastic Galerkin discretization of the log-normal isotropic diffusion problem. Math. Models Methods Appl. Sci. **20**(2), 237–263 (2010)
46. Gittelson, C.J.: Representation of Gaussian fields in series with independent coefficients. IMA J. Numer. Anal. **32**(1), 294–319 (2012)
47. Gittelson, C.J.: An adaptive stochastic Galerkin method for random elliptic operators. Math. Comp. **82**(283), 1515–1541 (2013)
48. Gittelson, C.J.: Adaptive wavelet methods for elliptic partial differential equations with random operators. Numer. Math. **126**(3), 471–513 (2014)

49. Guignard, D., Nobile, F.: A posteriori error estimation for the stochastic collocation finite element method. SIAM J. Numer. Anal. **56**(5), 3121–3143 (2018)
50. van Harten, W., Scarabosio, L.: Exploiting locality in sparse polynomial approximation of parametric elliptic pdes and application to parameterized domains. arXiv Preprint arXiv:2308.06188 (2023)
51. Herrmann, L., Lang, A., Schwab, C.: Numerical analysis of lognormal diffusions on the sphere. Stoch. Partial Differ. Equ. Anal. Comput. **6**(1), 1–44 (2018)
52. Herrmann, L., Schwab, C.: Multilevel quasi-Monte Carlo integration with product weights for elliptic PDEs with lognormal coefficients. ESAIM Math. Model. Numer. Anal. **53**(5), 1507–1552 (2019)
53. Herrmann, L., Schwab, C.: QMC integration for lognormal-parametric, elliptic PDEs: local supports and product weights. Numer. Math. **141**(1), 63–102 (2019)
54. Hoang, V.H., Schwab, C.: N-term Wiener chaos approximation rate for elliptic PDEs with lognormal Gaussian random inputs. Math. Models Methods Appl. Sci. **24**(4), 797–826 (2014)
55. Kazashi, Y.: Quasi-Monte Carlo integration with product weights for elliptic PDEs with log-normal coefficients. IMA J. Numer. Anal. **39**(3), 1563–1593 (2019)
56. Lang, A., Schwab, C.: Isotropic Gaussian random fields on the sphere: Regularity, fast simulation and stochastic partial differential equations. The Annals of Applied Probability **25**(6), 3047 – 3094 (2015)
57. Longo, M.: Adaptive quasi-Monte Carlo finite element methods for parametric elliptic PDEs. Journal of Scientific Computing **92**(1), 19 (2022)
58. Luschgy, H., Pagès, G.: Expansions for Gaussian processes and Parseval frames. Electron. J. Probab. **14**, no. 42, 1198–1221 (2009)
59. Marinucci, D., Peccati, G.: Random fields on the sphere, *London Mathematical Society Lecture Note Series*, vol. 389. Cambridge University Press, Cambridge (2011). Representation, limit theorems and cosmological applications
60. Matérn, B.: Spatial variation, *Lecture Notes in Statistics*, vol. 36, second edn. Springer-Verlag, Berlin (1986). With a Swedish summary
61. Meyer, Y.: Wavelets and operators, *Cambridge Studies in Advanced Mathematics*, vol. 37. Cambridge University Press, Cambridge (1992). Translated from the 1990 French original by D. H. Salinger
62. Mugler, A., Starkloff, H.J.: On the convergence of the stochastic Galerkin method for random elliptic partial differential equations. ESAIM Math. Model. Numer. Anal. **47**(5), 1237–1263 (2013)
63. Narcowich, F.J., Petrushev, P., Ward, J.D.: Localized tight frames on spheres. SIAM J. Math. Anal. **38**(2), 574–594 (2006)
64. Potthoff, J.: Sample properties of random fields. II. Continuity. Commun. Stoch. Anal. **3**(3), 331–348 (2009)
65. Roynette, B.: Mouvement brownien et espaces de Besov. Stochastics Stochastics Rep. **43**(3-4), 221–260 (1993)
66. Schwab, C., Gittelson, C.J.: Sparse tensor discretizations of high-dimensional parametric and stochastic PDEs. Acta Numer. **20**, 291–467 (2011)
67. Wood, A.T.A., Chan, G.: Simulation of stationary Gaussian processes in $[0, 1]^d$. J. Comput. Graph. Statist. **3**(4), 409–432 (1994)

Nonlinear compressive reduced basis approximation for multi-parameter elliptic problem

Hassan Ballout,
Yvon Maday and
Christophe Prud'homme

Abstract Reduced basis methods for approximating the solutions of parameter-dependant partial differential equations (PDEs) are based on learning the structure of the set of solutions - seen as a manifold $\mathcal{S}$ in some functional space - when the parameters vary. This involves investigating the manifold and, in particular, understanding whether it is close to a low-dimensional affine space. This leads to the notion of Kolmogorov N-width that consists of evaluating to which extent the best choice of a vectorial space of dimension N approximates $\mathcal{S}$ well enough. If a good approximation of elements in $\mathcal{S}$ can be done with some well-chosen vectorial space of dimension N – provided N is not too large – then a "reduced" basis can be proposed that leads to a Galerkin type method for the approximation of any element in $\mathcal{S}$. In many cases, however, the Kolmogorov N-width is not so small, even if the parameter set lies in a space of small dimension yielding a manifold with small dimension. In terms of complexity reduction, this gap between the small dimension of the manifold and the large Kolmogorov N-width can be explained by the fact that the Kolmogorov N-width is linear while, in contrast, the dependency in the parameter is, most often, non-linear. There have been many contributions aiming at reconciling these two statements, either based on deterministic or AI approaches. We investigate here further a new paradigm that, in some sense, merges these two aspects: the nonlinear compressive reduced basis approximation. We focus on a simple multiparameter problem and illustrate rigorously that the complexity associated with the approximation of the solution to the parameter dependant PDE is directly related to the number of parameters rather than the Kolmogorov N-width.

Hassan Ballout
Cemosis, IRMA UMR 7501, University of Strasbourg, CNRS, e-mail: hassan.ballout@etu.unistra.fr

Yvon Maday
Sorbonne Université, CNRS, Université Paris Cité, Laboratoire Jacques-Louis Lions (LJLL), F-75005 Paris, France e-mail: yvon.maday@sorbonne-universite.fr

Christophe Prud'homme
Cemosis, IRMA UMR 7501, Université de Strasbourg, CNRS, e-mail: prudhomme@unistra.fr

R. DeVore, A. Kunoth (eds.), *Multiscale, Nonlinear and Adaptive Approximation II*,
https://doi.org/10.1007/978-3-031-75802-7_4

1 Introduction

From the very beginning, the notion of "learning" is closely associated with — if not at the root of — the reduced basis methods and the complexity reduction techniques for the approximation of the solutions of parameter-dependant partial differential equations (PDEs) written as :
For some value of the parameter $\mu \in \mathcal{P}$ (the set of all parameters), find the solution $u_\mu \in X$ such that

$$\mathfrak{R}(u_\mu; \mu) = 0, \tag{1}$$

where X is some appropriate Hilbert space, chosen for the above problem to be well-posed.

In opposition to generic, multipurpose discretization methods for PDEs like finite volume, finite element, and spectral methods ... that can be used for any type of problem, the reduced basis methods are tuned to a certain class of problems of interest, the features of this problem are learned and digested, and the direct application to another problem makes no sense. These methods are based on a two phases approach :

- an "*off-line phase*", that can be (and generally is) quite long and during which the learning of $\mathcal{S} = \{u_\mu; \mu \in \mathcal{P}\}$ is performed
- an "*on-line phase*", much (and generally very much) faster, during which the solution(s) u_μ for a new (set of) parameter(s) is (are) approximated.

One preliminary step in the — *off-line* — learning process consists in investigating the set $\mathcal{S}$ of all solutions – seen as a manifold in X – that consists in understanding whether the manifold is close to a low-dimensional affine space. This is related to the notion of Kolmogorov N-width that aims at evaluating the extent to which the best choice of an N-dimensional vector space X_N approximates $\mathcal{S}$ sufficiently well. The discovery of such an accurate N-dimensional vector space X_N is not straightforward and getting the best one may even be impossible. Various algorithms (POD, SVD, Greedy [29, 17]), have been proposed to determine alternative quasi-optimal X_N, and performed in the off-line phase based on the data of many instances in $\mathcal{S}$. Then, provided that such a quasi-optimal X_N has been identified, a "reduced" basis method can be deployed — in the *on-line phase* — under the form, e.g. of a Galerkin formulation in X_N for the approximation of the solution u_μ for some given value μ of the parameter. This low complexity approximation works for a large variety of parameter-dependant problems and provides fast, even, superfast approximation methods, with a complexity that scales as $O(N^3)$.

In many instances however, even if the set of parameters lies in a space of small dimension $\mathcal{P} \subset \mathbb{R}^P$ ($P \in \mathbb{N}$), the Kolmogorov N-width of $\mathcal{S}$ is not so small (even if $\mathcal{S}$ can then be considered as a manifold with dimension $\leqslant P$). This conceptual mismatch, between the manifold dimension and the complexity Kolmogorov width can be understood by the fact that the Kolmogorov width is a linear concept whereas the dependency of the solution u_μ in the parameter μ is, most often highly non-linear.

However, it is true that the intrinsic complexity of the manifold is, heuristically at least, related to the dimension of $\mathcal{P}$.

Several authors have proposed nonlinear approximation/compression approaches to deal with this problem. There are essentially 2 wide classes of approaches: Lagrangian and Eulerian.

Lagrangian approaches (also called registration methods see [32]) involve a map φ_μ so that the set of all $v_\mu = u_\mu \circ \varphi_\mu$ defines a manifold better suited to linear compression methods than the original $\mathcal{S}$ with thus a faster decay of the Kolmogorov width. Examples of Lagrangian approaches have been proposed in [18, 23, 6, 24, 30, 14]. Eulerian approaches have a more nonlinear root: some approaches are based on a partition of the parameter domain in an adaptive way [12], others rely on basis updates [22, 7, 13, 25], or on a Grasmannian learning [1, 34, 31] or finally based on artificial learning methods, some involving convolutional autoencoders, [20, 33, 19, 15] or Operator Inference [21].

In these last references, the original notion of "learning" of the reduced basis methods is hence pushed further to the extent of investigating the nonlinearity of the manifold. This learning process can be used to the point of not even referring to the problem (1) and at the end propose a method that, to any input parameter $\mu \in \mathcal{P}$, proposes – as an output – an approximation to u_μ. Even if this is shown to work in those publications, this approach has some obvious drawbacks :

- the size of the database for the learning stage, needs to be quite large in order to apprehend the complexity of $\mathcal{S}$
- there is no evaluation of the residual of equation (1) that would allow to state
  - if the problem is well or not approximated
  - what to do in order to improve a non-sufficient accuracy

this is why, following [2, 16], the authors in [3, 9] have proposed to "help" the statistical approaches by using intermediate models and still ground the approximation of the approximated resolution of problem (1).

In this paper, we investigate further this idea applied to a simple multiparameter problem and show, in particular, that, indeed, the complexity associated with the approximation of the solution to the parameter-dependent PDE is related directly to the number of parameters rather than the Kolmogorov N-width.

The paper is organized as follows: Section 2 presents the problem setting, detailing the multiparameter elliptic problem and the physical model used for our analysis. Section 3 provides a comprehensive discussion on the analysis of the solution manifold, introducing the classical reduced basis methods and extending the discussion to our proposed nonlinear compressive approach. Section 4 explores the application of machine learning techniques for nonlinear approximation within the reduced basis framework, including detailed discussions on the regression models used for the recovery of higher RB modes from lower ones. Section 5 illustrates the efficacy of the proposed method through a series of numerical experiments, demonstrating the advantages of our approach in terms of accuracy and computational efficiency. Finally, Section 6 concludes the paper with a summary of our findings and suggestions for future research directions.

The numerical experiments presented in the paper are conducted using the Feel++ library [27] to produce the finite element method (FEM) computations as well as the reduced basis using proper orthogonal decomposition (POD). For the machine learning aspects of our methodology, we utilize the scikit-learn framework [26].

2 Problem setting: the multiparameter problem

We consider the problem of designing a thermal fin to remove heat from a surface effectively, see *e.g.* [28]. The two-dimensional fin, shown in Figure 1, consists of a vertical central **post** and say N_{fins} horizontal **subfins**; the fin conducts heat from a prescribed uniform flux source at the root, $\mathbf{\Gamma}_{\text{root}}$, through the large-surface-area subfins to surrounding flowing air. The fin is characterized by a ($P = N_{fins} + 1$)-component parameter vector $\mu = (\mu_1, \mu_2, \ldots, \mu_P)$, where $\mu_i = k^i, i = 1, \ldots, N_{fins}$, and $\mu_P = Bi$; μ may take on any value in a specified design set $\mathcal{P} \subset \mathbb{R}^P$.

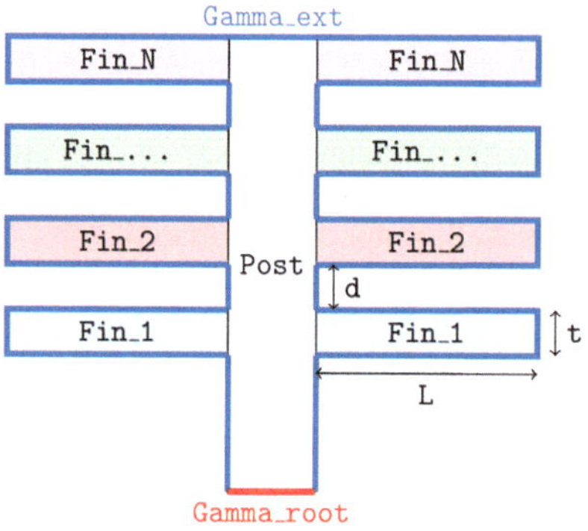

Fig. 1: Schematic of a thermal fin showing the structure and important elements, see also [28].

Here k^i is the thermal conductivity of the i-th subfin (normalized relative to the post conductivity $k^0 = 1$); and Bi is the Biot number, a nondimensional heat transfer coefficient reflecting convective transport to the air at the fin surfaces (larger Bi means better heat transfer). For example, suppose we choose a thermal fin with $N_{fins} = 4$, $k^1 = 0.4$, $k^2 = 0.6$, $k^3 = 0.8$, $k^4 = 1.2$, and $Bi = 0.1$; for this particular configuration $\mu = (0.4, 0.6, 0.8, 1.2, 0.1)$, which corresponds to a single point in the set of all possible configurations $\mathcal{P}$.

We are interested in the design of this thermal fin, and we thus need to look at certain outputs or cost-functionals of the temperature as a function of μ. For our output T_{root}, we choose the average steady-state temperature of the fin root normalized by the prescribed heat flux into the fin root. The particular output chosen relates directly to the cooling efficiency of the fin — lower values of T_{root} imply better thermal performance. The steady-state temperature distribution within the fin, u_μ, is governed by the elliptic partial differential equation

$$-\nabla.[\mathbf{k}\nabla]u_\mu = 0 \ \text{ in } \Omega, \tag{2}$$

which implies in particular that

$$-k^i \Delta u = 0 \text{ in } \Omega^i,\ i = 0, \ldots, N_{fins}, \tag{3}$$

where Δ is the Laplacian operator, and u_i refers to the restriction of u to Ω^i. Here Ω^i is the region of the fin with conductivity k^i, $i = 0, \ldots, N_{fins}$; Ω^0 thus the central post, and Ω^i, $i = 1, \ldots, N_{fins}$, corresponds to the subfins. The entire fin domain is denoted $\Omega = \bigcup_{i=0}^{N_{fins}} \Omega^i$; the boundary is denoted $\mathbf{\Gamma}$. The continuity of temperature and heat flux at the conductivity-discontinuity interfaces is also implicitly set on $\Gamma^i_{int} \equiv \partial\Omega^0 \cap \partial\Omega^i$, $i = 1, \ldots, N_{fins}$, where $\partial\Omega^i$ denotes the boundary of Ω^i, we have on Γ^i, $i = 1, \ldots, N_{fins}$:

$$\begin{aligned} [u] &= 0, \\ \left[k^i \frac{\partial u}{\partial n^i}\right] &= 0, \end{aligned} \tag{4}$$

where n^i is the outward normal on $\partial\Omega^i$. Finally, we introduce a Neumann flux boundary condition on the fin root:

$$-k^0 \frac{\partial u}{\partial n^0} = -1 \text{ on } \Gamma_{\text{root}}. \tag{5}$$

which models the heat source; and a Robin boundary condition:

$$-k^i \frac{\partial u}{\partial n^i} = Bi\, u^i \text{ on } \Gamma^i_{\text{ext}},\ i = 1, \ldots, N_{fins}, \tag{6}$$

which models the convective heat losses. Here Γ^i_{ext} is that part of the boundary of Ω^i exposed to the flowing fluid; note that $\bigcup_{i=0}^{N_{fins}} \Gamma^i_{\text{ext}} = \Gamma \setminus \Gamma_{\text{root}}$. The average temperature at the root, $T_{\text{root}}(\mu)$, can then be expressed as $\ell^O(u_\mu)$, where

$$\ell^O(v) = \int_{\Gamma_{\text{root}}} v. \tag{7}$$

Note that $\ell(v) = \ell^O(v)$ for this problem (compliant problem).

In order to proceed with a Galerkin type approximation of the solutions to problem (3), it is now written under a parametric variational problem : Given $\mu \in \mathcal{P}$, find $u_\mu \in X$ such that:

$$a(u_\mu, v; \mu) = f(v; \mu), \quad \forall v \in X \tag{8}$$

here $a(.,.;\mu)$ is thus a symmetric elliptic bilinear form.

3 Analysis of the solution manifold

3.1 Kolmogorov width - Classical RBM approximation

The Kolmogorov m-width of $\mathcal{S}$ is defined as

$$d_m(\mathcal{S}) = \inf_{\dim(X_m) \leqslant m} \max_{v \in \mathcal{S}} \min_{w \in X_m} \|v - w\|_X. \tag{9}$$

As was recalled in the introduction, this describes how well the manifold can be approximated by an ideally selected (and usually out of reach) m-dimensional space. A slow decay of the Kolmogorov m-width means that a large number of basis functions is needed to achieve a good approximation, which compromises the efficiency of our reduced basis method.

The determination of the optimal spaces that realize the minimum is most of the times an open problem and has been circumvented, from a practical point of view, within the frame of classical reduced basis methods, particularly Proper Orthogonal Decomposition (POD) that aim to approximate the solution manifold by a well-chosen (quasi-optimal) N-dimensional vector space.

For the sake of completeness, let us recall the definition of the POD (or SVD) method. We start by introducing a training set $\mathcal{P}_h = \{\mu_1, \mu_2, ..., \mu_M\} \subset \mathcal{P}$ of M parameter values, with M chosen large enough to "represent well enough" the set of parameters $\mathcal{P}$.

For each parameter value $\mu_i \in \mathcal{P}_h$, we approximate the corresponding solution $u_{\mu_i} \in \mathcal{S}$ with, – say – an accurate enough finite element method (hence the presence of the index h), that we denote as u_{h,μ_i}. Then, we construct the correlation matrix $C \in \mathbb{R}^{M \times M}$ as follows:

$$C_{ij} = \langle u_{h,\mu_i}, u_{h,\mu_j} \rangle_X \quad \forall i, j \in \{1, 2, ..., M\} \tag{10}$$

where $\langle ., . \rangle_X$ denotes the inner product in X. We consider the eigenvalue-eigenvector pairs (λ_i, V_i) of the correlation matrix C such that:

$$C\, V_i = \lambda_i\, V_i, \quad \forall i \in \{1, 2, ..., M\} \tag{11}$$

and:

$$\lambda_1 \geqslant \lambda_2 \geqslant ... \geqslant \lambda_M \tag{12}$$

The POD basis functions ζ_i corresponding to the eigenvectors V_i are given by:

$$\zeta_i = \frac{1}{\lambda_i} \sum_{j=1}^{M} (V_i)_j\, u_{h,\mu_j} \quad \forall i \in \{1, 2, ..., M\} \tag{13}$$

where $(V_i)_j$ denotes the j-th component of V_i.

The key property of this basis is that if one truncates the POD basis to the first N functions $\{\zeta_1, \zeta_2, ..., \zeta_N\}$, they span the optimal N-dimensional subspace X_N. This optimality is in the sense of minimizing the following error measure:

$$\sum_{\mu \in \mathcal{P}_h} \inf_{v_N \in X_N} \|u_{h,\mu} - v_N\|_X^2 \tag{14}$$

over all N-dimensional subspaces X_N of $X_M = \text{span}\{u_{h,\mu} \mid \mu \in \mathcal{P}_h\}$. In practice, we choose only $N \ll M$ modes corresponding to the largest N eigenvalues λ_i. This reveals a drawback of the POD method: it requires solving $M \gg N$ truth problems in order to explore the solution manifold. An alternative iterative method is the Greedy method.

For our multiparameter problem, with 8 subfins ($N_{fins} = 8$) and a piecewise linear finite element functions space of dimension of $N_h \approx 10000$. We checked the quality of the finite element solution previously described for $N_{\text{fins}} = 8 (P = 9)$ for the Kolmogorov width estimation. The average L_2 error norm of the finite element solutions described previously with respect to a finer mesh and quadratic finite element instead of linear is about 3.10^{-3} in average with a maximum value of 5.10^{-3} over the sample of parameters used.

We show in Figure 2 the number of POD modes, N, needed to reach a given accuracy, ϵ_{POD}, as a function of the parameter space dimension, P. We can see that the Kolmogorov N-width highly depends on P. More precisely, as P increases, the decay of the Kolmogorov N-width becomes slower (even if it can be described as fast). Note that the notion of ϵ_{POD} is based on the Relative Information Content, and $N(\epsilon)$ is chosen as the smallest value that satisfies:

$$I(N) = \frac{\sum_{i=1}^{N} \lambda_i}{\sum_{i=1}^{M} \lambda_i} \geqslant 1 - \epsilon^2.$$

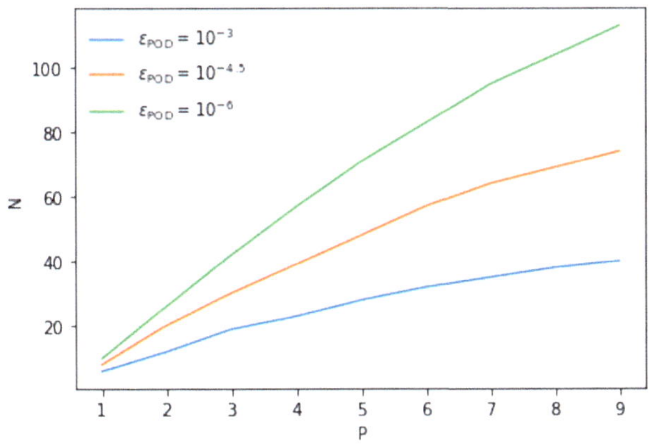

Fig. 2: Number of reduced basis functions as a function of the parameter dimension for different accuracies

Following this purely POD-based analysis to estimate the Kolmogorov width, we can achieve a reduced basis approximation using a POD-Galerkin method.

We thus defined the reduced space $X_N = \text{span}\{\zeta_1, \ldots, \zeta_N\} \subset X$. Given a parameter value μ in $\mathcal{P}$, we seek a reduced basis approximation u_μ^N in X_N such that:

$$a(u_\mu^N, v_N; \mu) = f(v_N; \mu), \quad \forall v_N \in X_N \tag{15}$$

This reduced basis approximation u_μ^N is expressed as follows:

$$u_\mu^N = \sum_{i=1}^{N} \hat{u}_{i,\mu} \zeta_i \tag{16}$$

And by testing the previous formulation with $v_N = \zeta_i$ for $i = 1, \ldots, N$, we obtain the coordinates vector of u_μ^N as the solution of the following linear system:

$$\mathbf{A}_N^\mu \mathbf{u}_N^\mu = \mathbf{f}_N^\mu \tag{17}$$

where $\mathbf{A}_N^\mu = \{a(\zeta_i, \zeta_j; \mu)\}_{i,j=1}^N$ and $\mathbf{f}_N^\mu = \{f(\zeta_i; \mu)\}_{i=1}^N$. Finally, we can evaluate the reduced basis output as:

$$s_N(\mu) = (\mathbf{u}_N^\mu)^T \mathbf{f}_N^\mu \tag{18}$$

In Figure 3, we illustrate the decay of both energy and output errors with respect to N for $P = 1$ and $P = 9$. By reducing the dimension from N_h to N, we obtained a speed-up factor of 1000 for $P = 1$ ($N = 10$) with a mean relative energy error (between the reduced method approximation and the finite element solution) of 10^{-6}, and a speed-up factor of 100 for $P = 9$ ($N = 114$) with a relative energy error of 10^{-5}. Additionally, we observe that the output error aligns with the square of the energy error, which is a classic result for compliant problems.

From now on, we use a coarser grid with linear piecewise finite element and $N_{\text{fins}} = 5$. The average L_2 error norm with respect to the fine grid setting with quadratic piecewise finite elements is then in average $7e-3$ and the maximum value $1e-2$ over the sample of parameters used.

The reduced basis space is constructed using POD such that $\epsilon_{POD} = 10^{-6}$. Note that, of course, the overall error with respect to the exact solution is, in our case, dominated by the finite element error. In contrast, the subsequent errors in our nonlinear compressive reduced basis are computed with respect to the finite element approximation.

3.2 Linear and nonlinear notions of complexity reduction

In order to investigate the nonlinear complexity reduction, let us remind that the frame of complexity reduction is classically described by a pair of continuous mappings, the encoder

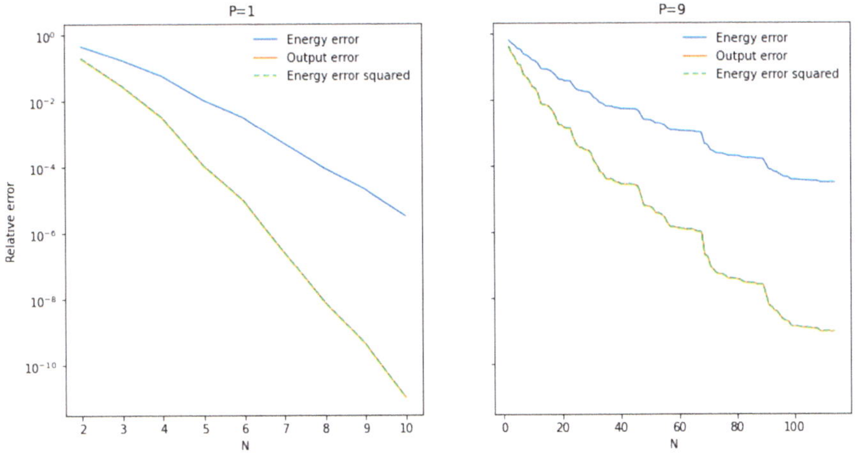

Fig. 3: Mean relative errors vs number of POD modes for $P = 1$(left) and $P = 9$(right)

$$E : X \longrightarrow \mathbb{R}^m,$$

and the decoder

$$D : \mathbb{R}^m \longrightarrow X.$$

These two mappings are tuned so that the quantity

$$\max_{v \in \mathcal{S}} \|v - D(E(v))\|_X \,, \qquad \text{is minimized under some constraints,} \tag{19}$$

and m here aims at reflecting the complexity of $\mathcal{S}$.

For instance, the already mentioned Kolmogorov width $d_m(\mathcal{S})_X$ assumes that both E and D are linear mappings.

Even if these two notions should not be mixed up, in order to reconcile the dimension P of the manifold $\mathcal{S}$ and the complexity reduction it is natural to think at alternative definitions of the complexity (see e.g. [10, 11] and also [8] for the link with Gelfand width) that leads to the idea of enlarging the class where the encoder and decoder stay (see [9] for an extended presentation in this frame). This is somehow the idea explored in references [19, 15], where the notion of auto-encoder is used to minimize the information needed to reconstruct the identity in $\mathcal{S}$, as a combination of two neural networks $\mathcal{E}$ and $\mathcal{D}$ (i.e. $Id = \mathcal{D} \circ \mathcal{E}$), and where the size of the layer at the output of $\mathcal{E}$ and the input of $\mathcal{D}$ is minimized.. looking thus to identify m. In these publications, neural networks are therefore used to discover $\mathcal{E}$ and $\mathcal{D}$, and further investments are made in the field of machine learning, with two warnings: the amount of data required to train the associated networks, and the lack of understanding underlying machine learning.

In the opposite direction, we can refer to a, still timely but much older problem of "parameter reconstruction" or "parameter identification", and ask what minimal knowledge on the elements v of $\mathcal{S}$ is required to identify the value of parameters μ in $\mathcal{P}$ such that $v = u_\mu$. Indeed, assuming that we have such a reconstruction $\mathcal{R} : v \in \mathcal{S} \mapsto \mu \in \mathcal{P}$, then the identity mapping in $\mathcal{S}$: $Id_{\mathcal{S}}$ would involve the solution operator $\mu \in \mathcal{P} \mapsto u_\mu \in \mathcal{S}$ so that $Id_{\mathcal{S}} = v \mapsto u_{\mathcal{R}(v)}$.

Our approach stands in the middle of these two extreme encoder-decoder paradigms and relies on the two notions. The first one is the *sensing numbers* $s_m(\mathcal{S})$ where E in (19) is still assumed to be linear but D is assumed to be nonlinear. In other words, they can be defined as

$$s_m(\mathcal{S}) := \inf_{D,\ell_1,\dots,\ell_m} \max_{v\in\mathcal{S}} \|v - D(\ell_1(v),\dots,\ell_m(v))\|_X, \tag{20}$$

where the infimum is taken over all choices of linear functionals $\ell_1,\dots,\ell_m \in X'$ and decoding map D. Then, referring to the parameter reconstruction/identification, we can raise the question: what are the best linear mappings $\ell_1,\dots,\ell_m$ that represent well $\mathcal{S}$, or, if we remember that we are in a Hilbert setting, from Riesz identification between X and X', the question becomes: what are the best elements in X: $v_1,\dots,v_m$ that are the more aligned with $\mathcal{S}$ (in order to have the largest scalar product with any element of $\mathcal{S}$), so that, for any $w \in \mathcal{S}$ and any j, $1 \leqslant j \leqslant m$: $\ell_j(w) = \langle v_j, w\rangle_X$ are the best linear functional appearing above? This is actually what is considered in the POD framework within the notion of Relative Information Content: the first modes on the POD approach are the ones that contain in the vectorial space spanned by $\mathcal{S}$ the largest amount or information. This naturally leads to considering that the first POD functions $v_j = \zeta_j$ are the best elements that carry the information on solutions in $\mathcal{S}$. This is what led us in [2, 3, 9] to propose the "anszatz" that a candidate for the optimal encoder in (20) consists in the moments associated with those POD modes.

In order to evaluate this claim, let us understand to which extent – and how many of them – these first POD moments allow to recover the parameters and can define a reconstruction $\mathcal{R} : v \in \mathcal{S} \mapsto \mu \in \mathcal{P}$, that is written under the form $\mathcal{R} = \mathcal{D} \circ E$ where the encoder E is linear and defined as being the set of moments

$$v \mapsto \{\langle \zeta_i, v\rangle_X\}_{i=1,..,m} \tag{21}$$

Actually, in dimension $P = 1$ (only the Biot number is varying) the following figure 4 graphically illustrates the simple dependence of the Biot number as a function of the first moment of the solution u_{Bi}, which is the sought for Decoder $\mathcal{D}$. The (Lipschitz) continuity of u with respect to Bi and, thus, the continuity of the decoder D, is a classical result (see e.g. [29] Sect.5.3 for the proof).

3.3 Sensing number and Parameter identification

This being said, in higher dimensions, we have to investigate the nonlinear decoder. As is presented in [2, 3, 9], where polynomial and ensemble learning regression methods (like a random forest) can be used to define the decoder.

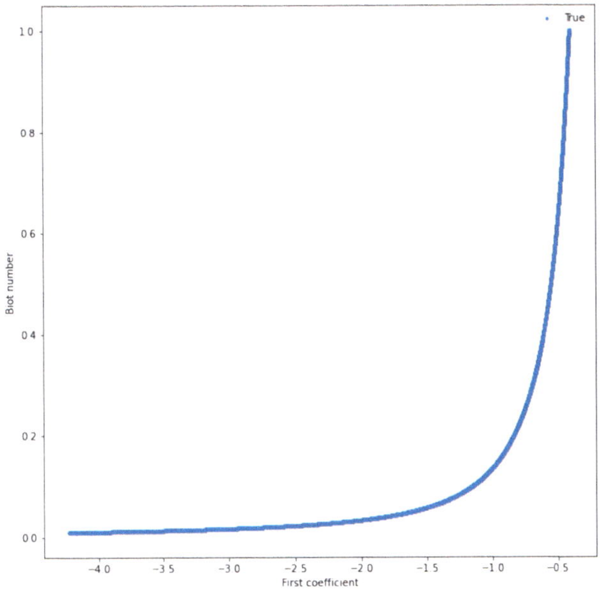

Fig. 4: The Biot number as a function of the first RB coefficient

Back with the 1-dimensional problem ($P = 1$), the dependence of $\mu = Bi$ with respect to the first coefficient displayed in Figure 4 can very well be approximated as a spline regression problem. By increasing the number of knots (n_{knots}) and the degree of the model, spline regression can achieve machine precision. Using a training dataset size of 10^7, decision trees and random forests also deliver excellent accuracy, with mean relative errors reaching up to 10^{-7}. However, polynomial models perform very weakly on this problem and do not achieve the same level of precision.

In higher-dimensional problems ($P > 1$), the regression problem becomes harder, and some algorithms suffer from the curse of dimensionality, as is well reported in the literature see [5]. In our case, the reconstruction of the parameter using decision trees (or random forests) gives "acceptable" results up to $P = 3$ (a mean relative error of 10^{-3} on the Biot number reconstruction and 10^{-2} for K^0 and K^1). For the same reason, for $P > 3$, more advanced models need to be employed to achieve a reasonable reconstruction of the parameters.

We can conclude from these initial results that indeed $s_m(\mathcal{S}) = 0$ for $m \geqslant P$, at least for $P = 1, 2, 3$, which sets the connection between the manifold dimension and the complexity of the manifold written in terms of sensing number rather than Kolmogorov m width.

To finish, we have also numerically investigated the stability of the inverse problem for $P = 1, 2, 3$ for the abovementioned regressors. The inverse problem is unstable whenever we use the polynomial regressors and their input data is outside the training domain; otherwise, in all other cases, it is stable. The inverse problems associated with spline, decision tree, and random forest regressors are, on the contrary, stable, including for input data outside the training domain.

4 Recovering the high RB modes from the low ones

After this analysis on the inverse/identification problem, let us come back to the approximation of the functions in $\mathcal{S}$. One of the fundamental steps in the nonlinear compressive reduced basis method is the regression task, specifically training a model that accurately predicts the high RB coefficients from the first n independent ones. Both accuracy and performance (prediction cost for approximating the high modes from the low ones) are crucial in this step. Accuracy is vital because any error in this step introduces a corresponding error in the final RB solution. Performance is equally important, as it significantly impacts online performance, especially if the prediction is used iteratively. Another fundamental consideration is the stability of these approximations. An unstable model is more likely to cause divergence in the associated solvers.

4.1 $P = 1$

Starting with the 1-dimensional case ($P = 1$), we consider the problem of predicting the last $N - n$ coefficients from the first n coefficient(s). Following the discussion in sections 3.2 and 3.3, it is natural to expect that $n = 1$ is sufficient for the reconstruction of high modes. The value of N is determined such that $\epsilon_{POD} = 10^{-6}$, which gave us $N(\epsilon) = 9$.

For this regression task, we tested several classical ML algorithms, namely decision tree, random forest, spline, and polynomial regression. The training dataset size is 10^7 for decision tree and random forest models and 10^5 for polynomial and spline models. In Figure 6, we can see that splines provide the best accuracy compared to other models. The accuracy of polynomial regression improves with increasing degree due to the oscillatory behavior of the higher RB coefficients (see Figure 5). The spline model, tuned with hyperparameters (knots=100, degree=10), achieves the best accuracy due to its local adaptability, giving it a significant advantage over polynomials.

Next, we consider increasing values of n for $P = 1$. The results are depicted in Figure 7. We observe that decision trees (and similarly random forests) do not improve, while polynomial regressors gain in accuracy. This result is expected and can be explained by the fact that the first coefficient is the only independent one. Increasing n does not provide additional information but implicitly increases the degree of polynomial models, which explains the improvement in accuracy.

While this approach appears to work well with polynomials, it has significant drawbacks. Firstly, increasing n significantly raises the prediction cost, given by

$\binom{n+d}{d}$ where d is the degree of the polynomial. Secondly, it introduces substantial instability in the model (see Figure 8), making it impractical for the online phase where a nonlinear system is solved iteratively.

4.2 $P > 1$

We now consider the case $P = 2$. The number of POD modes $N(\epsilon_{POD}) = 25$ where $\epsilon_{POD} = 10^{-6}$. Figure 9 shows that for $n = P = 2$, decision trees and random forests perform better than polynomials; however, we do not achieve the same accuracy as for $P = 1$. Similar to $P = 1$, increasing n improves the accuracy of polynomial models but leads to instability.

For higher parameter dimensions ($P > 2$), as seen in section 3.1, the multivariate regression task becomes more complex. Achieving acceptable accuracy using these

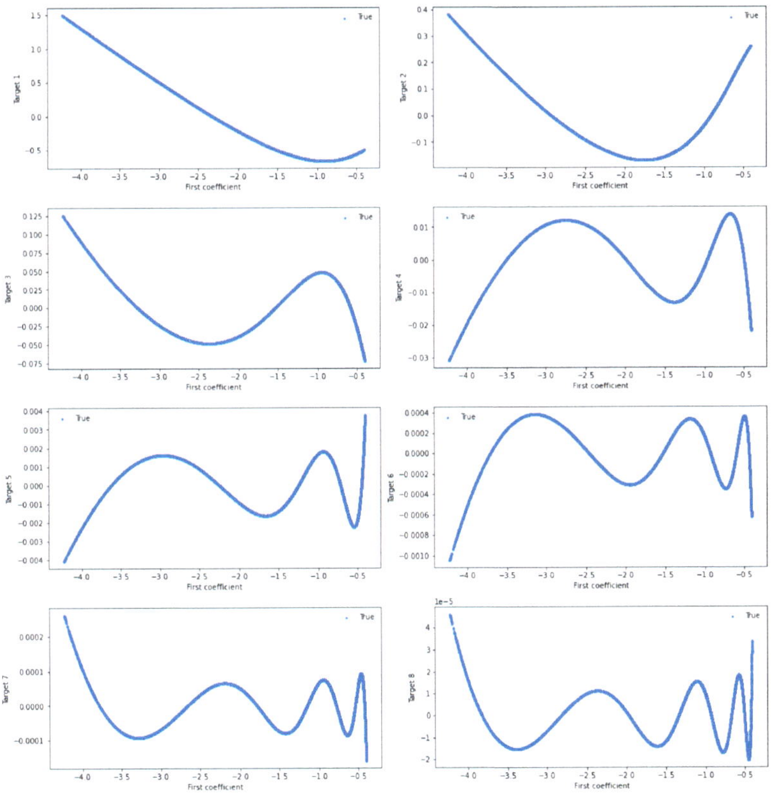

Fig. 5: Higher coefficients as a function of the first coefficient for $P = 1$ and $n = 1$

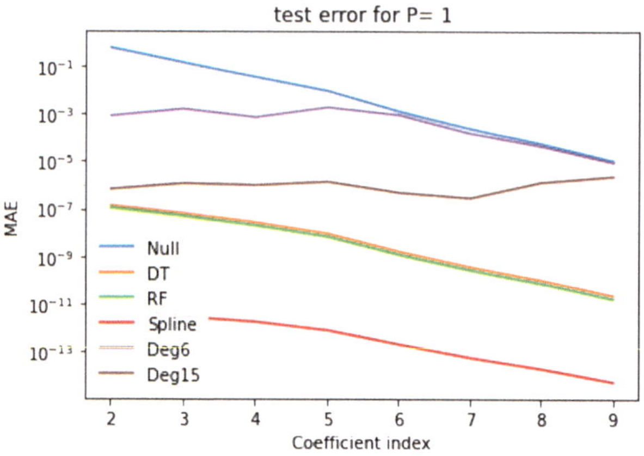

Fig. 6: Mean absolute error on a test set for different ML models for $P = 1$ and $n = 1$.

classic machine-learning algorithms is challenging and improved approaches need be involved.

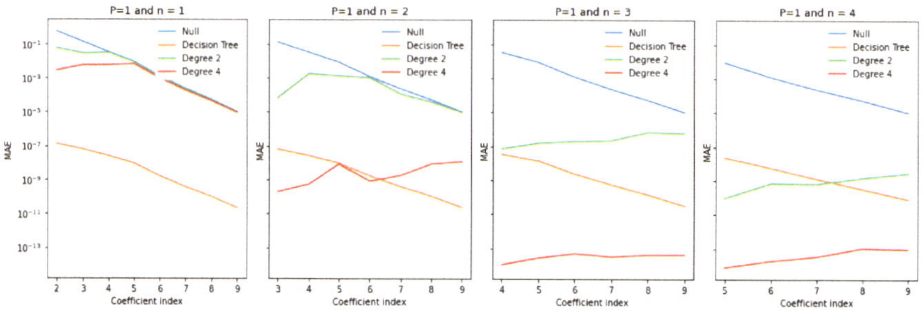

Fig. 7: Mean absolute error for different number of features n and $P = 1$

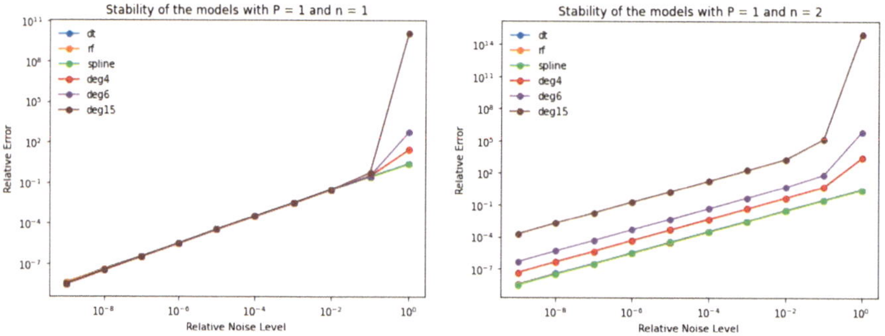

Fig. 8: Stability of models for $P = 1$, $n = 1$ (left panel) and $n = 2$ (right panel).

5 Nonlinear compressive RBM in action

Once we have identified proper (linear) encoder and (nonlinear) decoder to approximate elements in $\mathcal{S}$, we can use this paradigm to approximate the solution of problem (1) for new values of μ. This is done by writing, e.g. : Find $u_\mu^{n,N}$ in X_N that is written as

$$u_\mu^{n,N} = \sum_{i=1}^{n} \hat{u}_{i,\mu}\zeta_i + \sum_{j=n+1}^{N} \hat{u}_j(\{\hat{u}_{i,\mu}\}_{i=1,\ldots,n})\zeta_j \tag{22}$$

such that

$$\Pi_n\Big[\Re(u_\mu^{n,N};\mu)\Big] = 0, \tag{23}$$

where Π_n is some rank n operator.

In this equation, the degress of freedom are the n values $\{\hat{u}_{i,\mu}\}_{i=1,\ldots,n}$, and the unknown $\hat{u}_j(\{\hat{u}_{i,\mu}\}_{i=1,\ldots,n})$ for $j = n+1,\ldots,N$ are functions of the degrees of freedom. In our case, we choose a Galerkin approach and choose Π_n to be the projection on X_n.

This problem is nonlinear and iterative methods are advocated to solve it.

The problem is quite simple, and we have implemented a Picard iteration procedure (for a larger problem, a Newton or quasi-Newton method should be implemented). This gives rise to the following (where k is the iteration index and where we leave all dependencies in μ, n and N).

$$\sum_{i=1}^{n} \hat{u}_i^{k+1}\zeta_i = \sum_{i=1}^{n} \hat{u}_i^k\zeta_i - \gamma^k\Pi_n\Big[\Re(u^k;\mu)\Big] \tag{24}$$

with

Fig. 9: Mean absolute error on a test set for different ML models for $P = 2$

$$u^k = \sum_{i=1}^{n} \hat{u}_i^k \zeta_i + \sum_{j=n+1}^{N} \hat{u}_j(\{\hat{u}_i^k\}_{i=1,..,n})\zeta_j \tag{25}$$

To present the results, in Section 4.1, we showed that for $P = 1$, the best regression accuracy is achieved using the spline model. Applying the Nonlinear Compressive Method with $n = 1$ yields the results shown in Table 1. We observe that the nonlinear reconstruction of high modes significantly reduces the error compared to the classical RB method with only $n = 1$ mode. It is also evident that regression accuracy plays a crucial role in the final results, and with splines, we can achieve nearly the same accuracy as RB with $N = 9$ modes.

Model	**Relative Energy Error**	**Relative Output Error**
RB (with 9 modes)	9.39×10^{-7}	1.50×10^{-12}
RB (with 1 mode)	4.68×10^{-1}	2.89×10^{-1}
Degree 4 regression	1.17×10^{-2}	1.31×10^{-3}
Degree 6 regression	2.91×10^{-3}	6.48×10^{-4}
Degree 15 regression	3.41×10^{-6}	4.70×10^{-7}
Spline	9.39×10^{-7}	2.24×10^{-11}
Decision tree	9.91×10^{-7}	4.72×10^{-8}
Random forest	9.78×10^{-7}	3.70×10^{-8}

Table 1: Relative energy and output Errors for P=1

For $P = 2$, a mean energy error of 10^{-4} can be achieved using a decision tree model, compared to 10^{-1} with 2 modes and 10^{-6} with 25 modes using the classical RBM. In this case and for $P > 2$, it is essential to employ more robust nonlinear solvers alongside advanced regression models to achieve better accuracy.

We want finally and consider the potential economy from the pure reduced basis method to the nonlinear compressive reduced basis approximation.

The plain reduced basis method involves :

- as already indicated in the introduction, for a linear problem: the inversion of a full $N \times N$ matrice, i.e. $O(N^3)$ operations
- for nonlinear problems, thanks to hypereductions methods like EIM [4], only an addition (and negligible) contribution like $O(N^2)$

On the other hand, this nonlinear compressive RBM involves

- for a linear problem : the inversion of a $n \times n$ matrix hence $O(n^3)$ operations and for each iteration $k = 1, .., K$ the evaluation of the high modes with respect to the n first ones, hence $O(n^2 + (N - n)^2)$ complexity
- and additional $(KO(N^2))$, again if the original problem is nonlinear.

which of course rapidly came in favor of this latter approach.

6 Conclusion and perspective

As a preamble to this conclusion, we would like to point out that the proposed approach can be considered from two angles. The first, and the one we prefer, is that of approximating elements u_μ of $\mathcal{S}$: given some data that are linear evaluations of $u_\mu \in \mathcal{S}$ (the first n coefficients of u_μ in a reduced basis), how can we construct a good approximation of u_μ by authorizing non-linear constructions (of the next $N-n$ coefficients of u_μ in the reduced basis) from the data? We're well within the "sensing numbers" framework and this approximation paradigm is used to determine the first n coefficients of u_μ.

The other way of looking at the approach is to think in terms of μ and not u_μ. We then look directly at the problem (23), which is a nonlinear problem in the degrees of freedom $\{\hat{u}_{i,\mu}\}_{i=1,\dots,n}$ and the linearity of the encoder is hidden. We want to stress that the situation is similar if we consider a classical RBM approximation for a nonlinear problem. the resulting system is nonlinear in the component of the solution, while the foundation of the approach is properly linear and based on Kolmogorov's N-width.

Going back to the purpose of this paper, we have demonstrated on the low dimension $P = 1, 2, 3$ at least, that we can indeed compute through a non-linear construction $N-n$ coefficients from the first n coefficients associated with the more important modes. The critical offline ingredient is the regressor fitted to reconstruct the $N-n$ coefficients online with the Picard iterations. A few remarks are necessary. Several regressors have been used, — decision trees and random forests, polynomials are various degrees and splines — and compared. As P increases, the regressor accuracy decreases, or the regressor is not available, such as the spline which gave the best results for $P = 1$. Clearly, more advanced regressors must be used to support higher dimensions. The online procedure will also need to be studied in more detail concerning the robustness and computational cost. Finally, comparing the present work with other generation strategies for the reduced basis space would be interesting, *e*.g. greedy.

Acknowledgements This work has been partially funded by the European Research Council (ERC) under the European Union's Horizon 2020 research and innovation program (grant No 810367), project EMC2 (YM), and the France 2030 NumPEx Exa-MA (ANR-22-EXNU-0002) project managed by the French National Research Agency (ANR).

The authors want to thank Gérard Biau, Anthony Nouy and Agustin Somacal for interesting discussions during the preparation of this manuscript.

References

1. D Amsallem and C Farhat. Interpolation method for adapting reduced-order models and application to aeroelasticity. AIAA journal, 46(7):1803–1813, (2008).
2. Barnett, Joshua and Farhat, Charbel, Quadratic approximation manifold for mitigating the Kolmogorov barrier in nonlinear projection-based model order reduction, Journal of Computational Physics, **464**, 111348 (2022)

3. Barnett, Joshua, Charbel Farhat, and Yvon Maday, Neural-network-augmented projection-based model order reduction for mitigating the Kolmogorov barrier to reducibility, Journal of Computational Physics **492** 112420 (2023).
4. Barrault, Maxime, Yvon Maday, Ngoc Cuong Nguyen, and Anthony T. Patera. An "empirical interpolation" method: application to efficient reduced-basis discretization of partial differential equations. Comptes Rendus Mathematique 339, no. 9, 667–672 (2004)
5. Biau, Gérard, Analysis of a random forests model, The Journal of Machine Learning Research, Volume 13 Issue 1 1063–1095 (2012)
6. Cagniart Nicolas, Maday Yvon, and Stamm Benjamin. Model order reduction for problems with large convection effects. In Contributions to Partial Differential Equations and Applications, 131-150. Springer, (2019).
7. K Carlberg. Adaptive h-refinement for reduced-order models. International Journal for Numerical Methods in Engineering, 102(5):1192–1210, (2015).
8. Cohen, Albert, Wolfgang Dahmen, and Ronald DeVore. Compressed sensing and best k-term approximation. Journal of the American mathematical society 22, no. 1 211–231 (2009). Harvard
9. Cohen, Albert, Farhat, Charbel, Maday, Yvon and Somacal, Agustin, Nonlinear compressive reduced basis approximation for PDE's, Comptes Rendus. Mécanique, **351**-S1, pp 1–18, (2023)
10. DeVore, Ronald A. Nonlinear approximation. Acta numerica **7** 51-150 (1998). Harvard
11. DeVore, Ronald, Boris Hanin, and Guergana Petrova. Neural network approximation. Acta Numerica **30** 327-444 (2021): . Harvard
12. Jens L Eftang, David J Knezevic, and Anthony T Patera. An "hp" certified reduced basis method for parametrized parabolic partial differential equations. Mathematical and Computer Modelling of Dynamical Systems, 17(4):395–422, (2011).
13. Etter, Philip A., and Kevin T. Carlberg. "Online adaptive basis refinement and compression for reduced-order models via vector-space sieving." Computer Methods in Applied Mechanics and Engineering 364 (2020):
14. Ferrero, Andrea, Tommaso Taddei, and Lei Zhang. "Registration-based model reduction of parameterized two-dimensional conservation laws." Journal of Computational Physics 457 111068 (2022).
15. Fresca, Stefania, Luca Dedé, and Andrea Manzoni, A comprehensive deep learning-based approach to reduced order modeling of nonlinear time-dependent parametrized PDEs, Journal of Scientific Computing **87** 1–36 (2021) .
16. Geelen, Rudy, Stephen Wright, and Karen Willcox. Operator inference for non-intrusive model reduction with quadratic manifolds. Computer Methods in Applied Mechanics and Engineering **403** 115717 (2023).
17. Hesthaven, Jan, Rozza, Gianluigi and Stamm, Benjamin, Certified Reduced Basis Methods for Parametrized Partial Differential Equations (2016)
18. Angello Iollo and Damiano Lombardi. Advection modes by optimal mass transfer. Physical Review E, 89(2):022923, (2014).
19. Lee, Kookjin, and Kevin T. Carlberg, Model reduction of dynamical systems on nonlinear manifolds using deep convolutional autoencoders, Journal of Computational Physics **404** 108973 (2020).
20. K Kashima. Nonlinear model reduction by deep autoencoder of noise response data. In 2016 IEEE 55th Conference on Decision and Control (CDC), pages 5750–5755. IEEE (2016).
21. Kramer, Boris, Benjamin Peherstorfer, and Karen E. Willcox. "Learning nonlinear reduced models from data with operator inference." Annual Review of Fluid Mechanics **56** 521–548 (2024).
22. Maday, Yvon, and Benjamin Stamm. Locally adaptive greedy approximations for anisotropic parameter reduced basis spaces. SIAM Journal on Scientific Computing 35, no. 6, 2417-A2441 (2013)
23. Rambod Mojgani and Maciej Balajewicz. Arbitrary Lagrangian Eulerian framework for efficient projection-based reduction of convection dominated nonlinear flows. In APS Division of Fluid Dynamics Meeting Abstracts (2017).

24. Mario Ohlberger and Stephan Rave. Nonlinear reduced basis approximation of parameterized evolution equations via the method of freezing. Comptes Rendus Mathématique, 351(23-24):901–906, (2013).
25. B Peherstorfer. Model reduction for transport-dominated problems via online adaptive bases and adaptive sampling. SIAM Journal on Scientific Computing 42, no. 5 A2803-A2836 (2020).
26. Pedregosa et al., Scikit-learn: Machine Learning in Python, JMLR 12, pp. 2825–2830, (2011)
27. Prud'homme, Christophe et al. (2023). feelpp/feelpp: Feel++ V111 preview 9. Zenodo. https://doi.org/10.5281/zenodo.7326971
28. Prud'homme, Christophe, Rovas, Dimitrios, Veroy, Karen, Machiels, Luc, Maday, Yvon, Patera Anthony T. and Turinici Gabriel, Reliable Real-Time Solution of Parametrized Partial Differential Equations: Reduced-Basis Output Bound Methods J. Fluids Eng. Mar **124**(1): 70-80 (2002)
29. Quarteroni, Alfio, Manzoni, Andrea and Negri, Federico, Reduced basis methods for partial differential equations: An introduction 1–263 (2015).
30. Tommaso Taddei, Simona Perotto, and Alfio Quarteroni. Reduced basis techniques for nonlinear conservation laws. ESAIM: Mathematical Modelling and Numerical Analysis, 49(3):787–814, (2015).
31. Polack, Etienne, Geneviève Dusson, Benjamin Stamm, and Filippo Lipparini. "Grassmann extrapolation of density matrices for Born-Oppenheimer molecular dynamics." Journal of Chemical Theory and Computation 17, no. 11 6965–697 (2021) .
32. Tommaso Taddei. A registration method for model order reduction: data compression and geometry reduction. SIAM Journal on Scientific Computing, 42(2):A997–1027, (2020)
33. Kim, Youngkyu, Youngsoo Choi, David Widemann, and Tarek Zohdi. "Efficient nonlinear manifold reduced order model." arXiv preprint arXiv:2011.07727 (2020).
34. R Zimmermann, B Peherstorfer, and K Willcox. Geometric subspace updates with applications to online adaptive nonlinear model reduction. SIAM Journal on Matrix Analysis and Applications, 39(1):234–261, (2018).

Sparse Besov Space Analysis of Representations in Machine Learning

Ido Ben Shaul and
Shai Dekel

Abstract We present a survey of a unified approach that analyzes representation spaces in signal processing, classic machine learning and deep learning using familiar concepts of sparsity and smoothness spaces. The theory is especially suited for high dimensional spaces. It is known that it is not possible to characterize the approximation spaces of deep learning models using classic smoothness spaces [19]. Furthermore, many problems solved by deep learning are high dimensional where classical function spaces such as the isotropic Besov spaces are somewhat inadequate. Here, we shed some light on this problem by analyzing the dynamics of sparse Besov function smoothness of representations across the layers of a deep neural network, during and after training. We justify our approach by extensive experiments demonstrating that in well-performing trained networks, the sparse Besov smoothness of the training set, measured in the corresponding hidden layer feature map representation, increases from layer to layer.

1 Introduction

In recent years there has been growing interest and much research on the topic of the 'mathematics of deep learning'. While deep learning (DL) exhibits overwhelming effectiveness in many challenging problems in computer vision and natural language processing, the deep neural networks themselves are considered by many as mysterious black boxes. Thus, there is ongoing significant research efforts to find a theory to explain this success using statistics, dynamical systems, compositionality classes, approximation theory and more (e.g. [20, 32, 19, 29, 2]). However,

Ido Ben Shaul
School of Mathematical Sciences, Tel Aviv University, e-mail: ido.benshaul@gmail.com

Shai Dekel
School of Mathematical Sciences, Tel Aviv University e-mail: shaidekel6@gmail.com

R. DeVore, A. Kunoth (eds.), *Multiscale, Nonlinear and Adaptive Approximation II*,
https://doi.org/10.1007/978-3-031-75802-7_5

most of the proposed theories do not present a unified approach to the three pillars of data processing which are: signal processing, classic machine learning models (e.g. decision trees, random forest, gradient boosting) and deep learning. Quite often data scientists struggle to understand when does deep learning provide an advantage over the simpler machine learning models (e.g. tree-based models). There is some folklore knowledge that DL does not provide any advantage for most tabular data, but this phenomena is not well understood [31]. In this work we advocate to model datasets as samples of functions and then estimate the weak type smoothness of these functions using sparse tree-based Besov spaces. We also characterize this type of smoothness using an equivalent form of sparsity of wavelet decompositions of tree representations. This allows us to link the well established approximation theory of signal theory [16, 17], the theory of classical tree-based machine learning models [13, 18] and the approximation theoretical mechanisms of DL. We advocate that in cases where the input function has low 'weak-type' smoothness, the role of a feed-forward neural network is to perform automatic feature engineering which improves the smoothness of the representations as we go deeper through the layers of the network. In Figure 1 we see a visualization of the feature space of two famous datasets: Titanic and MNIST. The Titanic dataset is composed of tabular data where the passenger features are: age, gender, class of ticket, port of boarding, etc. and then the response variable is the binary survival outcome. The MNIST dataset is composed of 28×28 grayscale images, each containing a hand-written digit and each labeled: '0' to '9'. Although the feature spaces of the samples of the two datasets are high-dimensional, they are depicted colored with their label using a nonlinear dimension reduction algorithm. Roughly speaking, the tabular Titanic dataset has more cluster structures in the feature space, while the feature space of MNIST seems very unstructured. We argue that while DL is not needed for the Titanic dataset, the role of DL for the MNIST dataset is to unravel the feature space and improve it layer by layer, where at each layer clustering improves as shown in Figure 2. Consequently, the smoothness of the dataset improves across layers in the corresponding representation spaces.

Let us provide more details for this process: Assume we are presented with a set of gray-scale images of dimension $\sqrt{n_0} \times \sqrt{n_0}$ with L class labels. Assume further that a DL network has been successfully trained to classify these images with relatively high accuracy. This allows us to extract the representation of each image in each of the hidden layers. To create a representation at layer 0, we concatenate the $\sqrt{n_0}$ rows of pixel values of each image, to create a vector of dimension n_0. We also normalize the pixel values to the range $[0, 1]$. Since we advocate a function-theoretical approach, we transform the class labels into vector-values in the space $\mathbb{R}^L$ by assigning the l-th label to the l-th standard basis vector, so the first class to $(1, 0, \cdots)$, the second class to $(0, 1, 0, \cdots)$, etc. Thus, the images are considered as samples of a function $f_0 : [0, 1]^{n_0} \rightarrow \mathbb{R}^L$. Typically, for non-tabular data, such as in computer vision problems, there is no hope that there exists geometric clustering of the classes in this initial pixel feature space and that f_0 has sufficient 'weak-type' smoothness. Thus, a transform into a different feature space is needed. We associate with each k-th layer of a DL network, a function $f_k : [0, 1]^{n_k} \rightarrow \mathbb{R}^L$ where the samples are vectors

Fig. 1: Visualization of input feature spaces: MNIST dataset (left) and Titanic dataset (right)

Fig. 2: MNIST: improved clustering across layer representation

created by normalizing and concatenating the feature maps computed from each of the images. Interestingly enough, although the series of functions f_k are embedded in different dimensions n_k, through the simple normalizing of the features, our method is able to assign smoothness indices to each layer that are comparable. We claim that for well performing networks, the representations in general 'improve' from layer to layer and that our method mathematically quantifies the phenomena that is rendered in Figure 2.

The paper is organized as follows: In Section 2 we review our sparse approximation machinery for high dimensional datasets which is the Wavelet Decomposition of Random Forest (RF) [18]. As we shall see, equivalences between wavelet sparsity and sparse Besov semi-norms, also serve as our applied smoothness analysis machinery. In Section 3 we present the required sparse geometric function space theoretical background. Since we will be comparing different representations over different spaces of different dimensions, we ensure the theory is 'dimension-free', in the sense that certain sparsity and smoothness forms do not depend on the dimension [3]. In Section 4 we present some basic DL architectures and discuss the problem of the characterization of DL approximation spaces. In Section 5 we show how to apply the theory in practice. Specifically, we use a high dimensional sparse generalization of the equivalence of wavelet sparsity formulation and sparse Besov semi-norms. This allows us to numerically estimate a sparse Besov 'weak-type' smoothness index of a given function in any representation space (e.g hidden layer). Section 6 provides experimental results that demonstrate how our theory is able to explain empirical findings in various scenarios.

2 Wavelet decomposition of Random Forests

To overcome the challenge of analyzing the smoothness of high dimensional representation spaces of datasets (e.g. inner layers in DL), we apply the construction of wavelet decompositions of Random Forests [18]. The Random Forest (RF) [5, 9, 21] introduced by Breiman [6, 7] as a machine learning algorithm, is in fact a powerful adaptive sparse piecewise polynomial approximation algorithm for high dimensional problems. The forest overcomes the 'greedy' nature and high variance of a single decision tree. The geometric wavelets [13, 18] used to decompose the forest allow sparse representations and have many properties that are shared with the classical wavelets [11, 26] that are not computationally feasible in high dimensions (in dimension $n = 100,000$, there are $2^{100,000} - 1$ classic tensor wavelet 'types'). When combined, the wavelet decomposition of the RF unravels the sparsity of the underlying function and establishes an order of the RF nodes from 'important' components to 'negligible' noise. Therefore, the method provides a better understanding of any constructed RF. As we shall see, our motivation to use these wavelet decompositions is the equivalence between sparse Besov smoothness over forests and the wavelet sparsity (see (18)).

We begin with an overview of single trees. In statistics and machine learning the construction is called a Decision Tree or the Classification and Regression Tree (CART) [8, 1, 5, 21]. Assume we are given a real-valued function $f \in L_p([0,1]^n)$ or a discrete dataset $\{x_i \in [0,1]^n, f(x_i)\}_{i\in I}$ (the generalization to any convex bounded domain $\Omega_0 \subset \mathbb{R}^n$ is trivial). The goal is to compute an adaptive piecewise polynomial approximation of f, even for large dimensions n (potentially in the range of hundreds of thousands). To this end, we subdivide the initial domain $\Omega_0 := [0,1]^n$ into two subdomains, e.g. by intersecting it with a hyper-plane. The goal is to find a subdivision that approximately minimizes a given cost function. This subdivision process then continues recursively on the subdomains until some stopping criterion is met, which in turn, determines the leaves of the tree. We now describe one instance of the cost function which is related to minimizing variance. At each stage of the subdivision process, at a certain node of the tree, the algorithm finds, for the convex domain $\Omega \subset \mathbb{R}^n$ associated with the node:

(i) A partition by an hyper-plane into two convex subdomains $\Omega', \Omega'', \Omega' \cup \Omega'' = \Omega$,
(ii) Two multivariate polynomials $Q_{\Omega'}, Q_{\Omega''} \in \Pi_{r-1}(\mathbb{R}^n)$, of fixed (typically low) total degree r − 1.

The partition and the polynomials are chosen to minimize the following quantity

$$\|f - Q_{\Omega'}\|^p_{L_p(\Omega')} + \|f - Q_{\Omega''}\|^p_{L_p(\Omega'')} . \tag{1}$$

For the theory that follows, we require that the polynomials $Q_{\Omega'}, Q_{\Omega''}$ provide local near best approximation on their respective subdomains. In applications, where the dataset is discrete, consisting of feature vectors $x_i \in [0, 1]^n, i \in I$, with given values $f(x_i)$, a discrete functional is minimized over all partitions $\Omega' \cup \Omega'' = \Omega$

$$\sum_{x_i \in \Omega'} |f(x_i) - Q_{\Omega'}(x_i)|^p + \sum_{x_i \in \Omega''} |f(x_i) - Q_{\Omega''}(x_i)|^p. \tag{2}$$

Observe that for any given subdividing hyperplane, the approximating polynomials in (2) can be uniquely determined for $p = 2$, by least square minimization. For the order $r = 1$, the approximating polynomials are nothing but the mean of the function values over each of the subdomains

$$Q_{\Omega'}(x) = C_{\Omega'} = \frac{1}{\#\{x_i \in \Omega'\}} \sum_{x_i \in \Omega'} f(x_i),$$

$$Q_{\Omega''}(x) = C_{\Omega''} = \frac{1}{\#\{x_i \in \Omega''\}} \sum_{x_i \in \Omega''} f(x_i).$$

In many applications of decision trees, the high-dimensionality of the data does not allow to search through all possible subdivisions. As in our experimental results, one may restrict the subdivisions to the class of hyperplanes aligned with the main axes. In contrast, there are cases where one would like to consider more advanced form of subdivisions, where they take certain hyper-surface form or even non-linear forms through kernel Support Vector Machines. Our paradigm of wavelet decompositions can support in principle all of these forms.

Random Forest (RF) is a popular machine learning tool that collects decision trees into an ensemble model. The trees are constructed independently in a diverse fashion and prediction is done by a voting mechanism among all trees. A key element [6], is that large diversity between the trees reduces the ensemble's variance. There are many RFs variations that differ in the way randomness is injected into the model, e.g bagging, random feature subset selection and the partition criterion. Bagging [7] is a method that produces partial replicates of the training data for each tree. A typical approach is to randomly select for each tree a certain percentage of the training set (e.g. 80%).

Additional methods to inject randomness can be achieved at the node partitioning level. For each node, we may restrict the partition criteria to a small random subset of the feature coordinates. A typical selection is to search for a partition from a random subset of $\sqrt{n}$ features [6]. This technique is also useful for reducing the amount of computations when searching the appropriate partition for each node. Bagging and random feature selections are not mutually exclusive and can be used together.

For $j = 1, ..., J$, one constructs a tree $\mathcal{T}_j$, based on a subset of the data, X^j. One then provides a weight (score) w_j to the tree $\mathcal{T}_j$, based on the estimated performance of the tree, where $\sum_{j=1}^{J} w_j = 1$. In the supervised learning, one typically uses the remaining data points $x_i \notin X^j$ to evaluate the performance of $\mathcal{T}_j$. For simplicity, we will mostly consider in this paper the choice of uniform weights $w_j = 1/J$. For any point $x \in [0, 1]^n$, the approximation associated with the jth tree, denoted by $\tilde{f}_j(x)$, is computed by finding the leaf $\Omega \in \mathcal{T}_j$ in which x is contained and then evaluating $\tilde{f}_j(x) := Q_\Omega(x)$, where Q_Ω is the corresponding polynomial associated with the decision node Ω. One then assigns the RF approximate value to any point $x \in [0.1]^n$ by

$$\tilde{f}(x) = \sum_{j=1}^{J} w_j \tilde{f}_j(x).$$

As already discussed, in classification problems, each input training point $x_i \in [0, 1]^n$ is assigned with a class $Cl(x_i)$. To convert the problem to the 'functional' setting described above, one assigns to the l-th class the standard basis vector $e_l \in \mathbb{R}^L$, $1 \leqslant l \leqslant L$. In this case, if we choose approximation using constants ($r = 1$), then the calculated mean over any subdomain Ω is in fact a point $\mathbf{E}_\Omega \in \mathbb{R}^L$. Obviously, any vector value $v = (v_1, ..., v_L) \in R^L$ can be mapped back to a class, along with an estimated confidence level, by calculating $\arg\max_{i=1}^{L} v_i$.

Next, we recall the construction of a wavelet decomposition of a forest [18]. Let Ω' be a child of Ω in a tree $\mathcal{T}$, i.e. $\Omega' \subset \Omega$ and Ω' was created by a partition of Ω. Denote by $\mathbf{1}_{\Omega'}$, the indicator function over the child domain Ω', i.e. $\mathbf{1}_{\Omega'}(x) = 1$, if $x \in \Omega'$ and $\mathbf{1}_{\Omega'}(x) = 0$, if $x \notin \Omega'$. We use the polynomial approximations $Q_{\Omega'}, Q_\Omega \in \Pi_{r-1}(\mathbb{R}^n)$, computed by the local minimization (1) and define

$$\psi_{\Omega'}(x) := \psi_{\Omega'}(f)(x) := \mathbf{1}_{\Omega'}(x)\left(Q_{\Omega'}(x) - Q_\Omega(x)\right), \tag{3}$$

as the geometric wavelet associated with the subdomain Ω' and the function f, or the given discrete dataset $\{x_i, f(x_i)\}_{i \in I}$. Each wavelet $\psi_{\Omega'}$, is a 'local difference' component that belongs to the detail space between two levels in the tree, a 'low resolution' level associated with Ω and a 'high resolution' level associated with Ω'. Also, the wavelets (3) have the 'zero moments' property, i.e., if the response variable is sampled from a polynomial of degree $r - 1$ over Ω, then our local scheme will compute $Q_{\Omega'}(x) = Q_\Omega(x) = f(x)$, $\forall x \in \Omega$, and therefore $\psi_{\Omega'} = 0$. We also define $\psi_{[0,1]^n} := Q_{[0,1]^n}$.

Under certain mild conditions on the tree $\mathcal{T}$ and the function f, we have by the nature of the wavelets, the 'telescopic' sum of differences

$$f = \sum_{\Omega \in \mathcal{T}} \psi_\Omega, \qquad \psi_{\Omega_0} := Q_{\Omega_0}. \tag{4}$$

For example, (4) holds in L_p-sense, $1 \leqslant p < \infty$, if $f \in L_p([0, 1]^n)$ and for any $x \in [0, 1]^n$ and series of domains $\Omega_l \in \mathcal{T}$, each on a level l, with $x \in \Omega_l$, we have that $\lim_{l \to \infty} \operatorname{diam}(\Omega_l) = 0$.

The norm of a wavelet supported on a child Ω' of Ω is computed by

$$\|\psi_{\Omega'}\|_p^p = \int_{\Omega'} (Q_{\Omega'}(x) - Q_\Omega(x))^p \, dx.$$

For the case $r = 1$, where $Q_\Omega(x) = C_\Omega$ and $Q_{\Omega'}(x) = C_{\Omega'}$ this simplifies to

$$\|\psi_{\Omega'}\|_p^p = |C_{\Omega'} - C_\Omega|^p \, |\Omega'|, \tag{5}$$

where $|\Omega'|$ denotes the volume of Ω'. Observe that for $r = 1$ and $p = 2$, the subdivision process for partitioning a node by minimizing (1) is equivalent to maximizing the sum of squared norms of the wavelets that are formed in that partition (see [18]).

Recall that our approach is to convert classification problems into a 'functional' setting by assigning the class labels to vector values in $\mathbb{R}^L$. In such cases of vector-valued functions, choosing $r = 1$, the wavelet $\psi_{\Omega'} : \mathbb{R}^n \to \mathbb{R}^L$ is

$$\psi_{\Omega'}(x) = \mathbf{1}_{\Omega'}(x) \left(\mathbf{E}_{\Omega'} - \mathbf{E}_\Omega\right),$$

and its norm is given by

$$\|\psi_{\Omega'}\|_p^p = \|\mathbf{E}_{\Omega'} - \mathbf{E}_\Omega\|_{l_2(\mathbb{R}^L)}^p \, |\Omega'|, \tag{6}$$

where for $\mathbf{v} \in \mathbb{R}^L, \|\mathbf{v}\|_{l_2} := \sqrt{\sum_{i=1}^L v_i^2}$.

Using any given weights assigned to the trees, we obtain a wavelet representation of the entire RF

$$\tilde{f}(x) = \sum_{j=1}^{J} \sum_{\Omega \in \mathcal{T}_j} w_j \psi_\Omega(x). \tag{7}$$

The theory tells us that, just as in the classical case [14], sparse approximation is achieved by reordering the wavelet components based on their norm [18]

$$w_{j(\Omega_{k_1})} \left\|\psi_{\Omega_{k_1}}\right\|_p \geqslant w_{j(\Omega_{k_2})} \left\|\psi_{\Omega_{k_3}}\right\|_p \geqslant \cdots, \tag{8}$$

with the notation $\Omega \in \mathcal{T}_j \Rightarrow j(\Omega) = j$. Thus, the adaptive M-term approximation of a RF is

$$f_M(x) := \sum_{m=1}^{M} w_{j(\Omega_{k_m})} \psi_{\Omega_{k_m}}(x). \tag{9}$$

Observe that, contrary to most existing tree pruning techniques [23], where each tree is pruned separately, the above approximation process applies a 'global' pruning strategy where the significant components can come from any node of any of the trees at any level. For simplicity, one could choose $w_j = 1/J$, and obtain

$$f_M(x) = \frac{1}{J} \sum_{m=1}^{M} \psi_{\Omega_{k_m}}(x). \tag{10}$$

Fig. 3 depicts an M-term (10) selected from an RF ensemble. The red colored nodes illustrate the selection of the M wavelets with the highest norm values from the entire forest. Observe that they can be selected from any tree at any level, with no connectivity restrictions.

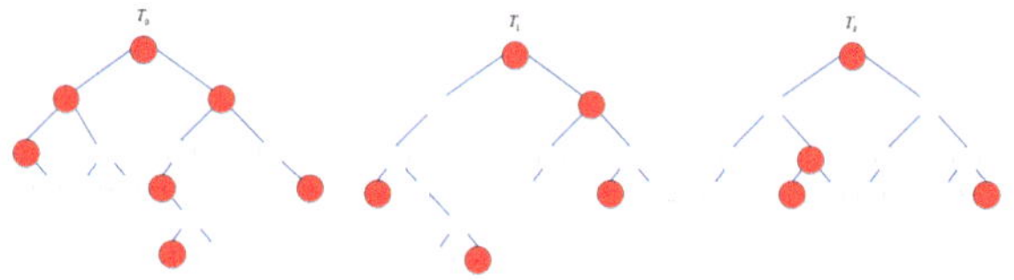

Fig. 3: Selection of an M-term approximation from the entire forest.

3 Geometric multivariate function space theory

An important research area of approximation theory, pioneered by Pencho Petrushev, is the characterization of adaptive geometric approximation algorithms by generalizations of the classic isotropic Besov spaces to sparse Besov-type spaces [10, 13, 22] that are more adequate for unstructured data in high-dimensions. We begin by defining the 'weak-type' sparse Besov smoothness of a function, subject to the geometry of a single (possibly adaptive) tree To this end, recall that for a function $f \in L_\tau(\Omega)$, $0 < \tau \leqslant \infty$, $h \in \mathbb{R}^n$ and $r \in \mathbb{N}$, we have the r-th order difference operator

$$\Delta_h^r(f,x) = \Delta_h^r(f,\Omega,x) := \sum_{k=0}^{r}(-1)^{r+k}\binom{r}{k} f(x+kh),$$

where we assume the segment $[x, x+rh]$ is contained in Ω. Otherwise, we set $\Delta_h^r(f,\Omega,x) = 0$. The modulus of smoothness of order r is defined by

$$\omega_r(f,t)_\tau := \sup_{|h|\leqslant t} \left\|\Delta_h^r(f,\Omega,\cdot)\right\|_{L_\tau(\Omega)}, \quad t > 0,$$

where for $h \in \mathbb{R}^n$, $|h|$ denotes the norm of h. We also denote

$$\omega_r(f,\Omega)_\tau := \omega_r\left(f, \frac{diam(\Omega)}{r}\right)_\tau. \tag{11}$$

We also recall a subset of the classic Besov spaces defined by functions in $L_\tau(\Omega)$, $0 < \tau < \infty$, for which

$$|f|_{\mathcal{B}_\tau^\alpha(\Omega)} := \left(\int_0^\infty \left(t^{-\alpha}\omega_r(f,\Omega,t)_\tau\right)^\tau \frac{dt}{t}\right)^{1/\tau} < \infty, \tag{12}$$

with $\alpha > 0$, $r \geqslant \lfloor \alpha \rfloor + 1$. The (quasi-)norm is then defined by

$$\|f\|_{\mathcal{B}_\tau^\alpha(\Omega)} := \|f\|_{L_\tau(\Omega)} + |f|_{\mathcal{B}_\tau^\alpha(\Omega)}.$$

A useful equivalent form of the Besov semi-norm uses dyadic cubes. Assume for simplicity that $\Omega = \mathbb{R}^n$ or $\Omega = [0,1]^n$. We denote by Q the dyadic cubes partitions of Ω. In the case of the unit cube, the initial level of Q contains $\Omega = [0,1]^n$, which is then partitioned into 2^n sub-cubes, etc. It is well known that

$$|f|_{\mathcal{B}_\tau^{n\alpha,r}} \sim \left(\sum_{Q \in Q} \left(|Q|^{-\alpha} \omega_r (f, Q)_\tau \right)^\tau \right)^{1/\tau}, \tag{13}$$

This leads to the following generalization to adaptive sparse smoothness.

Definition 3.1 For $\alpha < 1/\tau$, $r \in \mathrm{N}$, $f \in L_\tau([0,1]^n)$, and tree $\mathcal{T}$, we define the associated sparse tree Besov smoothness in $\mathcal{B}_\tau^{\alpha,r}(\mathcal{T})$, by

$$|f|_{\mathcal{B}_\tau^{\alpha,r}(\mathcal{T})} := \left(\sum_{\Omega \in \mathcal{T}} \left(|\Omega|^{-\alpha} \omega_r (f, \Omega)_\tau \right)^\tau \right)^{1/\tau}, \tag{14}$$

where $|\Omega|$ denotes the volume of Ω.

In cases where we wish to approximate in the p-norm, $0 < p < \infty$, we shall typically set $1/\tau = \alpha + 1/p$. The higher the index α for which (14) is finite, the smoother the function is. It is easy to see that the classical Besov spaces are a special case, where the tree is constructed (non-adaptively) by partitioning over dyadic knots such that at levels which are multiples of n one obtains dyadic cubes. For these special dyadic trees $\mathcal{T}_Q$, one has $\mathcal{B}_\tau^\alpha(\mathcal{T}_Q) \sim \mathcal{B}_\tau^{\alpha n}$, where the latter space is the classic Besov semi-norm defined by (12).

We note that we do not impose the condition $r > \alpha$ for (3.1) as we have done for (12). In fact, as we shall see below, for most part we will use $r = 1$. We remind the reader that in certain configurations of classical Besov spaces $\mathcal{B}_\tau^{\alpha,r}$, where $r < \alpha$ and $1 \leqslant \tau \leqslant \infty$, we get a trivial space [15] (e.g. polynomials of degree $r-1$ when Ω is a segment and $\{0\}$ for $\Omega = \mathbb{R}$). However, this is not the case here, since whenever $\alpha \geqslant 1$, we immediately have $\tau < 1$. For example, any $f(x) := \mathbf{1}_{\tilde{\Omega}}(x)$, where $\tilde{\Omega} \subset \mathbb{R}^n$ is a compact domain with a smooth boundary, satisfies $f \in \mathcal{B}_\tau^{\alpha,r}(\mathcal{T}_Q)$, for any $r \geqslant 1$ and $\alpha < 1/\tau$.

For a given forest $\mathcal{F} = \{\mathcal{T}_j\}_{j=1}^J$ and weights $w_j = 1/J$, the α sparse Besov semi-norm associated with the forest is

$$|f|_{\mathcal{B}_\tau^{\alpha,r}(\mathcal{F})} := \frac{1}{J} \left(\sum_{j=1}^J |f|^\tau_{\mathcal{B}_\tau^{\alpha,r}(\mathcal{T}_j)} \right)^{1/\tau}. \tag{15}$$

Definition 3.2 Given a (possibly adaptive) forest representation, we define the sparse Besov smoothness index of f by the maximal index α for which (15) is finite.

Remark It is known that different geometric approximation schemes are characterized by different flavors of Besov-type smoothness. In this work, for example, all of our experimental results compute smoothness of representations using tree subdivisions along the main n axes. This restriction may lead, in general, to potentially lower Besov smoothness of the underlying function and lower sparsity of the wavelet representation. Yet, the theoretical definitions and results of this paper can also apply to more generalized schemes where, for example, tree partitions are performed using arbitrary hyper-planes. In such a case, the smoothness index of a given function may increase.

Next, for a given tree $\mathcal{T}$ and parameter $0 < \tau < p$, we denote the τ-sparsity of the tree by

$$N_\tau(f, \mathcal{T}) = \left(\sum_{\Omega \neq [0,1]^n, \Omega \in \mathcal{T}} \|\psi_\Omega\|_p^\tau \right)^{1/\tau}. \tag{16}$$

Let us further denote the τ-sparsity of a forest $\mathcal{F}$, by

$$\begin{aligned} N_\tau(f, \mathcal{F}) &:= \frac{1}{J} \left(\sum_{j=1}^{J} \sum_{\Omega \neq [0,1]^n, \Omega \in \mathcal{T}_j} \|\psi_\Omega\|_p^\tau \right)^{1/\tau} \\ &= \frac{1}{J} \left(\sum_{j=1}^{J} N_\tau(f, \mathcal{T}_j)^\tau \right)^{1/\tau}. \end{aligned}$$

Under mild conditions on the partitions (see remark after (4) and condition (19)), the theory of [13] proves the equivalence

$$|f|_{\mathcal{B}_\tau^{\alpha,r}(\mathcal{T})} \sim N_\tau(f, \mathcal{T}). \tag{17}$$

This implies that there are constants $0 < C_1 < C_2 < \infty$, that depend on parameters such as α, p, n, r and ρ in condition (19) below, such that

$$C_1 |f|_{\mathcal{B}_\tau^{\alpha,r}(\mathcal{T})} \leqslant N_\tau(f, \mathcal{T}) \leqslant C_2 |f|_{\mathcal{B}_\tau^{\alpha,r}(\mathcal{T})}.$$

Here, we assume $1/\tau := \alpha + 1/p$ and that the wavelets are local differences of near-best local polynomial approximations from the space of polynomials of degree $r-1$ (see (1)). Therefore, we also have for the forest model

$$|f|_{\mathcal{B}_\tau^{\alpha,r}(\mathcal{F})} \sim N_\tau(f, \mathcal{F}). \tag{18}$$

Next, we present a simple invariance property of the smoothness analysis under higher dimensional embedding.

Lemma 3.1 *[3] Let $f : [0,1]^n \to \mathbb{R}^L$, $f \in L_p([0,1]^n)$ and let $\mathcal{F}$ be a forest approximation of f. For any $m \geqslant 0$, let $\tilde{x} = (x, 0, \ldots, 0) \in [0,1]^{n+m}$, $x \in [0,1]^n$. Let us further define $\tilde{f}(\tilde{x}) := f(x)$. Next, denote by $\tilde{\mathcal{F}}$ a forest defined over $[0,1]^{n+m}$ which is the natural extension of $\mathcal{F}$, using the same trees with same partitions over* *the first n coordinates. Then, for $r = 1$ and any $\tau > 0$,*

$$N_\tau\left(\tilde{f}, \tilde{\mathcal{F}}\right) = N_\tau\left(f, \mathcal{F}\right).$$

Remark Note that the above invariance property also holds if the additional m redundant features are intertwined with the n significant features that are 'detected' and used for tree subdivisions during the construction of the random forest.

In the setting in which we wish to apply our function theoretical approach, we are comparing smoothness of representations over different layers of DL networks. This implies that we are analyzing and comparing the smoothness of a set of functions f_k, each over a different representation space of a different dimension n_k. This is, in some sense, non-standard in function space theory, where the space, or at least the dimension, over which the functions have their domain is typically fixed. Specifically, observe that the equivalence (18) depends on the dimension n of the feature space. To this end, we add to the theory 'dimension-free' analysis for the case $r = 1$.

We begin with a Jackson-type estimate for the degree of the adaptive wavelet forest approximation, which we keep 'dimension free' for the case $r = 1$.

Theorem 3.1 *[3] Let $\mathcal{F} = \left\{\mathcal{T}_j\right\}_{j=1}^J$ be a forest. Assume there exists a constant $0 < \rho < 1$, such that for any domain $\Omega \in \mathcal{F}$ on a level l and any domain $\Omega' \in \mathcal{F}$, on the level $l + 1$, with $\Omega \cap \Omega' \neq \emptyset$, we have*

$$\left|\Omega'\right| \leqslant \rho \left|\Omega\right|. \tag{19}$$

For any $r \geqslant 1$, denote formally $f = \sum_{\Omega \in \mathcal{F}} w_{j(\Omega)}\psi_\Omega$, and assume that $N_\tau(f, \mathcal{F}) < \infty$, where

$$\frac{1}{\tau} = \alpha + \frac{1}{p}.$$

Then, for the M-term approximation (9) we have for $r = 1$

$$\sigma_M(f) := \|f - f_M\|_p \leqslant C\left(p, \alpha, \rho\right) J M^{-\alpha} N_\tau(f, \mathcal{F}). \tag{20}$$

and for $r > 1$

$$\sigma_M(f) := \|f - f_M\|_p \leqslant C\left(p, \alpha, \rho, r, n\right) J M^{-\alpha} N_\tau(f, \mathcal{F}). \tag{21}$$

Using the equivalence (18), we get for any $r \geqslant 1$

$$\sigma_M(f) \leqslant C\left(p, \alpha, \rho, n\right) J M^{-\alpha} \left|f\right|_{\mathcal{B}_\tau^{\alpha, r}(\mathcal{F})},$$

which is not a 'dimension-free' Jackson estimate, as the one will show below for $r = 1$ (see (24)).

Next, to allow our smoothness analysis to be 'dimension free' we modify the modulus of smoothness (11) for $r = 1$ and use the following form of 'averaged modulus'

Definition 3.3 For a function $f : \Omega \to \mathbb{R}^L$ we define

$$w_1(f,\Omega)_\tau := \left(\int_\Omega \|f(x) - \mathbf{E}_\Omega\|_{l_2(\mathbb{R}^L)}^\tau \, dx\right)^{1/\tau}, \tag{22}$$

where $\mathbf{E}_\Omega$ is the average of f over Ω.

It is well known that averaged forms of the modulus are equivalent to the form (11), but with constants that depend on the dimension. However, replacing (11) with (22) allows us to produce 'dimension-free' analysis. We use (22) to define

$$|f|_{\tilde{\mathcal{B}}_\tau^\alpha(\mathcal{T})} := \left(\sum_{\Omega\in\mathcal{T}} \left(|\Omega|^{-\alpha} w_1(f,\Omega)_\tau\right)^\tau\right)^{1/\tau},$$

$$|f|_{\tilde{\mathcal{B}}_\tau^\alpha(\mathcal{F})} := \frac{1}{J}\left(\sum_{j=1}^J |f|_{\tilde{\mathcal{B}}_\tau^\alpha(\mathcal{T}_j)}^\tau\right)^{1/\tau}.$$

This allows to show

Theorem 3.2 *[3] Let $f : [0,1]^n \to \mathbb{R}^L$. Then the following equivalence holds for the case $r = 1$,*

$$|f|_{\tilde{\mathcal{B}}_\tau^\alpha(\mathcal{F})} \sim N_\tau(f,\mathcal{F}), \tag{23}$$

where $1/\tau = \alpha + 1/p$, and the constants of equivalence depend on α, τ, ρ, but not n.

The equivalence (23) together with (20) imply that for $r = 1$ we do have a 'dimension-free' Jackson estimate

$$\sigma_M(f) \leqslant C(p,\alpha,\rho)\, J M^{-\alpha} |f|_{\tilde{\mathcal{B}}_\tau^\alpha(\mathcal{F})}. \tag{24}$$

4 Deep Learning approximation models

4.1 Deep Learning Architectures

Neural network architectures can be highly complex, but in this paper we focus on three simple models:

(i) Multilayer Perceptron (MLP) - An MLP network is a forward feed network $\tilde{f} : \mathbb{R}^n \to \mathbb{R}^L$

$$\tilde{f} = \Phi(\sigma_K \circ T_K \circ \ldots \sigma_1 \circ T_1), \tag{25}$$

where each T_k is an affine linear transformation, $T_k : \mathbb{R}^{n_k} \to \mathbb{R}^{n_{k+1}}$, $T_k(x_k) = M_k x_k + b_k$, and the functions σ_k are pointwise nonlinear activation functions. One of the most popular nonlinear activation functions is the Rectified Linear Unit (ReLU)

$$\mathrm{ReLU}(x) := \begin{cases} 0, & -\infty < x < 0, \\ x, & 0 \leqslant x < \infty. \end{cases} \tag{26}$$

Observe that an MLP network (25), with σ_k =ReLU, $1 \leqslant k \leqslant K$, imposes a piecewise linear approximation over polyhedral subdomains in the input feature space.

Typically, in classification problems, the last layer is of dimension L and the function Φ in (25) is the softmax

$$\Phi(x_1, \cdots, x_L) = \frac{1}{\sum_{i=1}^{L} e^{x_i}} (e^{x_1}, e^{x_2}, \cdots, e^{x_L}),$$

which 'normalizes' the output of the network to be a vector whose components are in the range [0, 1] and sum up to one. This design facilitates training the network to output the probabilities of a sample to belong to each of the L classes.

(ii) Convolution Neural Network (CNN) - A convolutional layer is in fact a special case of an MLP layer, where the linear transformation M_k is broken up to a finite number of compactly supported convolutional operations on the input to the layer. This is useful if we are expecting 'local' correlations in the input and in some cases allows to reduce the number of network weights substantially. Applying convolutions requires to maintain the shape of the input signal through the layers. So for example, if the input is an image, the layers will be three dimensional, as a stack of two-dimensional 'feature maps' and the convolutions applied to the input layer k will be three dimensional. The dimension of the stack of two-dimensional 'feature maps' in the output layer $k + 1$ is equal to the number of convolutions applied between the layers, where each convolution produces a 'feature map'.

(iii) Multiplicative Neural Network - Multiplicative networks provide a platform for a polynomial type of nonlinearity which has been found to be extremely powerful in the form of attention mechanisms as part of the Transformer network architectures [28]. Here we describe one useful variant which can be regarded as a multiplicative network of degree 2. Let $x_k \in \mathbb{R}^{n_k}$ be an input vector to the k-th layer. For each component $1 \leqslant i \leqslant n_{k+1}$ of the layer output $x_{k+1} \in \mathbb{R}^{n_{k+1}}$, one selects as part of the architecture two predetermined indices of the input space $1 \leqslant j_1(i), j_2(i) \leqslant n_k$, such that for $1 \leqslant i \leqslant n_{k+1}$

$$x_{k+1}(i) = A_k(i) x_k(j_1(i)) x_k(j_2(i)) + \sum_{j=1}^{n_k} M_k(i, j) x_k(j) + b_k(i),$$

where the vector $A_k \in \mathbb{R}^{n_{k+1}}$ is an additional set of network parameters. In this variant, there is no additional nonlinearity applied as part of the transformation between layers. The multiplicative structure of degree 2 is extremely useful in efficient representation of multivariate polynomials, leading to superior approximation properties over regular MLP networks (see Theorem 4.1 below). Also, one can regard one of the inputs to the multiplication, for example $x_k(j_1(i))$, as an adaptive context, or attention for the second input $x_k(j_2(i))$. This can be regarded as an adaptive generalization of MLP, where the weights depend on the input.

4.2 Deep Learning Approximation Spaces

One of the main goals of approximation theory is the characterization of the approximation spaces of a given approximation algorithm using function spaces [15]. To this day, the characterization of deep learning approximation is still out of reach [19]. It is certainly possible to prove various Jackson-type theorems that bound the degree of approximation using smoothness norms such as Sobolev,Besov, mixed-smoothness, etc. Under certain conditions it is even possible to remove the well known 'curse of dimensionality' from these estimates. Let us review two such results. We recall the Sobolev space $W_p^r(\Omega)$, $1 \leqslant p \leqslant \infty$, $\Omega \subseteq \mathbb{R}^n$, consisting of functions with distributional derivatives up to order r satisfying

$$\|f\|_\Omega := \sum_{|\alpha| \leqslant r} \|\partial^\alpha f\|_{L_p(\Omega)} < \infty.$$

The semi-norm is given by

$$|f|_{W_p^r(\Omega)} = |f|_{r,p} := \sum_{|\alpha|=r} \|\partial^\alpha f\|_{L_p(\Omega)}.$$

In this survey we are focused on cases where the dimension n is large, since even the simplest classification network trained on a computer vision dataset consisting of very small images of size 32×32, produces inner layer dimensions of up to $n = 16,000 - 64,000$ neurons. However, when trying to approximate functions from the unit ball of $W_p^r([0,1]^n)$, using neural networks, one encounters the 'curse of dimensionality'. That is, in the worst case, approximation with error $\varepsilon > 0$, requires a network of size $\sim \varepsilon^{-n/r}$ (e.g. there are lower bounds [34, 33]). In contrast, the following result demonstrates, that with additional mild conditions, the size of the network may depend linearly or quadratically on the dimension n (see definitions in Subsection 4.1). These results are useful in cases where $n >> r$, that is, when the dimension is relatively higher than the given smoothness.

Theorem 4.1 *[4] Let $n \geqslant 2$, $r \in \mathbb{N}$, $f \in W_2^r(\mathbb{R}^n)$ such that $\|\hat{f}\|_{L_1(\mathbb{R}^n)} < \infty$, where $\hat{f}$ is the Fourier transform of f and $0 < \varepsilon \leqslant 1$. Then there exist:*

(i) A deep MLP network $\tilde{f}_{MLP}$ with ReLU activations, of depth $O(n^2\varepsilon^{-2/r})$ and $O(n^2\varepsilon^{-(2+2/r)})$ neurons, such that

$$\|f - \tilde{f}_{MLP}\|_{L_2([0,1]^n)} \leqslant c \max(|f|_{r,2}^2, \|\hat{f}\|_1)\varepsilon.$$

(ii) A deep multiplicative network $\tilde{f}_{MUL}$ *of depth* $O(n\varepsilon^{-1/r})$ *and* $O(n\varepsilon^{-(2+1/r)})$ *neurons, such that*

$$\|f - \tilde{f}_{MUL}\|_{L_2([0,1]^n)} \leqslant c \max(|f|_{r,2}^2, \|\hat{f}\|_1)\varepsilon.$$

Returning to our discussion of the characterization of the approximation spaces,the main difficulty lies with the inverse embedding. It is not trivial to conclude the smoothness of a given function from the behavior of the degree of DL approximation. Let us provide the well-known counter-example of the 'sawtooth' functions.

Let $\Delta_1 : \mathbb{R} \to [0, 1]$ be defined by

$$\Delta_1(x) := \begin{cases} 2x, & 0 \leqslant x < 1/2, \\ -2x + 2, & 1/2 \leqslant x \leqslant 1, \\ 0, & \text{else}. \end{cases} \tag{27}$$

We then define $\Delta_j := \Delta_{j-1} \circ \Delta_1$, for $j \geqslant 2$. The sawtooth function Δ_j has 2^{j-1} 'teeth', as depicted in Figure 4. The function Δ_1 has the following representation

$$\begin{aligned} \Delta_1(x) &:= 2(x)_+ - 2(2x-1)_+ + 2(x-1)_+ \\ &= \sigma(2x) - 2\sigma(2x-1) + \sigma(2x-2), \end{aligned}$$

where σ is the ReLU nonlinear activation. Therefore, it can be realized using a network block composed of two MLP layers

$$x \to (2x, 2x, 2x) + (0, -1, -2) \to \sigma \to (x_1, x_2, x_3) \to x_1 - 2x_2 + x_3.$$

This implies that Δ_j, $j \geqslant 1$, can be realized as a composition of j blocks, which is a network of depth $2j$.

It can be shown [19] that for any $\alpha < 1/\tau$

$$|\Delta_j|_{\mathcal{B}_\tau^\alpha} \sim 2^{j\alpha}, \tag{28}$$

This is a pro-typical example of a function that, one the one hand, has a compositional structure, that allows it to be represented by a relatively shallow network of depth $\sim j$ and yet its Besov smoothness semi-norm is exponentially larger with j. That is, when initially inspecting the smoothness of the input function, it is not obvious that it can be approximated or, as in this special case, even realized, by a relatively shallow network.

In going further, let us denote by $E_N(f)_p$ the degree of approximation of a neural network with width of size $\sim n$ and depth of size N of a function $f \in L_p([0,1]^n)$. Let us recall the approximation space $\mathcal{A}_\infty^\gamma(L_p)$, defined by functions in L_p for which

$$|f|_{\mathcal{A}_\infty^\gamma(L_p)} := \sup_N N^\gamma E_N(f)_p < \infty.$$

Since Δ_j can be realized by a network of width 3 and $2j$ layers, we have that

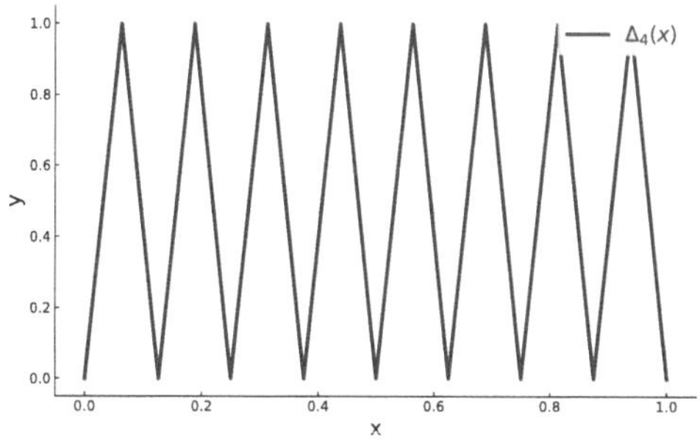

Fig. 4: Sawtooth function Δ_4

$$E_N(\Delta_j)_p \sim \begin{cases} 1, & N < j/2, \\ 0, & N \geqslant j/2. \end{cases}$$

Therefore

$$|\Delta_j|_{\mathcal{A}^\gamma_\infty(L_p)} \sim j^\gamma.$$

Combining this observation with (28), we see that we do not have a continuous embedding of $\mathcal{B}^\alpha_\tau$ in $\mathcal{A}^\gamma_\infty(L_p)$ for any values of $\alpha, \gamma > 0$.

Yet, with the realization of $f_0 := \Delta_j$ as a neural network, we have that $|f_0|_{\mathcal{B}^\alpha_\tau} \sim 2^{j\alpha}$ and then as we proceed trough the layers, the Besov semi-norm of the inner layer representations decreases exponentially, since $f_2 = \Delta_{j-1}$, $f_4 = \Delta_{j-2}$, etc. That is, $|f_2|_{\mathcal{B}^\alpha_\tau} \sim 2^{(j-1)\alpha}$, $|f_4|_{\mathcal{B}^\alpha_\tau} \sim 2^{(j-2)\alpha}$, etc.

Observe that one can easily construct more examples of such compositional 'sawtooth-type' functions. For example, let $1 < A < B < \infty$ and for each $1 \leqslant i \leqslant j$, pick a random parameter $a_i \in [A, B]$. Next define the piecewise linear 'tooth' with support in $[0, 1]$ realized by

$$\Delta_{a_i}(x) := a_i(x)_+ - \frac{a_i}{a_i - 1}(a_i x - 1)_+ + \frac{a_i}{a_i - 1}(x - 1)_+, \quad 1 \leqslant i \leqslant j.$$

One has $\Delta_{a_i}(a_i^{-1}) = 1$. Then, the 'sawtooth-type' function

$$\tilde{\Delta} := \Delta_{a_1} \circ \cdots \circ \Delta_{a_j},$$

also has 2^{j-1} teeth and also satisfies $|\tilde{\Delta}|_{\mathcal{B}^\alpha_\tau} \sim 2^{j\alpha}$. The sawtooth with uniform knots Δ_j corresponds to the choices $a_i = 2$, $1 \leqslant i \leqslant j$. Furthermore, $\tilde{\Delta}$ can also be realized using a network of depth $\sim j$, with the Besov semi-norm of the inner layer representation spaces also decaying exponentially. All this, with constants that further depend on A, B.

In the general case, every piecewise linear function on $[0, 1]$ over N (possibly non-uniform) knots can be realized by a network of fixed width and $\sim N$ layers [12, 34].

Furthermore, one can provide a relatively simple and stable NN construction where the Besov smoothness of the inner layer representations is guaranteed to decrease at a certain given rate.

Theorem 4.2 *[3] Let $f : [0,1] \to \mathbb{R}$ be a continuous piecewise linear function with N knots and let $\alpha < 1/\tau$. Then, there exists a neural network realization of f of width 5 and depth $N+3$, where the Besov semi-norms of the inner layer representations $f_k : \Omega \to \mathbb{R}$, $\Omega := [0,1]^4 \times [-\|f\|_\infty, \|f\|_\infty]$, $2 \leqslant k \leqslant N+2$, are bounded by*

$$|f_k|_{\mathcal{B}^\alpha_\tau(\Omega)} \leqslant c(\alpha, \tau, \|f\|_\infty)(N-k+2)^{1/\tau}. \tag{29}$$

5 Smoothness analysis of the representation layers in deep learning networks

We now explain how the theory presented in Section 3 is used to estimate the 'weak-type' smoothness of an inner representation layer. Recall from the introduction that we create a representation of images at layer 0 by concatenating the $\sqrt{n_0}$ rows of pixel values of each grayscale image, to create a vector of dimension n_0 (or $3 \times n_0$ for a color image). We also normalize the pixel values to the range $[0,1]$. We then transform the class labels into vector-values in the space $\mathbb{R}^L$ by assigning each label l, the standard basis vector $e_l \in \mathbb{R}^L$ (see Section 2). Thus, the images are considered as samples of a function $f_0 : [0,1]^{n_0} \to \mathbb{R}^L$. In the same manner, we associate with each k-th layer of a DL network, a function $f_k : [0,1]^{n_k} \to \mathbb{R}^L$, where n_k is the number of features/neurons at the k-th layer. The samples of f_k are obtained by applying the network on the original images up the given k-th layer. For example, in a convolution layer, we capture the representations after the cycle of convolution, non-linearity and pooling. We then extract vectors created by normalizing and concatenating the feature map values corresponding to the images. Recall that although the functions $\{f_k\}$ are embedded in different dimensions $\{n_k\}$, through the simple normalizing of the features, our 'dimension free' analysis is able to assign smoothness indices to each layer that are comparable.

Next we describe how we estimate the smoothness of each function f_k using the method of [3]. We compute a random forest $\mathcal{F}_k$ over the samples of f_k with the choice $r = 1$ and then apply the wavelet decomposition of the RF (see Section 2). We work under the assumption that at each layer, we have a sufficient number of samples relative to k-th feature space dimension n_k and that we construct a forest $\mathcal{F}_k$ which is diverse enough to capture the 'geometry' of f_k.

With $p = 2$, for each k, the goal is to numerically estimate

$$\tau_k^* := \inf_{0<\tau<2} \{\tau | \quad N_\tau(f_k, \mathcal{F}_k) < \infty\}.$$

Based on the 'dimension free' equivalence (23) for $r = 1$, this gives an estimate for critical (optimal) smoothness of each f_k

$$\alpha_k^* = \frac{1}{\tau_k^*} - \frac{1}{2}. \tag{30}$$

It is obvious that for any $f \in L_2([0,1]^n)$ and random forest $\mathcal{F}$ built adaptively to fit it, the univariate function $N_\tau(f,\mathcal{F})$, as function of τ is monotonically non-increasing. Yet, in discrete setting, there is no value $0 < \tau < 2$, however small, for which $N_\tau(f,\mathcal{F}) = \infty$. So the estimate of a 'critical' τ^* is a numeric estimate for where the derivative (as a function of τ)

$$\frac{d}{d\tau}N_\tau(f,\mathcal{F}) = \frac{1}{J}\frac{d}{d\tau}\left(\sum_{j=1}^{J}\sum_{\Omega\neq\Omega_0,\Omega\in\mathcal{T}_j}\|\psi_\Omega\|_2^\tau\right)^{1/\tau},$$

crosses some lower threshold.

First, we compute a series of samples $N_{\tau_i}(f,\mathcal{F})$, for a set of discrete evenly spaced samples $\{\tau_i\}_{i=1}^M$, $0 < \tau_i < 2$. We then approximate, using the samples, the numerical derivatives

$$N_i' \approx \frac{d}{d\tau}N_{\tau_i}(f,\mathcal{F}), \quad 1 \leqslant i \leqslant M.$$

We use the angles of the numerical derivatives

$$\theta_i := \arctan(N_i'),$$

to estimate the transition index τ^* which is associated with an infinite derivative, or equivalently, an angle of $-\pi/2$. To this end, we use two meta parameters: $\varepsilon_{\text{low}}, \varepsilon_{\text{high}}$, and define

$$S := \{\tau_i : -\frac{\pi}{2} + \varepsilon_{\text{low}} \leqslant \theta_i \leqslant -\frac{\pi}{2} + \varepsilon_{\text{high}}\}.$$

We then approximate the transition index by

$$\tau^* \approx -\varepsilon_{\text{low}} + \frac{1}{|S|}\sum_{\tau_i\in S}\tau_i.$$

A demonstration is shown in Figure 5.

6 Experimental Results: Smoothness analysis across DL representation layers

Our goal is to show how the theory laid out in this paper enables to look into the 'black box' of a neural network and analyze its performance. For all the experiments presented in this section, 3 different neural networks were trained using the same training set, each time with different initialization seeds. Then, for each of the 3 networks, we applied the numerical computation of the representation layers'

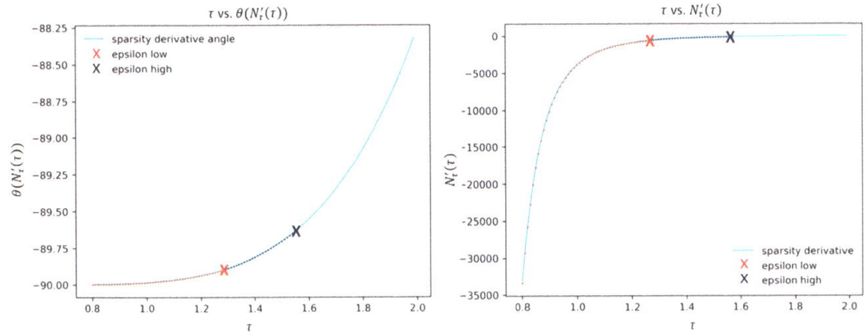

Fig. 5: Example of plots of N'_τ and $\theta(N'_\tau)$

smoothness (30) using wavelet decompositions of RFs of three trees each, with maximal depth of 15 layers. The hyper parameters of Section 5 were selected as $\varepsilon_{\text{low}} = 0.1$, $\varepsilon_{\text{high}} = 0.4$. The smoothness estimates for each layer were averaged over the 3 trained models. It is important to note that in the results below, the function space smoothness estimates are performed using only the training dataset. Then, testing of the accuracy of the models is performed using out of bag testing dataset. Nevertheless, as we shall see below, there is a strong correlation between the rate of improvement of smoothness across the model's intermediate layers and its performance in the testing phase.

6.1 Analysis of a classic well-performing convolutional neural network

We start off with a simple example of a relatively simple dataset and an appropriate and well functioning neural network architecture. In this example we use MNIST-1D [27], a well known univariate equivalent of the classic MNIST image dataset. In MNIST-1D, as opposed to MNIST, the signals are intertwined in the input space, and are then much less separable. We apply to it a classic convolutional network with 7 layers (see Subsection 4.1). The question is, how does the representation smoothness improve across layers and throughout the training phase? To this end we monitor for each k-th layer, the smoothness α_k^* (see (30)) and record it every 10 epochs of the training phase. The results are shown in figure 6. While obviously the smoothness of the input dataset at layer 0 is constant throughout the training, it is clear that the smoothness improves as we progress forward through the intermediate layers and through the training process. Also notice that while the smoothness of the intermediate layers is estimated using only the training data, it highly correlates with the results on the testing data depicted on the right hand side.

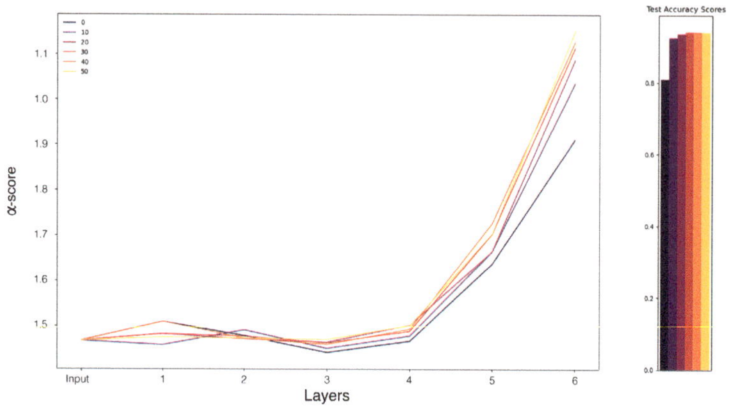

Fig. 6: Smoothness improvement through intermediate layers and the training phase

6.2 Analysis of a 'problematic' neural network architecture

In most cases, where there is a problem with the chosen network architecture, we can only hopelessly observe the bad testing result. This is especially frustrating in cases where the problematic architecture is still able to overfit the training data and produce a low training loss. To demonstrate this, we use the well known Fashion-MNIST image dataset and create two types of classification models. The first model is based on a 'sensible' architecture of convolutional residual networks [24]. For the second model, we deliberately construct an architecture that does not make sense from a computer-theoretical perspective. Specifically, we construct a network where the types of layers alternate between convolutional and MLP. Thus, every convolutional layer is followed by reshaping of the 3D feature space (sequence of 2D feature maps) into a vector and then an MLP layer. In Figure 7 we see the analysis of the two models. We see that the 'bad' architecture does not provide representation improvement across the inner layers which results in poor testing performance.

6.3 Analysis of the performance of nonlinear activation functions

In Subsection 4.1 we reviewed the role of the nonlinear activation functions and provided some examples. Here we use again the MNIST-1D dataset and fix a convolutional neural network as the classification model, where the only hyper-parameter we change for each run is the activation function. Suppose we are trying to create a new activation function, by using a simple step function:

$$\sigma_H(x) = \begin{cases} 1, & \text{if } x > 0. \\ 0, & \text{otherwise.} \end{cases}$$

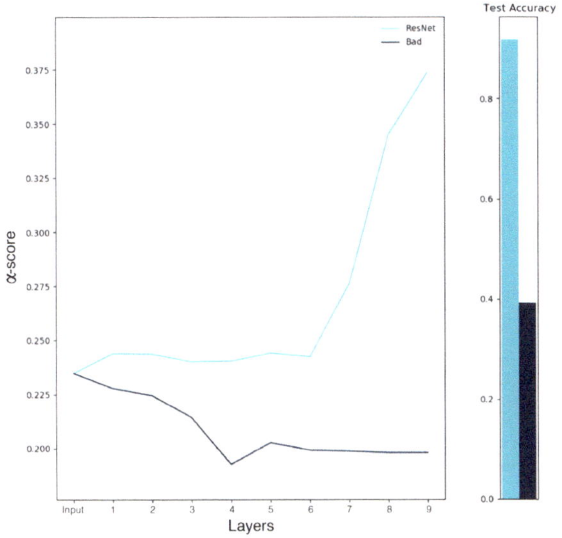

Fig. 7: Analysis of Fashion-MNIST classification models: a 'good' ResNet architecture and a 'bad' alternating architecture.

At face value, this seems a potentially problematic activation function since it does not output any magnitude values, only simple activation of '0' or '1'. Other non-linearities we analyze here are the ReLU (26) its variants Leaky ReLU and GELU and the Sigmoid. Figure 8 shows the dominance of the ReLU activation over other activations. The ReLU variants - Leaky ReLU and GELU are relatively close in performance. The Sigmoid is indeed far lower in terms of the behavior of α^*. Again, we see a clear correlation of the architectural smoothness space analysis using the training data and the testing performance of the different activations.

References

1. E. Alpaydin, *Introduction to machine learning*, MIT Press, 2004.
2. Y. Bengio, A. Courville and P. Vincenty, *Representation Learning: A Review and New Perspectives*, IEEE Transactions on Pattern Analysis and Machine Intelligence 8 (2013), 1798–1828.
3. I. Ben-Shaul, S. Dekel and O. Elisha, *Spars Besov space analysis of deep learning representation layers in high dimensions* Pure and Applied Functional Analysis, to appear.
4. I. Ben-Shaul, T. Galanti and S. Dekel, *Exploring the approximation capabilities of multiplicative neural networks for smooth functions*, Transactions of Machine Learning 2023.
5. G. Biau and E. Scornet, *A random forest guided tour*, TEST 25 (2016), 197–227.
6. L Breiman, *Random forests*, Machine Learning 45 (2001), 5–32.
7. L. Breiman, *Bagging predictors*, Machine Learning 24 (1996), 123–140.
8. L. Breiman, J. Friedman, C. Stone and R. Olshen, *Classification and Regression Trees*, Chapman and Hall/CRC, 1984.

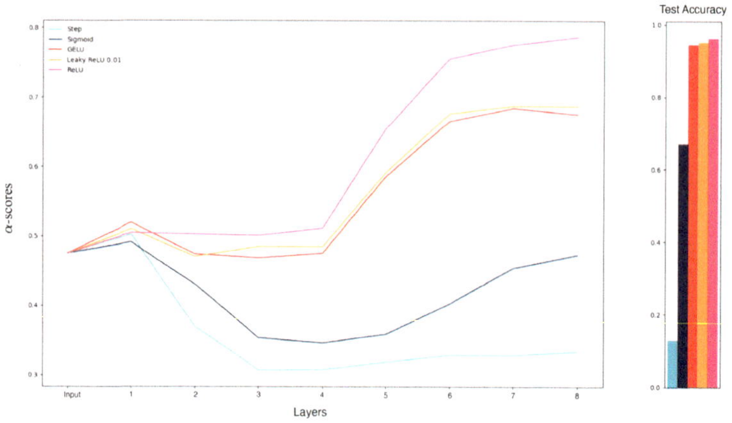

Fig. 8: Analysis of MNIST-1D classification models: Different nonlinear activation functions.

9. A. Criminisi, J. Shotton and E. Konukoglu, *Forests for Classification, Regression, Density Estimation, Manifold Learning and Semi-Supervised Learning*, Microsoft Research technical report, report TR-2011-114, 2011.
10. W. Dahmen, S. Dekel and P. Petrushev, *Two-level-split decomposition of anisotropic Besov spaces*, Constructive approximation 31 (2001), 149-194.
11. I. Daubechies, *Ten lectures on wavelets*, CBMS-NSF Regional Conference Series in Applied Mathematics,1992.
12. I. Dubechies, R. DeVore, S. Foucart, B. Hanin and G. Petrova, *Nonlinear approximation and (deep) ReLU networks*, Constructive approximation 55 (2022), 127-172.
13. S. Dekel and D. Leviatan, *Adaptive multivariate approximation using binary space partitions and geometric wavelets*, SIAM Journal on Numerical Analysis 43 (2005), 707–732.
14. R. DeVore, *Nonlinear approximation*, Acta Numerica 7 (1998), 51–150.
15. R. DeVore and G. Lorentz, *Constructive approximation*, Springer Science and Business, 1993.
16. R. DeVore, B. Jawerth and B. Lucier, *Image compression through wavelet transform coding*, IEEE transactions on information theory 38 (1992), 719-746.
17. D.Donoho, M. Vetterli, R DeVore and I. Daubechies *Data compression and harmonic analysis*, IEEE Transactions on information theory 44 (1998), 2435-2476.
18. O. Elisha and S. Dekel, *Wavelet decompositions of Random Forests - smoothness analysis,sparse approximation and applications*, Journal of machine learning research 17 (2016), 1-38.
19. R. Gribonval, G. Kutyniok, M. Nielsen and F. Voigtlaender, *Approximation spaces of deep neural networks*, Constructive Approximation 55 (2022), 259-367.
20. P. Grohs and G. Kutyniok (ed.), *Mathematical aspects of deep learning*, Cambridge University press, 2022.
21. T. Hastie, R. Tibshirani and J. Friedman, *The elements of statistical learning*, Springer, 2009.
22. B. Karaivanov and P. Petrushev, *Nonlinear piecewise polynomial approximation beyond Besov spaces*, Applied and computational harmonic analysis 15 (2003), 177-223.
23. V. Kulkarni and P. Sinha, *Pruning of Random Forest classifiers: A survey and future directions*, In International Conference on data science and engineering, 64-68, 2012.
24. K. He, X. Zhang, S. Ren and Jian Sun, *Deep Residual Learning for Image Recognition*, IEEE Conference on Computer Vision and Pattern Recognition 2016, 770-778.

25. W. Loh, *Classification and regression trees*, Wiley Interdisciplinary Reviews: Data Mining and Knowledge Discovery 1 (2011), 14-23.
26. S. Mallat, *A Wavelet tour of signal processing, 3rd edition (the sparse way)*, Acadmic Press, 2009.
27. Sam Greydanus, *Scaling down Deep Learning*, arXiv:2011.14439, 2020.
28. M. Phuong and M. Hutter, *Formal algorithms for transformers*, DeepMind report, arXiv:2207.09238 (2022).
29. T. Poggio, H. Mhaskar, L. Rosasco, B. Miranda and Q. Liao, *Why and when can deep-but not shallow-networks avoid the curves of dimensionality: a review*, International Journal of Automation and computing 14 (2017), 503-519.
30. P. Salembier and L. Garrido, Binary partition tree as an efficient representation for image processing, segmentation, and information retrieval, *IEEE transactions on image processing*9:561-576, 2000.
31. R. Schwartz-Ziv and A. Armon, *Tabular data:deep learning is not all you need*, Information Fusion 81 (2022), 84-90.
32. R. Schwartz-Ziv and N. Tishbi, *Opening the black box of Deep Neural Networks via Information*, http://arxiv.org/abs/1703.00810.
33. J. Siegel, *Optimal Approximation Rates for Deep ReLU Neural Networks on Sobolev and Besov Spaces*, JMLR 24 (2023), 1-52.
34. D. Yarotsky, *Error bounds for approximations with deep ReLU networks*, Neural Networks 94 (2017), 103-114.

Joint Denoising and Line Distortion Correction for Raster-Scanned Image Series

Benjamin Berkels and Peter Binev

Abstract The problem of noise in a general data acquisition procedure can be resolved more accurately if it is based on a model that describes well the distortions of the data including both spatial and intensity changes. The focus of this article is the modeling of the position distortions during sequential data acquisitions. A guiding example is the data obtained by Scanning Transmission Electron Microscopy (STEM) and High Angular Annular Dark Field (HAADF) data, in particular. The article discusses different models of the position noise and their numerical implementations comparing some computational results.

1 General Data Distortion Model

The process of recovering information from data usually involves several steps. A standard presumption is to assume that the data set d is a result of applying an operator Φ to an ideal set $\mathcal{P}$ describing the phenomenon. The goal is to recover $\mathcal{P}$ from d by inverting Φ. Often, the input and output spaces of Φ are different, e.g., in the case Φ is a "parameter-to-state" map in which the goal is to recover from d the set of parameters $\mathcal{P}$ that determine it. The mathematical problem of inverting the "parameter-to-state" map is well understood in case the data is clean from any distortions typically accompanying the data acquisition process. This article considers the case of having an operator Φ with the same input and output spaces Ω in which the goal is to recover a "clean" version c of the data set d. In this case, Φ is a *distortion* operator acting "state-to-state". We consider this as a standalone problem, although anticipate the scenario of it being the first step of the two-step

Benjamin Berkels
Aachen Institute for Advanced Study in Computational Engineering Science (AICES), RWTH Aachen University, Germany, e-mail: berkels@aices.rwth-aachen.de

Peter Binev
University of South Carolina, Columbia, SC 29208, USA, e-mail: binev@math.sc.edu

R. DeVore, A. Kunoth (eds.), *Multiscale, Nonlinear and Adaptive Approximation II*,
https://doi.org/10.1007/978-3-031-75802-7_6

inversion problem recovering c from d which is followed by the "state-to-parameter" map recovering $\mathcal{P}$ from c.

The choice of the data space Ω depends on the nature of the distortion process described by the operator Φ. One typical example is the case in which Ω is the space of images. The continuous version of it is the space Ω of all bounded functions $f : [0,1]^2 \to \mathbb{R}_+^n$, where $\mathbb{R}_+$ is the set of non-negative real numbers. The most studied situation is $n = 1$ and we will concentrate on it. A straightforward generalization of the methods often works for $n > 1$ but the problems become more challenging in some settings for $n > 1$, in particular for multimodal data that might even require replacing the range $\mathbb{R}_+^n$ of f with a more intricate set. Other choices of Ω that can be treated similarly include bounded functions $f : [0,1]^3 \to \mathbb{R}_+^n$ representing data sets in 3D space and $f : [0,1]^4 \to \mathbb{R}_+^n$ for 4D space/time. The only significant problem in such generalizations is the implementation of the algorithms that have to manage the increased amounts of data, memory, and computations.

An abstract problem formulation involves a data function $g := \Phi \circ f$ that is assumed to be received after the application of the operator Φ that is unknown or at least has a stochastic component. In reality, we do not have g but some data points that can be treated as the values $\hat{g}$ of some functionals applied to g. Note that $\hat{g}$ belongs to a finite dimensional space while Ω is infinite dimensional. On the other hand, f is also approximated by some $\hat{f}$ that belongs to a chosen class of approximating functions that is finite dimensional. In addition, the operator Φ is also unknown and has to be approximated by an operator $\hat{\Phi}$ based on a model depending on finite number of parameters. The result of these approximations is a discretized version $\hat{g}$ of $\hat{\Phi} \circ \hat{f}$ that we have to compare to the actual data d.

The problem of interest is typical for sequential data acquisition in which the data is scanned at different positions p_j on a given lattice, assuming that $j \in J$ for a finite index set J. The process is time-dependent meaning that both f and p_j change with time. It is often the case that f does not change significantly during scanning and its changes can be attributed to a general component. Then the assumption is that f does not change during the data acquisition but instead of sampling f at p_j the data is received by sampling f at distorted positions $\hat{p}_j$.

We measure the discrepancy using a distance functional $\mathrm{Dist}_{\mathrm{dat}}(d, \hat{g})$ that is often a square of the ℓ_2-norm of the difference of the numerical values. The process of finding $\hat{f}$ involves a regularized minimization of this distance. A more complicated distance functional is used in Subsection 2.2.5. In the following, we derive it from an assumption of Poisson noise on the input data. Recalling the Poisson distribution $P_\lambda(k) = \frac{\lambda^k}{k!}\exp(-\lambda)$, given a true intensity $u \geqslant 0$, the probability to observe the intensity $v \in \mathbb{N}^0$ is given by

$$P(v|u) = P_u(v) = \frac{u^v}{v!}\exp(-u)$$

Thus, given an observation v, the solution u should maximize the probability $P(v|u)$. Therefore, we assign $-\log P(v|u)$ as cost for u given v, i.e.

$$\begin{aligned}\mathrm{Dist}_{\mathrm{dat}}(u, v) &= -\log(P(v|u)) = -\log \frac{u^v}{v!} + u \\ &= u - v \log u + \underbrace{\log(v!)}_{\text{independent of } u}\end{aligned}$$

Thus, we get $\mathrm{Dist}_{\mathrm{dat}}(u, v) = u - v \log u$. Note that this approach can be applied to other noise models as well, as long as the distribution is known.

Both, the operator $\hat{\Phi}$ and the approximating function $\hat{f}$, can be forced to belong to appropriate smoothness classes. The preferred way to express this enforcement is through two other functionals. The first one is motivated by the assumption that the distortion $\hat{\Phi}$ should be close to the identity mapping and we have a way to measure this closeness. Thus, it is (a power of) a distance $\mathrm{Dist}_{\mathrm{op}}(\hat{\Phi}, I)$ between $\hat{\Phi}$ and the identity I in a chosen metric. The second one is usually (a power of) a norm $\|\hat{f}\|_Y$ in a chosen Banach space. Often $\|\hat{f}\|_Y$ is replaced by a semi-norm or a quasi-norm that better expresses the nature of f. It should be noted that $\mathrm{Dist}_{\mathrm{op}}(\hat{\Phi}, I)$ and $\|\hat{f}\|_Y$ are acting on discrete objects inspired by the corresponding continuous models.

To simplify the notation, we recognize that g, f, and Φ are not available to us and will use the same notation for their approximations identifying $\hat{g}$ with g, $\hat{f}$ with f, and $\hat{\Phi}$ with Φ from now on. However, we will keep separate notations for p_j and $\hat{p}_j$.

We find the approximations of f and Φ using regularized minimization process with a *loss function*

$$\mathcal{L} = \mathcal{L}(\mathcal{P}_f, \mathcal{P}_\Phi) = \mathrm{Dist}_{\mathrm{dat}}(d, \Phi \circ f) + \lambda \, \mathrm{Dist}_{\mathrm{op}}(\Phi, I) + \mu \|f\|_Y \, , \tag{1}$$

where $\mathcal{P}_f$ and $\mathcal{P}_\Phi$ are the parameters determining the approximations of f and Φ, while $\lambda \geqslant 0$ and $\mu \geqslant 0$ are tuning parameters weighting the regularization terms for f and Φ. Sometimes, one or even both of the constants λ and μ are set to zero. This effectively leaves only implicit regularization imposed via the approximation procedure and restrictions on the sets $\mathcal{P}_f$ and/or $\mathcal{P}_\Phi$.

The simplest interpretation of the above model assumes acquisition of a single frame of data at a rectangular set P of positions p_j. Then the minimization of the loss function (1) provides good results only in cases of small distortions and regular data patterns. However, the problems of interest rarely possess either of these features. To process more involved problems one needs to obtain additional information about the distortions. One way of gathering this information is to acquire several frames from the same function f assuming that f itself does not change significantly from frame to frame. This will allow a comparison of the distortions for each individual frame and provide a basis for a better estimation of the approximations of these distortions. In the next section, we consider different models that take advantage of the multi-frame setup.

2 Distortion Estimation in the Multi-Frame Setup

Let K be the number of frames gathered in the experiment. Each frame is a discrete approximation of the ideal image f. While the intention is to sample f at the same set P of positions p_j for $j \in J$, in reality f is sampled at different positions $\hat{p}_j^{[k]}$ for $j \in J$ at different frames $k = 1, 2, ..., K$. Each frame is determined by the set of distorted positions $P^{[k]} := \{\hat{p}_j^{[k]}, j \in J\}$ corresponding to the discretization $g_k := f(P^{[k]})$.

The general model must be slightly modified to account for the multiple frames. The operator Φ is now acting from Ω to Ω^K and the regularization through $\mathrm{Dist}_{\mathrm{op}}$ should compare Φ with I^K instead of I. The data $d = \{d_k\}_{k=1}^K$ is also treated as K-dimensional entity. In addition, instead of assuming that Φ is such that $\Phi \circ f = g$ as in the general model, we will focus on its position distortion part and will assume that $\Phi \circ f = (g_k)_{k=1}^K$ anticipating that the estimation of the other distortions will be handled by the regularization term $\mathrm{Dist}_{\mathrm{dat}}(d, \Phi \circ f)$ by penalizing the distances between d_k and g_k. We associate the operator Φ with a vector of operators $(\phi_k)_{k=1}^K$ that are transforming $P^{[k]}$ back to P yielding $g_k \circ \phi_k = f$. In some sense $(\phi_k)_{k=1}^K$ represents the inverse operator of Φ. To state the updated form of (1), we need to redefine the two distance functionals for Φ as sums of individual distance functionals $\mathrm{Dist}_{\mathrm{op}}^{\diamond}$ and $\mathrm{Dist}_{\mathrm{dat}}^{\diamond}$ for each frame and ϕ_k reminding the reader that often for the formulation of a loss function the distance functionals are powers of an actual distance measure. We set

$$\mathrm{Dist}_{\mathrm{op}}(\Phi, I^K) := \sum_{k=1}^{K} \mathrm{Dist}_{\mathrm{op}}^{\diamond}(\phi_k, I)$$

and

$$\mathrm{Dist}_{\mathrm{dat}}(d, \Phi \circ f) := \sum_{k=1}^{K} \mathrm{Dist}_{\mathrm{dat}}^{\diamond}(d_k \circ \phi_k, f).$$

The general form of a loss function for the multi-frame setup is

$$\mathcal{L} = \mathcal{L}(\mathcal{P}_f, \mathcal{P}_\Phi) = \sum_{k=1}^{K} \Big(\mathrm{Dist}_{\mathrm{dat}}^{\diamond}(d_k \circ \phi_k, f) + \lambda\, \mathrm{Dist}_{\mathrm{op}}^{\diamond}(\phi_k, I)\Big) + \mu \|f\|_Y. \tag{2}$$

Depending on the settings, the minimization of the loss function could be a very complicated process. Therefore, simplifying the dependencies in the formulation could be beneficial. In general, $\mathrm{Dist}_{\mathrm{op}}(\Phi_k, I)$ in (1) could depend on f since $\Phi : \Omega \to \Omega$ is defined for Ω being a set of functions. However, in (2) we define ϕ_k as a transformation between sets of positions avoiding a potential dependence on f.

There are several possible applications of the models described by the loss function (2). One of the most interesting is scanning transmission electron microscopy (STEM). In the next subsection, we consider the specifics in the STEM applications.

2.1 Atomic Scale Data Processing in Scanning Transmission Electron Microscopy (STEM)

STEM images are usually two-dimensional rectangular arrays of counts of the numbers of electrons that have hit the detector(s) during the short period of time when the electron beam has been positioned at the corresponding spot. The interactions of the electrons from the beam with the atoms of the specimen result in changing the trajectories of some electrons. The standard detectors are *bright field* (BF) detecting the electrons whose trajectories are not changed substantially, *annular bright field* (ABF) detecting the electrons diverted by a small angle, *annular dark field* (ADF) detecting the electrons diverted by a moderate angle, and *high angle annular dark field* (HAADF) detecting the electrons diverted by a high angle. HAADF is the most used STEM acquisition mode due to the so-called Z-contrast allowing to better distinguish atoms with different number of protons. The examples in this article are for HAADF data at atomic scale resolution. As discussed in the introduction, one can also consider a vector data output ($n > 1$), e.g., by combining the data of BF, ABF, ADF, and/or HAADF or even consider a two-dimensional array of detectors resulting in so-called 4D-STEM data (see, e.g. [7]). While the vector data would result in better estimation of the distortions, the associated algorithms are more complicated and their computational complexity is higher.

The electron beam is interacting with the specimen within a small volume around the central axis of the beam. The density of the interaction is higher closer to the axis. It is usually modeled by a 2D Gaussian distribution on the planes perpendicular to the axis although in actuality it is much more complicated. We identify the axis with its intersection point p with the xy-plane of the specimen. The data collected for the corresponding volume of interaction is the (approximate) value of $f(p)$. The standard scanning routine moves the electron beam in x-direction for a given number of equal steps h_x at fixed y-position and then changes the y-position by a step h_y and starts scanning from the initial x-position. The scanning continues for a given number of steps in y-direction. The electron beam is positioned by incremental changes of an electromagnetic field guiding it. Different environmental factors can influence this electromagnetic field and distort the intended position of the beam - the higher the resolution, the bigger the relative distortions. In atomic scale resolution, these position distortions are the main obstacle of receiving higher precision. This is the main motivation for developing acquisition procedures and numerical algorithms for estimating the position distortions. The beam is shut out during the repositioning but might not be completely steady in maintaining the same position during the acquisition process of a single data point (or data vector). This would change somewhat the density of interaction within the volume corresponding to p creating data that is deviating from $f(p)$. Such a deviation is expected to be more pronounced for the first point with a new y-position due to the larger repositioning from the previously scanned point. As our goal is the estimation of the position distortions, this and other value distortions will be handled by the regularization terms of the loss function.

2.2 Models and algorithms for estimating the position distortion in HAADF-STEM data

We consider five particular models. The first two are based on continuous models that are discretized and the others are discrete by design. In Subsection 2.2.5 we introduce a new discrete model for joint denoising and distortion correction.

2.2.1 Iterative approximating algorithm based on the estimation of the position distortions of pairs of consecutive frames

The method introduced in [1, 8] does not consider a global minimization procedure for a given loss function like (2) but suggests an iterative process consisting of several steps. The input is the data for the frames d_k, $k = 1, \ldots, K$ and the output is the approximation f calculated together with position distortions ϕ_k, $k = 1, \ldots, K$. The key ingredient of the method is the minimization process that for given two functions u and v defined on a domain $\mathcal{D}$ finds a transformation ϕ for which the functions $u(p)$ and $v \circ \phi(p) := v(\phi(p))$ are as close as possible in terms of minimizing the loss function

$$\mathcal{L}^\star = \mathcal{L}^\star(\mathcal{P}_\phi) = \mathrm{Dist}_{\mathrm{dat}}(u, v \circ \phi) + \lambda R(\phi) \tag{3}$$

among all vector-valued piecewise bilinear functions $\phi : \mathcal{D} \to \mathcal{D}$ with a positive semi-definite Jacobian $\mathcal{J}(\phi)$. The similarity measure $\mathrm{Dist}^\diamond_{\mathrm{dat}}$ is based on the normalized cross correlation and is defined by

$$\mathrm{Dist}^\diamond_{\mathrm{dat}}(u, v) := -\frac{1}{|\mathcal{D}|} \int_{\mathcal{D}} \frac{u - \bar{u}}{\sigma_u} \, \frac{v - \bar{v}}{\sigma_v} \, \mathrm{d}p \,, \tag{4}$$

where as usual $\bar{u} := (1/|\mathcal{D}|) \int_{\mathcal{D}} u(p) \, \mathrm{d}p$ is the mean value of u over the domain $\mathcal{D}$ and $\sigma_u := \sqrt{(1/|\mathcal{D}|) \int_{\mathcal{D}} |u(p) - \bar{u}|^2 \, \mathrm{d}p}$ is the standard deviation of u. The regularization term $R(\phi)$ is the squared integrated Frobenius norm of the difference between the Jacobian $\mathcal{J}(\phi)$ of ϕ and the identity I:

$$R(\phi) := \int_{\mathcal{D}} \|\mathcal{J}(\phi) - I\|_F^2 \mathrm{d}p. \tag{5}$$

It is corresponding to the term $\mathrm{Dist}^\diamond_{\mathrm{op}}(\phi_k, I)$ in (2). The parameter λ is used as a normalization that makes the two terms in (3) of similar magnitude.

In the above description, it is assumed that both functions u and v are defined in the same domain $\mathcal{D}$. If this is not the case and we have that u is defined in $\mathcal{D}_u$ and v is defined in $\mathcal{D}_v$, then we can use $\mathcal{D} := \mathcal{D}_u \cap \mathcal{D}_v$. However, we can opt for using $\mathcal{D} := \mathcal{D}_u \cup \mathcal{D}_v$ by first calculating the mean values $\bar{u} := (1/|\mathcal{D}_u|) \int_{\mathcal{D}_u} u(p) \, \mathrm{d}p$ and $\bar{v} := (1/|\mathcal{D}_v|) \int_{\mathcal{D}_v} v(p) \, \mathrm{d}p$ and then extending u and v by their mean values at the points they are not defined. Note that the product under the integral in (4) is zero for $p \notin \mathcal{D}_u \cap \mathcal{D}_v$.

A key feature of the algorithm is using that the distortion between two consecutive frames is close to the identity and it can be used as initial approximation. The loss function (3) is defined for continuous functions and thus we need to define functional analogs g_k of the discrete data d_k. It can be done in different ways but the most convenient one is to define g_k to be the piecewise bilinear function on a rectangular grid interpolating the data d_k at the integer points. As a preprocessing step of the iterative procedure, we determine the transformations $\phi = \phi_{k+1,k}$ that make the functions $u = g_k$ and $v \circ \phi = g_{k+1} \circ \phi_{k+1,k}$ as close as possible by minimizing (3) with initial approximation $\phi = I$. The practical implementation of the minimization is rather involved. One of the issues that should be addressed is the near-periodicity of the data that creates several local minima. This necessitates involvement of multiresolution processing and estimation of the rigid registration component in the procedure. Details can be found in [1, 8].

The outline of the general algorithm is as follows:

Algorithm: Iterative approximation using general bilinear distortions

Input: g_k for $k = 1, \ldots, K$ and $\phi_{k+1,k}$ for $k = 1, \ldots, K-1$;

Initialization: set the initial guess $f^{[0]} = g_1$ and $\phi_1 = I$;

repeat

  estimate ϕ_1 by the approximate minimizer ϕ of (3) with $u = f^{[0]}$ and $v = g_1$ using as initial guess the last version of ϕ_1;

  consecutively for $k = 2, \ldots, K$ estimate ϕ_k by the approximate minimizer ϕ of (3) with $u = f^{[0]}$ and $v = g_k$ using as initial guess $\phi_{k,k-1} \circ \phi_{k-1}$;

  calculate f as the average (e.g., median or mean) of the functions $g_k \circ \phi_k$ for $k = 1, \ldots, K$;

  if *the error estimator* $\|f - f^{[0]}\|$ *is below a prescribed tolerance or other stopping criterion is satisfied (number of iterations, time of computation, etc.).* ;

  then

    break;

  else

    set $f^{[0]} := f$;

  end

There can be different variants of this algorithm depending on the type of average used and the way the domains $\mathcal{D}_k$ of each of the frames are treated. Usually, the median averaging produces better results. The standard approach is to consider the intersection of the domains $\mathcal{D} := \cap_{k=1}^{K} \mathcal{D}_k$ but it is often beneficial to work with weighted norms over the union $\cup_{k=1}^{K} \mathcal{D}_k$ assigning higher weights on the subregions belonging to more domains $\mathcal{D}_k$.

2.2.2 Iterative approximating algorithm with global regularization

The iterative algorithm proposed in Subsection 2.2.1 attempts to approximate well the solution of the global minimization problem based on (2). It performs the minimization of (3) frame by frame for a fixed function $f = f^{[0]}$ that is updated at each iteration by an average of $g_k \circ \phi_k$ for $k = 1, ..., K$. The averaging process can be adjusted to yield a function f with specific properties which are usually imposed by the term $\mu\|f\|_Y$ in (2). Using weighted averaging can balance well the terms $\text{Dist}^{\diamond}_{\text{dat}}(f, g_k \circ \phi_k)$ in the corresponding sum in (2) making it close to optimal. However, the sum $\sum_{k=1}^{K} \lambda \, \text{Dist}^{\diamond}_{\text{op}}(\phi_k, I) = \lambda \sum_{k=1}^{K} R(\phi_k)$ might stay far from optimal for significant number of iterations.

The algorithm proposed in [2] adds a step at the end of each iteration that improves substantially the optimality of $\sum_{k=1}^{K} R(\phi_k)$ leading to very good results after a few iterations. The idea is to introduce a local change of variables ψ replacing f with $f \circ \psi$ and ϕ with $\phi \circ \psi$ in the sums of the loss function. The values of terms $\text{Dist}^{\diamond}_{\text{dat}}(f \circ \psi, g_k \circ \phi_k \circ \psi)$ are only slightly perturbed since they are calculated by (4) with integration over $\psi(\mathcal{D})$ instead of $\mathcal{D}$ and $\psi(p) \approx p$. The ψ is calculated by minimizing $\sum_{k=1}^{K} \|\phi_k \circ \psi - \text{id}_{\mathcal{D}}\|^2_{L^2}$ for the estimated distortions ϕ_k at the current iteration, where $\text{id}_{\mathcal{D}}$ is the identity mapping on $\mathcal{D}$. Then f is updated as an average of the functions $g_k \circ \phi_k \circ \psi$ for $k = 1, ..., K$. As already mentioned, this modification of the algorithm from Subsection 2.2.1 converges very fast to the desired solution and noticeably reduces the bias of the reconstruction f towards the coordinate system of g_1, which is due to the alternating minimization scheme used in [1] combined with using g_1 as initial guess for f.

2.2.3 Smart align

In [5], the authors suggest a non-rigid registration methodology and a software package called Smart Align. The procedure uses two types of images, a static reference image and a set of "moving" images. The reference image is calculated initially as an average of the rigidly aligned frames. Once the moving images are updated, it is recalculated as an average of them to start the next iteration of the procedure. The original frames are the starting state of the moving images each of which is gradually modified by a gradient descent process utilizing the difference of the gradients of a moving image and the reference image to detect the position varying offsets between them. This results in obtaining an incremental calculation field that is improved through implementing constrains motivated by the specifics of the STEM images. The software presents solutions to some practical situations like processing frames with beam damage. Smart Align presents an alternative approach to the loss function minimization discussed in the previous subsections providing a good algorithmic solution to the distortion correction problems.

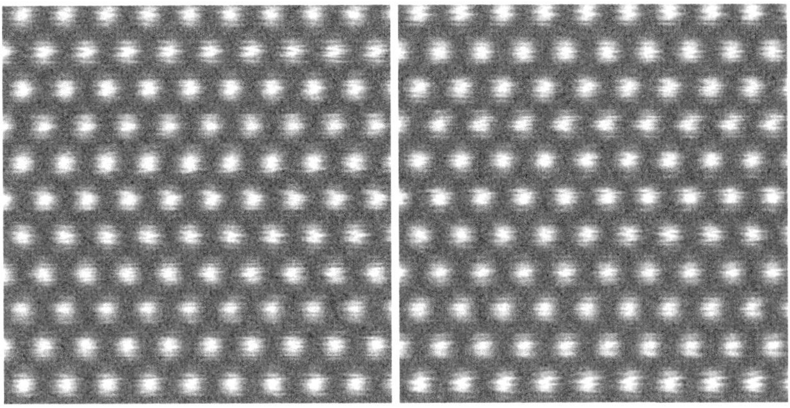

d_1 d_{64}

Fig. 1: First and last image of the synthetic STEM images series used for the numerical performance evaluation.

d_1 d_{64}

Fig. 2: First and last image of the subset of the HAADF STEM GaN $[11\bar{2}0]$ image series from [8] used for the numerical performance evaluation here.

2.2.4 Joint denoising and distortion correction with a bump fit image model

In [3], a Bayesian method is derived that, for a given STEM image series, jointly estimates the distortion in each image and reconstructs the underlying image of the material. For this discrete model, the data $d = \{d_k\}_{k=1}^K$ is assumed to be given as K noisy STEM pixel images of size $M \times N$ each, i.e., $d_k \in \mathbb{R}^{M \times N}$, $k = 1, \ldots, K$, cf. Figure 1. Here, the pixel values are considered to correspond to a Cartesian pixel grid discretizing the image domain $\mathcal{D} = [0, 1]^2 \subset \mathbb{R}^2$. The coordinates of the pixels are denoted by $x_{ij} \in \mathcal{D}$, $(i, j) \in \{1, \ldots, M\} \times \{1, \ldots, N\}$.

The sought denoised, continuous image $f : \mathcal{D} \to \mathbb{R}$ is modelled as a sum of parametrized Gaussians, i.e.

$$f[\mathcal{P}](x) = \sum_{l=1}^{L} b[c_l](x - y_l) + o\,,$$

where L denotes the number of atoms, b is a parameterized Gaussian, c_l and y_l are parameters and the center of the Gaussian for atom l, respectively, and o is the background intensity. $\mathcal{P} = (L, y_1, \ldots, y_L, c_1, \ldots, c_L, o)$ is the vector of all parameters.

The deformation aligning f with d_k is split into a global sample drift in terms of a rigid body motion consisting of a rotation $R_k \in SO(2)$ and a translation $v_k \in \mathbb{R}^2$ (using the positions of d_1 as reference coordinate system) and the local displacement/shift $s_{ij}^k \in \mathbb{R}^2$ for each pixel i, j. Note that this shift is linked to a scanline and thus it is inherently discrete in y-direction. Thus, the sought image f approximates the discrete image d_k via

$$f[\mathcal{P}](R_k(x_{ij} + s_{ij}^k) + v_k) \approx (d_k)_{ij}\,.$$

The objective functional is

$$\begin{aligned}&\mathcal{L}^K[\mathbf{s}^1, \ldots, \mathbf{s}^K, \mathcal{P}, (R_k, v_k)_{k=1}^K]\\ &\quad= \sum_{k=1}^{K} \left[\sum_{i=1}^{M} \sum_{j=1}^{N} \mathrm{Dist}_{\mathrm{dat}}\Big(f[\mathcal{P}](R_k(x_{ij} + s_{ij}^k) + v_k), (d_k)_{ij}\Big) + R_1[\mathbf{s}^k] + R_2[\mathbf{s}^k] \right],\end{aligned}$$

where with a specific scaling factor Λ, Δt being the time to scan one pixel, and ΔT being the time between scanning two consecutive lines

$$R_1[\mathbf{s}] = \frac{\Lambda}{2} \left[\frac{|s_{11}|^2}{\Delta t} + \sum_{j=2}^{N} \left(\frac{|s_{1j} - s_{M(j-1)}|^2}{\Delta T} + \sum_{i=2}^{M} \frac{|s_{ij} - s_{(i-1)j}|^2}{\Delta t} \right) \right]$$

for $\mathbf{s} \in (\mathbb{R}^2)^{M \times N}$, which results from the assumption that sample undergoes a Brownian motion [3]. For numerical reasons,

$$R_2[\mathbf{s}] = \frac{1}{2} \sum_{i=1}^{M} \sum_{j=1}^{N} \left(\nu_{\mathrm{hor}} |(s_{ij})_1|^2 + \nu_{\mathrm{vert}} |(s_{ij})_2|^2 \right)$$

is further added. Here, $(s_{ij})_1$ and $(s_{ij})_2$ denote the horizontal and vertical component of $s_{ij} \in \mathbb{R}^2$ and $\nu_{\mathrm{hor}}, \nu_{\mathrm{vert}} > 0$ are scaling factors. We would like to point out that the input data $(d_k)_{ij}$ is not interpolated at all. Due to the nature of the STEM acquisition process, interpolation in vertical direction, i.e. between different scanlines, must be avoided when trying to estimate the discontinuous scanline shifts.

The numerical optimization of this model is also rather involved. In particular, getting an initial guess for $\mathcal{P}$ that contains all atoms visible in the reconstructed image is challenging. For details on this, we refer to [3]. Further down, we describe the other aspects of the minimization strategy, since they will be used for the new model we propose here.

2.2.5 A new model - Joint denoising and distortion correction with a spline based image model

In order to generalize [3] to images that cannot be described as sum of Gaussians, we propose to use a new model for the sought denoised continuous image $f : \mathcal{D} \to \mathbb{R}$. It is parametrized using the Cartesian product space of 1D cubic B-splines. More specifically, let $\varphi_1, \dots, \varphi_n$ be a basis of the cubic B-splines on $[0, 1]$ with equidistant knots [4]. In other words, the value of f at a position $x = (x_1, x_2) \in \mathcal{D}$ is given by

$$f[\mathcal{P}](x_1, x_2) = \sum_{k,l=1}^{n} p_{kl}\varphi_k(x_1)\varphi_l(x_2),$$

where $\mathcal{P} = (p_{kl})_{k,l=1}^{n}$ is the parameter vector.

The resulting objective function is numerically minimized using a Trust Region solver [6]. Due to the non-convex nature of this minimization problem, a careful minimization strategy is necessary. To this end, we follow a similar strategy to the one described in [3], which considers a similar approach where u is not represented using splines, but as a sum of Gaussians, one Gaussian for each atom visible in the image domain. $\mathbf{s}^1, \dots, \mathbf{s}^K, \mathcal{P}$ are each initialized with 0. The global rotations and translations $(R_k, v_k)_{k=1}^{K}$ are initialized using the strategy of [1], but just with a rigid deformation model instead of a non-rigid one. Then, $\mathcal{P}$ is updated, by fitting $f[\mathcal{P}]$ to the average of the rigidly aligned input images, i.e. by minimizing

$$\sum_{i=1}^{M}\sum_{j=1}^{N} \mathrm{Dist}_{\mathrm{dat}}\left(f[\mathcal{P}](x_{ij}), \frac{1}{K}\sum_{k=1}^{N} g_k\left(R_k^{-1}(x_{ij}) + v_k^{-1}\right)\right),$$

with respect to $\mathcal{P}$ using the BFGS Quasi Newton algorithm [6]. Here, R_k^{-1} and v_k^{-1} denote the rotation and translation corresponding to the inverse of the rigid transformation given by R_k and v_k, g_k is the piecewise bilinear interpolation of d_k. With this rough but reasonable guess for $f[\mathcal{P}]$, the objective function is minimized with respect to $\mathcal{P}$ and $(R_k, v_k)_{k=1}^{K}$ using Trust Region. From here on, the global rigid alignment parameters $(R_k, v_k)_{k=1}^{K}$ are considered final and thus fixed. Then, $\mathcal{L}^K[\mathbf{s}^1, \dots, \mathbf{s}^K, \mathcal{P}]$ is minimized, while the shifts $\mathbf{s}^k$ are constrained to be constant in each scan line, i.e. $s_{ij}^k = s_{lj}^k$ for all $i, l \in \{1, \dots, M\}$, $j \in \{1, \dots, N\}$ using Trust Region. Finally, $\mathcal{L}^K[\mathbf{s}^1, \dots, \mathbf{s}^K, \mathcal{P}]$ is minimized using Trust Region.

Note that the initialization of this model is simpler than the one for [3]. For the latter, one needs to know the number of atoms and a good initial guess of the position of each atom. In [3], this is obtained by using a special convex surrogate model that treats the detection of the atom as a deconvolution problem with a known kernel (a Gaussian representing a single atom / atomic column).

3 Numerical results

For the numerical evaluation of the performance of the proposed approach, we created a series consisting of 64 synthetic STEM images. Here the image is modelled as a sum of Gaussians located on a regular grid, intensity noise is added using a Poisson noise model, location noise is added by assuming that the sample is undergoing a Brownian motion. The same approach is used for the generation of synthetic test images in [3]. Figure 1 shows the first and the last image from this series. This synthetic data set was modelled to resemble the characteristics of the HAADF STEM GaN $[11\bar{2}0]$ image series used in [8], cf. Figure 2.

To this synthetic dataset, the non-rigid reconstruction [1, 8] (denoted by "NRR"), the extended non-rigid reconstruction with bias correction [3] (denoted by "NRR+") and the method proposed here (denoted by "JUD", run with $\nu_{\text{hor}} = 25.9$, $\nu_{\text{vert}} = 71.4$, $\frac{\Lambda}{2\Delta t} = 50.0$, $\Delta T = 1000\Delta t$, $n = 2^6$) are applied.

The quality of the reconstruction here is measured in terms of the so-called *precision*. In electron microscopy, this refers to the precision with which atoms can be located and is computed on images assumed to show a perfect crystal. To compute it, all atoms in a given image are individually fitted using a parameterized Gaussian bump. Then all pairwise distances of neighboring atoms in x and y-direction are computed. Finally, the standard deviation of these distances are computed, resulting in a precision in x-direction and a precision in y-direction. The overall precision is defined as the square root of the sum of the squares of the precision in x and y-direction. Since the image is supposed to show a perfect crystal, all pairwise distances in one direction should be the same. In this sense, the precision implicitly measures how well atom centers can be located. Figure 3 illustrates the pairwise distances used for the precision. x-neighboring atoms are connected with blue arrows, y-neighboring atoms with yellow arrows. To convert the precision from pixels to picometer, the size a pixel in picometer is estimated as follows. The average of all x-distances and all y-distances in pixels is computed from all reconstructions. Combined with the expected physical distances for GaN $[11\bar{2}0]$ in x and y-directions of 276.174 pm and 518.5 pm, respectively, this allows to compute a pixel size in pm in x and y-direction. Their average is finally used as conversion factor. A very similar conversion logic was used in [8]. The left plot in Figure 4 shows the resulting precision for the different methods depending on the number of input images. To

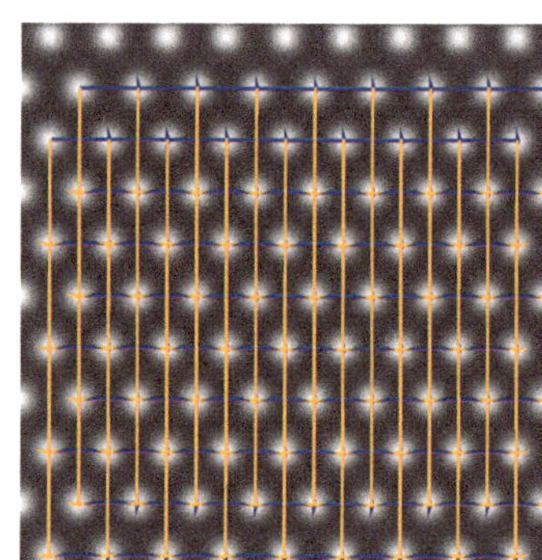

Fig. 3: Distances used to compute the precision in x and y-direction.

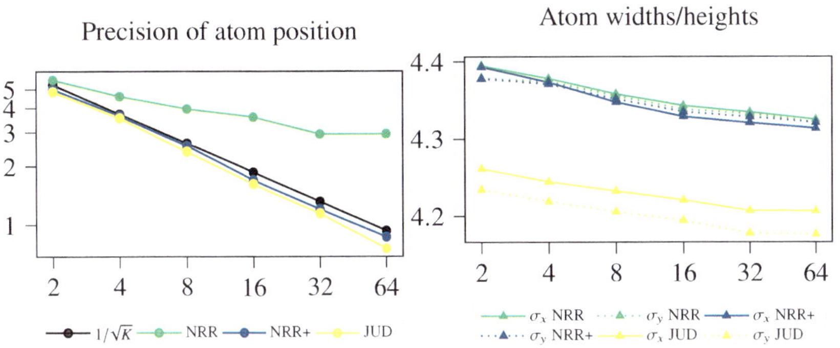

Fig. 4: Left: Precision plot for the synthetic test data set for different methods. The horizontal axis shows the number of images used for the reconstruction, the vertical axis the precision value in picometer. Right: Estimated atom sizes in pixels for a synthetic test data set.

compute the precision for a given number of images, the image series is split into non-overlapping series to generate multiple samples. For instance, for two images, the series of 64 images is split into 32 series with 2 images each, each method is applied to each of the 32 small series and the resulting precision values are averaged to get the precision for two images. The only exemption here is the case of one image. Here only every other image is used, i.e., the methods are applied to 32 series with one image each. As to be expected, the precision of all methods improves with the number of images. However, the scaling is different. Both "NRR+" and the proposed method scale as one over the square root of the number of images, "NRR" scales noticeable worse. While the proposed method shows a slightly better precision than "NRR+", both methods are equally suited when it just comes to locating the atom centers. Looking beyond just the precision, this picture changes though. The right plot in Figure 4 shows the average sizes of the Gaussians used to compute the precision. Note that the size of the Gaussian used to create the test data is 4.25 pixels. Thus, the proposed method more accurately reconstructs this dataset than the other methods.

Figure 5 shows the reconstructed images when the whole series of 64 images is used. Side by side, the images look very similar, even though the corresponding precision varies strongly between "NRR" on the one hand, and "NRR+" and the proposed method on the other hand, cf. Figure 4, left. Even though the latter two are rather similar in terms of precision (with the proposed method still outperforming "NRR+"), the proposed method reconstructs the intensity values of the ground truth considerably more accurately than "NRR+". This becomes visible when comparing the intensity of the ground truth to the reconstruction, cf. Figure 6. Note that some rows and columns at the boundaries have been cropped to avoid boundary artifacts. It is apparent that the intensity errors of "NRR+" are not only larger, but that "NRR+" has not fully corrected the scanline artifacts, which is not surprising, since this

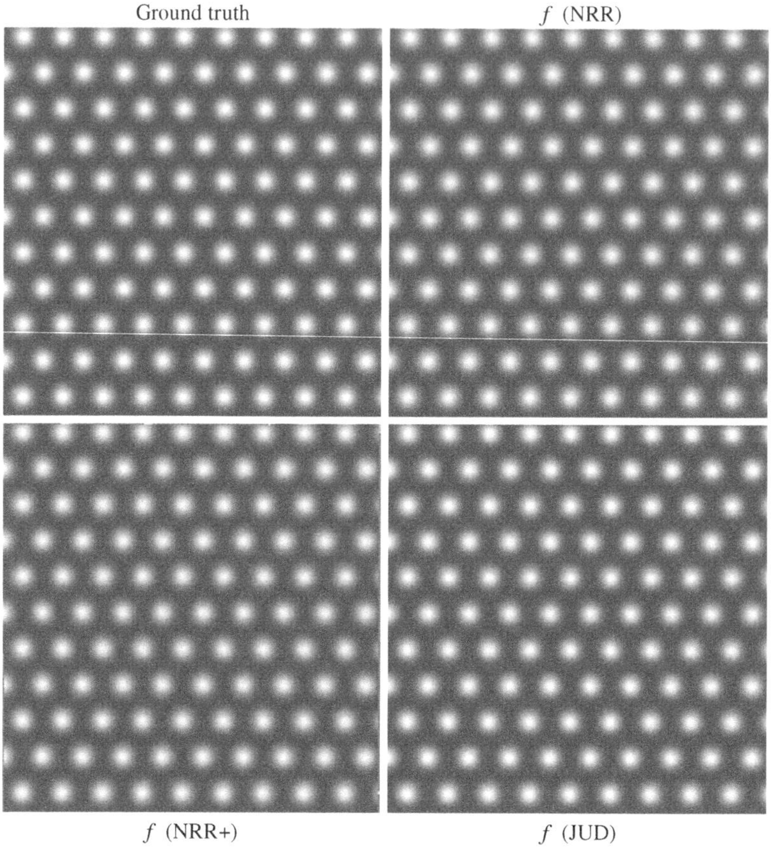

Fig. 5: Ground truth image the synthetic STEM image series was created from and the three reconstructions obtained with the different methods using the whole series of 64 images.

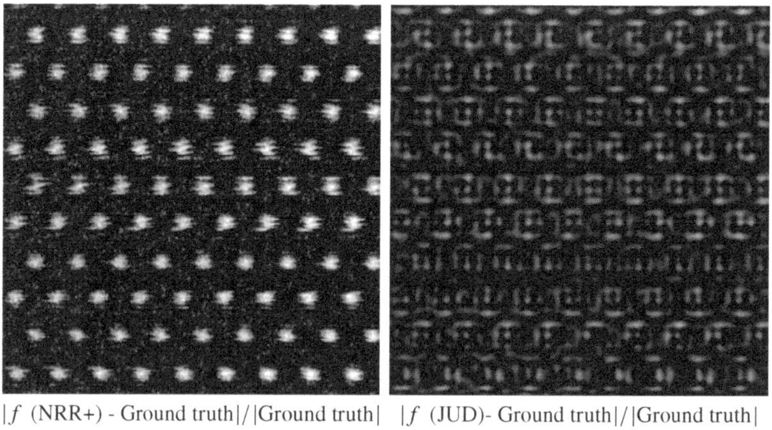

Fig. 6: Pixel-wise relative error of the ground truth image to the reconstructed images from the bottom row of Figure 5. 0 ▬ 5.5%

method does not allow for discontinuous displacements and thus cannot fully correct those.

We also applied the three methods on a series of 64 real STEM images. This are the first 64 images of the 512 HAADF STEM GaN $[11\bar{2}0]$ image series used in [8]. Figure 2 shows the first and the last of these 64 images. Figure 7 shows the resulting precision in picometer, where the same unit conversion logic as for the synthetic data was used with the same physical reference distances. The relative performance of the different methods is similar to the synthetic data case. The differences however are much smaller and the precision decreases slower than one over the square root of the number of images, which indicates that all of these models still miss some relevant effects encountered in real STEM data. Note that the precision numbers here are not comparable to the ones listed in [8], since the reconstructed image for a given number of images were computed very differently. In [8], series of 512 images were considered and an entire series was aligned and averaged following the algorithm from [1]. Then, the first n of the resulting aligned images were averaged and used to compute the precision value for n frames. Figure 8 illustrates the reconstructions using all images for the different methods. Like in the synthetic case (cf. Figure 5), the reconstructions look very similar. However, since no ground truth is available, we cannot illustrate the effect of the improved scanline distortion correction for the real data case as in the synthetic data case.

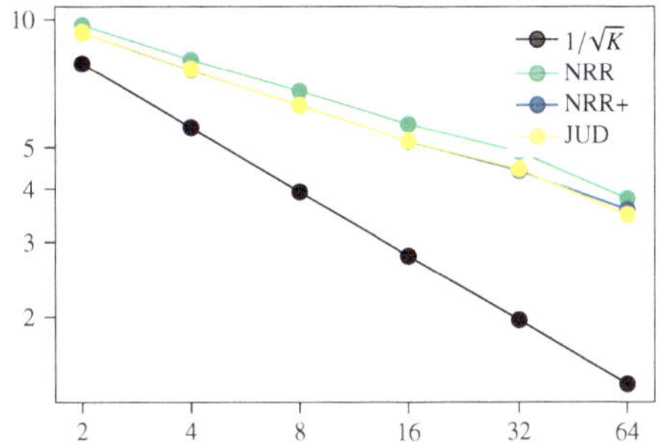

Fig. 7: Precision plot for the HAADF STEM GaN $[11\bar{2}0]$ image series.

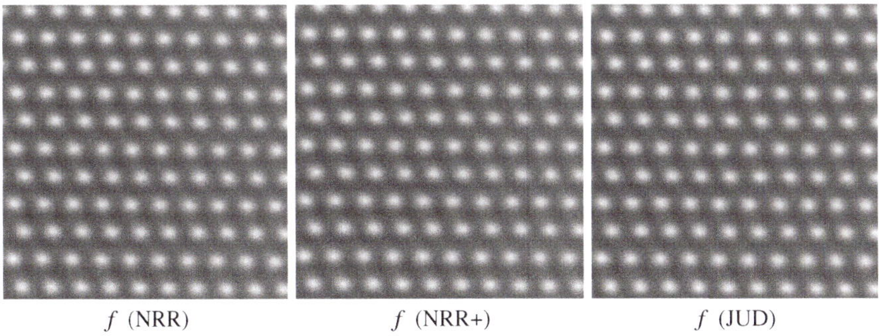

Fig. 8: The three reconstructions obtained with the different methods using the whole 64 HAADF STEM GaN $[11\bar{2}0]$ image series.

Acknowledgements The work of B. Berkels was supported by the German research foundation (DFG) within the Collaborative Research Centre SFB 1394 "Structural and Chemical Atomic Complexity—From Defect Phase Diagrams to Materials Properties" (Project ID 409476157) in Project A04. The work of P. Binev was supported by the NSF grants DMS 2038080 and DMS 2245097.

References

1. Berkels, B., Binev, P., Blom, D.A., Dahmen, W., Sharpley, R.C., Vogt, T.: Optimized imaging using non-rigid registration. Ultramicroscopy **138**, 46–56 (2014). DOI 10.1016/j.ultramic.2013.11.007
2. Berkels, B., Liebscher, C.H.: Joint non-rigid image registration and reconstruction for quantitative atomic resolution scanning transmission electron microscopy. Ultramicroscopy **198**, 49–57 (2019). DOI 10.1016/j.ultramic.2018.12.016
3. Berkels, B., Wirth, B.: Joint denoising and distortion correction of atomic scale scanning transmission electron microscopy images. Inverse Problems **33**(9), 095002 (2017). DOI 10.1088/1361-6420/aa7b94
4. de Boor, C.: A practical guide to splines, vol. 27. Springer verlag New York (1978)
5. Jones, L., Yang, H., Pennycook, T.J., Marshall, M.S.J., Van Aert, S., Browning, N.D., Castell, M.R., Nellist, P.D.: Smart align - a new tool for robust non-rigid registration of scanning microscope data. Advanced Structural and Chemical Imaging **1** (2015). DOI 10.1186/s40679-015-0008-4
6. Nocedal, J., Wright, S.J.: Numerical Optimization, 2 edn. Springer (2006)
7. Ophus, C.: Four-dimensional scanning transmission electron microscopy (4d-stem): From scanning nanodiffraction to ptychography and beyond. Microscopy and Microanalysis **25**(3), 563–582 (2019). DOI 10.1017/S1431927619000497
8. Yankovich, A.B., Berkels, B., Dahmen, W., Binev, P., Sanchez, S.I., Bradley, S.A., Li, A., Szlufarska, I., Voyles, P.M.: Picometre-precision analysis of scanning transmission electron microscopy images of platinum nanocatalysts. Nature Communications **5** (2014). DOI 10.1038/ncomms5155

The Approximation of Cauchy-Stieltjes and Laplace-Stieltjes Functions

Dietrich Braess and Wolfgang Hackbusch

Abstract The treatment of fractional diffusion operators leads to the rational approximation of Cauchy–Stieltjes functions. Similarly the computation of some high-dimensional integrals in quantum chemistry leads to the approximation of Lebesgue–Stieltjes functions by exponential sums. In both cases the exponential rate of approximation is simply computed via Gauss' arithmetic-geometric iteration and an approximation procedure of the square root function. Surprisingly the convergence rates are different in the two cases although the derivations of the two results are very similar.

1 Introduction and Historical Remarks

Two approximation problems have recently got interest due to new applications. The approximation of the function $x^{-\alpha}$, $\alpha > 0$, by *exponential sums* helps to compute some high-dimensional integrals via sums and products of one-dimensional integrals [11], [13], [27]. Differential equations with fractional elliptic differential operators occur in the treatment of materials with a non-local constitutive law. Here the approximation of $x^{-\alpha}$, $0 < \alpha < 1$, by *rational functions* is used with high precision (cf. [3], [4], [6], [14], [15], [23, §14], [26]).

Both approximation problems are related to Zolotarev's famous problems from 1877: given two separate intervals on the real line, rational functions of order n are desired that are small on one interval and their inverses are small on the second interval. The solution of the problem with symmetrical intervals is often determined

Dietrich Braess
Mathematisches Institut, Ruhr-Universität Bochum, 44780 Bochum, Germany,
e-mail: `Dietrich.Braess@rub.de`

Wolfgang Hackbusch
Max-Planck-Institut *Mathematik in den Naturwissenschaften*, Inselstr. 22, 04103 Leipzig, Germany,
e-mail: `wh@mis.mpg.de`

R. DeVore, A. Kunoth (eds.), *Multiscale, Nonlinear and Adaptive Approximation II*,
https://doi.org/10.1007/978-3-031-75802-7_7

by the rational approximation of the sign function, and the results are expressed in terms of elliptic functions, but we prefer the solution via a weighted approximation of the square root function. This approach is more appropriate for the applications that were the reason for the revival of the interest in this approximation problem. The basic tool is a variant of Newton's method for computing square roots of positive real numbers.

In the second part of the last century a rich theory of the rational approximation was created. Quantitative and numerical results for the degree of approximation were established for various special functions, e.g., they helped to disprove the $\frac{1}{9}$-conjecture for the rational approximation of the exponential function [18]. Cauchy–Stieltjes functions, however, were disregarded in this context, since the mentioned results in terms of elliptic integrals were available.

When recently new interest arose due to new applications, it was clear that a similar behaviour for the approximation of $x^{-\alpha}$ by exponential sums and by rational functions was expected, but no numerical results for a comparison existed. We expected that the asymptotic decay of the error is the same for the two approximation problems. It turned out, however, that the approximation error of exponential sums and of rational functions correspond to approximation errors of the square root function on intervals of *different sizes*. Therefore, the decay of the approximation error is faster for the rational approximation than for the treatment with exponential sums.

The asymptotic law of the degree of approximation can be obtained by a few steps of Gauss' arithmetic-geometric process and therefore by very simple calculations. The well-known fast convergence of the arithmetic-geometric iteration implies that the asymptotic rules provide very good results for the applications we have in mind. It is clear that approximations with a relative error (much) smaller than 10^{-4} are here required. From Theorem 4.1 we conclude that the deviation of the values for finite n from the asymptotics is of the order 10^{-8}. As a consequence, the error bounds according to the asymptotic law that are calculated with the data in the small Table 3 are sufficient for the actual applications.

There are more facts which are emphasised in approximation theory, but are rather misleading in the applications under consideration. If a Cauchy-Stieltjes function is considered on a finite interval without singularity at the boundary of the interval, then the asymptotic behaviour of the approximation error has the form $ce^{-\beta n}$ with appropriate parameters c and β. On the other hand, a law of the form $ce^{-\sqrt{\alpha n}}$ holds for infinite intervals.

When elliptic differential operators are involved, then usually you have to cover a larger spectrum, say with a ratio of the order 10^5 between the largest and the lowest eigenvalue. Thus one may expect that an error of the form $ce^{-\beta n}$ is observed only for extremely large n while for mid-sized n the behaviour is closer to that of infinite intervals. Lemma 2.2, however, will tell us that the approximation on $[1, 159]$ is the same as on the mentioned large interval if we double the number n. The very fast convergence of the arithmetic-geometric iteration implies that the behaviour is closer to the situation on small intervals.

For completeness, we mention that different approaches are required, when the relative error of the approximation by exponential sums[1] is considered as in the treatment of elliptic equations in very high dimensions [33].

The paper is structured as follows. Sect. 2 treats the exact error of the best relative approximation of the square root function on finite intervals. The required tools from Gauss' arithmetic-geometric mean and complete elliptic integrals are provided in Appendix A. Sect. 3 investigates the approximation of Laplace–Stieltjes functions by exponential sums by using results from the previous section. The sharpness of the estimates for finite intervals is elucidated by numerical examples. Sect. 4 is concerned with the rational approximation of Cauchy–Stieltjes functions.

2 Best Rational Approximation of the Square Root in a Weighted Norm

2.1 Setting of the Problem

A powerful tool for the analysis of the best approximation of the associated Stieltjes functions by rational functions or by exponential sums on an interval $[a,b]$, $0 < a < b \leqslant \infty$, are the results about the approximation of the square root function with respect to the weighted norm

$$|||f|||_{[a,b]} := \sup_{a \leqslant x \leqslant b} \left| \frac{f(x)}{\sqrt{x}} \right|. \tag{1}$$

The set of rational functions of degree (m,n) is denoted by $R_{m,n} := \{\frac{p_m}{q_n} : \text{degree}(p_m) \leqslant m \text{ and degree}(q_n) \leqslant n\}$. The best approximation $u_{m,n} \in R_{m,n}$ is the minimiser of

$$\tilde{E}_{m,n}([a,b]) := \min_{v \in R_{m,n}} |||\sqrt{x} - v|||_{[a,b]}. \tag{2}$$

In particular, we consider the case $n = m - 1$. The unique best approximation exists and is characterised by the *equi-oscillation property*, i.e., there is an alternant $a = \alpha_0 < \alpha_1 < \ldots < \alpha_{2m} \leqslant b$ with

$$\frac{\sqrt{\alpha_\nu} - v(\alpha_\nu)}{\sqrt{\alpha_\nu}} = (-1)^\nu \, \tilde{E}_{m,n}([a,b]) \qquad \text{for } 0 \leqslant \nu \leqslant 2m.$$

Remark 2.1 Given $m, n \in \mathbb{N}_0$, $\tilde{E}_{m,n}([a,b])$ only depends on the ratio $\frac{b}{a}$, i.e., for $\lambda \in (0,\infty)$

$$\tilde{E}_{m,n}([a,b]) = \tilde{E}_{m,n}([\lambda a, \lambda b]).$$

If u is the best approximation in $[a,b]$, then $u_\lambda(x) := \sqrt{\lambda}\, u(x/\lambda)$ is the best approximation in the interval $[\lambda a, \lambda b]$.

[1] The page [20] also contains best approximations by exponential sums with respect to the relative error.

The next theorem may certainly be traced back to Zolotarev [34] in 1877.

Theorem 2.1 *Let $m \geqslant 1$. Then the error of the best relative approximation of $f(x) = \sqrt{x}$ is bounded by*

$$\tilde{E}_{m,m-1}([a,b]) \leqslant 4\exp[-2m\,\mu(\kappa)] \quad \textit{with } \kappa = \tilde{E}_{0,0}([a,b]) = \frac{\sqrt{b}-\sqrt{a}}{\sqrt{b}+\sqrt{a}}, \tag{3a}$$

where

$$\mu(\kappa) := \frac{\pi\,\mathbf{K}'(\kappa)}{2\,\mathbf{K}(\kappa)} \tag{3b}$$

is the Grötzsch ring function.

The elliptic integrals $\mathbf{K}(\kappa)$ and $\mathbf{K}'(\kappa)$ are explained in Appendix A.1.

2.2 Proof of Theorem 2.1

The proof of the theorem is based on two different relations between $\tilde{E}_{m,m-1}$ and $\tilde{E}_{2m,2m-1}$ stated in Lemmata 2.1 and 2.2. The following lemma holds for all $m \in \mathbb{N}_0 = \mathbb{N} \cup \{0\}$. In the case of $m = 0$, $\tilde{E}_{m,m-1}([a,b])$ becomes $\tilde{E}_{0,0}([a,b])$ and $\tilde{E}_{2m,2m-1}([a,b])$ becomes $\tilde{E}_{1,0}([a,b])$.

Lemma 2.1 *For all $m \in \mathbb{N}_0$ and all $0 < a < b < \infty$ the approximation errors satisfy*

$$\tilde{E}_{2m,2m-1}([a,b]) = \left(\frac{\tilde{E}_{m,m-1}([a,b])}{1+\sqrt{1-\tilde{E}^2_{m,m-1}([a,b])}}\right)^2. \tag{4}$$

Proof. The proof is illustrated in Fig. 1. Let $u_{m,m-1}$ be the best relative approximation to the square root function in $R_{m,m-1}$ and $(x_0, x_1, \ldots, x_{2m})$ be the associated alternant of length $2m+1$. The scaled function $v := \{1 - (\tilde{E}_{m,m-1}([a,b]))^2\}^{-1/2} u_{m,m-1}$ satisfies

$$-\sigma \leqslant \frac{\sqrt{x}-v(x)}{\sqrt{x}+v(x)} \leqslant \sigma \text{ for } x \in [a,b] \text{ where } \left(\frac{1+\sigma}{1-\sigma}\right)^2 = \frac{1+\tilde{E}_{m,m-1}([a,b])}{1-\tilde{E}_{m,m-1}([a,b])}. \tag{5}$$

It is used to form $w := \frac{1}{2}\left(v + \frac{x}{v}\right) \in R_{2m,2m-1}$ (cf. the second graph in Fig. 1). The function w can be considered as the result of one Newton step $t \mapsto \frac{1}{2}\left(t + \frac{x}{t}\right)$ for the solution of the equation $t^2 - x = 0$. The error function $w - \sqrt{x}$ is known to be nonnegative. Consider the scaled function $u_{2m,2m-1} := (1-\sigma)\,w$. Now the error function has an alternant $(x_0, \xi_1, x_1, \xi_2, \ldots, x_{2m})$ of length $4m+1$ (cf. the third graph in Fig. 1). The zeros ξ_j of w are the new points. The equi-oscillation property implies that $u_{2m,2m-1}$ is the best approximation in $R_{2m,2m-1}$ with the error $\tilde{E}_{2m,2m-1}([a,b]) = \sigma^2$. Expressing σ in terms of $\tilde{E}_{m,m-1}([a,b])$ according to (5) we obtain (4). □

More details of this proof are found in [12], [8, (5.42)], [11, §3.2].

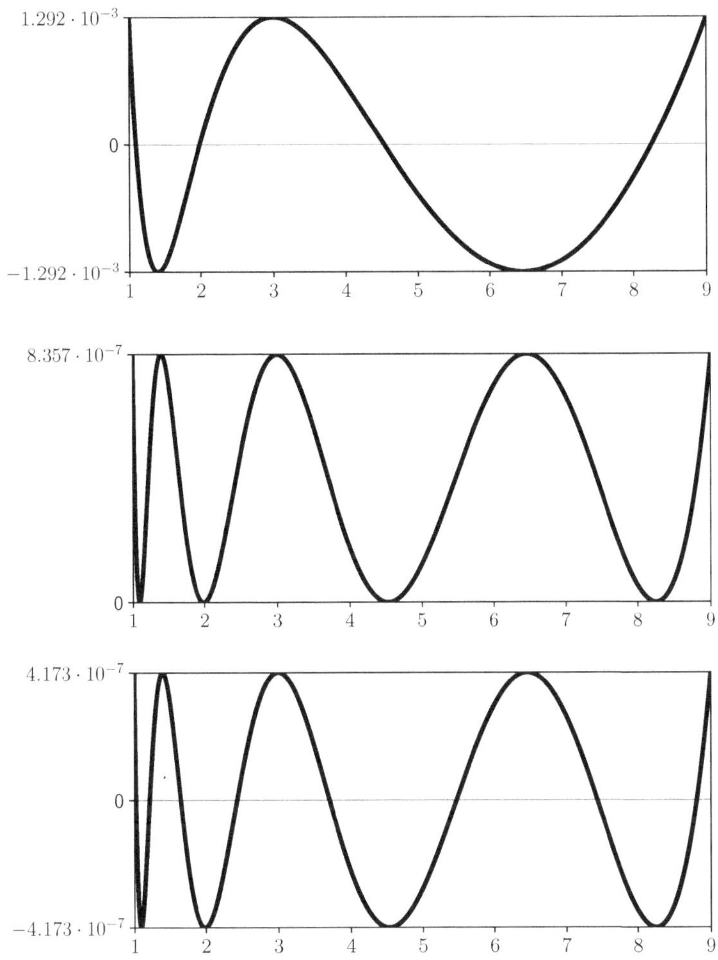

Fig. 1: Relative errors (1) for $u_{2,1}$, (2) for the auxiliary function w after a Newton step, (3) for $u_{4,3}$.

The second lemma compares $\tilde{E}_{m,m-1}$ and $\tilde{E}_{2m,2m-1}$ for different intervals.

Lemma 2.2 *Let $\{a_k, b_k\}$ be a double sequence generated by the arithmetic-geometric process (27):*

$$a_{k+1} = \sqrt{a_k b_k}, \quad b_{k+1} = \frac{a_k + b_k}{2}$$

with $0 < a_0 < b_0$. Then

$$\tilde{E}_{m,m-1}([a_{k+1}^2, b_{k+1}^2]) = \tilde{E}_{2m,2m-1}([a_k^2, b_k^2]), \quad m \in \mathbb{N},\ k \in \mathbb{N}_0.$$

According to Remark 2.1, an equivalent statement is

$$\tilde{E}_{m,m-1}([1,r_{k+1}^2]) = \tilde{E}_{2m,2m-1}([1,r_k^2]) \quad \text{with } r_k := b_k/a_k \qquad (k \in \mathbb{N}_0). \tag{6}$$

Proof. The function

$$u_1(x) := (a_k b_k)^{-1/2} \frac{x + a_k b_k}{a_k + b_k}$$

is a multiple of the best relative approximation to $\sqrt{x}$ from $R_{1,0}$. The function $x \mapsto x/u_i^2(x)$ maps the interval $[a_k^2, b_k^2]$ twice onto the interval $[a_{k+1}^2, b_{k+1}^2]$ as elucidated by the special values in the following table:

x	a_k^2	$a_k b_k$	b_k^2
$\dfrac{x}{u_1^2(x)}$	$a_k b_k$	$\left(\dfrac{a_k + b_k}{2}\right)^2$	$a_k b_k$

Let $p/q \in R_{m,m-1}$ be the best relative approximation to $\sqrt{x}$ for the interval $[a_{k+1}^2, b_{k+1}^2]$. Then

$$\frac{P(x)}{Q(x)} := u_1(x) \frac{p(x/u_1^2(x))}{q(x/u_1^2(x))} \in R_{2m,2m-1}$$

provides the best relative approximation to $\sqrt{x} = u_1(x)\sqrt{x/u_1^2(x)}$ on the original interval with the same norm for the relative error as p/q on the smaller interval. Indeed, an alternant of length $4m+1$ is generated from the alternant of length $2m+1$, since $x \mapsto x/u_1^2(x)$ is monotone on the subintervals $[a_k^2, a_k b_k]$ and $[a_k b_k, b_k^2]$, respectively. Hence, P/Q is a best approximation from $R_{2m,2m-1}$, and the proof is complete. □

Combining (6) with (4), we obtain the recursion

$$\tilde{E}_{m,m-1}([1,r_{k+1}^2]) = \left(\frac{\tilde{E}_{m,m-1}([1,r_k^2])}{1 + \sqrt{1 - \tilde{E}_{m,m-1}^2([1,r_k^2])}} \right)^2. \tag{7}$$

In the following we consider (forward) *Landen sequences* $\{x_k : k \in \mathbb{N}_0\}$ defined by a starting value $x_0 \geqslant 1$ and the rule

$$x_{k+1} = \mathcal{L}(x_k) := \frac{1}{2}\left(\sqrt{x_k} + \frac{1}{\sqrt{x_k}}\right). \tag{8}$$

Accordingly, an *inverse Landen sequence* $\{y_k : k \in \mathbb{N}_0\}$ is defined by the inverse of the function $\mathcal{L}$,

$$y_{k+1} = \mathcal{L}^{-1}(y_k) = \left(y_k + \sqrt{y_k^2 - 1}\right)^2. \tag{9}$$

These sequences will be studied in Appendix A.2. From (9) we conclude that $4y_{k+1} \leqslant (4y_k)^2$. This is the reason for the fact that a factor of 4 or 1/4 is found in many inequalities, e.g., in (10), (14), or (31).

We will complete the proof of Theorem 2.1 for intervals of the form $[a^2, b^2]$. Following Remark 2.1 we focus on intervals of the form $[1, R^2]$, $R > 1$.

Remark 2.2 (a) The sequence $\eta_k := 1/\tilde{E}_{m,m-1}([1, r_k^2])$ defined by (7) is an inverse Landen sequence.
(b) The sequence r_k defined in Lemma 2.2 satisfies the rule of the forward Landen sequence (8), while the induced sequence

$$\lambda_k := \frac{r_k + 1}{r_k - 1} \qquad \text{(equivalent to } r_k = \frac{\lambda_k + 1}{\lambda_k - 1}\text{)}$$

yields an inverse Landen sequence.

The asymptotic behaviour of *inverse Landen sequences* is provided in Appendix A.2: The parameters of the exponential decay of the Landen sequences $\{\lambda_k\}$ and $\{\eta_k\}$ are denoted in (10) as ω and Ω, respectively:

$$\begin{aligned} \lambda_k &= \tfrac{1}{4}\omega^{2^k}(1 + o(1)) \to \infty, \ \omega := \exp\left[\mu(\tfrac{1}{\lambda_0})\right] \text{ with } \tfrac{1}{\lambda_0} = \tfrac{b_0 - a_0}{b_0 + a_0}, \\ \eta_k &= \tfrac{1}{4}\Omega^{2^k}(1 + o(1)) \to \infty, \ \Omega := \exp\left[\mu(E_0)\right] \text{ with } E_0 = \tilde{E}_{m,m-1}([1, r_0^2]). \end{aligned} \tag{10}$$

We recall that $\mu(\cdot)$ is the Grötzsch ring function defined in (3b).

Next we consider the Padé approximation $r_{\text{Padé}} = p/q \in R_{m,m-1}$ of $f(z) := \sqrt{1+z}$ at the origin

$$p(x) = \frac{1}{2}\left[(1 + \sqrt{1+x})^{2m} + (1 - \sqrt{1+x})^{2m}\right],$$
$$q(x) = \frac{1}{2}\left[(1 + \sqrt{1+x})^{2m} - (1 - \sqrt{1+x})^{2m}\right]/\sqrt{1+x}.$$

The error may be written in the form

$$\sqrt{1+x} - \frac{p(x)}{q(x)} = -\frac{(1 - \sqrt{1+x})^{2m}}{q(x)} = \frac{x^{2m}}{q(x)(1 + \sqrt{1+x})^{2m}}.$$

Now we obtain the bound $|\sqrt{1+x} - r_{\text{Padé}}(1+x)| \leqslant C_1|x|^{2m}$ for $|x| \leqslant 1/2$ and some $C_1 > 0$. The length of the interval $[1, r_k^2]$ is

$$r_k^2 - 1 = (r_k - 1)(r_k + 1) = (r_k + 1)^2/\lambda_k = (4 + o(1))/\lambda_k \to 0.$$

There is a k^* so that $r_k^2 - 1 \leqslant 1/2$ for $k \geqslant k^*$. This proves

$$1/\eta_k = \tilde{E}_{m,m-1}([1, r_k^2]) \leqslant |||\sqrt{x} - r_{\text{Padé}}|||_{[1,r_k^2]} \leqslant C_1\lambda_k^{-2m} \quad \text{for } k \geqslant k^*.$$

Hence, $\eta_k \geqslant \frac{1}{C_1}\lambda_k^{2m}$ holds for $k \geqslant k^*$. Using the asymptotic laws in (10) we obtain

$$\frac{1}{4}\Omega^{2^k} \geqslant \left(\frac{\omega^{2^k}}{4C_1}\right)^{2m}(1 + o(1)) \quad \text{for } k \geqslant k^*.$$

Applying the 2^k-th root and exploiting the fact that $\sqrt[2^k]{c} \to 1$ for any positive constant c we conclude that

$$\Omega \geqslant \omega^{2m}.$$

Recalling (10) once more, we see that the last inequality is equivalent to

$$\mu(E_0) \geqslant 2m\mu(1/\lambda_0). \tag{11}$$

Now the rule (29) for the elliptic integral yields

$$\log\left(\frac{4}{E_0}\right) \geqslant \mu(E_0) \geqslant 2m\mu(1/\lambda_0)\,.$$

Exponentiation and the expressions for E_0 and λ_0 in (10) yield

$$\tilde{E}_{m,m-1}([a_0^2, b_0^2] \leqslant 4\exp\left[-2m\mu\left(\frac{b_0 - a_0}{b_0 + a_0}\right)\right],$$

and the proof of Theorem 2.1 for any interval $[a^2, b^2]$ is complete.

2.3 Computational Scheme for the Exponential Convergence Rate

For $m = 2^k$ ($k \in \mathbb{N}_0$) the solutions $u_{m,m-1}$ of problem (2) as well as their approximation errors $\tilde{E}_{m,m-1}([a,b])$ can be computed explicitly by the following recursion. The start is given by the best constant approximation $u_{0,0} \in R_{0,0}$ in the interval $[a,b]$ and its error:

$$u_{0,0} = \frac{2\sqrt{ab}}{\sqrt{a}+\sqrt{b}}, \qquad \tilde{E}_{0,0}([a,b]) = \frac{\sqrt{b}-\sqrt{a}}{\sqrt{b}+\sqrt{a}}. \tag{12}$$

Now (4) yields $\tilde{E}_{2^{k-1},2^{k-1}-1}([a,b])$ for all $k \geqslant 1$. The inverse values $\lambda_0 := \frac{1}{\tilde{E}_{0,0}([a,b])}$ and $\lambda_k := 1/\tilde{E}_{2^{k-1},2^{k-1}-1}([a,b])$ belong to an inverse Landen iteration. As in Section 2.2 it follows that $\omega = \lim\,(\lambda_k)^{1/2^k} = \exp[-2m\,\mu(\kappa)]$ with $\kappa := \frac{\sqrt{b}-\sqrt{a}}{\sqrt{b}+\sqrt{a}}$ is the exponential convergence rate in (3a). $\mu(\cdot)$ is defined via the elliptic integrals, but the numerical computation uses the inverse Landen iteration as explained below.

Let $m = 2^{k-1}$. The previous construction shows

$$\tilde{E}_{m,m-1}([a,b]) = 1/\lambda_k \leqslant 4\omega_0^{-2m} = 4\omega_1^{-m},$$

where λ_{-k} and ω_0, ω_1 are computed by the scheme

$$\boxed{\begin{array}{ccccccccc} \lambda_0 = \frac{\sqrt{b}+\sqrt{a}}{\sqrt{b}-\sqrt{a}} & \mapsto & \lambda_1 & \mapsto & \ldots & \mapsto & \lambda_{k-1} & \mapsto & \lambda_k \\ & & & & & & & & \downarrow\times 4 \\ \omega_0 & \leftarrow & \omega_1 & \leftarrow & \ldots & \leftarrow & \omega_{k-1} & \leftarrow & \omega_k \end{array}}\,. \tag{13}$$

Here $\mapsto$ denotes one step of the inverse Landen iteration (9), while $\leftarrow$ denotes an application of the square root function.

The following example refers to the interval $[a,b] = [1, 3.24]$ and $k = 3$ (i.e., $m = 4$):

$$\boxed{\begin{array}{ccccccc} \lambda_0 = 3.5 & \mapsto & 46.978\,713\,764 & \mapsto & 882.599\,807 & \mapsto & 3.115\,929\,660 \times 10^8 \\ & & & & & & \downarrow\times 4 \\ 13.707\,427\,407 & \leftarrow & 187.893\,566\,105 & \leftarrow & 35\,303.992\,184 & \leftarrow & 1.246\,371\,864 \times 10^9 \end{array}}$$

We recall that the statement in (3a) follows from $\frac{1}{\lambda_k} = 4\omega_0^{-2m} \leqslant 4\exp[-2m\mu(\frac{1}{\lambda_0})]$, $m = 2^{k-1}$ (cf. also Lemma 4.2). The rate of convergence is $\omega_1 = \omega_0^2 \leqslant$

$\exp[-2\,\mu(\frac{1}{\lambda_0})]$. Estimate (31) shows that

$$\frac{\exp[\mu(1/\lambda_0)]}{\omega_0} = 1 - \frac{2}{m\lambda_k^2} + \text{lower order terms.}$$

If $\lambda_k > 10^8$ as in the example above, the values of ω_0 and $\exp[\mu(1/\lambda_0)]$ coincide as double precision numbers.

Hence, given $\lambda_0 > 1$, the value $\exp[\mu(1/\lambda_0)]$ is obtained by the computational scheme (13) with k chosen such that $\lambda_k \geqslant 10^8$.

2.4 Large Intervals

The approximation on the infinite interval $[1, \infty)$ with respect to the standard supremum norm will be studied later. Because of $\tilde{E}_{m,m-1}([1, \infty)) = \infty$, the approximation is not possible for the weighted norm (1). Instead we consider a large finite interval $[a^2, b^2]$ with $b \gg a$.

We recall (12), $\frac{1}{\lambda_0} = \kappa_0 = \tilde{E}_{0,0}([a^2, b^2]) = \frac{b-a}{b+a}$ and apply the Landen iteration $\frac{1}{\lambda_1} = \kappa_1 = \frac{2\sqrt{\kappa_0}}{1+\kappa_0}$. The complementary modulus is given by $\kappa_1' = \sqrt{1-\kappa_1^2} = \frac{1-\kappa_0}{1+\kappa_0} = \frac{a}{b}$. Using (29), we obtain

$$\mu(\kappa_0) = \frac{\pi \mathsf{K}'(\kappa_0)}{2\mathsf{K}(\kappa_0)} = \frac{\pi \mathsf{K}'(\kappa_1)}{\mathsf{K}(\kappa_1)} = \frac{\pi \mathsf{K}(\kappa_1')}{\mathsf{K}'(\kappa_1')} = \frac{\pi^2 \mathsf{K}(a/b)}{\pi \mathsf{K}'(a/b)} \geqslant \frac{\pi^2/2}{\log(4b/a)}.$$

Inserting this inequality into Theorem 2.1, we obtain

$$\tilde{E}_{m,m-1}([a^2, b^2]) \leqslant 4 \exp\left[\frac{-m\,\pi^2}{\log(4\,b/a)}\right]. \tag{14}$$

The bound (14) will be used in the next sections for treating approximation problems on infinite intervals.

2.5 Blaschke Product

In 1877 Zolotarev [34] considered the minimisation problem

$$Z_k(A, B) = \inf_{r_k \in R_{k,k}} \frac{\sup\{|r_k(x)| : x \in A\}}{\inf\{|r_k(x)| : x \in B\}}. \tag{15}$$

Our results apply to the symmetrical case where $A = [a, b]$, $B = -A = [-b, -a]$, and $0 < a < b < \infty$.

Assume that $u = p/q$ is the best approximation in $R_{k,k-1}$ and in the interval $[a^2, b^2]$. Construct $v = \tilde{p}(x)/q(x)$ as in the proof of Lemma 2.1. The considerations in the proof yield

$$\left|\frac{\tilde{p}(x)/q(x) - \sqrt{x}}{\tilde{p}(x)/q(x) + \sqrt{x}}\right| \leqslant \sigma \quad \text{for } x \in [a^2, b^2]$$

with σ defined in (5). Let $x = z^2$ and $P_{2k}(z) := p(z^2) - zq(z^2)$. The *Blaschke product* $r_{2k}(z) := P_{2k}(z)/P_{2k}(-z) \in R_{2k,2k}$ satisfies

$$|r_{2k}(z)| \leqslant \sigma \text{ for } z \in [a,b], \qquad |r_{2k}(z)| = 1 \text{ for } \mathrm{Re}\,\mathfrak{e}(z) = 0 \text{ or } |z| \to \infty. \tag{16}$$

Note that r_{2k} has $2k$ zeros in $[a,b]$: $r_{2k}(z) = \prod_{j=1}^{2k} \frac{z-z_j}{z+z_j}$. The connection with the Zolotarev numbers $Z_k(A,B) = \sigma^2$ is described in [12].

3 Approximation of Completely Monotone Functions by Exponential Sums

3.1 Exponential Sums

The set of exponential sums of order n is denoted by

$$\mathfrak{E}_n := \Big\{u_n(x) = \sum_{\nu=1}^{n} \alpha_\nu \mathrm{e}^{-t_\nu x} : \alpha_\nu, t_\nu \in \mathbb{R}\Big\}.$$

Recently the approximation by exponential sums has got new interest, since the computation of some high-dimensional integrals in chemistry was facilitated by the use of exponential sums (cf. [1, 9, 11, 13, 22, 27, 32]). An application to a pde with a fractional time derivative is given in [19]. The exponential sums for the approximation of $x^{-\alpha}$ can also be used to approximate $L^{-\alpha}$, in particular, if L is given in tensor form (cf. [24, §9.8.2.6]).

The above mentioned d-dimensional integrals have the form

$$\int_B \frac{g(x)\,\mathrm{d}x}{\|x-y\|^{2\alpha}} \quad \text{where } g(x) = g_1(x_1)g_2(x_2)\ldots g_d(x_d),\ B = [a_1,b_1]\times\ldots\times[a_d,b_d]$$

and $\|x-y\|^2 = (x_1-y_1)^2 + (x_2-y_2)^2 + \cdots + (x_d-y_d)^2$. The Coulomb or Newton potential corresponds to $\alpha = 1/2$ and $d = 3$. Approximating the negative power $x^{-\alpha}$ by an exponential sum $u_n \in \mathfrak{E}_n$ we obtain a sum of products of one-dimensional integrals:

$$\int_B \alpha_k g(x) \mathrm{e}^{-\beta_k \|x-y\|^2} \mathrm{d}x = \alpha_k \prod_{i=1}^{d} \int_{a_i}^{b_i} g(x_i)\, \mathrm{e}^{-\beta_k (x_i-y_i)^2} \mathrm{d}x_i\,. \tag{17}$$

In the case of $y \notin B$, the error estimates are obvious. However, in quantum chemistry the standard integration domain of the Coulomb potential is $\mathbb{R}^d$, i.e., the singularity at $x = y$ is always present. Nevertheless, we get the following estimate.

Lemma 3.1 *Consider a source term $f \in L^1(\mathbb{R}^d) \cap L^\infty(\mathbb{R}^d)$ with $\|f\|_{L^1(\mathbb{R}^d)} \leqslant C_1$, $\|f\|_{L^\infty(\mathbb{R}^d)} \leqslant C_\infty$ and $d \geqslant 2$. Let E_n be the error of the best approximation of $x^{-1/2}$ in $[1,\infty)$ by an n-term exponential sum. Then there is an exponential sum $u_n \in \mathfrak{E}_n$ such that*

$$\left\| \int_{\mathbb{R}^d} \left[\frac{f(x)}{\|x-\bullet\|} - u_n(\|x-\bullet\|^2) f(x) \right] \mathrm{d}x \right\|_{L^\infty(\mathbb{R}^d)} \leqslant \frac{d\sqrt{\pi}}{d-1} \sqrt[d]{\frac{2}{\Gamma(d/2)} C_1^{d-1} C_\infty}\; E_n^{\frac{d-1}{d}}.$$

Proof. Split the integral into $I_\eta := \int_{\|x-\bullet\|\leqslant\eta}\dots$ and $I_\infty := \int_{\|x-\bullet\|\geqslant\eta}\dots$. Since $\|y\|^{-1} > u_n(\|y\|^2)$ for $\|y\| \leqslant \eta$ and the later choice of u_n, the first integral is

$$\left|\int_{\|y\|\leqslant\eta} f(x-y)\left[\frac{1}{\|y\|} - u_n(\|y\|^2)\right]\mathrm{d}y\right| \leqslant C_\infty \int_{\|y\|\leqslant\eta} \frac{\mathrm{d}y}{\|y\|} = C_\infty\,\mu(\Omega)\int_0^\eta t^{d-2}\mathrm{d}t,$$

where $\mu(\Omega) = 2\pi^{d/2}/\Gamma(\frac{d}{2})$ is the measure of the surface of the d-dimensional unit sphere. Hence, $I_\eta \leqslant 2C_\infty\,\pi^{d/2}\frac{\eta^{d-1}}{d-1}/\Gamma(\frac{d}{2})$.

The approximation error $E_n := E_n(x^{-1/2}, [1,\infty))$ on the interval $[1,\infty)$ is shown in Table 2. A simple scaling argument yields $E_n(x^{-1/2}, [\eta^2,\infty)) = E_n/\eta$. For $\|y\| \geqslant \eta$, the estimate $1/\|y\| - u_n(\|y\|^2) \leqslant E_n/\eta$ holds for the optimal u_n on $[\eta^2,\infty)$ and leads to $I_\infty \leqslant C_1 E_n/\eta$.

The sum $2C_\infty\,\pi^{\frac{d}{2}}\frac{\eta^{d-1}}{d-1}/\Gamma(\frac{d}{2}) + \frac{C_1 E_n}{\eta}$ is minimal for the choice $\eta^d = \frac{C_1 E_n \Gamma(\frac{d}{2})}{2C_\infty\,\pi^{d/2}}$ and yields the bound stated above. □

3.2 Completely Monotone Functions

Here we will consider best approximations in the sense of Chebyshev, i.e., the error is to be minimised with respect to the (non-weighted) supremum norm $\|\cdot\|_{[a,b]}$ on a given interval $[a,b] \subset (0,\infty)$. A unique best exponential sum of order n exists for functions with the following property, and here also error bounds are available describing the asymptotic behaviour very well.

Definition 3.1 A function $f : (0,\infty) \to \mathbb{R}$ is said to be completely monotone (or a Laplace–Stieltjes function) if

$$f(x) = \int_0^\infty \mathrm{e}^{-tx}\mathrm{d}\mu(t)$$

holds for some nonnegative measure $\mathrm{d}\mu$ and the integral is absolutely convergent.

The negative powers are Laplace–Stieltjes functions due to

$$x^{-\alpha} = \frac{1}{\Gamma(\alpha)}\int_0^\infty t^{\alpha-1}\mathrm{e}^{-tx}\mathrm{d}t \qquad \text{for } \alpha > 0. \tag{18}$$

Theorem 3.1 *Given a Laplace–Stieltjes function f and $2n$ points $0 < x_1 < x_2 < \dots < x_{2n}$. Then there exists an exponential sum*

$$v(x) = \sum_{\nu=1}^{n} \alpha_\nu \mathrm{e}^{-t_\nu x}, \quad \alpha_\nu > 0, \quad 0 < t_1 < t_2 < \dots < t_n,$$

which satisfies the interpolation conditions

$$v(x_k) = f(x_k) \quad \textit{for } k = 1, 2, \dots, 2n.$$

If, in addition, f is continuous at $x = 0$, also $v(0) < f(0)$ holds.

A proof can be found in [8].

3.3 Approximation Error

Results on the weighted approximation in $R_{n,n-1}$ for the squared interval $[a^2, b^2]$ are useful in the study of the approximation by exponential sums in the interval $[a, b]$.

Theorem 3.2 *Let f be a Laplace–Stieltjes function with finite value $f(0)$ and $0 < a < b$.. Then there exists an exponential sum v_n of order n such that*

$$|f(x) - v_n(x)| \leqslant 2\sigma f(0) \quad \text{for } x \in [a, b],$$

where

$$\left(\frac{1+\sigma}{1-\sigma}\right)^2 = \frac{1 + \tilde{E}_{n,n-1}([a^2, b^2])}{1 - \tilde{E}_{n,n-1}([a^2, b^2])}.$$

Proof. Let $u_{n,n-1}$ be the best relative approximation on the squared interval $[a^2, b^2]$ in $R_{n,n-1}$. The error function $u_{n,n-1} - \sqrt{x}$ as well as the Blaschke product $r_{2n} \in R_{2n,2n}$ in (16) have $2n$ zeros in $[a^2, b^2]$. According to (16), r_{2n} has a maximum of 1 on the boundary of the right half-plane. Let $v_n \in \mathfrak{E}_n$ be the interpolant of f at the $2n$ zeros of P_{2n}. We know from Theorem 3.1 and the behaviour of Laplace–Stieltjes functions that

$$|f(z)| \leqslant f(0), \quad |v_n(z)| \leqslant v_n(0) \leqslant f(0) \quad \text{for } \mathrm{Re}\,\mathfrak{e}(z) \geqslant 0.$$

Hence,

$$\left|\frac{f(z) - v_n(z)}{r_{2n}}\right| \leqslant 2f(0) \quad \text{at the boundary of the right half-plane.}$$

The given function f, the approximating exponential sum, and the quotient $(f - v_n)/P_{2n}$ are analytic functions in the right half-plane of $\mathbb{C}$. The maximum principle in complex analysis completes the proof:

$$|f(z) - v_n(z)| \leqslant 2f(0)\,|r_{2n}| \leqslant 2f(0)\,\sigma \quad \text{for } z \in [a, b]. \qquad \square$$

We define the error of the best approximation on the interval $[a, b]$ by

$$E_n(f, [a, b]) := \min_{v_n \in \mathfrak{E}_n} \|f - v_n\|_{[a,b]}\,.$$

Theorem 3.2 provides the result

$$E_n(f, [a, b]) \leqslant 2\ f(0)\,\tilde{E}_{n,n-1}(\sqrt{x}, [a^2, b^2])$$

and compares therefore the approximation in $[a, b]$ with the rational approximation of the square root function on the larger interval $[a^2, b^2]$. We emphasise that Theorem 3.2 can be applied to any $n \in \mathbb{N}$, it is not an asymptotic law. Combining Theorems 2.1 and 3.2, we get

Corollary 3.1 *Let f be a Laplace–Stieltjes function. Then there exists an exponential sum $v_n \in \mathfrak{E}_n$ such that*

$$E_n(f, [a, b]) \leqslant 8f(0)\,\exp\left[-2n\,\mu\left(\tfrac{b-a}{b+a}\right)\right] \qquad \text{for } n \in \mathbb{N}.$$

f	$1/x$		$1/\sqrt{x}$	
n	$E_n(f, [1, 10])$	$\dfrac{E_n}{\sqrt{n} \cdot 14.54611^{-n}}$	$E_n(f, [1, 10])$	$\dfrac{E_n}{14.54611^{-n}}$
1	$8.556 \cdot 10^{-2}$	1.2446	$9.057 \cdot 10^{-2}$	1.3174
2	$8.752 \cdot 10^{-3}$	1.3094	$6.096 \cdot 10^{-3}$	1.2898
3	$7.145 \cdot 10^{-4}$	1.2696	$4.165 \cdot 10^{-4}$	1.2819
4	$5.577 \cdot 10^{-5}$	1.2484	$2.856 \cdot 10^{-5}$	1.2786
5	$4.243 \cdot 10^{-6}$	1.2357	$1.960 \cdot 10^{-6}$	1.2764
6	$3.173 \cdot 10^{-7}$	1.2271	$1.346 \cdot 10^{-7}$	1.2750
7	$2.344 \cdot 10^{-8}$	1.2208	$9.251 \cdot 10^{-9}$	1.2747
8	$1.716 \cdot 10^{-9}$	1.2160	$6.356 \cdot 10^{-10}$	1.2740

Table 1: Numerical results for $1/x$ (left) and $1/\sqrt{x}$ (right) on the interval $[1, 10]$

Example 3.1 In Table 1 we consider the approximation of the Laplace–Stieltjes functions $f(x) = 1/x$ and $1/\sqrt{x}$ by exponential sums on the interval $[1, 10]$. According to Theorem 3.2 we compare the approximation errors with $\tilde{E}_{n,n-1}([1, 100])$ for the interval with the squared ratio b/a. The corresponding rate 14.54611 is obtained by the scheme (13):

$$\boxed{\begin{array}{l} \lambda_0 = \frac{11}{9} \mapsto 3.705435 \mapsto 52.90208 \mapsto 11192.52 \mapsto 5.010902 \cdot 10^8 \\ \hfill \downarrow \times 4 \\ \qquad 14.54611 \leftarrow 211.5894 \leftarrow 44770.09 \leftarrow 2.004361 \cdot 10^9 \end{array}}\ .$$

Since the given functions are singular at $x = 0$, an adaptation of Theorem 3.2 is required. The original problem is equivalent to the approximation of $g(x) := f(x+1/n)$ on the interval $[1-1/n, 10-1/n]$. Since the function $\mu(\cdot)$ is differentiable, we may take the shift of the interval into account by a factor of $1 + c/n$ with some constant c. We note that $(1+c/n)^n \leqslant e^c$. Since the shifted function satisfies $f(0) = n$, we obtain the upper bound

$$E_n(1/x, [1, 10]) \leqslant c\, n\, \omega^{-n} \quad \text{where } \omega = 14.54611.$$

The numerical results in Table 1 show the better behaviour

$$E_n(1/x, [1, 10]) \approx c\sqrt{n}\, \omega^{-n} \quad \text{and} \quad E_n(1/\sqrt{x}, [1, 10]) \approx c\, \omega^{-n}$$

with constants c about 1.2. The important observation is that the rate $\omega = 14.54611$ obtained from the rational approximation errors $\tilde{E}_{n,n-1}([1, 100])$ is also the rate of the exponential approximation, but now in the interval $[1, 10]$.

3.4 Infinite Intervals

We consider the function $f(x) = x^{-\alpha}$, $0 < \alpha \leqslant 1$, and its best approximation u_n in $[1, r]$. Since $x^{-\alpha} \longrightarrow 0$ and $u_n \longrightarrow 0$ for $x \longrightarrow \infty$, it follows that there is a point $r > 1$ such that $|x^{-\alpha} - u_n| \leqslant \|x^{-\alpha} - u_n\|_\infty$ for all $x \geqslant r$. If we choose r as the smallest number with this property, we have $r^{-\alpha} - u_n(r) = \|x^{-\alpha} - u_n\|_\infty$. There exists an alternant of length $2n + 1$ in the subinterval $[1, r]$ and $E_n(x^{-\alpha}, [1, r]) = E_n(x^{-\alpha}, [1, \infty))$. Moreover, $x^{-\alpha} - u_n(x)$ has already $2n$ zeros between the points of the alternant and thus no zero for $x > r$. Hence, $x^{-\alpha} - u_n(x) < x^{-\alpha} < r^{-\alpha}$ holds for $x > r$.

Figure 2 shows the best approximation of $1/x$ in the interval $[1, \infty)$. Here, the critical value is $r^* \approx 146.8$.

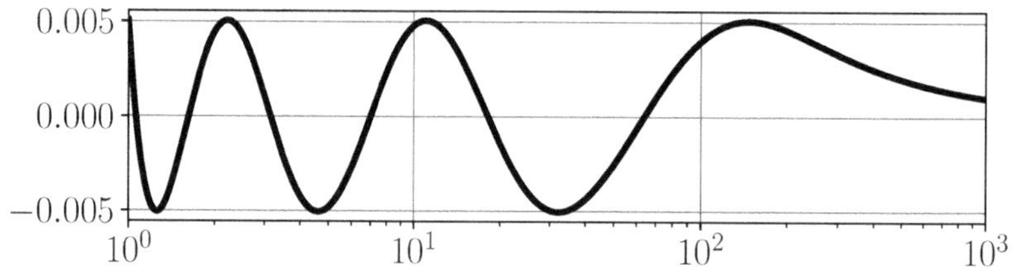

Fig. 2: Error function for the approximation of $1/x$ and $n = 3$ on $[1, \infty)$

For a given r we apply a shift of $1/2$ and consider the function $g(x) := (\frac{1}{2} + x)^{-\alpha}$ on the interval $[1/2, r - 1/2]$ instead of $f(x) = x^{-\alpha}$ on $[1, r]$. We use (14) with $b/a = 2r - 1$ to obtain

$$E_n(x^{-\alpha}, [1, r]) = E_n(g, [1/2, r - 1/2]) \leqslant 2 \cdot 2^\alpha 4 \exp\left[\frac{-n\pi^2}{\log(8r - 4)}\right].$$

Setting $8r - 4 = Ce^{\pi\sqrt{n/\alpha}}$, we have

$$E_n(x^{-\alpha}, [1, r]) \leqslant 8 \cdot (2C)^\alpha \, \mathrm{e}^{-\pi\sqrt{\alpha n}} \, (1 + o(1)).$$

The approximation error $f - v_n$ has no zero for $x > r$. Hence we conclude that $f(x) - v(x) < f(x) < f(r) = r^{-\alpha} < (r - 1/2)^{-\alpha} = (8/C)^\alpha \mathrm{e}^{-\pi\sqrt{\alpha n}}$ for $x > r$. Choosing $C = 2^{1-3/(2\alpha)}$, both parts can be estimated by

$$E_n(x^{-\alpha}, [1, \infty)) \leqslant 2^{2\alpha+3/2} \exp\left[-\pi\sqrt{\alpha n}\right] (1 + o(1)). \tag{19}$$

The numerical results in Table 2 show that the factors in the exponentially decreasing terms differ from those of our theoretical bound in (19) by a factor of $\sqrt{2}$. The values are based on the numerically computed best approximations v_n for $1 \leqslant n \leqslant 50$. An extensive list of best approximations and their coefficients can be found on the web [20].

The behaviour of the error function for the approximation of $1/x$ and large x is elucidated in Figure 2. Denote the largest point of the alternant of the best

f	$1/x$		$1/\sqrt{x}$	
n	E_n	$E_n \frac{e^{\pi\sqrt{2n}}}{\log(2+n)}$	E_n	$E_n\, e^{\pi\sqrt{n}}/4$
1	$8.55641 \cdot 10^{-2}$	6.62	$1.39861 \cdot 10^{-1}$	0.8091
2	$1.78498 \cdot 10^{-2}$	6.89	$4.08681 \cdot 10^{-2}$	0.8687
3	$5.05162 \cdot 10^{-3}$	6.90	$1.55559 \cdot 10^{-2}$	0.8974
5	$6.42813 \cdot 10^{-4}$	6.82	$3.29743 \cdot 10^{-3}$	0.9267
10	$1.31219 \cdot 10^{-5}$	6.67	$1.85217 \cdot 10^{-4}$	0.9553
15	$6.31072 \cdot 10^{-7}$	6.62	$2.01096 \cdot 10^{-5}$	0.9673
20	$4.79366 \cdot 10^{-8}$	6.60	$3.08291 \cdot 10^{-6}$	0.9740
30	$6.18824 \cdot 10^{-10}$	6.61	$1.32130 \cdot 10^{-7}$	0.9815
40	$1.55388 \cdot 10^{-11}$	6.64	$9.26374 \cdot 10^{-9}$	0.9858
50	$5.99210 \cdot 10^{-13}$	6.68	$8.90145 \cdot 10^{-10}$	0.9885

Table 2: Numerical results for $1/x$ (left) and $1/\sqrt{x}$ (right) on $[1, \infty)$

approximation in $[1, \infty)$ by r_n. It is the smallest value such that

$$E_n(x^{-1}, [1, r_n]) = E_n(x^{-1}, [1, \infty)).$$

The value r chosen in the proof of (19) yields $rE_n(x^{-1}, [1, \infty)) \leqslant 1 + o(1)$, while the numerical results from the tables in [20] show that

$$0.741 \leqslant r_n \cdot E_n(x^{-1}, [1, \infty)) \leqslant 0.742.$$

3.5 Application to the Møller–Plesset Approach

A typical application in quantum chemistry is the Møller–Plesset approach (cf. [1, 32]). The critical expression is a quotient

$$\frac{\text{numerator}}{\sum_j \varepsilon_j}$$

containing a sum of energies ε_j in the denominator. The division by $\sum_j \varepsilon_j$ couples all variables and leads to a high-dimensional problem of high complexity. As in the case of (17) the approximation of $1/x$ by an exponential sum leads to $1/\sum_j \varepsilon_j \approx \prod_j \exp(-t_\nu \varepsilon_j)$ and allows a decoupling. As a consequence the computational complexity is reduced.

The approximation is needed for a *finite* interval $[E_{\min}, E_{\max}]$. In the case of benzene the ratio is $E_{\max}/E_{\min} \approx 278$ (for more details see [11, 5.2]). If $n \leqslant 3$, the approximation on the interval $[1, 278]$ is the same as on the infinite interval $[1, \infty)$ (cf. Figure 2). The asymptotic rate for large n is given by ω^{-n} with $\omega = 4.084286$.

4 Approximation of Cauchy–Stieltjes Functions by Rational Functions

4.1 Cauchy–Stieltjes Functions

In the study of differential equations with nonlocal differential operators, the equations with fractional elliptic differential operators have recently attracted interest [14, 15]. The fractional operators are defined via the eigenvalues and eigenfunctions. Let $\{u_j\}_{j=1}^{\infty}$ be a complete set of eigenfunctions of the Laplacian,

$$-\Delta u_j = \lambda_j u_j, \quad j = 1, 2, 3, \dots$$

and $0 < \alpha < 1$, then the power α of the Laplace operator is defined by

$$(-\Delta)^{\alpha} u_j = \lambda_j^{\alpha} u_j, \quad j = 1, 2, 3, \dots .$$

The solution of a partial differential equation by a computation of all significant eigenfunctions, however, is too time-consuming for a numerical treatment. The rational approximation of the negative power

$$z^{-\alpha} \approx \sum_{\nu=1}^{n} \frac{c_\nu}{1 + t_\nu z} \in R_{n-1,n}$$

leads to an approximation $\tilde{u}$ of $u = (-\Delta)^{-\alpha} g$ defined by

$$\tilde{u} = \sum_{\nu=1}^{n} c_\nu v_\nu, \quad \text{where } (1 - t_\nu \Delta) v_\nu = g, \quad \nu = 1, 2, \dots, n.$$

Parabolic differential equations with a fractional time differential operator have been treated similarly by Khristenko–Wohlmuth [26], where rational approximations are applied to convolution integrals.

When we deal with rational functions, in general we have to take care of poles. Fortunately, the mentioned applications refer to a class of functions well behaving with respect to interpolation and approximation by the rational functions in $R_{n-1,n}$.

Definition 4.1 A function $f : (0, \infty) \to \mathbb{R}$ is said to be a Cauchy–Stieltjes function if

$$f(x) = \int_0^{\infty} \frac{\mathrm{d}\mu(t)}{1 + tx}$$

holds for some nonnegative measure $\mathrm{d}\mu$ and the integral is absolutely convergent.

Negative powers have this property due to the Balakrishnan formula [3],

$$x^{-\alpha} = \pi^{-1} \sin(\pi\alpha) \int_0^{\infty} \frac{t^{\alpha-1}}{1 + tx} \mathrm{d}t, \quad 0 < \alpha < 1. \tag{20}$$

4.2 Approximation by Rational Functions

As in the proof of Theorem 3.2 it follows that a Cauchy–Stieltjes function has an interpolant $p/q \in R_{n-1,n}$ satisfying the following estimate.

Remark 4.1 Let f be a Cauchy–Stieltjes function. Then there exists a rational function $v_{n-1,n} \in R_{n-1,n}$ such that

$$|f(x) - v_{n-1,n}(x)| \leqslant 2\,\tilde{E}_{n,n-1}([a^2, b^2])\, f(0) \qquad \text{for } x \in [a, b].$$

Therefore the approximation error $E_{n,n-1}(f, [a, b])$ of the best approximation in $R_{n-1,n}$ is bounded by

$$E_{n,n-1}(f, [a, b]) \leqslant 2\,\tilde{E}_{n,n-1}([a^2, b^2])\, f(0). \tag{21}$$

The error estimate (21) is the analogon of the error bound in Theorem 3.2 that turned out to be optimal for the approximation of Lebesgue–Stieltjes functions by exponential sums. Surprisingly, here the approach only provides a suboptimal exponential decay of the error. The correct exponential behaviour is given by

$$\tilde{E}_{n,n-1}([a, b])$$

and not by the approximation on the interval $[a^2, b^2]$.

We will prove it for the Cauchy–Stieltjes function $x^{-1/2}$. The extension to all Cauchy–Stieltjes functions will be based on the result by Ganelius [16] and Gončar [17] that the asymptotic behaviour is the same for all Cauchy–Stieltjes functions whose support of the defining measure is not degenerate. In addition, we confirm the statement by numerical computations.

Gončar's studies use results on so-called condensers in the framework of harmonic analysis (cf. [2]). Rational functions minimising (15) play there an important rule.

4.3 Best Approximations of $x^{-1/2}$

When the approximations of $x^{-\alpha}$ in the intervals $[a, b]$ and $[\lambda a, \lambda b]$, $\lambda > 0$, are compared, the uniform norm depends on the scaling factor λ,

$$E_{m,n}(x^{-\alpha}, [\lambda a, \lambda b]) = \lambda^{-\alpha} E_{m,n}(x^{-\alpha}, [a, b]). \tag{22}$$

Let $r(x)$ be the best approximation in $[a, b]$. Then the best approximation r_λ in $[\lambda a, \lambda b]$ is given by $r_\lambda(x) := \lambda^{-\alpha} r(x/\lambda)$.

Lemma 4.1 *Let $0 < a < b$. For sufficiently large $n \geqslant n_0(b/a)$ we have*

$$\frac{1}{2}\sqrt{a}\,\tilde{E}_{n,n-1}(\sqrt{x}, [a, b]) \leqslant E_{n-1,n}(x^{-1/2}, [a, b]) \leqslant 2\sqrt{b}\,\tilde{E}_{n,n-1}(\sqrt{x}, [a, b]). \tag{23}$$

Proof. By Remark 2.1 and (22) it is sufficient to prove (23) for the interval $I_R = [1/\sqrt{R}, \sqrt{R}]$ with $R := b/a$. Here $\|\cdot\|_\infty$ denotes the maximum norm on I_R.

By Theorem 2.1 the approximation error $\tilde{E}_{n,n-1}(\sqrt{x}, I_R)$ decreases exponentially. Therefore there exists a number $n_0 := n_0(R) = n_0(b/a)$ such that

$$\tilde{E}_{n,n-1}(\sqrt{x}, I_R) \leqslant \frac{1}{4\sqrt{R}} \leqslant \frac{1}{2} \text{ holds for all } n \geqslant n_0. \tag{24}$$

This condition is helpful for obtaining lower bounds of the approximating rational functions.

(a) Let $n \geqslant n_0$, and let $\tilde{u}_{n,n-1} \in R_{n,n-1}$ be the best$_\surd$ approximation of $x^{1/2}$ in I_R. We conclude from (24) that we have in I_R,

$$\tilde{u}_{n,n-1}(x) \geqslant \sqrt{x} - \sqrt{x}\frac{|\sqrt{x} - \tilde{u}_{n,n-1}|}{\sqrt{x}} \geqslant \sqrt{x}\left(1 - \tilde{E}_{n,n-1}(\sqrt{x}, I_R)\right) \geqslant \frac{1}{2}\sqrt{x}.$$

Since $1/u_{n,n-1} \in R_{n-1,n}$, it follows that

$$\begin{aligned} E_{n-1,n}(x^{-1/2}, I_R) \leqslant \|1/\tilde{u}_{n,n-1} - x^{-1/2}\|_\infty &= \left\|\frac{\tilde{u}_{n,n-1} - \sqrt{x}}{\tilde{u}_{n,n-1}\cdot\sqrt{x}}\right\|_\infty \\ &\leqslant 2\sqrt[4]{R}\,\tilde{E}_{n,n-1}(\sqrt{x}, I_R). \end{aligned} \tag{25}$$

(b) Let $n \geqslant n_0$, and let $u_{n-1,n} \in R_{n-1,n}$ be the best approximation of $x^{-1/2}$ in I_R. The conditions (24) and (25) imply

$$E_{n-1,n}(x^{-1/2}, I_R) \leqslant 2\sqrt[4]{R}/(4\sqrt{R}) = 1/(2\sqrt[4]{R}).$$

Together with $x^{-1/2} \geqslant 1/\sqrt[4]{R}$ in I_R we conclude that

$$u_{n-1,n}(x) \geqslant x^{-1/2} - \|x^{-1/2} - u_{n-1,n}\|_\infty \geqslant \frac{1}{\sqrt[4]{R}} - \frac{1}{2\sqrt[4]{R}} = \frac{1}{2\sqrt[4]{R}}.$$

Since $1/u_{n-1,n} \in R_{n,n-1}$, it follows that

$$\begin{aligned} \tilde{E}_{n,n-1}(\sqrt{x}, I_R) \leqslant \|\sqrt{x} - 1/u_{n-1,n}\|_\surd &= \left\|\frac{u_{n-1,n} - x^{-1/2}}{u_{n-1,n}}\right\|_\infty \\ &\leqslant 2\sqrt[4]{R}\left\|u_{n-1,n} - x^{-1/2}\right\|_\infty = 2\sqrt[4]{R}\,E_{n-1,n}(x^{-1/2}, I_R). \end{aligned}$$

(c) According to (22), the transformation of I_R to $[a, b]$ induces the scaling factor $\sqrt[4]{ab}$, and the proof of (23) is complete. □

Corollary 4.1 *If* $0 < a < b < \infty$, *then*

$$\lim_{n\to\infty} E_{n,n-1}(x^{-\alpha}, [a, b])^{1/n} = \exp\left[-2\mu\left(\tfrac{\sqrt{b}-\sqrt{a}}{\sqrt{b}+\sqrt{a}}\right)\right],$$

Numerical Observation. For $R > 1$, the approximation errors $\tilde{E}_{n,n-1}([1/\sqrt{R}, \sqrt{R}\,])$ and $E_{n-1,n}(x^{-1/2}, [1/\sqrt{R}, \sqrt{R}\,])$ seem to be asymptotically equal for increasing n, e.g.,

$$E_{4,5}(x^{-1/2}, [1/10, 10]) = \mathbf{6.142\,215\,473}\,182 \cdot 10^{-6},$$
$$\tilde{E}_{5,4}([1/10, 10]) = \mathbf{6.142\,215\,479}\,162 \cdot 10^{-6}.$$

Together with (22), the mentioned observation indicates fast convergence of

$$\lim_{n\to\infty} \frac{E_{n-1,n}(x^{-1/2}, [a,b])}{\tilde{E}_{n,n-1}([a,b])} = (ab)^{-1/4}.$$

The behaviour for general $0 < \alpha < 1$ seems to be

$$E_{n-1,n}(x^{-\alpha}, [a,b]) = (ab)^{-\alpha/2}\,(4 + o(1))\sin(\alpha\pi)\,\exp\left[-2n\,\mu\left(\tfrac{\sqrt{b}-\sqrt{a}}{\sqrt{b}+\sqrt{a}}\right)\right].$$

The inverse powers $x^{-\alpha}$, $0 < \alpha < 1$, are Stieltjes function with respect to the approximation by exponential sums and by rational functions. In both cases the degree of approximation decays exponentially. The numbers in Table 3 elucidate that it occurs with different rates.

approximation by								
exponential sums	$b/a =$	2	5	10	$10\sqrt{10}$	100	1000	10^4
rational functions	$b/a =$	4	25	100	1000	10^4	10^6	10^8
$\exp[\pi\mathbf{K}/\mathbf{K}']$		135.85	27.27	14.546	7.6845	5.1928	3.287	2.538

Table 3: Characteristic parameter for the exponential decay of the approximation error as a function of the size of the interval $[a,b]$ measured by b/a.

4.4 Infinite Intervals

The asymptotic behaviour of the approximation on the infinite interval $[1,\infty)$ may be directly obtained from Stahl's famous result [29, (7.34)]:

$$E_{n,n}(x^{\alpha}, [0,1]) = 4^{\alpha+1}\,|\sin\pi\alpha|\,\mathrm{e}^{-2\pi\sqrt{\alpha n}}\,(1 + o(1)) \quad \text{as } n \to \infty,\ \alpha > 0.$$

Replacing x by $1/x$, we obtain

$$E_{n,n}(x^{-\alpha}, [1,\infty)) = 4^{\alpha+1}\,|\sin\pi\alpha|\,\mathrm{e}^{-2\pi\sqrt{\alpha n}}\,(1 + o(1)) \quad \text{as } n \to \infty,\ \alpha > 0. \quad (26)$$

Here, the factor $\sqrt{n}$ in the exponent is typical. Note that for finite intervals $E_{n-1,n}(x^{-\alpha}, [a,b])$ decays like $O(\mathrm{e}^{-cn})$, $c > 0$, i.e., n is linear in the exponent.

Example 4.1 At a first glance, an interval with $b/a = 10^5$ may be considered as a very large interval. This impression, however, is not correct due to the fast convergence of the arithmetic-geometric process. A Landen iteration maps the interval with $b/a = 10^5$ onto an interval with $b/a \approx 159$. By (6) the approximation with $2n$ terms on the large interval has the same error as the approximation with n terms on the interval with $b/a \approx 159$. The exponential decay of the approximation error is the same for all $\alpha \in (0,1)$. There is a difference, of course, whether the approximation is done on the interval $[1, 10^5]$ or on $[10^{-5}, 1)$. The associated scaling factor is $10^{5\alpha}$. Moreover, a weak dependence on α due to the shift of the singularity may be observed.

Remark 4.2 A similar splitting technique as used in Sect. 3.4, based on the estimate in Lemma 4.1 leads to the suboptimal inequality

$$E_{n-1,n}(x^{-1/2}, [1,\infty)) \leqslant 2^{5/4} \cdot \exp\left(-\pi\sqrt{n/2}\right)\left(1 + O(1/\sqrt{n}\right).$$

The following table shows the approximation errors $E_{n-1,n}(x^{-1/2}, [1,\infty))$. r^* is the smallest value such that the best approximation in $[1, r^*]$ coincides with that in $[1,\infty)$.

n	r^*	$E_{n-1,n}(x^{-1/2}, [1,\infty))$	$\frac{E_{n-1,n}}{8e^{-\pi\sqrt{2n+0.349}}}$	$\sqrt{r^*} \cdot E_{n-1,n}$
1	$5.20115 \cdot 10^1$	$6.483485 \cdot 10^{-2}$	0.999 6	0.467 58
2	$1.64233 \cdot 10^3$	$1.142363 \cdot 10^{-2}$	1.000 0	0.462 95
3	$2.51305 \cdot 10^4$	$2.919186 \cdot 10^{-3}$	1.000 0	0.462 77
4	$2.56515 \cdot 10^5$	$9.136775 \cdot 10^{-4}$	1.000 0	0.462 75
5	$2.00836 \cdot 10^6$	$3.265330 \cdot 10^{-4}$	1.000 0	0.462 75
⋮	⋮	⋮	⋮	⋮
19	$2.64684 \cdot 10^{14}$	$2.844357 \cdot 10^{-8}$	1.000 0	0.462 75
20	$7.20727 \cdot 10^{14}$	$1.723702 \cdot 10^{-8}$	1.000 0	0.462 75
21	$1.91496 \cdot 10^{15}$	$1.057470 \cdot 10^{-8}$	1.000 0	0.462 75
22	$4.97325 \cdot 10^{15}$	$6.561871 \cdot 10^{-9}$	1.000 0	0.462 75

The table shows that $(r^*)^{-1/2}$ and $E_{n-1,n}(x^{-1/2}, [1,\infty))$ are asymptotically proportional. The decay of the approximation error is very accurately described by[2]

$$E_{n-1,n}(x^{-1/2}, [1,\infty)) \approx 8\,\exp(-\pi\sqrt{2n + 0.349}\,).$$

[2] Note that

$$8\exp(-\pi\sqrt{2n+0.349}) = 4^{\alpha+1}\,|\sin\pi\alpha|\,e^{-2\pi\sqrt{\alpha n}}\,(1+o(1))$$

for $\alpha = 1/2$ (cf. (26)). The approximation errors $E_{n,n}(x^{1/2}[0,1]) = E_{n,n}(x^{-1/2}, [1,\infty))$ are listed in Varga–Carpenter [31, Table 2.4].

Appendix A: Gauss' Arithmetic-Geometric Mean

When Rutishauser [28] pointed out that Heron's method can be improved by a rescaling, he referred to Zolotarev's problems [34], and so already to a connection with Gauss' arithmetic-geometric iteration.

A.1 Some Elliptic Integrals

At an early age, Gauss became enamoured in a sequential procedure that is now known as the arithmetic-geometric process; see e.g. [7, 16]. Given two numbers $0 < a_0 \leqslant b_0$, one successively takes the arithmetic mean and the geometric mean,

$$a_{j+1} = \sqrt{a_j b_j}, \qquad b_{j+1} = \frac{1}{2}(a_j + b_j). \tag{27}$$

Since $\{a_j\}$, $\{b_j\}$ stay bounded and $b_j/a_j \to 1$ (see Sect. A.2), there is a common limit

$$m(a_0, b_0) := \lim_{j\to\infty} a_j = \lim_{j\to\infty} b_j$$

called the *arithmetic-geometric mean* of a_0 and b_0 (cf. [7, §1.1]). As shown below, it can be expressed in terms of the complete elliptic integral

$$I(a,b) := \int_0^\infty \frac{\mathrm{d}t}{\sqrt{(a^2+t^2)(b^2+t^2)}}. \tag{28}$$

The crucial observation for establishing the relation between $m(a,b)$ and $I(a,b)$ is that $I(a,b)$ is invariant under the transformation $(a,b) \mapsto (a_1, b_1) = (\sqrt{ab}, \frac{a+b}{2})$. For a proof substitute $t = \frac{1}{2}(x - \frac{ab}{x})$. As x varies from 0 to ∞, the variable t increases from $-\infty$ to ∞. Moreover,

$$\mathrm{d}t = \frac{x^2+ab}{2x^2}\,\mathrm{d}x, \quad t^2 + \left(\frac{a+b}{2}\right)^2 = \frac{(x^2+a^2)(x^2+b^2)}{4x^2}, \quad t^2 + ab = \frac{(x^2+ab)}{4x^2}.$$

Hence,

$$I(a_1,b_1) = \frac{1}{2}\int_{-\infty}^\infty \frac{\mathrm{d}t}{\sqrt{(a_1^2+t^2)(b_1^2+t^2)}} = \int_0^\infty \frac{\mathrm{d}x}{\sqrt{(a^2+x^2)(b^2+x^2)}} = I(a,b)$$

yields the invariance. Let $m = m(a,b)$, and set $a_0 = a$, $b_0 = b$. By induction it follows that $I(a_0,b_0) = I(a_j,b_j)$ for all j, and by continuity $I(a_0,b_0) = I(m,m)$. Obviously, $I(m,m) = \int_0^\infty \frac{\mathrm{d}t}{m^2+t^2} = \frac{\pi}{2m}$, and we conclude that [16]

$$m(a,b) = \frac{\pi}{2I(a,b)}.$$

The involved *elliptic functions* K and K' are defined by $\mathsf{K}'(\kappa) := I(\kappa, 1)$ and

$$\mathsf{K}'(\kappa) = \mathsf{K}(\kappa') \qquad \text{for } \kappa^2 + \kappa'^2 = 1.$$

The Grötzsch ring function $\mu(\kappa) = \frac{\pi}{2}\mathbf{K}'(\kappa)/\mathbf{K}(\kappa)$ satisfies $\mu(\kappa) \leqslant \log(\frac{4}{\kappa})$ (cf. [12] or [7, pp. 8-9]), hence

$$\exp(\mu(\kappa)) \leqslant \frac{4}{\kappa}. \tag{29}$$

A.2 The Landen Iteration

The arithmetic-geometric mean $m(a, b)$ is usually not determined via the integral (28). On the contrary, elliptic integrals are computed via the arithmetic-geometric iteration, since its convergence is extremely fast. Set $\rho_j := b_j/a_j$. The mapping $\rho \mapsto \frac{1}{2}(\sqrt{\rho} + 1/\sqrt{\rho})$ is called *Landen iteration* and appears in

$$\rho_{j+1} = \frac{a_j + b_j}{2\sqrt{a_j b_j}} = \frac{1+\rho_j}{2\sqrt{\rho_j}} = \frac{1}{2}\left(\sqrt{\rho_j} + \frac{1}{\sqrt{\rho_j}}\right). \tag{30}$$

n	λ_n
−4	$6.825 \cdot 10^{14}$
−3	$1.306 \cdot 10^{7}$
−2	1807.087
−1	21.26669
0	2.414214
1	1.098684
2	1.001107
3	$1 + 1.531 \cdot 10^{-7}$
4	$1 + 2.930 \cdot 10^{-13}$

Table 4: Landen transformation with $\lambda_0 = 1 + \sqrt{2}$

We are interested in the *inverse Landen iteration* $\rho_{j+1} \mapsto \rho_j$ defining ρ_j for $j = 0, -1, -2, \ldots$. We set $\lambda_j := \rho_{-j}$. The sequence $\{\lambda_j : j \in \mathbb{N}_0\}$ is defined by (9):

$$\lambda_{j+1} = \left(\lambda_j + \sqrt{\lambda_j^2 - 1}\right)^2 \qquad (j \in \mathbb{N}_0).$$

The (forward) Landen sequence $\{\rho_j\}$ converges quadratically to $\lim_{j\to\infty}\rho_j = 1$, whereas the sequence $\{\lambda_j\}$ of the inverse iteration starting with $\lambda_0 > 1$ diverges quadratically. The very fast convergence/divergence is illustrated in Table 4. Taking the 2^j-th root of λ_j, we obtain a convergent sequence as shown in the next lemma.

Lemma 4.2 *Define $\{\lambda_j\}$ by the inverse Landen iteration starting with $\lambda_0 = 1/\kappa_0 > 1$. Then the limit*

$$\lim_{j\to\infty} (4\lambda_j)^{1/2^j} = \omega := \exp[\mu(\kappa_0)]$$

exists and satisfies

$$4\lambda_j\,(1 - o(1)) \leqslant \omega^{2^j} \leqslant 4\lambda_j \quad \text{as } j \to \infty.$$

Concerning the proof we refer to [12]. Therein also the following sharper estimates are shown.

Theorem 4.1 *Let λ_j, $\kappa_0 = 1/\lambda_0$, and ω as in Lemma 4.2. Then*

$$\frac{1}{4}\omega^{2^j} \leqslant \lambda_j \leqslant \frac{1}{4}\omega^{2^j} + 4\,\omega^{2^{-j}}$$

holds. Moreover, the asymptotic behaviour is

$$\lambda_j = \frac{1}{4}\omega^{2^j} + \omega^{-2^j} + O(\omega^{-3\cdot 2^j}) \quad \text{as } j \to \infty. \tag{31}$$

Equation (31) explains the accurate values obtained by the scheme (13).

Appendix B: Approximation by Sinc Quadrature

An alternative technique for producing approximations on semi-infinite intervals is the use of the sinc quadrature.

Assume that a function $\varphi(x)$ is represented by an integral $\int_I f(t;x)\mathrm{d}t$, where for fixed t the integrand $f(t;x)$ is a rational function of x or, respectively, exponential in x. Then any standard quadrature method $Q_I(f(\cdot;x)) = \sum_k \omega_k f(t_k;x)$ with weights ω_k and quadrature points t_k produces a rational function or, respectively, an exponential sum approximating $\varphi(x)$. The supremum norm of $\varphi(x) - Q_I(f(\cdot;x))$ in an interval J requires a uniform estimate of the quadrature error for all parameters $x \in J$.

Since the standard sinc quadrature applies to integrals over $(-\infty, \infty)$, we need a substitution $t = \psi(z)$ mapping $(-\infty, \infty)$ onto I to obtain $\int_I f(t;x)\mathrm{d}t = \int_{-\infty}^{\infty} f(\psi(z);x)\psi'(z)\mathrm{d}z$. The exact integration of the *sinc interpolation* of the integrand $F(z;x) := f(\psi(z);x)\psi'(z)$ with step size $h > 0$ yields the result (cf. [30], [23, p. 466])

$$T(F,h)(x) := h \sum_{k=-\infty}^{\infty} F(kh;x).$$

Next, the infinite summation is replaced by the finite sum

$$T_{M,N}(F,h)(x) := h \sum_{k=-M}^{N} F(kh;x).$$

The number of terms is

$$n = N + M + 1. \tag{32}$$

The approximation error $\eta(x) := \varphi(x) - T_{M,N}(F,h)(x)$ is split into three parts:

$$\eta_0(x) := \varphi(x) - T(f,h)(x), \quad \eta_1(x) := h \sum_{k=-\infty}^{-M-1} F(kh;x), \quad \eta_2(x) := h \sum_{k=N+1}^{\infty} F(kh;x).$$

The parameters h, N, M are chosen such that η_0, η_1, η_2 are equilibrated (in the following this means that all terms have the same exponential behaviour).

The error term $\eta_0(x)$ depends on the analytic properties of $F(\cdot, x)$ (for fixed x) in the strip $\mathfrak{D}_d = \{\zeta \in \mathbb{C} : |\mathrm{Im}\,\mathfrak{m}(\zeta)| < d\}$. In addition, d must the chosen so that the boundary integral

$$\|F\|_{\mathfrak{D}_d} := \int_{\partial\mathfrak{D}_d} |F(\zeta;x)|\,\mathrm{d}\zeta = \int_{-\infty}^{\infty} |F(\xi + \mathrm{i}d;x)|\,\mathrm{d}\xi + \int_{-\infty}^{\infty} |F(\xi - \mathrm{i}d;x)|\,\mathrm{d}\xi$$

is finite. Then the following estimate holds (cf. [30, pp. 144f], [23, p. 466]):

$$|\varphi(x) - T(F,h)(x)| \leqslant \exp(-2\pi d/h)\ \|F\|_{\mathfrak{D}_d}\,. \tag{33}$$

η_1 and η_2 depend on the decay of $F(\xi;x)$ for real $\xi \to -\infty$ and $\xi \to \infty$, respectively.

B.1 Rational approximation of $x^{-\alpha}$

A candidate for the integral $\int_I f(t;z)\mathrm{d}t$ is Balakrishnan's formula (20) with the interval $I = [0,\infty)$ and $f(t;x) = \frac{\sin(\pi\alpha)}{\pi}\frac{t^{\alpha-1}}{1+tx}$. The substitution $t = \mathrm{e}^{\xi}$ yields[3]

$$x^{-\alpha} = \frac{\sin(\pi\alpha)}{\pi}\int_{-\infty}^{\infty} F(\xi;x)\mathrm{d}\xi \qquad \text{with } F(\xi;x) := \frac{\exp((\alpha-1)\xi)}{\exp(-\xi)+x}. \tag{34}$$

Since $F(\xi;x)$ is singular at $\xi = -\log(x) - \mathrm{i}\pi$, the parameter d is bounded by $d \leqslant \pi$. As $\|f\|_{\mathfrak{D}_\pi} = \infty$, but $\|f\|_{\mathfrak{D}_d} < \infty$ for $d < \pi$, we choose $d := \pi - \varepsilon$ for some $\varepsilon > 0$ so that

$$\eta_0 \leqslant \exp(-2\pi\,(\pi-\varepsilon)\,/h)\;\|f\|_{\mathfrak{D}_{\pi-\varepsilon}}.$$

Equilibrating the rates of η_0 with

$$\eta_1(x) \leqslant \eta_1(1) < \int_{-\infty}^{-Mh} F(\xi;1)\mathrm{d}\xi < \frac{\sin(\pi\alpha)}{\pi}\int_0^{\mathrm{e}^{-Mh}} t^{\alpha-1}\mathrm{d}t = \frac{\sin(\pi\alpha)}{\pi}\frac{\mathrm{e}^{-Mh\alpha}}{\alpha},$$

$$\eta_2(x) \leqslant \eta_2(1) < \int_{Nh}^{\infty} F(\xi;1)\mathrm{d}\xi < \frac{\sin(\pi\alpha)}{\pi}\int_{\mathrm{e}^{Nh}}^{\infty} t^{\alpha-2}\mathrm{d}t = \frac{\sin(\pi\alpha)}{\pi}\frac{\mathrm{e}^{(\alpha-1)Nh}}{1-\alpha},$$

we obtain $(1-\alpha)\,N = \alpha M = 2\pi^2h^{-2}$.

The number of terms and the degree of p_n is $n = N + M + 1$ (cf. (32)). Using

$$n = M + N + 1 = \left(\frac{1}{1-\alpha} + \frac{1}{\alpha}\right)2\pi^2h^{-2} + 1 \approx \left(\frac{1}{\alpha} + \frac{1}{1-\alpha}\right)2\pi^2h^{-2} = \frac{2\pi^2h^{-2}}{\alpha\,(1-\alpha)},$$

we conclude that the asymptotic behaviour of the step size is

$$h = \pi\sqrt{\tfrac{2}{n\alpha(1-\alpha)}}.$$

Given h, M should be an integer close to $2\pi^2h^{-2}/\alpha$, similarly for N. Neglecting ε, these values show that the total error is about

$$O\left(\exp(-\sqrt{2n\alpha\,(1-\alpha)}\,\pi)\right).$$

In the particular case of $\alpha = \frac{1}{2}$ we have $N = M$. A careful analysis yields the following result.

Lemma 4.3 *For $\alpha = 1/2$ the sinc quadrature $T_N(F,h)$ with F from (34) and $h = 2\pi/\sqrt{N}$ is a rational function in $R_{2N,2N+1}$ with the error estimate*

$$\left|z^{-1/2} - T_N(F,h)\right| \leqslant \exp\left(-(\pi-\varepsilon)\sqrt{N}\right)\|F\|_{\mathfrak{D}_{\pi-\varepsilon}} + \frac{4}{\pi}\exp\left(-\pi\sqrt{N}\right) \qquad (\varepsilon > 0).$$

[3] Sinc quadratures are e.g. applied in [4], [6], and [26].

The right-hand side is almost $O(\exp(-\pi\sqrt{N})$. This behaviour is supported by the following numerical test.

N	$\left\|x^{-1/2}-T_N(F,h)(x)\right\|_{[1,\infty)}$	$e^{-\pi\sqrt{N}}$	$\left\|x^{-1/2}-T_N(F,h)(x)\right\|_{[1,\infty)}/e^{-\pi\sqrt{N}}$
10	1.5783×10^{-4}	4.8469×10^{-5}	3.26
20	2.4704×10^{-6}	7.9127×10^{-7}	3.12
30	1.0288×10^{-7}	3.3651×10^{-8}	3.06
40	7.0874×10^{-9}	2.3492×10^{-9}	3.02

A comparison with the table in Subsection 4.4 shows that the best approximation of degree $n = 22$ yields a better result than the sinc approach with $2N + 1 = 81$ terms.

B.2 Approximation by Exponential Sums

Starting from the representation of $x^{-\alpha}$, $0 < \alpha < 1$, by the integral (18), we apply the substitution $t = e^{\xi}$ and obtain

$$x^{-\alpha} = \frac{1}{\Gamma(\alpha)} \int_{-\infty}^{\infty} \exp\left(\alpha\xi - xe^{\xi}\right) d\xi.$$

The errors η_0, η_1, η_2 are defined as in Sect. B.1. The integrand decays doubly exponentially as $\xi \to +\infty$, while the behaviour for $\xi \to -\infty$ is like $e^{-\alpha|\xi|}$. Again we use the sinc approximation

$$T_{M,N}(f,h)(x) := h \sum_{k=-M}^{N} f(kh;x) \qquad \text{for } f(\xi,x) = \frac{1}{\Gamma(\alpha)} \exp\left(\alpha\xi - xe^{\xi}\right)$$

(cf. [33, (1.5)][4]) with appropriate M and N. Since $\eta_0 \leqslant e^{-\pi(\pi-\varepsilon)/h} \|f\|_{\mathfrak{D}_{(\pi-\varepsilon)/2}}$ (for all $\varepsilon > 0$), we choose the step size $h = \pi/\sqrt{\alpha M}$ and obtain the estimate $\eta_1 < \frac{1}{\alpha\Gamma(\alpha)} e^{-\alpha M h} = \frac{1}{\alpha\Gamma(\alpha)} e^{-\pi\sqrt{\alpha M}}$. Choosing $N := \left\lceil \log(\pi\sqrt{M\alpha})\frac{\sqrt{M\alpha}}{\pi} \right\rceil$, a careful analysis yields $\eta_2 \leqslant c\,(M\alpha)^{\frac{\alpha-1}{2}} \exp(-\pi\sqrt{M\alpha})$ with $c \approx \pi^{1+\alpha}$.

$T_{M,N}(f,h)(x)$ is a sum of $n = M + N + 1$ exponential terms. Expressing M as a function of n, we obtain

$$\exp(-\pi\sqrt{M\alpha}) \lessapprox \exp\left(-\pi\sqrt{n\alpha}\right) \cdot \left(\pi\sqrt{n\alpha}\right)^{\alpha/2} \lessapprox \exp\left(-(\pi-\varepsilon)\sqrt{n\alpha}\right)$$

and

$$\left|x^{-1/2} - T_{M,N}(f,h)(z)\right| \lessapprox O\left(\exp\left(-(\pi-\varepsilon)\sqrt{n\alpha}\right)\right).$$

[4] In [33] the *relative* approximation error should be small. Best approximations for this task are provided in [20].

The following numerical results correspond to $\alpha = \frac{1}{2}$, $n = N + M + 1$, the supremum norm in $[1, \infty)$, and the choice $\pi - \varepsilon = 3.1$:

n	M	N	$\|\cdot^{-1/2} - T_{M,N}(f,h)\|$	$e^{-(\pi-\varepsilon)\sqrt{n/2}}$	$\|\cdot^{-1/2} - T_{M,N}(f,h)\| \times e^{(\pi-\varepsilon)\sqrt{n/2}}$
20	17	2	2.6×10^{-4}	1.05×10^{-5}	4.7
50	45	4	7.5×10^{-7}	1.86×10^{-7}	4.0
100	92	7	2.1×10^{-9}	3.02×10^{-10}	7.0

Table 2 shows that the best approximation of degree $n = 50$ yields a better result than the sinc method with $n = 100$. In fact, even $n = 47$ is better (see [20]).

Eventually we summarise the exponential behaviour of the error estimates and of the observed numerical results in the case of the approximation of $x^{-\alpha}$ in the infinite interval $[1, \infty)$. Here we use the abbreviations ES=exponential sums, RF=rational functions, BA=best approximation, sinc=sinc method:

method	error estimate		observed rate
ES / BA	(19):	$\exp[-\pi\sqrt{\alpha n}\,]$	$\exp[-\pi\sqrt{2\alpha n}\,]$
ES / sinc	Sect. B.2:	$\exp\left[-(\pi-\varepsilon)\sqrt{\alpha n}\,\right]$	$\exp\left[-(\pi-\varepsilon)\sqrt{\alpha n}\,\right]$ for $\alpha=\frac{1}{2}$
RF / BA	Stahl, (26):	$\exp[-2\pi\sqrt{\alpha n}\,]$	$\exp[-2\pi\sqrt{\alpha n}\,]$
"	Remark 4.2:	$\exp[-\pi\sqrt{\alpha n}\,]$ for $\alpha = \frac{1}{2}$	"
RF / sinc	Sect. B.1:	$\exp[-(\pi-\varepsilon)\sqrt{2n\,\alpha(1-\alpha)}]$	$\exp[-\pi\sqrt{n}\,]$ for $\alpha = \frac{1}{2}$

Appendix C: Computation of Best Rational Approximations

To exploit the excellent approximation properties of rational function, one must be able to compute the best approximations. In the following we consider best approximations in $R_{n-1,n}$. An appropriate representation of $r \in R_{n-1,n}$ is

$$r(x) = \sum_{j=1}^{n} \frac{a_\nu}{x + b_\nu}. \tag{35}$$

In the following, f denotes a Cauchy–Stieltjes function. It is known that the best approximation of f satisfies $a_\nu, b_\nu > 0$. Therefore, the evaluation of (35) at $x > 0$ is stable. Note that $\{-b_\nu\}$ are the poles of r.

Given points $0 < x_1 < \ldots < x_{2n}$ and values $y_\nu = f(x_\nu)$ $(1 \leqslant \nu \leqslant 2n)$, the uniquely existing interpolating function $r \in R_{n-1,n}$ with $r(x_\nu) = y_\nu$ can be determined by Newton's method (together with a damping strategy), provided that appropriate starting values are available. The required functional matrix is explicitly computable.

Let $[a, b]$ be the interval in which f is to be approximated. The iterative solution of the best approximation problem requires an *admissible* function $r_0 \in R_{n-1,n}$ for the start, i.e., there must be $2n$ zeros ξ_μ of the error $f - r_0$:

$$0 < a < \xi_1 < \xi_2 < \cdots < \xi_{2n} < b.$$

For instance, one may choose some ξ_μ and solve the interpolation problem $r_0(\xi_\mu) = f(\xi_\mu)$. As a result the error $\varepsilon := f - r_0$ is oscillating:

$$\varepsilon(x) > 0 \text{ in } [a, \xi_1) \cup (\xi_2, \xi_3) \cup \cdots \cup (\xi_{2n}, b], \quad \varepsilon(x) < 0 \text{ in } (\xi_1, \xi_2) \cup (\xi_3, \xi_4) \cup \cdots .$$

The corresponding alternant $(\alpha_0, \alpha_1, \ldots, \alpha_{2n})$ is defined by: $\varepsilon(\alpha_0) = \max\{\varepsilon(x) : x \in [a, \xi_1)\}$, $\varepsilon(\alpha_1) = \min\{\varepsilon(x) : x \in (\xi_1, \xi_2)\}, \ldots$ An efficient method for computing the alternant is as follows. For interior intervals (ξ_i, ξ_{i+1}) determine the extremal value on a discrete grid $\{\xi_i + jh : 1 \leqslant j \leqslant m - 1\}$ for $h := (\xi_{i+1} - \xi_i)/m$ (e.g. with $m = 10$). Let the discrete extremum be attained at $\xi^* := \xi_i + j^* h$ and compute the parabola interpolating at $\xi^* - h, \xi^*, \xi^* + h$. Its extremum ξ^{**} is a good approximation of α_i. If necessary, this process can be iterated. In the case of the interval $(\xi_{2n}, b]$ one has to take care since $\alpha_{2n} = b$ may hold.

The rational functions r_0 and $\varepsilon = f - r_0$ are uniquely determined by the tuple $\boldsymbol{\xi} := (\xi_\mu)_{1 \leqslant \mu \leqslant 2n}$. Therefore, formally, $r = r(x; \boldsymbol{\xi})$ and $\varepsilon(x; \boldsymbol{\xi})$ can be considered as functions of x and the parameters $\boldsymbol{\xi}$.

The equi-oscillation property leads to the equations

$$F_j(\boldsymbol{\xi}) := \varepsilon(\alpha_{j-1}; \boldsymbol{\xi}) + \varepsilon(\alpha_j; \boldsymbol{\xi}) = 0 \qquad (1 \leqslant j \leqslant 2n).$$

The nonlinear system $\mathbf{F}(\boldsymbol{\xi}) := \left(F_j(\boldsymbol{\xi})\right)_{1 \leqslant j \leqslant 2n} = 0$ for the $2n$ values of $\boldsymbol{\xi}$ has a unique solution (the best approximation) and can be solved by Newton's method leading to the next iterate r_1. The required functional matrix may be approximated via divided differences. In the case of $\partial \mathbf{F}(\boldsymbol{\xi})/\partial \xi_j$, the value $\mathbf{F}(\boldsymbol{\xi}')$ for $\boldsymbol{\xi}'_j = \boldsymbol{\xi}_j + \delta$ and $\boldsymbol{\xi}'_i = \boldsymbol{\xi}_i$ for $i \neq j$ is needed. For its computation a new interpolation problem $r'(\boldsymbol{\xi}'_j) = f(\boldsymbol{\xi}'_j)$ has to be solved by a secondary Newton iteration and the alternant of $\varepsilon' = f - r'$ need be determined. Note that the new iterate r_1 is again admissible.

Assume that the best approximation $r^*_{[1,b]}$ is determined for some interval $[1, b]$ ($a = 1$ w.l.o.g., cf. Remark 2.1). For another interval we can use the continuation method: if b' is close enough to b, $r^*_{[1,b]}$ is a good starting value of Newton's method solving the problem in $[1, b']$. More details about the continuation method are described in [25], where the completely analogous situation of best approximations by exponential sums is considered.

A more severe problem is the computation of the transition from $n - 1$ to n. After computing best approximations $r^*_{n-2,n-1}$ in various intervals, we need a first best approximation $r^*_{n,n-1}$ for some interval $[1, b]$. Its iterative computation requires an admissible starting function $r \in R_{n-1,n}$ which can be obtained by various approaches.

First approach: Set $r = p/q$, $q(x) = x^n + \ldots$. For given interpolation points $\boldsymbol{\xi}_j$ solve the linear system $p(\boldsymbol{\xi}_j) - f(\boldsymbol{\xi}_j)q(\boldsymbol{\xi}_j) = 0$ $(1 \leqslant j \leqslant 2n)$. To obtain the representation (35), one has to determine the zeros of q.

Second approach: The direct computation of the interpolating function represented by (35) is described by [5, Algorithm 3.1]. It requires the solution of an eigenvalue problem.

The third approach, used by the second author, is explained by the following example. Given the situation that the best approximations $r^*_{20,21}$ for $[1, 10^{15}]$ is known, we want to obtain an admissible function $r \in R_{21,22}$ for $n = 22$. The coefficients a_ν, b_ν of (35) for $r^*_{20,21}$ are

ν	1	2	...	19	20	21
a_ν	0.3338	0.4258	...	$2.54_{10}5$	$1.51_{10}6$	$2.20_{10}7$
b_ν	0.0066	0.0705	...	$8.44_{10}10$	$1.84_{10}12$	$1.42_{10}14$

A rough extrapolation of these values for $\nu = 22$ yields $a_{22} = 2.5_{10}8$ and $b_{22} = 1_{10}16$. Set $\hat{r}(x) := a_{22}/(x + b_{22})$ and define $\varphi(x;\lambda) := \frac{1}{\sqrt{x}} - \lambda\hat{r}(x)$. $r^*_{20,21}$ is the best approximation of $\varphi(\cdot;0)$. Using the continuation technique, we obtain a best approximation $r^{**}_{20,21}$ for $\lambda = 1$ with zeros of $r^{**}_{20,21} - \varphi(\cdot;1)$ at ξ_j $(1 \leqslant j \leqslant 42)$. Note that $r := r^{**}_{20,21} + \hat{r} \in R_{21,22}$ has the representation (35). Checking the values of r, we find further zeros $\xi_{43} \approx 6.4_{10}14$ and $\xi_{44} \approx 7.3_{10}16$. Set $\boldsymbol{\xi} = (\xi_j)_{1\leqslant j\leqslant 44}$. The underlying interval is taken as $[1, 5_{10}17]$. Now, $r(\cdot, \boldsymbol{\xi})$ is an admissible function in $R_{21,22}$ which can be used as starting value of the damped Newton method.

The web page [21] contains the coefficients and approximation errors of the best approximations of $x^{-1/2}$ for various intervals and orders $0 \leqslant n \leqslant 22$.

References

1. J. Almlöf: *Elimination of energy denominators in Møller-Plesset perturbation theory by a Laplace transform approach.* Chem. Phys. Lett. **176** (1991), 319–320
2. Th. Bagby: *The modulus of a plane condenser.* J. Math. and Mech. **17** (1967), 315–329
3. A. V. Balakrishnan: *Fractional powers of closed operators and the semigroups generated by them.* Pacific J. Math. **10** (1960), 419–437
4. L. Banjai, J. M. Melenk, and C. Schwab: *Exponential convergence of hp-FEM for spectral fractional Laplacian in polygons.* Numer. Math. **153** (2020), 1–47
5. B. Beckermann, J. Bisch, and R, Luce: *On the rational approximation of Markov functions, with applications to the computation of Markov functions of Toeplitz matrices.* ArXiv:2106.05098
6. A. Bonito and J. Pasciak: *Numerical approximation of fractional powers of elliptic operators.* Math. Comp. **84** (2015), 2083–2110
7. J. M. Borwein and P. B. Borwein: *Pi and the AGM.* John Wiley & Sons, 1987
8. D. Braess: *Nonlinear Approximation Theory.* Springer-Verlag, Berlin, 1986
9. D. Braess: *Asymptotics for the approximation of wave functions by exponential sums.* J. Approximation Theory **83** (1995), 93–103
10. D. Braess and W. Hackbusch: *Approximation of* $1/x$ *by exponential sums in* $[1,\infty)$. IMA J. Numer. Anal. **25** (2005), 685–697
11. D. Braess and W. Hackbusch: *On the efficient computation of high-dimensional integrals and the approximation by exponential sums.* In: Multiscale, nonlinear and adaptive approximation (R. DeVore, A. Kunoth, eds.), Springer Berlin 2009, pp. 39-74

12. D. Braess and W. Hackbusch: *On the Approximation of Stieltjes Functions by Exponential Sums and Rational Functions with Applications to Partial Differential Equations.* MIS Preprint 10/2024, Max-Planck-Institut Leipzig, 2024
13. E. Cancès, M. Defranceschi, W. Kutzelnigg, C. Le Bris, and Y. Maday: *Computational quantum chemistry: a primer.* in 'Handbook of Numerical Analysis', **X**, pp. 3–270, C. Le Bris (ed.), Elsevier, Amsterdam 2003
14. T. Danczul, C. Hofreiter, and J. Schöberl: *A unified rational Krylov method for elliptic and parabolic fractional diffusion problems.* Numer. Linear Algebra Appl. (2023), 685–697
15. M. Faustmann, C. Marcati, J.M. Melenk, C. Schwab: *Exponential convergence of hp-FEM for the integral fractional Laplacian in polygons.* arXiv:2209.11468v1
16. T. Ganelius , W. K. Hayman, and D. J. Newman: *Lectures on approximation and value distribution.* Montréal: Presses de l'Université de Montréal, 1982
17. A. A. Gončar: *On the speed of rational approximation of some analytic functions,* Math. USSR Sbornik **24** (1978), 131–145
18. A. A. Gončar and E. A. Rakhmanov: *Equilibrium distributions and the rate of rational approximation of analytic functions,* Math. USSR Sbornik **62** (1989), 305-348
19. Guanglian Li: *Wavelet-based edge multiscale parareal algorithm for subdiffusion equations with heterogeneous coefficients in a large time domain.* J. Comput. Appl. Math. **440** (2024), 115608
20. W. Hackbusch: Webpages `www.mis.mpg.de/scicomp/EXP_SUM`
21. W. Hackbusch: Webpages `www.mis.mpg.de/scicomp/RationalAppr`
22. W. Hackbusch: *Approximation of* $1/\|x-y\|$ *by exponentials for wavelet applications.* Computing, **76** (2006), 359–366
23. W. Hackbusch: *Hierarchical Matrices: Algorithms and Analysis.* Springer-Verlag, Heidelberg, 2015
24. W. Hackbusch: *Tensor Spaces and Numerical Tensor Calculus.* 2nd edition. Springer-Verlag, 2019
25. W. Hackbusch: *Computation of best* L^∞ *exponential sums for* $1/x$ *by Remez' Algorithm.* Computing and Visualization in Science **20** (2019) 1-11
26. U. Khristenko and B.Wohlmuth: *Solving time-fractional differential equations via rational approximation.* IMA J. Numer. Anal.**43** (2023), 1263–1290
27. W. Kutzelnigg: *Theory of the expansion of wave functions in a Gaussian basis.* Int. J. of Quantum Chemistry **51** (1994), 447–463
28. R. Rutishauser: *Betrachtungen zur Quadratwurzeliteration.* Monatshefte Math. **67** (1963), 452–464
29. H. R. Stahl: *Best uniform rational approximation of* x^α *on* $[0,1]$. Acta Math. **190** (2003), 241–306
30. F. Stenger: *Numerical Methods Based of Sinc and Analytic Functions.* Springer-Verlag, New York 1993
31. R. S. Varga and A. J. Carpenter: *Some numerical results on best uniform rational approximation of* x^α *on* $[0,1]$. Numer. Alg. **2** (1992), 171-186
32. A. Takatsuka, S. Ten-no, and W. Hackbusch: *Minimax approximation for the decomposition of energy denominators in Laplace-transformed Møller-Plesset perturbation theories.* J. Chem. Phys. **129** (2008), 044112
33. H. Yserentant: *An iterative method for the solution of Laplace-like equations in high and very high space dimensions.* Numer. Math. **156** (2024), 777–811
34. E. I. Zolotarev: *Application of elliptic functions to questions of functions deviating least and most from zero* (Russian). Zap. Imp. Akad. Nauk (1877). St. Petersburg 30 no. 5; reprinted in collected works II, pp. 1–59. Izdat, Akad. Nauk SSSR, Moscow 1932

Approximating Partial Differential Equations without Boundary Conditions

Andrea Bonito and Diane Guignard

Abstract We consider the problem of numerically approximating the solutions to an elliptic partial differential equation (PDE) for which the boundary conditions are lacking. To alleviate this missing information, we assume to be given measurement functionals of the solution. In this context, a near optimal recovery algorithm based on the approximation of the Riesz representers of these functionals in some intermediate Hilbert spaces is proposed and analyzed in [2]. Inherent to this algorithm is the computation of H^s, $s > 1/2$, inner products on the boundary of the computational domain. We take advantage of techniques borrowed from the analysis of fractional diffusion problems to design and analyze a fully practical near optimal algorithm not relying on the challenging computation of H^s inner products.

1 Introduction

In many applications, a lack of full information on boundary conditions arises for various reasons [25, 11, 12]. For example, the correct physics might not be fully understood [21, 23], the boundary values are not accessible [15] or they must be appropriately modified in numerical schemes [14, 22]. In that cases, measurements can be used to alleviate the missing information; see for instance [15, 20, 2].

We consider the model elliptic partial differential equation (PDE)

$$-\Delta u = f \quad \text{in } \Omega, \tag{1}$$

Andrea Bonito
Department of Mathematics, Texas A&M University, College Station, TX 77845, USA, e-mail: bonito@tamu.edu

Diane Guignard
Department of Mathematics and Statistics, University of Ottawa, ON K1N 6N5, Canada, e-mail: dguignar@uottawa.ca

R. DeVore, A. Kunoth (eds.), *Multiscale, Nonlinear and Adaptive Approximation II*,
https://doi.org/10.1007/978-3-031-75802-7_8

satisfied by the function $u \in H^1(\Omega)$ of interest on a bounded polygonal domain $\Omega \subset \mathbb{R}^d$ $(d = 2, 3)$ and where $f \in H_0^1(\Omega)^*$, the dual of $H_0^1(\Omega)$. Because the influence of the forcing term f is marginal, we assume for the purpose of this introduction that $f \equiv 0$, i.e., the function of interest u is harmonic in Ω

$$-\Delta u = 0 \quad \text{in } \Omega. \tag{2}$$

Notice that (2) must be supplemented by suitable boundary conditions to uniquely determine u. In lack of such boundary conditions, we assume to have access to measurements of u, namely to the vector $\omega \in \mathbb{R}^m$ given by

$$\omega = \lambda(u) := (\lambda^1(u), \lambda^2(u), \ldots, \lambda^m(u)) \in \mathbb{R}^m, \tag{3}$$

where for $j = 1, \ldots, m$, λ^j are known functionals in $H^1(\Omega)^*$, the dual of $H^1(\Omega)$. Still, this additional information does not sufficiently narrow down the set of functions u satisfying the PDE (2) and the measurements (3). The last ingredient is a model class assumption (prior knowledge), which restricts the possible candidates to a compact subset of $H^1(\Omega)$. The model class assumption in our context are functions $v \in H^1(\Omega)$ with traces on $\Gamma := \partial\Omega$ satisfying $v|_\Gamma \in K_\Gamma^s$ for some $s > 1/2$, where

$$K_\Gamma^s := U(H^s(\Gamma)) := \{g \in H^s(\Gamma) : \|g\|_{H^s(\Gamma)} \leqslant 1\}. \tag{4}$$

To characterize further the model class, we denote by

$$\mathcal{H}^s(\Omega) := \left\{u \in H^1(\Omega) : u \text{ satisfies (2) and } u|_\Gamma \in H^s(\Gamma)\right\} \subset H^1(\Omega), \tag{5}$$

the space of harmonic functions with traces in $H^s(\Gamma)$. The model class is the unit ball in $\mathcal{H}^s(\Omega)$, i.e.,

$$K^s := U(\mathcal{H}^s(\Omega)),$$

and we gather all the information available on the function to be recovered in

$$K_\omega^s := \{u \in K^s : \lambda(u) = \omega\}. \tag{6}$$

The *optimal recovery problem* is to identify a function in K_ω^s that would be the best representative of all K_ω^s. More precisely, given $\omega \in \mathbb{R}^m$ and $s > 1/2$, it consists of finding $u^* \in H^1(\Omega)$ that minimizes the worst-case recovery error, namely

$$u^* \in \operatorname*{arg\,min}_{v \in H^1(\Omega)} \sigma(v; K_\omega^s), \quad \text{where} \quad \sigma(v; K_\omega^s) := \sup_{u \in K_\omega^s} \|u - v\|_{H^1(\Omega)}. \tag{7}$$

It is well-known [19] that the solution to this problem is unique and $u^* \in H^1(\Omega)$ is the center of the smallest closed ball in $H^1(\Omega)$ containing K_ω^s (the Chebyshev ball). Moreover, we have that $\sigma(u^*; K_\omega^s) = R(K_\omega^s)$ is the radius of the Chebyshev ball which shall serve as a benchmark for the performances of practical algorithms used to construct approximations of u^*. Equipped with the standard $H^s(\Gamma)$ inner product, $\mathcal{H}^s(\Omega)$ is an Hilbert space. This together with the fact that K^s is the unit ball in $\mathcal{H}^s(\Omega)$, we must have that u^* is the element of K_ω^s with minimal norm [19]. This crucial property offers a computationally favorable expression

$$u^* = \sum_{i=1}^{m} U_i \phi^i, \tag{8}$$

where $\phi^i \in \mathcal{H}^s(\Omega)$ is the Riesz representer of the functional λ^i over $\mathcal{H}^s(\Omega) \subset H^1(\Omega)$, i.e.,

$$\langle \phi^i, v \rangle_{\mathcal{H}^s(\Omega)} = \lambda^i(v), \qquad \forall v \in \mathcal{H}^s(\Omega). \tag{9}$$

The coefficients $\mathbf{U} := (U_i)_{i=1}^m \subset \mathbb{R}^m$ are determined by the relation

$$G\mathbf{U} = \omega,$$

where $G = (g_{ij})_{i,j=1}^m$ with $g_{ij} := \lambda^i(\phi^j)$. Since $\phi^i \in \mathcal{H}^s(\Omega)$, it is harmonic and thus $\phi^i = E\psi^i$, where $E : H^{1/2}(\Gamma) \to H^1(\Omega)$ is the harmonic extension operator (see (32)) and $\psi^i \in H^s(\Gamma)$ satisfies

$$\langle \psi^i, v \rangle_{H^s(\Gamma)} = (\lambda^i \circ E)(v), \qquad \forall v \in H^s(\Gamma). \tag{10}$$

The algorithm proposed in [2] takes advantage of the representation (8), where ϕ^i and ψ^i are replaced by their respective finite element approximations.

We reiterate that the algorithm in [2] is fully implementable provided a finite element approximation of (9) can be constructed. While standard finite element methods can be used for integral s, the approximation of the PDE (10) for nonintegral s is computationally challenging in presence of the nonlocal $H^s(\Gamma)$ inner product. We focus on the case $1/2 < s < 1$ since the general case, modulo additional technicalities, can be obtain by recurrence as described in Section 5. For $1/2 < s < 1$, we advocate the use of the following inner product on $H^s(\Gamma)$

$$\langle \mathcal{L}^{s/2} v, \mathcal{L}^{s/2} w \rangle_{L^2(\Gamma)}, \qquad \forall v, w \in H^s(\Gamma), \tag{11}$$

where $\mathcal{L} := (I - \Delta_\Gamma)$ and $\mathcal{L}^{s/2}$ denotes its fractional powers, see Section 2.2. The advantage of using (11) as inner product is that (10) becomes a fractional diffusion problem

$$(I - \Delta_\Gamma)^s \dot{\psi}^i = \lambda^i \circ E. \tag{12}$$

Hereafter we use the dot notation to indicate quantities defined using the alternate inner product (12).

The purpose of this work is to design a finite element method for its approximation and accounts for its effect in the analysis of the optimal recovery algorithm. The approximation of fractional diffusion problems is an active research topic. We refer to the reviews [3, 17] and the reference therein for details. In this work, we follow the strategy initiated in [8] later refined in [9, 7]. Compared to the setting considered in the previous references, the fractional diffusion problem (12) is set on a Lipschitz closed hyper-surfaces Γ and involves a forcing term $\lambda^i \circ E$ that is not necessarily in $L^2(\Gamma)$. Nevertheless, the strategy proposed in [8] relies on the following integral representation of $\dot{\psi}^i$ in (12)

$$\dot{\psi}^i = \frac{\sin(\pi s)}{\pi} \int_{-\infty}^{\infty} e^{(1-s)y}(e^y I + \mathcal{L})^{-1}(\lambda^i \circ E)\mathrm{d}y \tag{13}$$

and consists of employing a sinc quadrature for the indefinite integral and a finite element method for the approximation of each integrand at the quadrature points. The resulting numerical method, see Algorithm 3, is straightforward to implement as it only require a finite element solver for standard reaction-diffusion problems. Once the approximation $\dot{\psi}^i_{k,h}$ is computed, the approximation $\dot{\phi}^i_{k,h}$ is defined as the discrete harmonic extension of $\dot{\psi}^i_{k,h}$; see Section 3.1.

The error analysis performed in this work is inspired from [5], see also [6], where fractional diffusion problems on closed smooth manifolds with a specific forcing term (white noise) are addressed. Theorem 4.1 guarantees that

$$\|\dot{\phi}^i - \dot{\phi}^i_{k,h}\|_{H^1(\Omega)} \in O(e^{-\frac{\pi^2}{k}} + h^q),$$

i.e., the error in computing the Riesz representers is exponentially decreasing as the sinc quadrature parameter k decreases and exhibits an algebraic decay rate h^q in terms of the finite element resolution characterized by the mesh size h, where $q > 0$ depends on various parameters. This result is used in Corollary 4.1 to provide a recovery error estimate for the output of Algorithm 1

$$\|u - u_{k,h}\|_{H^1(\Omega)} \leqslant R(\dot{K}^s_\omega) + O(e^{-\frac{\pi^2}{k}} + h^q).$$

In Section 2, we introduce the optimal recovery algorithm (Algorithm 1) from [2] and justify expression (12) along with the integral representation (37) of $\dot{\psi}^i$. The approximation of the latter is discussed in Section 3. We propose a *practical* algorithm (Algorithm 3), which build on the *theoretical* algorithm (Algorithm 2), but provides an efficient way to approximate the Riesz representers when s is non-integral without computing demanding $H^s(\Gamma)$ inner products. The performance of the practical algorithm is analyzed in Section 4, which contains the main results of this work (Theorem 4.1 and Corollary 4.1). We illustrate numerically the performance of Algorithm 1 and analyze the effect of the parameters s, m and h on the recovery error. We postpone to the appendix results related to the approximation of fractional diffusion problems, which may not be readily available in the literature but whose proofs are similar to existing ones.

From now on, C denotes a generic constant independent of the discretization parameters (i.e., the mesh size h and the sinc quadrature spacing k) and we use the notation $a \lesssim b$ to indicate that $a \leqslant Cb$.

2 Preliminaries

2.1 The optimal recovery algorithm

In this section, we recall the algorithm proposed [2] along with its properties. We return to the PDE (1) with $f \in H_0^1(\Omega)^*$ and decompose u as

$$u = u_f + u_{\mathcal{H}}, \tag{14}$$

where $u_f \in H_0^1(\Omega)$ solves

$$-\Delta u_f = f \qquad \text{in } \Omega \tag{15}$$

and $u_{\mathcal{H}} \in U(\mathcal{H}^s(\Omega))$. Since $u_f \in H_0^1(\Omega)$ is uniquely determined by (15), the recovery procedure only targets $u_{\mathcal{H}}$. Therefore, without loss of generality, we assume from now on that $f = u_f = 0$.

Algorithm 1: Recovery Algorithm - simplified from [2]

Require: $\varepsilon > 0$
Ensure : $\widehat{u}$
◇ Compute an approximation $\widehat{u}_f$ to the function $u_f \in H_0^1(\Omega)$ in (15) such that $\|u_f - \widehat{u}_f\|_{H^1(\Omega)} \leqslant \varepsilon$
◇ Set $\widehat{\omega} := \omega - \lambda(\widehat{u}_f)$
◇ For each $j \in \{1, 2, \ldots, m\}$, compute an approximation $\widehat{\phi}^j$ to the Riesz representer ϕ^j in (9) such that $\|\phi^j - \widehat{\phi}^j\|_{H^1(\Omega)} \leqslant \varepsilon$
◇ Set $\widehat{G} = (\widehat{g}_{ij})_{ij=1}^m$ with $\widehat{g}_{ij} = \lambda^i(\widehat{\phi}^j)$
◇ Find the coefficients $\widehat{\mathbf{U}} = (\widehat{U}_i)_{i=1}^m$ satisfying $\widehat{G}\widehat{\mathbf{U}} = \widehat{\omega}$ ▹ Use Moore-Penrose inverse of $\widehat{G}$
◇ Set $\widehat{u} = \sum_{i=1}^m \widehat{U}_i \widehat{\phi}^i + \widehat{u}_f$

We have already pointed out that Algorithm 1 is fully implementable provided approximations of the Riesz representers can be computed. The purpose of this work is to design and analyze an efficient algorithm for the approximation of the Riesz representers.

Theorem 3.1 in [2] guarantees that there is a constant Λ depending on the measurements λ^i and their Riesz representers ϕ^i, $i = 1, \ldots, m$, on m and ω as well as the constant C_s in

$$\|v\|_{H^1(\Omega)} \leqslant C_s \|v\|_{\mathcal{H}^s(\Omega)}, \qquad \forall\, v \in \mathcal{H}^s(\Omega), \tag{16}$$

such that

$$\|u - \widehat{u}\|_{H^1(\Omega)} \leqslant R(K_\omega^S) + \Lambda\varepsilon. \tag{17}$$

In particular, for ε sufficiently small, we have

$$\|u - \widehat{u}\|_{H^1(\Omega)} \leqslant C(\varepsilon) R(K_\omega^S), \tag{18}$$

where $C(\varepsilon) \to 1^+$ as $\varepsilon \to 0^+$. When the output of a recovery algorithm satisfies (18), we say that the algorithm is near-optimal.

2.2 The space $H^s(\Gamma)$ and the model class assumption

Let $H^0(\Gamma) := L^2(\Gamma)$ be the space of square integrable functions over Γ with inner product and associated norm

$$\langle v, w\rangle_{L^2(\Gamma)} := \int_\Gamma vw, \qquad \|v\|^2_{L^2(\Gamma)} := \langle v, v\rangle_{L^2(\Gamma)}, \qquad \forall\, v, w \in L^2(\Gamma).$$

We shall frequently use without mentioning the identification of $L^2(\Gamma)$ with its dual. Also, define

$$H^1(\Gamma) := \{v \in L^2(\Gamma) : \ |\partial_i v|_{L^2(\Gamma)} < \infty,\ \forall\, i = 1, ..., d\},$$

where ∂_i are the components of the tangential gradient ∇_Γ. It is an Hilbert space when equipped with the inner product

$$\langle v, w\rangle_{H^1(\Gamma)} := \langle v, w\rangle_{L^2(\Gamma)} + \sum_{i=1}^{d} \langle \partial_i v, \partial_i w\rangle_{L^2(\Gamma)}, \qquad \|v\|^2_{H^1(\Gamma)} := \langle v, v\rangle_{H^1(\Gamma)}.$$

Its dual space is denoted $H^{-1}(\Gamma)$ and $\langle ., .\rangle$ stands for the duality pairing between $H^{-1}(\Gamma)$ and $H^1(\Gamma)$.

Sobolev fractional spaces on surfaces can be defined using an intrinsic norm, see for instance [2]. However, it will be more convenient to use a spectral characterization of these fractional spaces. In this aim, we define $\mathcal{T} : H^{-1}(\Gamma) \to H^1(\Gamma)$, where for $g \in H^{-1}(\Gamma)$, $w := \mathcal{T} g$ is the unique function in $H^1(\Gamma)$ satisfying

$$\int_\Gamma wv + \int_\Gamma \nabla_\Gamma w \cdot \nabla_\Gamma v = \langle g, v\rangle, \qquad \forall\, v \in H^1(\Gamma). \tag{19}$$

The operator $\mathcal{L} := I - \Delta_\Gamma : H^1(\Gamma) \to H^{-1}(\Gamma)$ is the inverse of $\mathcal{T}$.

Let $\{(\tau_j, e_j)\}_{j=1}^\infty \subset \mathbb{R} \times H^1(\Gamma)$ be the eigenpairs of $\mathcal{L}$, ordered and normalized so that the eigenvalues and eigenfunctions satisfy

$$1 = \tau_1 < \tau_2 \leqslant \tau_3 \leqslant \cdots, \qquad \text{and} \qquad \langle e_i, e_j\rangle_{L^2(\Gamma)} = \delta_{ij} \quad \forall\, i, j \geqslant 1.$$

For any $g \in L^2(\Gamma)$ we thus have

$$g = \sum_{j=1}^{\infty} g_j e_j, \qquad g_j := \langle g, e_j\rangle_{L^2(\Gamma)}, \tag{20}$$

which offers an alternate characterization of $L^2(\Gamma)$. This motivates the introduction of $\dot{H}^p(\Gamma)$, $p \geqslant 0$, as the space of functions in $L^2(\Gamma)$ for which

$$\|g\|^2_{\dot{H}^p(\Gamma)} = \langle g, g\rangle_{\dot{H}^p(\Gamma)} < \infty, \tag{21}$$

where

$$\langle f, g\rangle_{\dot{H}^p(\Gamma)} := \sum_{j=1}^{\infty} \langle f, e_j\rangle_{L^2(\Gamma)} \langle g, e_j\rangle_{L^2(\Gamma)} \tau_j^p < \infty. \tag{22}$$

We identify the duality product $\langle .,. \rangle$ between $\dot{H}^{-p}(\Gamma)$ and $\dot{H}^p(\Gamma)$ with the $L^2(\Gamma)$ scalar product so that

$$\left\langle \sum_{j=1}^{\infty} c_j e_j, \sum_{j=1}^{\infty} d_j e_j \right\rangle := \sum_{j=1}^{\infty} c_j d_j \qquad \text{for} \quad \sum_{j=1}^{\infty} d_j e_j \in \dot{H}^p(\Gamma).$$

This leads to the following definitions for negative indices

$$\dot{H}^{-p}(\Gamma) := \left\{ \left\langle \sum_{j=1}^{\infty} c_j e_j, . \right\rangle : \ \{c_j\}_{j=1}^{\infty} \subset \mathbb{R}, \ \sum_{j=1}^{\infty} c_j^2 \tau_j^{-p} < \infty \right\}, \quad p > 0. \tag{23}$$

We refer to [9] for more details.

The regularity properties of the operators $\mathcal{T}$ and $\mathcal{L}$ are critical to assess the equivalence of the spaces $\dot{H}^q(\Gamma)$ with the standard intermediate spaces $H^q(\Gamma)$. We make the following assumption.

Assumption. *We assume that there is $\alpha_\Gamma > 0$ such that the operator $\mathcal{T}$ has elliptic regularity pickup index α_Γ. That is, $\mathcal{T}$ is an isomorphism from $H^{-1+q}(\Gamma)$ to $H^{1+q}(\Gamma)$ for any $q \in [0, \alpha_\Gamma]$.*

When Assumption 2.2 holds, a similar argument that the one used in Proposition 4.1 in [8], see also [9], guarantees that for $q \in [-1-\alpha_\Gamma, 1+\alpha_\Gamma]$ we have

$$\dot{H}^q(\Gamma) \cong H^q(\Gamma) \tag{24}$$

with equivalent norms. We denote by d_s and D_s the constants only depending on s and Γ such that

$$d_s\|v\|_{\dot{H}^s(\Gamma)} \leqslant \|v\|_{H^s(\Gamma)} \leqslant D_s\|v\|_{\dot{H}^s(\Gamma)}, \qquad \forall v \in H^s(\Gamma). \tag{25}$$

The sets $\dot{K}^s_\Gamma$, $\dot{K}^s$, and $\dot{K}^s_\omega$ associated with the recovery problem are as their counterparts without the dot symbol but using the norm $\|.\|_{\dot{H}^s(\Gamma)}$ instead of $\|.\|_{H^s(\Gamma)}$. With these notations, the model class assumption reads

Assumption. *There exists $s > 1/2$ such that the function u to be recovered belongs to $\dot{K}^s_\omega$.*

With this notation, the recovery estimate (17) reads

$$\|u - \hat{u}\|_{H^1(\Omega)} \leqslant R(\dot{K}^s_\omega) + \dot{\Lambda}\varepsilon, \tag{26}$$

for some constant $\dot{\Lambda}$ depending on the measurements λ^i and their Riesz representers $\dot{\phi}^i$, $i = 1, ..., m$ in $\dot{H}^s(\Gamma)$, on m and ω as well as the constant C_s in (16). Moreover, we have

$$cR(\dot{K}^s_\omega) \leqslant R(K^s_\omega) \leqslant CR(\dot{K}^s_\omega),$$

where the constants c, C depends on the equivalence constants in (25).

2.3 The role of fractional diffusion

In this section, we further exploit the structure of the norm $\|.\|_{\dot{H}^s(\Gamma)}$ to deduce a characterization using the (spectral) fractional powers of $\mathcal{L}$. In turns, this justify the fractional diffusion formulation (9) for the Riesz representers.

Recall that for $1/2 < s < 1$, the spaces $\dot{H}^s(\Gamma)$ and $H^s(\Gamma)$ are the same. Also, $\{(\tau_j, e_j)\}_{j=1}^\infty$ denotes the eigenpairs of $\mathcal{L}$ and define for $\beta \in (-1, 1)$ and $q \in \mathbb{R}$, $\mathcal{L}^\beta : \dot{H}^q(\Gamma) \to \dot{H}^{q-2\beta}(\Gamma)$ by

$$\mathcal{L}^\beta v := \sum_{j=1}^\infty \tau_j^\beta v_j e_j, \qquad \forall v = \Big\langle \sum_{j=1}^\infty v_j e_j, . \Big\rangle \in \dot{H}^q(\Gamma). \tag{27}$$

Thanks to the Lax-Milgram theory and the equivalence (24),

$$\mathcal{L}^\beta : H^q(\Gamma) \to H^{q-2\beta}(\Gamma) \tag{28}$$

is an isomorphism provided that

$$q \in [-1 - \alpha_\Gamma + 2\beta, 1 + \alpha_\Gamma] \quad \text{for } \beta \in (0, 1),$$

and

$$q \in [-1 - \alpha_\Gamma, 1 + \alpha_\Gamma + 2\beta] \quad \text{for } \beta \in (-1, 0).$$

Moreover, in view of (22), we have for $q \in [-1 - \alpha_\Gamma, 1 + \alpha_\Gamma]$

$$\langle v, w \rangle_{\dot{H}^q(\Gamma)} = \langle \mathcal{L}^{\frac{q}{2}} v, \mathcal{L}^{\frac{q}{2}} w \rangle_{L^2(\Gamma)} = \langle \mathcal{L}^q v, w \rangle, \qquad v, w \in H^q(\Gamma) \tag{29}$$

and

$$\|v\|_{\dot{H}^q(\Gamma)} = \|\mathcal{L}^{\frac{q}{2}} v\|_{L^2(\Gamma)}, \qquad v \in H^q(\Gamma). \tag{30}$$

We equip $\mathcal{H}^s(\Omega)$ with the $\dot{H}^s(\Gamma)$ inner product and define the Riesz representer $\dot{\phi}^i \in \mathcal{H}^s(\Omega)$ of the functional $\lambda^i : \mathcal{H}^s(\Omega) \to \mathbb{R}$ by the relation

$$\langle \dot{\phi}^i, v \rangle_{\dot{H}^s(\Gamma)} = \lambda^i(v), \qquad \forall v \in \mathcal{H}^s(\Omega). \tag{31}$$

Since $\dot{\phi}^i$ is harmonic, it reads $\dot{\phi}^i = E\dot{\psi}^i$, where E is the harmonic extension operator defined for $g \in H^{1/2}(\Gamma)$ as the unique function $w = Eg \in H^1(\Omega)$ satisfying

$$w|_\Gamma = g \qquad \text{and} \qquad \int_\Omega \nabla w \cdot \nabla v = 0, \qquad \forall\, v \in H^1_0(\Omega). \tag{32}$$

It is well-known that $E : H^{1/2}(\Gamma) \to H^1(\Omega)$ is continuous and we denote by C_E the positive constant such that

$$\|E(v)\|_{H^1(\Omega)} \leqslant C_E \|v\|_{H^{1/2}(\Gamma)}, \qquad \forall\, v \in H^{1/2}(\Gamma). \tag{33}$$

Returning to (31) and recalling (29), we find that the trace $\dot{\psi}^i \in H^s(\Gamma)$ of $\dot{\phi}^i$ satisfies

$$\langle \mathcal{L}^s \dot{\psi}^i, \eta\rangle = (\lambda^i \circ E)(\eta), \qquad \forall\, \eta \in H^s(\Gamma). \tag{34}$$

Note that $\lambda^i \in H^1(\Omega)^*$ implies that

$$\mu^i := \lambda^i \circ E \in H^{-1/2}(\Gamma) \tag{35}$$

with

$$\begin{aligned}\|\mu^i\|_{H^{-1/2}(\Gamma)} &= \sup_{v \in H^{1/2}(\Gamma)\setminus\{0\}} \frac{\lambda^i(E(v))}{\|v\|_{H^{1/2}(\Gamma)}} \leqslant \|\lambda^i\|_{H^1(\Omega)^*} \sup_{v \in H^{1/2}(\Gamma)\setminus\{0\}} \frac{\|E(v)\|_{H^1(\Omega)}}{\|v\|_{H^{1/2}(\Gamma)}} \\ &\leqslant C_E \|\lambda^i\|_{H^1(\Omega)^*}.\end{aligned} \tag{36}$$

Whence, $\dot{\psi}^i = \mathcal{L}^{-s}\mu^i \in H^{\min(2s-\frac{1}{2}, 1+\alpha_\Gamma)}(\Gamma) \subset H^s(\Gamma)$, $1/2 < s < 1$, and the Balakrishnan formula [1], see also Proposition 3.1 in [5], offers an integral representation of $\dot{\psi}^i$

$$\begin{aligned}\dot{\psi}^i = \mathcal{L}^{-s}\mu^i &= \frac{\sin(\pi s)}{\pi} \int_{-\infty}^{\infty} e^{(1-s)y}(e^y I + \mathcal{L})^{-1}\mu^i \mathrm{d}y \\ &= \frac{\sin(\pi s)}{\pi} \int_{-\infty}^{\infty} e^{(1-s)y} w(y) \mathrm{d}y,\end{aligned} \tag{37}$$

where $w(y) \in H^1(\Gamma)$ solves

$$(e^y + 1)\int_\Gamma w(y)v + \int_\Gamma \nabla_\Gamma w(y) \cdot \nabla_\Gamma v = \mu^i(v), \qquad \forall\, v \in H^1(\Gamma). \tag{38}$$

3 Computation of the Riesz representers

The key ingredient in the recovery algorithm (Algorithm 1) is the approximation for $i = 1, \ldots, m$ of the Riesz representer $\dot{\phi}^i$ via its trace $\dot{\psi}^i$ satisfying (37). We present two algorithms serving different purposes. The *theoretical algorithm* is based on the representation (34) and assumes as in [2] that an exact computation of the nonlocal $\dot{H}^s(\Gamma)$ inner product can be performed. Instead, the *practical algorithm* build from the theoretical algorithm and accounts for the approximation of the $\dot{\psi}^i$ without computing $\dot{H}^s(\Gamma)$ inner products. Before describing the two algorithms, we start with the introduction of the finite element tools. For the rest of this section, we simplify the notation and drop the superscript i indexing the measurements.

3.1 Finite element tools

Let $\{\mathcal{T}_h\}_{h>0}$ be a sequence of quasi-uniform and shape-regular subdivisions of $\overline{\Omega}$ made of simplices (resp. quadrilaterals/hexahedrons). Associated to each subdivision $\mathcal{T}_h$, we denote by $\{\varphi_1, \dots, \varphi_N\}$ and $\{\varphi_{N+1}, \dots, \varphi_{N+N_b}\}$ the Lagrange nodal basis functions [10] associated with interior and boundary nodes, respectively. We define

$$\mathbb{V}_h := \mathbb{W}_h \oplus \mathbb{V}_{b,h} \subset H^1(\Omega), \tag{39}$$

where

$$\begin{aligned}\mathbb{W}_h &:= \operatorname{span}\{\varphi_1, \varphi_2, \dots, \varphi_N\} \subset H_0^1(\Omega)\\ \mathbb{V}_{b,h} &:= \operatorname{span}\{\varphi_{N+1}, \varphi_{N+2}, \dots, \varphi_{N+N_b}\} \subset H^1(\Omega)\end{aligned}$$

and

$$\mathbb{T}_h := \mathbb{V}_{b,h}|_\Gamma = \mathbb{V}_h|_\Gamma = \operatorname{span}\{\varphi_{N+1}|_\Gamma, \varphi_{N+2}|_\Gamma, \dots, \varphi_{N+N_b}|_\Gamma\} \subset H^s(\Gamma),$$

since $\varphi_i|_\Gamma = 0$ for any $i \in \{1, 2, \dots, N\}$.

We can now define the discrete harmonic extension operator $E_h : \mathbb{T}_h \to \mathbb{V}_h$. For $g_h \in \mathbb{T}_h$, $w_h := E_h g_h \in \mathbb{V}_h$ is given by

$$w_h|_\Gamma = g_h \qquad \text{and} \qquad \int_\Omega \nabla w_h \cdot \nabla v_h = 0 \quad \forall\, v_h \in \mathbb{W}_h; \tag{40}$$

compare with (32). The stability of E_h follows from the stability (33) of E, refer to [2] for more details. In particular, there is a constant $D_E > 0$ independent of h such that

$$\|E_h g_h\|_{H^1(\Omega)} \leqslant D_E \|g_h\|_{H^{1/2}(\Gamma)}, \qquad \forall\, g_h \in \mathbb{T}_h. \tag{41}$$

We shall also need the Scott-Zhang projection $\Pi_h : L^2(\Gamma) \to \mathbb{V}_h$ introduced in [24] for Euclidean domains but whose definition and properties readily extend to the boundary of a polygonal domain like Γ. Its stability in $L^2(\Gamma)$ and $H^1(\Gamma)$ are standard and for $t \in [0, 1]$ we have

$$\|\Pi_h v\|_{H^t(\Gamma)} \lesssim \|v\|_{H^t(\Gamma)}, \qquad \forall\, v \in H^t(\Gamma). \tag{42}$$

In [13] the stability property (42) is extended to $t \in (1, 3/2)$ but with the Hilbert norms replaced by Besov norms. Nevertheless, we have for any $t \in (1, 3/2)$ and any

$\epsilon > 0$

$$\|\Pi_h v\|_{H^{t-\epsilon}(\Gamma)} \lesssim C(\epsilon)\|v\|_{H^t(\Gamma)}, \qquad \forall v \in H^t(\Gamma), \tag{43}$$

where $C(\epsilon) \to \infty$ as $\epsilon \to 0^+$. Moreover, for any $t_1 \in [0,1]$, $t_2 \in [0,2]$ such that $t_1 + t_2 \in [0,2]$, we have

$$\|(I - \Pi_h)v\|_{H^{t_1}(\Gamma)} \lesssim h^{t_2}\|v\|_{H^{t_1+t_2}(\Gamma)}, \qquad \forall v \in H^{t_1+t_2}(\Gamma). \tag{44}$$

We shall also need the discrete counter-part of the operator $\mathcal{L} := (I - \Delta_\Gamma)$, namely $\mathcal{L}_h : \mathbb{T}_h \to \mathbb{T}_h$, where for $w_h \in \mathbb{T}_h$, $g_h = \mathcal{L}_h w_h$ satisfies

$$\int_\Gamma g_h v_h = \int_\Gamma w_h v_h + \int_\Gamma \nabla_\Gamma w_h \cdot \nabla_\Gamma v_h, \qquad \forall v_h \in \mathbb{T}_h.$$

3.2 Theoretical algorithm

The finite element approximation of the Riesz representer $\dot{\phi} \in H^1(\Omega)$ is defined by $\dot{\phi}_h := E_h \dot{\psi}_h \in \mathbb{V}_h$, where $\dot{\psi}_h \in \mathbb{T}_h$ approximates $\dot{\psi}$ and is given by the relations

$$\langle \dot{\psi}_h, \eta_h \rangle_{\dot{H}^s(\Gamma)} = \mu_h(\eta_h) := (\lambda \circ E_h)(\eta_h), \qquad \forall \eta_h \in \mathbb{T}_h. \tag{45}$$

We have already pointed out that the computation of $\dot{\phi}_h = E_h \dot{\psi}_h$ using this approach is quite demanding for two reasons: for all $1 \leqslant i, j \leqslant N_b$ one must compute

1. exactly the nonlocal $H^s(\Gamma)$ inner product $\langle \varphi_{N+j}, \varphi_{N+i} \rangle_{\dot{H}^s(\Gamma)}$ leading to a full system;
2. the discrete harmonic extension $E_h(\varphi_{N+i}|_\Gamma)$.

The first issue is addressed in the next section. For the second issue, we follow [2] and consider the following saddle-point problem enforcing the harmonic constraint using Lagrange multipliers: find $(\dot{\phi}_h, \xi_h) \in \mathbb{V}_h \times \mathbb{W}_h$ such that

$$\begin{aligned} \langle \dot{\phi}_h, v_h \rangle_{\dot{H}^s(\Gamma)} + \textstyle\int_\Omega \nabla v_h \cdot \nabla \xi_h &= \lambda(v_h) \qquad && \forall v_h \in \mathbb{V}_h, \\ \textstyle\int_\Omega \nabla \dot{\phi}_h \cdot \nabla z_h &= 0 && \forall z_h \in \mathbb{W}_h. \end{aligned} \tag{46}$$

In [2], it is shown that the above saddle-point problem has a unique solution $(\dot{\phi}_h, \xi_h)$ and $\dot{\phi}_h = E_h \dot{\psi}_h$, where $\dot{\psi}_h \in \mathbb{T}_h$ solves (45). We now further explore the structure of (46) to reduce its complexity and write $\dot{\phi}_h = \dot{\phi}_h^0 + \dot{\phi}_h^b$, where $\dot{\phi}_h^0 \in \mathbb{W}_h$ and $\dot{\phi}_h^b \in \mathbb{V}_{b,h}$. Restricting the first equation of (46) to $v_h \in \mathbb{W}_h \subset \mathbb{V}_h$ yields

$$\int_\Omega \nabla \xi_h \cdot \nabla v_h = \lambda(v_h), \qquad \forall v_h \in \mathbb{W}_h, \tag{47}$$

which uniquely determine $\xi_h \in \mathbb{W}_h$. Then, restricting the same equation to $v_h \in \mathbb{V}_{b,h} \subset \mathbb{V}_h$ implies

$$\langle \dot{\phi}_h^b, v_h \rangle_{\dot{H}^s(\Gamma)} = \lambda(v_h) - \int_\Omega \nabla \xi_h \cdot \nabla v_h, \qquad \forall\, v_h \in \mathbb{V}_{b,h}, \tag{48}$$

which this time uniquely determine $\dot{\phi}_h^b \in \mathbb{V}_{b,h}$ and thus $\dot{\phi}_h^b|_\Gamma \in \mathbb{T}_h$. It remains to use the second equation in (46) to compute $\dot{\phi}_h^0 \in \mathbb{W}_h$ solving

$$\int_\Omega \nabla \dot{\phi}_h^0 \cdot \nabla v_h = -\int_\Omega \nabla \dot{\phi}_h^b \cdot \nabla v_h, \qquad \forall\, v_h \in \mathbb{W}_h. \tag{49}$$

We summarize this sequential algorithm in Algorithm 2.

Algorithm 2: Theoretical Algorithm for the approximation of $\dot{\phi}$

- ◇ Find $\xi_h \in \mathbb{W}_h$ satisfying (47)
- ◇ Find $\dot{\phi}_h^b \in \mathbb{V}_{b,h}$ satisfying (48)
- ◇ Find $\dot{\phi}_h^0 \in \mathbb{W}_h$ satisfying (49)
- ◇ Compute $\dot{\phi}_h = \dot{\phi}_h^0 + \dot{\phi}_h^b$

We end this section by noting that the sequential resolution of (47), (48), and (49) has the advantage of not requiring to assemble and store the (potentially large) matrix of size $(2N + N_b) \times (2N + N_b)$ associated to the saddle-point problem (46). More importantly, in contrast to the formulation (45), it does not require the expensive computation of the discrete harmonic extensions $E_h(\varphi_{N+i}|_\Gamma)$ for $i = 1, 2, \ldots, N_b$. However, it still assumes the exact computation of $H^s(\Gamma)$ inner products $\langle \varphi_{N+j}, \varphi_{N+i} \rangle_{\dot{H}^s(\Gamma)}$ to assemble the matrix associated to Problem (48). We address this aspect in the next section.

3.3 Practical algorithm

We restart from the Balakrishnan formula (37) for $\dot{\psi}$ and proceed as in [7, 6] for its approximation, thereby circumventing the computation of the $\dot{H}^s(\Gamma)$ inner product in (45). A sinc quadrature approximate the indefinite integral and a finite element method is advocated for the approximation of $w(y_l)$ at each quadrature point y_l.

Given $k > 0$, we set $y_l := lk$ for $l = -\mathtt{M}, \ldots, \mathtt{N}$, where the specific values for $\mathtt{M}, \mathtt{N} \in \mathbb{N}$ will be chosen later (see Lemma 4.1) to guarantee a quadrature exhibiting an exponential rate of approximation. The sinc approximation of $\dot{\psi}$ reads

$$\begin{aligned} \dot{\psi}_k := Q_k^{-s}(\mathcal{L})\mu &:= \frac{k \sin(\pi s)}{\pi} \sum_{l=-\mathtt{M}}^{\mathtt{N}} e^{(1-s)y_l} (e^{y_l} I + \mathcal{L})^{-1} \mu \\ &= \frac{k \sin(\pi s)}{\pi} \sum_{l=-\mathtt{M}}^{\mathtt{N}} e^{(1-s)y_l} w(y_l), \end{aligned} \tag{50}$$

where $w(y_l) \in H^1(\Gamma)$ solves (38). We further approximate $\dot{\psi}_k$ using a finite element method based on $\mathbb{T}_h$, namely we define

$$\dot{\psi}_{k,h} := \frac{k\sin(\pi s)}{\pi} \sum_{l=-\mathtt{M}}^{\mathtt{N}} e^{(1-s)y_l} w_h(y_l), \tag{51}$$

where for $y \in \mathbb{R}$, $w_h(y) \in \mathbb{T}_h$ is the solution to

$$(e^y + 1)\int_\Gamma w_h(y) v_h + \int_\Gamma \nabla_\Gamma w_h(y) \cdot \nabla_\Gamma v_h = \mu_h(v_h), \qquad \forall v_h \in \mathbb{T}_h, \tag{52}$$

where we recall that $\mu_h(v_h) = \lambda(E_h v_h)$ for $v_h \in \mathbb{T}_h$. Using the operator $\mathcal{L}_h$ defined in Section 3.1 and recalling that $L^2(\Gamma)$ is identified with its dual so that $\mu_h(v_h) = \int_\Gamma \mu_h v_h$ for all $v_h \in \mathbb{T}_h$, we have

$$\dot{\psi}_{k,h} = Q_k^{-s}(\mathcal{L}_h)\mu_h := \frac{k\sin(\pi s)}{\pi} \sum_{l=-\mathtt{M}}^{\mathtt{N}} e^{(1-s)y_l}(e^{y_l} I + \mathcal{L}_h)^{-1}\mu_h.$$

Once the approximation $\dot{\psi}_{k,h}$ is computed, the approximation of the Riesz representer $\dot{\phi}$ is then given by $\dot{\phi}_{k,h} := E_h \dot{\psi}_{k,h}$. In Section 4, we analyze the behavior of $\|\dot{\phi} - \dot{\phi}_{k,h}\|_{H^1(\Omega)}$. At this point, we emphasize again that (51) does not require the computation of $H^s(\Gamma)$ inner products; compare with (45). Also, notice that $\dot{\psi}_{k,h}$ is obtained by aggregating $\mathtt{N} + \mathtt{M} + 1$ standard approximation of reaction-diffusion problems.

As for the theoretical algorithm (Algorithm 2), we circumvent the computation of the N_b discrete harmonic extensions using the saddle-point framework. We write $\dot{\phi}_{k,h} = \dot{\phi}^0_{k,h} + \dot{\phi}^b_{k,h}$ with $\dot{\phi}^0_{k,h} \in \mathbb{W}_h$ and $\dot{\phi}^b_{k,h} \in \mathbb{V}_{b,h}$. Step 1 and 3 in Algorithm 2 remain unchanged while step 2 is replaced by

$$\dot{\phi}^b_{k,h}|_\Gamma = \frac{k\sin(\pi s)}{\pi} \sum_{l=-\mathtt{M}}^{\mathtt{N}} e^{(1-s)y_l} w_h(y_l), \tag{53}$$

where for $y \in \mathbb{R}$, $w_h(y) \in \mathbb{T}_h$ is given by

$$(e^y + 1)\int_\Gamma w_h(y) v_h + \int_\Gamma \nabla_\Gamma w_h(y) \cdot \nabla_\Gamma v_h = \lambda(v_h) - \int_\Omega \nabla \xi_h \cdot \nabla v_h, \quad \forall v_h \in \mathbb{V}_{b,h}, \tag{54}$$

instead of (52). The full practical algorithm is described in Algorithm 3.

The next lemma guarantees that the output $\dot{\phi}_{k,h}$ of Algorithm 3 is such that $\dot{\phi}_{k,h} = E_h \dot{\psi}_{k,h}$, where $\dot{\psi}_{k,h} \in \mathbb{T}_h$ satisfies (51).

Lemma 3.1 *Algorithm 3 produces the unique discrete harmonic function $\dot{\phi}_{k,h} \in \mathbb{V}_h$ whose trace $\dot{\phi}_{k,h}|_\Gamma$ satisfies* (51).

Proof. Comparing (51)-(52) with (53)-(54), it suffices to show that

Algorithm 3: Practical Algorithm for the approximations of $\dot{\phi}$

- ◇ Find $\xi_h \in \mathbb{W}_h$ satisfying (47)
- ◇ Find $\dot{\phi}^b_{k,h} \in \mathbb{V}_{b,h}$ satisfying (53)
- ◇ Find $\dot{\phi}^0_h \in \mathbb{W}_h$ satisfying (49) with ϕ^b_h replaced by $\dot{\phi}^b_{k,h}$
- ◇ Compute $\dot{\phi}_{k,h} = \dot{\phi}^0_h + \dot{\phi}^b_{k,h}$

$$\lambda(v_h) - \int_\Omega \nabla \xi_h \cdot \nabla v_h = \lambda(E_h(v_h|_\Gamma)) = \mu_h(v_h|_\Gamma) \tag{55}$$

for all $v_h \in \mathbb{V}_{b,h}$. Since $v_h - E_h(v_h|_\Gamma) \in \mathbb{W}_h$, relation (47) together with the definition of the discrete harmonic extension operator (40) yield

$$\lambda(v_h - E_h(v_h|_\Gamma)) = \int_\Omega \nabla \xi_h \cdot \nabla(v_h - E_h(v_h|_\Gamma)) = \int_\Omega \nabla \xi_h \cdot \nabla v_h,$$

which is (55) in disguised. From (55) we deduce that $\dot{\phi}_{b,h}|_\Gamma = \dot{\psi}_h$, where $\dot{\psi}_h \in \mathbb{T}_h$ solves (45). The proof is complete. □

The representation (51) is instrumental in the analysis provided below while Algorithm 3 is advocated in Section 5 to illustrate the performance of the method.

4 Error analysis for the Riesz representer and optimal recovery

As in the previous section, we drop the notation i indexing the measurements and let $\lambda \in H^1(\Omega)^*$ be a generic measurement with associated $\mu := \lambda \circ E$ which belongs to $H^{-1/2}(\Gamma)$ in view of (35).

The efficiency of the recovery algorithm (Algorithm 1) depends on the discrepancy between $\phi = E\psi$ with ψ the solution to the boundary problem (34) and $\dot{\phi}_{k,h} = E_h\dot{\psi}_{k,h}$ with $\dot{\psi}_{k,h}$ is the sinc-FE approximation defined in (51).

In this section, we derive an error estimate for $\|\phi - \dot{\phi}_{k,h}\|_{H^1(\Omega)}$ by decomposing the error into three parts

$$\|\phi - \dot{\phi}_{k,h}\|_{H^1(\Omega)} = \|E\psi - E_h\dot{\psi}_{k,h}\|_{H^1(\Omega)} \leqslant \mathrm{I} + \mathrm{II} + \mathrm{III}, \tag{56}$$

where

$$\mathrm{I} := \|E\psi - E\psi_k\|_{H^1(\Omega)}, \qquad \mathrm{II} := \|E\psi_k - E_h(\Pi_h\psi_k)\|_{H^1(\Omega)},$$
$$\text{and} \qquad \mathrm{III} := \|E_h(\Pi_h\psi_k) - E_h\dot{\psi}_{k,h}\|_{H^1(\Omega)}$$

correspond to the sinc quadrature error, the finite element error for the discrete harmonic extension, and the finite element error for ψ_k, respectively. We recall that Π_h denotes the Scott-Zhang operator defined in Section 3.1 and that we restrict our attention to the case $s \in (1/2, 1)$.

We bound each term separately in the following three subsections to derive the following result.

Theorem 4.1 *Let $1/2 < s < 1$, α_Γ be as in Assumption 2.2 and set*

$$\widetilde{\gamma} := \min(2s - 1, \alpha_\Gamma + 1/2, 2\alpha_\Gamma). \tag{57}$$

Let α_Ω be as in Assumption 4.2. For $i = 1, \ldots, m$, let $\lambda^i \in H^1(\Omega)^*$, $\dot{\phi}^i$ be given by (31) and $\dot{\phi}^i_{k,h}$ be the output of Algorithm 3. For every $0 < \epsilon < \widetilde{\gamma}$, there are constants $C(\epsilon)$ and C such that

$$\|\dot{\phi}^i - \dot{\phi}^i_{k,h}\|_{H^1(\Omega)} \leqslant C(\epsilon) h^{\min\{\widetilde{\gamma}-\epsilon,\alpha_\Omega\}} + Ce^{-\frac{\pi^2}{k}} \|\lambda^i\|_{H^1(\Omega)^*}, \tag{58}$$

where $C(\epsilon) \to \infty$ as $\epsilon \to 0^+$.

Proof. The proof follows upon combining Corollary 4.2, Lemma 4.3 and Lemma 4.4 below. □

Corollary 4.1 *Under the assumptions of Theorem 4.1, assume further that the model class Assumption 2.2 holds. Fix $0 < \epsilon < \widetilde{\gamma}$ and let*

$$\varepsilon = \left(C(\epsilon) h^{\min\{\widetilde{\gamma}-\epsilon,\alpha_\Omega\}} + Ce^{-\frac{\pi^2}{k}}\right) \max_{i=1,\ldots,m} \|\lambda^i\|_{H^1(\Omega)^*},$$

where $C(\epsilon)$ and C are the constants in Theorem 4.1. Then the output $\hat{u} = u_{k,h}$ of Algorithm 1 with input parameter ε and where the Riesz representers are computed using Algorithm 3 satisfies the near optimal recovery estimate

$$\|u - u_{k,h}\|_{H^1(\Omega)} \leqslant R(\dot{K}^s_\omega) + \dot{\Lambda}\left(C(\epsilon) h^{\min\{\widetilde{\gamma}-\epsilon,\alpha_\Omega\}} + Ce^{-\frac{\pi^2}{k}})\right) \max_{i=1,\ldots,m} \|\lambda^i\|_{H^1(\Omega)^*},$$

where $\dot{\Lambda}$ is the constant appearing in (26).

Proof. Recall that we assume that $u_f = f = 0$ so it suffices to combine the optimal recovery estimate (26) with the specific choice of ε and Theorem 4.1. □

4.1 The sinc quadrature error

We derive in this section an estimate for the sinc quadrature error, i.e., for I in (56). We take advantage of the stability (33) of the harmonic extension operator E and the following lemma.

Lemma 4.1 *Let $\lambda \in H^1(\Omega)^*$, $\dot{\psi} = \mathcal{L}^{-s}(\lambda \circ E)$ be as in (37) and $\dot{\psi}_k = Q^{-s}_k(\mathcal{L})(\lambda \circ E)$ be as in (50) with*

$$\mathtt{M} := \left\lceil \frac{\pi^2}{(1-s)k^2} \right\rceil \qquad \textit{and} \qquad \mathtt{N} := \left\lceil \frac{\pi^2}{\left(s-\frac{1}{2}\right)k^2} \right\rceil. \tag{59}$$

Then we have

$$\|\dot{\psi} - \dot{\psi}_k\|_{H^{1/2}(\Gamma)} \lesssim e^{-\frac{\pi^2}{k}} \|\lambda\|_{H^1(\Omega)^*},$$

where the hidden constant depends on $\max\{(1-s)^{-1}, (2s-1)^{-1}\}$.

Proof. We only give a sketch of the proof as it follows from [7, Theorem 3.2] (case $\lambda \circ E \in L^2(\Gamma)$) and [5, Proposition 5.1] (estimate for $\|\dot{\psi} - \dot{\psi}_k\|_{L^2(\Gamma)}$).

We set $\mu := \lambda \circ E$, which belong to $H^{-1/2}(\Gamma) \cong \dot{H}^{-1/2}(\Gamma)$ according to (35) and (24). Whence, we have $\dot{\psi} = \mathcal{L}^{-s}\mu \in \dot{H}^{2s-1/2}(\Gamma)$. Furthermore, we write

$$\|\dot{\psi} - \dot{\psi}_k\|_{\dot{H}^{1/2}(\Gamma)} = \|\mathcal{L}^{\frac{1}{4}}(\dot{\psi} - \dot{\psi}_k)\|_{L^2(\Gamma)} = \|\mathcal{L}^{\frac{1}{4}}(\mathcal{L}^{-s} - Q_k^{-s}(\mathcal{L}))\mathcal{L}^{\frac{1}{4}} v\|_{L^2(\Gamma)},$$

where $v := \mathcal{L}^{-1/4}\mu \in L^2(\Gamma)$. In view of the representation (37) of $\dot{\psi}$ and (50) of $\dot{\psi}_k$, we expand v in the basis of $L^2(\Gamma)$-orthonormal eigenfunctions $\{e_j\}_{j=1}^{\infty}$ according to (20) and deduce that for all $y \in \mathbb{R}$

$$\begin{aligned} e^{(1-s)y}\mathcal{L}^{\frac{1}{4}}(e^y I + \mathcal{L})^{-1}\mathcal{L}^{\frac{1}{4}} v &= e^{(1-s)y} \sum_{j=1}^{\infty} v_j \mathcal{L}^{\frac{1}{4}}(e^y I + \mathcal{L})^{-1}\mathcal{L}^{\frac{1}{4}} e_j \\ &= e^{(1-s)y} \sum_{j=1}^{\infty} v_j \tau_j^{\frac{1}{4}}(e^y + \tau_j)^{-1}\tau_j^{\frac{1}{4}} e_j, \end{aligned}$$

where $v_j := \langle v, e_j \rangle_{L^2(\Gamma)}$. This implies that

$$\|\dot{\psi} - \dot{\psi}_k\|_{\dot{H}^{1/2}(\Gamma)} \leqslant \left(\max_{\tau \geqslant \tau_1} |e(\tau)|\right) \|v\|_{L^2(\Gamma)} = \left(\max_{\tau \geqslant \tau_1} |e(\tau)|\right) \|\mu\|_{H^{-1/2}(\Gamma)}$$

with

$$e(\tau) := \int_{-\infty}^{\infty} g_\tau(y)\mathrm{d}y - k \sum_{l=-\mathtt{M}}^{\mathtt{N}} g_\tau(y_l), \quad g_\tau(y) := e^{(1-s)y}(e^y + \tau)^{-1}\tau^{\frac{1}{2}}.$$

The choice (59) for $\mathtt{M}$ and $\mathtt{N}$ together with Lemma 3.1 in [7], see also [18], guarantee that

$$e(\tau) \lesssim e^{-\frac{\pi^2}{k}} + e^{-(s-\frac{1}{2})\mathtt{N}k} + e^{-(1-s)\mathtt{M}k}$$

so that

$$\|\dot{\psi} - \dot{\psi}_k\|_{\dot{H}^{1/2}(\Gamma)} \lesssim \left(e^{-\frac{\pi^2}{k}} + e^{-(s-\frac{1}{2})\mathtt{N}k} + e^{-(1-s)\mathtt{M}k}\right) \|\mu\|_{H^{-1/2}(\Gamma)}.$$

The desired result follows upon using again the equivalence (24) and estimate (36) for $\|\mu\|_{H^{-1/2}(\Gamma)}$. □

Corollary 4.2 *Under the assumptions of Lemma 4.1, there holds*

$$\|E\psi - E\psi_k\|_{H^1(\Omega)} \lesssim e^{-\frac{\pi^2}{k}} \|\lambda\|_{H^1(\Omega)^*}. \tag{60}$$

Proof. This result directly follows from the stability (33) of the extension operator and Lemma 4.1. □

4.2 The finite element error for the harmonic extension

In this section, we derive an upper bound for the term II in (56). We start by estimating the discrepancy between Eg_h and $E_h g_h$ for $g_h \in \mathbb{T}_h$, which depends on the regularity of Eg_h for $g_h \in \mathbb{T}_h$. We make the following assumption.

Assumption. *There is $\alpha_\Omega > 0$ and a constant C such that for all $q \in [0, \alpha_\Omega]$ there holds*

$$Ev \in H^{1+q}(\Omega) \qquad \text{with} \qquad \|Ev\|_{H^{1+q}(\Omega)} \leqslant C\|v\|_{H^{1/2+q}(\Gamma)},$$

for all $v \in H^{1/2+q}(\Gamma)$.

As suggested by the notation, the value of α_Ω depends on Ω. For instance, for polygonal domains Ω with inner angles at most θ, we have that $\alpha_\Omega = \pi/\theta - \epsilon$ for all $\epsilon > 0$; see [16].

The approximation of the harmonic extension operator E by its discrete counterpart E_h is now assessed under Assumption 4.2.

Lemma 4.2 *Let α_Ω be as in Assumption 4.2. For any $g_h \in \mathbb{T}_h$ and any $\beta \in (1/2, 3/2)$ we have*

$$\|Eg_h - E_h g_h\|_{H^1(\Omega)} \lesssim h^{\min\{\beta-1/2, \alpha_\Omega\}} \|g_h\|_{H^\beta(\Gamma)}.$$

Proof. Lemma 4.3 in [2] guarantees that

$$\|Eg_h - E_h g_h\|_{H^1(\Omega)} \lesssim \inf_{v_h \in \mathbb{V}_h} \|Eg_h - v_h\|_{H^1(\Omega)}.$$

Notice that the functions in $\mathbb{T}_h$ belong to $W^{1,\infty}(\Gamma)$ and thus are in $W^{1,p}(\Gamma)$ for any $p \geqslant 1$. Consequently, we have $g_h \in H^\beta(\Gamma)$ for any $\beta < 3/2$ by the Sobolev embedding theorem. Whence, Assumption 4.2 implies that $Eg_h \in H^{1+\min\{\beta-1/2, \alpha_\Omega\}}(\Omega)$ and the desired result follows from a standard finite element approximation estimate. □

We now use the interpolation result (44), the stability estimates (42)-(43), and Lemma 4.2 to estimate II.

Lemma 4.3 *Let $1/2 < s < 1$, α_Γ be as in Assumption 2.2 and $\widetilde{\gamma}$ be as in (57). Let α_Ω be as in Assumption 4.2. Moreover for $\lambda \in H^1(\Omega)^*$ we set $\mu = \lambda \circ E$ and let $\dot{\psi}_k = Q_k^{-s}(\mathcal{L})\mu$ be given by (50) with $\mathtt{M}$, $\mathtt{N}$ as in (59). For any $0 < \epsilon < \widetilde{\gamma}$, there is a constant $C(\epsilon)$ such that $C(\epsilon) \to \infty$ as $\epsilon \to 0^+$ and*

$$\|E\dot{\psi}_k - E_h(\Pi_h \dot{\psi}_k)\|_{H^1(\Omega)} \leqslant C(\epsilon) h^{\min\{\widetilde{\gamma}-\epsilon, \alpha_\Omega\}} \|\lambda\|_{H^1(\Omega)^*}. \tag{61}$$

Proof. We first write $E\dot{\psi}_k - E_h(\Pi_h\dot{\psi}_k) = E\dot{\psi}_k - E(\Pi_h\dot{\psi}_k) + E(\Pi_h\dot{\psi}_k) - E_h(\Pi_h\dot{\psi}_k)$ so that in view of (33) we get

$$\begin{aligned} \mathrm{II} &\leqslant \|E\dot{\psi}_k - E(\Pi_h\dot{\psi}_k)\|_{H^1(\Omega)} + \|E(\Pi_h\dot{\psi}_k) - E_h(\Pi_h\dot{\psi}_k)\|_{H^1(\Omega)} \\ &\leqslant C_E\|\dot{\psi}_k - \Pi_h\dot{\psi}_k\|_{H^{1/2}(\Gamma)} + \|(E - E_h)(\Pi_h\dot{\psi}_k)\|_{H^1(\Omega)}. \end{aligned} \tag{62}$$

The approximation properties (44) of the Scott-Zhang projection with $t_1 = 1/2$ and $t_2 = \widetilde{\gamma} - \epsilon$, yield for the first term in (62)

$$\|\dot{\psi}_k - \Pi_h\dot{\psi}_k\|_{H^{1/2}(\Gamma)} \lesssim h^{\widetilde{\gamma}-\epsilon}\|\dot{\psi}_k\|_{H^{1/2+\widetilde{\gamma}-\epsilon}(\Gamma)}. \tag{63}$$

For the second term, we invoke Lemma 4.2 with $\beta = 1/2 + \widetilde{\gamma} - \epsilon < 3/2$ together with the stability estimate (43) to write

$$\begin{aligned} \|(E - E_h)(\Pi_h\dot{\psi}_k)\|_{H^1(\Omega)} &\lesssim h^{\min\{\widetilde{\gamma}-\epsilon,\alpha_\Omega\}}\|\Pi_h\dot{\psi}_k\|_{H^{1/2+\widetilde{\gamma}-\epsilon}(\Gamma)} \\ &\lesssim C(\epsilon)h^{\min\{\widetilde{\gamma}-\epsilon,\alpha_\Omega\}}\|\dot{\psi}_k\|_{H^{1/2+\widetilde{\gamma}-\epsilon/2}(\Gamma)}, \end{aligned} \tag{64}$$

where $C(\epsilon) \to \infty$ as $\epsilon \to 0^+$. Inserting (63), (64) in (62) we get

$$\mathrm{II} \lesssim C(\epsilon)h^{\min\{\widetilde{\gamma}-\epsilon,\alpha_\Omega\}}\|\dot{\psi}_k\|_{H^{1/2+\widetilde{\gamma}-\epsilon/2}(\Gamma)}.$$

To end the proof, it remains to note that the stability of $Q_k^{-s}(\mathcal{L})$ (Lemma 5.4 with $r = 1/2$ and $\beta = 1/2 + \widetilde{\gamma} - \epsilon$) implies that $\dot{\psi}_k = Q_k^{-s}(\mathcal{L})\mu \in H^{1/2+\widetilde{\gamma}-\epsilon/2}(\Gamma)$ and

$$\|\dot{\psi}_k\|_{H^{1/2+\widetilde{\gamma}-\epsilon/2}(\Gamma)} \lesssim C(\epsilon)\|\mu\|_{H^{-1/2}(\Gamma)} \lesssim C(\epsilon)\|\lambda\|_{H^1(\Omega)^*}. \tag{65}$$

□

4.3 The finite element error for $\dot{\psi}_k$

The following lemma provides an estimate for the last term III in (56).

Lemma 4.4 *Let* $1/2 < s < 1$, α_Γ *be as in Assumption 2.2 and* $\widetilde{\gamma}$ *be as in* (57). *Let* α_Ω *be as in Assumption 4.2. For* $\lambda \in H^1(\Omega)^*$, *we set* $\mu := \lambda \circ E$ *and* $\mu_h := \lambda \circ E_h$. *Let* $\dot{\psi}_k = Q_k^{-s}(\mathcal{L})\mu$ *be given by* (50) *with* $\mathtt{M}$ *and* $\mathtt{N}$ *as in* (59) *and* $\dot{\psi}_{k,h} = Q_k^{-s}(\mathcal{L}_h)\mu_h$ *be as in* (51). *For any* $0 < \epsilon < \widetilde{\gamma}$, *there is a constant* $C(\epsilon)$ *such that* $C(\epsilon) \to \infty$ *as* $\epsilon \to 0^+$ *and*

$$\|E_h(\Pi_h\dot{\psi}_k) - E_h\dot{\psi}_{k,h}\|_{H^1(\Omega)} \leqslant C(\epsilon)h^{\min\{\widetilde{\gamma}-\epsilon,\alpha_\Omega\}}\|\lambda\|_{H^1(\Omega)^*}. \tag{66}$$

Proof. From the stability (41) of the discrete harmonic extension, we deduce that

$$\|E_h(\Pi_h\dot{\psi}_k) - E_h\dot{\psi}_{k,h}\|_{H^1(\Omega)} \leqslant D_E\|\Pi_h\dot{\psi}_k - \dot{\psi}_{k,h}\|_{H^{1/2}(\Gamma)}.$$

Therefore, it remains to estimate the finite element error $\Pi_h\dot{\psi}_k - \dot{\psi}_{k,h}$ in the $H^{1/2}(\Gamma)$ norm. The approximation of $\dot{\psi}_k$ by $\dot{\psi}_{k,h}$ gives rise to two sources of error: the approximation of the source term $\mu = \lambda \circ E$ by $\mu_h = \lambda \circ E_h$ and the approximation of the reaction-diffusion problems using a finite element method. To separate the two errors, we introduce the intermediate function $\widetilde{\psi}_k := Q_k^{-s}(\mathcal{L})\widetilde{\mu}$, where $\widetilde{\mu} := \lambda \circ E_h \circ \Pi_h$ and Π_h is the Scott-Zhang projection. Note that proceeding as in (36) but using (41) as well as (42) with $t = 1/2$, we deduce that $\widetilde{\mu} \in H^{-1/2}(\Gamma)$ with

$$\|\widetilde{\mu}\|_{H^{-1/2}(\Gamma)} \lesssim \|\lambda\|_{H^1(\Omega)^*}. \tag{67}$$

We write

$$\begin{aligned}\|\Pi_h\dot{\psi}_k - \dot{\psi}_{k,h}\|_{H^{1/2}(\Gamma)} \leqslant &\|\Pi_h(\dot{\psi}_k - \widetilde{\psi}_k)\|_{H^{1/2}(\Gamma)} + \|\Pi_h\widetilde{\psi}_k - \widetilde{\psi}_k\|_{H^{1/2}(\Gamma)} \\ &+ \|\widetilde{\psi}_k - \dot{\psi}_{k,h}\|_{H^{1/2}(\Gamma)}\end{aligned} \tag{68}$$

and estimate each term separately. Note that for the last term, $\widetilde{\psi}_k$ and $\dot{\psi}_{k,h}$ are defined with the same data $\widetilde{\mu} = \lambda \circ E_h \circ \Pi_h$ because $\mu_h(v_h) = \widetilde{\mu}(v_h)$ for all $v_h \in \mathbb{T}_h$. Moreover, Lemma 4.3 ensures that $\dot{\psi}_k$ and $\widetilde{\psi}_k$ belong to $H^{1/2+\widetilde{\gamma}-\epsilon}(\Gamma)$. Whence, Lemma 5.3 and (67) guarantee that

$$\|\widetilde{\psi}_k - \dot{\psi}_{k,h}\|_{H^{1/2}(\Gamma)} \lesssim C(\epsilon)h^{\widetilde{\gamma}-\epsilon}\|\widetilde{\mu}\|_{H^{-1/2}(\Gamma)} \lesssim C(\epsilon)h^{\widetilde{\gamma}-\epsilon}\|\lambda\|_{H^1(\Omega)^*}.$$

The second term in (68) is an interpolation error. Proceeding with the same arguments leading to (65) and (63), we get

$$\|\Pi_h\widetilde{\psi}_k - \widetilde{\psi}_k\|_{H^{1/2}(\Gamma)} \lesssim C(\epsilon)h^{\widetilde{\gamma}-\epsilon}\|\widetilde{\mu}\|_{H^{-1/2}(\Gamma)} \lesssim C(\epsilon)h^{\widetilde{\gamma}-\epsilon}\|\lambda\|_{H^1(\Omega)^*}.$$

We now focus on the first term in (68) and invoke the stability (42) with $t = 1/2$ to obtain

$$\|\Pi_h(\dot{\psi}_k - \widetilde{\psi}_k)\|_{H^{1/2}(\Gamma)} \lesssim \|\dot{\psi}_k - \widetilde{\psi}_k\|_{H^{1/2}(\Gamma)}.$$

Moreover, Lemma 5.4 with $g = \mu - \widetilde{\mu}$, $\beta = 1/2$ and $r = \widetilde{\gamma} + 1/2 - \epsilon/2$ yields

$$\begin{aligned}\|\dot{\psi}_k - \widetilde{\psi}_k\|_{H^{1/2}(\Gamma)} &\lesssim \frac{1}{\epsilon}\|\mu - \widetilde{\mu}\|_{H^{-r}(\Gamma)} = \frac{1}{\epsilon}\sup_{v\in H^r(\Gamma)\setminus\{0\}} \frac{\lambda(Ev) - \lambda(E_h(\Pi_h v))}{\|v\|_{H^r(\Gamma)}} \\ &\lesssim \frac{1}{\epsilon}\|\lambda\|_{H^1(\Omega)^*} \sup_{v\in H^r(\Gamma)\setminus\{0\}} \frac{\|Ev - E_h(\Pi_h v)\|_{H^1(\Omega)}}{\|v\|_{H^r(\Gamma)}}.\end{aligned}$$

Now proceeding as in the proof of Lemma 4.3 we infer that

$$\|Ev - E_h(\Pi_h v)\|_{H^1(\Omega)} \lesssim C(\epsilon)h^{\min(r-1/2-\epsilon/2,\alpha_\Omega)}\|v\|_{H^r(\Gamma)}$$

and thus

$$\|\dot{\psi}_k - \widetilde{\psi}_k\|_{H^{1/2}(\Gamma)} \lesssim C(\epsilon)h^{\min(\widetilde{\gamma}-\epsilon,\alpha_\Omega)}\|\lambda\|_{H^1(\Omega)^*}.$$

The desired estimate follows upon using the estimates for each term in the right-hand side of (68). □

5 Numerical experiments

In this section, we illustrate the performances of the recovery algorithm (Algorithm 3). For the computational domain, we consider the L-shaped domain $\Omega = (-1,1)^2 \setminus \{[0,1)\times(-1,0]\}$.

We consider two functions to be recovered, namely

$$u_1(x,y) = e^x \cos(y) \tag{69}$$

and

$$u_2(x,y) = \widetilde{u}(r,\theta) = r^{\frac{2}{3}} \sin\left(\frac{2\theta}{3}\right), \qquad x = r\cos(\theta),\ y = r\sin(\theta), \tag{70}$$

referred to as the smooth and non-smooth solutions, respectively. Note that although u_1 is smooth, its trace belongs to any $H^s(\Gamma)$, $s < 3/2$ because Γ is not C^1. Also u_2 is only in $H^{5/3}(\Omega)$ and thus $u_2|_\Gamma \in H^{7/6}(\Gamma)$. In both cases, we have $f = 0$ as the functions are harmonic in Ω, and thus $u_f = 0$ in (14).

The measurements are Gaussian functionals mimicking point evaluations, that is of the form

$$\lambda(v;z) = \frac{1}{\sqrt{2\pi r^2}} \int_\Omega \exp\left(-\frac{|x-z|^2}{2r^2}\right) v(x)\mathrm{d}x, \qquad v \in H^1(\Omega), \qquad \text{with } r = 0.1. \tag{71}$$

The centers z are uniformly distributed in Ω: given an integer p controlling the number of points in each direction, we consider the centers

$$z_{i,j} = (-1 + i\delta_p, -1 + j\delta_p) \quad \text{for} \quad 1 \leqslant i,j \leqslant p, \quad \text{where} \quad \delta_p = \frac{2}{p+1},$$

as long as $z_{i,j} \in \Omega$. Table 1 reports the number of measurements for the values of p used in the numerical experiments.

p	2	3	4	5	6	7
m	3	5	12	16	27	33

Table 1: Number of measurements for the values of p used in the numerical experiments.

As mentioned earlier, the functions to be recover belong to $H^s(\Gamma)$ for $s < 3/2$ (smooth case) and $s = 7/6$ (rough case). In particular, both have $s > 1$. We now briefly describe how to extend our Algorithm 3 for $s > 1$. When $s = l + \overline{s}$ with $l \geqslant 1$ integral and $0 \leqslant \overline{s} < 1$, we obtain an approximation of $\psi = \mathcal{L}^{-s}\mu$ by proceeding iteratively as described in Algorithm 4.

Regarding the discretization parameters, we use uniform partitions of Ω made of $3 \cdot 2^{2n}$ quadrilaterals of side-length 2^{-n}, where $n = 2, \dots, 9$ is the refinement level,

Algorithm 4: Fractional diffusion problem for $s \geqslant 1$

Require: $s = l + \overline{s}$ with $l \geqslant 1$ integral and $0 \leqslant \overline{s} < 1$
Ensure : $\dot{\psi}_{k,h}$
◇ Set $\psi_h^{(0)} = \mu_h$
for $i = 1, 2, \ldots, l$ **do**
  ◇ Compute $\psi_h^{(i)} = \mathcal{L}_h^{-1} \psi_h^{(i-1)}$
end
if $\overline{s} > 0$ **then**
  ◇ Compute $\dot{\psi}_{k,h} = Q_k^{-\overline{s}}(\mathcal{L}_h)\psi_h^{(l)}$
else
  ◇ $\dot{\psi}_{k,h} = \psi_h^{(l)}$
end

corresponding to a mesh size of $h = \sqrt{2} \cdot 2^{-n}$. Also, in all experiments, the sinc quadrature parameters M and N are taken sufficiently large not to influence the results.

The recovery functions constructed by Algorithm 1 and associated to u_i, $i = 1, 2$, are denoted $\hat{u}_i$.

5.1 Performances of Algorithm 1 for the recovery of u

We analyze the behavior of the relative recovery error

$$e_i := \frac{\|u_i - \widehat{u}_i\|_{H^1(\Omega)}}{\|u_i\|_{H^1(\Omega)}}$$

for both the smooth ($i = 1$) and non-smooth ($i = 2$) for different values of s (regularity on the boundary), m (number of measurements), and n (refinement level). Note that we have $\|u_1\|_{H^1(\Omega)} \approx 2.648$ for the smooth solution and $\|u_2\|_{H^1(\Omega)} \approx 1.709$ for the non-smooth one.

In view of Theorem 4.1, the recovery error is expected to reduce as n increases until the optimal recovery threshold (Chebyshev radius $R(\dot{K}_\omega^s)$) is reached and as s increases (at least until the function to recover belongs to $\dot{K}_\omega^s$). We also expect that this threshold decreases as m increases. This is indeed what Table 2 (smooth case) and Table 3 (non-smooth case) indicate. Furthermore, we observe that the recovery error increase as m increase for n fixed. This effect is already documented in [2] and reflects the influence of the measurements in the constant $\dot{\Lambda}$ in Corollary 4.1. We also note that the relative recovery error does not seem to be affected by the smoothness of the function except that larger s can be chosen in the smooth case.

The influence of s in the recovered functions is illustrated in Figures 1 and 2 when using $m = 33$ measurements and $n = 6$. For small s the recovered solution fails to approximate accurately the exact solution near the boundary as indicated by the presence of spurious oscillations. However, the latter are reduced as s increases.

			$s=1$				$s=1.45$			
n	# cells	# dofs	$m=3$	$m=12$	$m=16$	$m=33$	$m=3$	$m=12$	$m=16$	$m=33$
2	48	65	0.71660	0.23765	0.95774	1264.02	0.63811	0.17785	1.06128	218.399
3	192	225	0.70115	0.25950	0.11432	0.53311	0.62669	0.18755	0.08289	0.60300
4	768	833	0.69910	0.26335	0.11677	0.093724	0.62505	0.18931	0.08058	0.08118
5	3072	3201	0.69838	0.26429	0.11762	0.08191	0.62445	0.18965	0.08020	0.06073
6	12288	12545	0.69813	0.26451	0.11785	0.08176	0.62423	0.18969	0.08010	0.06029
7	49152	49665	0.69803	0.26455	0.11790	0.08181	0.62415	0.18968	0.08008	0.06033

Table 2: Normalized recovery error e_1 for different refinement levels and different number of Gaussian measurements for $s = 1$ and $s = 1.45$ in the smooth solution case (69).

			$s=0.55$				$s=7/6$			
n	# cells	# dofs	$m=3$	$m=12$	$m=16$	$m=33$	$m=3$	$m=12$	$m=16$	$m=33$
2	48	65	0.74285	0.27451	1.38862	6871.97	0.34158	0.17675	1.49918	1047.31
3	192	225	0.70773	0.22492	0.19341	0.94360	0.34456	0.11568	0.12992	0.97837
4	768	833	0.69627	0.22385	0.17725	0.19350	0.34271	0.10777	0.07942	0.16295
5	3072	3201	0.69128	0.22290	0.17574	0.13210	0.34190	0.10417	0.07067	0.06218
6	12288	12545	0.68907	0.22238	0.17521	0.12761	0.34156	0.10265	0.06848	0.05098
7	49152	49665	0.68811	0.22214	0.17494	0.12694	0.34141	0.10203	0.06769	0.04937

Table 3: Normalized recovery error e_2 for different refinement levels and different number of Gaussian measurements for $s = 0.55$ and $s = 7/6$ in the non-smooth solution case (70).

Instead, Figures 3 and 4 depict the recovered functions for $s = 1$ and $n = 6$, but for different values of the number of measurements m.

Fig. 1: Recovery of the smooth solution u_1 in (69) with $m = 33$ Gaussian measurements. Left: $s = 0.55$ ($e_1 = 0.17372$), middle: $s = 1.45$ ($e_1 = 0.06029$), right: u_1.

We end this section with an open question: what can we expect from Algorithm 1 if the s chosen is larger than the regularity of the functions to be recovered? In Table 4, we report the recovery error for different values of m and s going beyond the limiting cases ($s < 3/2$ for u_1 and $s < 7/6$ for u_2). We observe that increasing s

Fig. 2: Recovery of the non-smooth solution u_2 in (70) with $m = 33$ Gaussian measurements. Left: $s = 0.55$ ($e_2 = 0.12761$), middle: $s = 7/6$ ($e_2 = 0.05098$), right: u_2.

Fig. 3: Recovery of the smooth solution u_1 in (69) with $s = 1$ and m Gaussian measurements. Left: $m = 3$ ($e_1 = 0.69813$), middle: $m = 12$ ($e_1 = 0.26451$), right: $m = 33$ ($e_1 = 0.08176$).

Fig. 4: Recovery of the non-smooth solution u_2 in (70) with $s = 1$ and m Gaussian measurements. Left: $m = 3$ ($e_2 = 0.40164$), middle: $m = 12$ ($e_2 = 0.11890$), right: $m = 33$ ($e_2 = 0.05865$).

beyond the value dictated by the regularity of u_i continues to yield smaller recovery errors but only until a certain limiting value after which the recovery errors increase. Whether this effect is a numerical artifact or can be predicted mathematically is left open for future studies.

	u_1 (smooth)		u_2 (non-smooth)	
s	$m = 3$	$m = 33$	$m = 3$	$m = 33$
0.55	0.85958	0.17372	0.68908	0.12761
0.66	0.80998	0.13560	0.59458	0.09864
0.75	0.77457	0.11460	0.53124	0.08277
1.00	0.69813	0.08176	0.40164	0.05865
7/6	0.66269	0.07059	0.34156	0.05098
1.25	0.64885	0.06671	0.31710	0.04845
1.50	0.61967	0.05915	0.26038	0.04382
1.75	0.60393	0.05512	0.22215	0.04170
1.95	0.59755	0.05325	0.20102	0.04093
2.00	0.59656	0.05290	0.19677	0.04082
5.00	0.61386	0.08286	0.15607	0.04479
10.00	0.61994	0.16415	0.15670	0.11683
20.00	0.62008	2.96580	0.15672	0.95191

Table 4: Normalized recovery error e_i for different values of s and number of Gaussian measurements for both the smooth ($i = 1$) and non-smooth ($i = 2$) cases. We use $n = 6$ for the refinement level of the mesh.

5.2 Performances of Algorithm 3 for the approximation of the Riesz representers

We end the numerical section with a discussion on the approximation of the Riesz representers using Algorithm 3 for two different choices of centers z in the Gaussian measurements (71). In absence of an exact solution, we compute the $H^1(\Omega)$ error between the output of Algorithm 3 and a reference discrete solution obtained using $n = 9$ refinements (786432 cells). The number of sinc quadrature points is large enough not to influence the approximations. The $H^1(\Omega)$ error for different values of the mesh size h is provided in Figure 5. In most cases, the observed convergence rate is approximately 0.75, which is marginally better than $2/3$. Note that for the L-shaped domain $\alpha_\Omega = 2/3 - \epsilon$ for any $\epsilon > 0$ and seems to dictate the convergence rate guaranteed by Theorem 4.1. The case $z = (-0.5, 0.5)$ and $s = 0.75$ is an outlier for which a rate closer to 1 is observed.

Acknowledgements AB is partially supported by the NSF Grant DMS-2409807. DG acknowledges the support provided by the Natural Science and Engineering Research Council (NSERC, grant RGPIN-2021-04311).

Appendix

This appendix contains several lemmas that cannot be directly found in the literature because of the low regularity of the source terms (measurements) as we all the

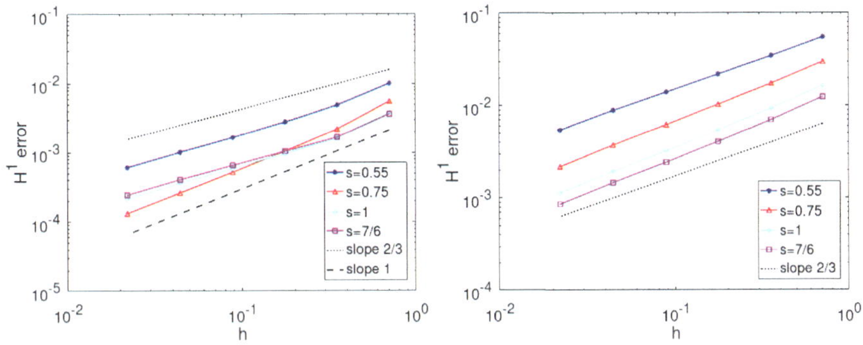

Fig. 5: Convergence of the Riesz representer for $n = 2, \ldots, 6$. The $H^1(\Omega)$ error is computed using a reference discrete solution ($n = 9$). Left: Gaussian measurement centered at $z = (-0.5, 0.5)$; right: Gaussian measurement centered at $z = (-0.1, 0.1)$.

boundary Γ considered in this work. However, the proofs presented in this appendix follows those provided in [6, 5]. Those that are only marginally different are omitted.

Given $g \in H^{-r}(\Gamma)$, $r < 1$, let $w = \mathcal{T} g \in H^1(\Gamma)$ be given by (19), i.e., w solves

$$a(w, v) := \langle w, v \rangle_{H^1(\Gamma)} = \int_\Gamma wv + \int_\Gamma \nabla_\Gamma w \cdot \nabla_\Gamma v = g(v), \qquad \forall\, v \in H^1(\Gamma). \tag{72}$$

Let $w_h \in \mathbb{T}_h$ be the finite element approximation to w, i.e., the solution to

$$a(w_h, v_h) = g(v_h), \qquad \forall\, v \in \mathbb{T}_h. \tag{73}$$

We start with estimates on the discrepancy $w - w_h$ including the two specific cases consistent with the estimates in [6, 5] under Assumption 2.2

- $\|w - w_h\|_{H^\beta(\Gamma)} \in O(h^{2-r-\beta})$ for $\beta \in [0, 1]$ and $\alpha_\Gamma = 1$;
- $\|w - w_h\|_{H^\beta(\Gamma)} \in O(h^{\alpha_\Gamma + \min\{\alpha_\Gamma, 1-\beta\}})$ for $\beta \in [0, 1]$ and $r = 0$.

Lemma 5.1 *Let α_Γ be as in Assumption 2.2. Let $g \in H^{-r}(\Gamma)$ for $r < 1$, and let w and w_h be given by* (72) *and* (73), *respectively. Then for any $\beta \in [0, 1]$ we have*

$$\|w - w_h\|_{H^\beta(\Gamma)} \lesssim h^{\gamma + \overline{\gamma}} \|g\|_{H^{-r}(\Gamma)},$$

where $\gamma := \min\{\alpha_\Gamma, 1 - r\}$ and $\overline{\gamma} := \min\{\alpha_\Gamma, 1 - \beta\}$.

Proof. We first prove the result for $\beta = 1$. Using the Galerkin orthogonality property

$$a(w - w_h, v_h) = 0, \qquad \forall\, v_h \in \mathbb{T}_h,$$

we have for $v_h \in \mathbb{T}_h$

$$\|w - w_h\|_{H^1(\Gamma)}^2 = a(w - w_h, w - w_h) = a(w - w_h, w - v_h) \leqslant \|w - w_h\|_{H^1(\Gamma)} \|w - v_h\|_{H^1(\Gamma)}$$

and thus

$$\|w - w_h\|_{H^1(\Gamma)} \leqslant \inf_{v_h \in \mathbb{T}_h} \|w - v_h\|_{H^1(\Gamma)}.$$

Now since $w \in H^{1+\gamma}(\Gamma)$, a standard finite element approximation estimate on Γ, see for instance [4], yields

$$\inf_{v_h \in \mathbb{T}_h} \|w - v_h\|_{H^1(\Gamma)} \lesssim h^\gamma \|w\|_{H^{1+\gamma}(\Gamma)}$$

and because $\gamma \leqslant \alpha_\Gamma$, Assumption 2.2 implies that

$$\|w\|_{H^{1+\gamma}(\Gamma)} \lesssim \|g\|_{H^{\gamma-1}(\Gamma)}.$$

Whence, we obtain

$$\|w - w_h\|_{H^1(\Gamma)} \lesssim h^\gamma \|g\|_{H^{\gamma-1}(\Gamma)} \lesssim h^\gamma \|g\|_{H^{-r}(\Gamma)},$$

and the case $\beta = 1$ is proved.

We deduce the general case $\beta \in (0, 1)$ from the case $\beta = 1$ using a duality argument. Let $z \in H^1(\Gamma)$ be the solution to

$$a(v, z) = \langle w - w_h, v\rangle_{\dot{H}^\beta(\Gamma)}, \qquad \forall v \in H^1(\Gamma).$$

Notice that $w - w_h \in H^\beta(\Gamma)$ and recall that $\langle w - w_h, v\rangle_{\dot{H}^\beta(\Gamma)} = \langle \mathcal{L}^\beta(w - w_h), v\rangle$. Whence, $z = \mathcal{T} g$ where $g = \mathcal{L}^\beta(w - w_h) \in H^{-\beta}(\Gamma)$. Thanks to Assumption 2.2 and the definition of $\overline{\gamma}$, we find that $z \in H^{1+\overline{\gamma}}(\Gamma)$. Choosing $v = w - w_h$ and using the Galerkin orthogonality property, we deduce that for any $v_h \in \mathbb{T}_h$ there holds

$$\|w - w_h\|_{\dot{H}^\beta(\Gamma)}^2 = a(w - w_h, z) = a(w - w_h, z - v_h) \leqslant \|w - w_h\|_{H^1(\Gamma)} \|z - v_h\|_{H^1(\Gamma)}.$$

Invoking again a standard finite element approximation estimate on Γ, we find that

$$\inf_{v_h \in \mathbb{T}_h} \|z - v_h\|_{H^1(\Gamma)} \lesssim h^{\overline{\gamma}} \|z\|_{H^{1+\overline{\gamma}}(\Gamma)} \lesssim h^{\overline{\gamma}} \|w - w_h\|_{\dot{H}^t(\Gamma)} \lesssim h^{\overline{\gamma}} \|w - w_h\|_{\dot{H}^\beta(\Gamma)},$$

for

$$t := -1 + \overline{\gamma} + 2\beta \leqslant \beta.$$

This together with the estimate previously obtained for $\beta = 1$ implies

$$\|w - w_h\|_{\dot{H}^\beta} \lesssim h^{\overline{\gamma}} \|w - w_h\|_{H^1(\Gamma)} \lesssim h^{\gamma+\overline{\gamma}} \|g\|_{H^{-r}(\Gamma)}.$$

The case $\beta \in [0, 1)$ is complete in view of the equivalence $\dot{H}^\beta(\Gamma) \cong H^\beta(\Gamma)$, see (24). □

Given $\widetilde{\mu} \in H^{-1/2}(\Gamma)$, we now derive an upper bound for the error between

$$\widetilde{\psi}_k := Q_k^{-s}(\mathcal{L})\widetilde{\mu} = \frac{k\sin(\pi s)}{\pi}\sum_{l=-\mathrm{M}}^{\mathrm{N}} \eta_l^{1-s} \underbrace{(\eta_l I + \mathcal{L})^{-1}\widetilde{\mu}}_{=:w(\eta_l)}$$

and its finite element approximation

$$\dot{\psi}_{k,h} := Q_k^{-s}(\mathcal{L}_h)\widetilde{\pi}_h\widetilde{\mu} = \frac{k\sin(\pi s)}{\pi}\sum_{l=-\mathrm{M}}^{\mathrm{N}} \eta_l^{1-s} \underbrace{(\eta_l I + \mathcal{L}_h)^{-1}\widetilde{\pi}_h\widetilde{\mu}}_{=:w_h(\eta_l)}. \tag{74}$$

Here $\widetilde{\pi}_h : H^{-1}(\Gamma) \to \mathbb{T}_h$ is a projection operator defined for $g \in H^{-1}(\Gamma)$ by the relations

$$\langle \widetilde{\pi}_h g, v_h\rangle_{L^2(\Gamma)} = \langle g, v_h\rangle, \qquad \forall\, v_h \in \mathbb{T}_h;$$

recall that we have identified $L^2(\Gamma)$ with its dual. In particular, if $g \in L^2(\Gamma)$ then $\widetilde{\pi}_h g = \pi_h g$ with $\pi_h : L^2(\Gamma) \to \mathbb{T}_h$ the $L^2(\Gamma)$ orthogonal projection operator.

To estimate the error $\|\widetilde{\psi}_k - \dot{\psi}_{k,h}\|_{H^{1/2}(\Gamma)}$, we state a result slightly more general than Lemma 4.8 in [6]. We omit the proof because it follows the one of Lemma 4.8 in [6].

Lemma 5.2 *Let α_Γ be as in Assumption 2.2 and let $g \in H^{-r}(\Gamma)$ with $r < 1$. Moreover, for $\eta \in (0, \infty)$, let $w(\eta) = (\eta I + \mathcal{L})^{-1} g$ and $w_h(\eta) = (\eta I + \mathcal{L}_h)^{-1}\widetilde{\pi}_h g$. Then for any $\delta \in [r, r+1]$ with $\delta < 1$ and any $\tau, \beta \in [0, 1]$ with $\tau + \beta \in [0, 1]$ we have*

$$\|\pi_h w(\eta) - w_h(\eta)\|_{H^\beta(\Gamma)} \lesssim \eta^{\frac{r-\delta-\tau}{2}} h^{\min\{\sigma_1,\sigma_2\}} \|g\|_{H^{-r}(\Gamma)}, \tag{75}$$

where

$$\sigma_1 := \min\{1-\delta, \alpha_\Gamma\} + 1 - \tau - \beta \quad \textit{and} \quad \sigma_2 := \min\{1-\delta, \alpha_\Gamma\} + \min\{1-\tau-\beta, \alpha_\Gamma\}.$$

In the specific case where $r = \beta = 1/2$, the estimate in Lemma 5.2 reads: for any $\delta \in [1/2, 1)$ and any $\tau \in [0, 1/2]$ we have

$$\|\pi_h w(\eta) - w_h(\eta)\|_{H^{1/2}(\Gamma)} \lesssim \eta^{\frac{1/2-\delta-\tau}{2}} h^{\min\{\sigma_1,\sigma_2\}} \|g\|_{H^{-1/2}(\Gamma)}, \tag{76}$$

where

$$\sigma_1 = \min\{1-\delta, \alpha_\Gamma\} + 1/2 - \tau \quad \text{and} \quad \sigma_2 = \min\{1-\delta, \alpha_\Gamma\} + \min\{1/2-\tau, \alpha_\Gamma\}. \tag{77}$$

We are now in position to derive an estimate for $\|\widetilde{\psi}_k - \dot{\psi}_{k,h}\|_{H^{1/2}(\Gamma)}$.

Lemma 5.3 *Let α_Γ be as in Assumption 2.2, $s \in (1/2, 1)$, and*

$$\widetilde{\gamma} := \min\{2s - 1, \alpha_\Gamma + 1/2, 2\alpha_\Gamma\}.$$

Then for $\widetilde{\mu} \in H^{-1/2}(\Gamma)$ and for $0 < \epsilon < \widetilde{\gamma}$, there is a constant $C(\epsilon)$ such that $C(\epsilon) \to \infty$ as $\epsilon \to 0^+$ and

$$\|\widetilde{\psi}_k - \dot{\psi}_{k,h}\|_{H^{1/2}(\Gamma)} \leqslant C(\epsilon) h^{\widetilde{\gamma}-\epsilon} \|\widetilde{\mu}\|_{H^{-1/2}(\Gamma)}.$$

Proof. Without loss of generality, we assume that $h \leqslant 1$. Let $\alpha^* := (\widetilde{\gamma} - \epsilon)/2$. We start by using triangle's inequality to decompose the error as

$$\|\widetilde{\psi}_k - \dot{\psi}_{k,h}\|_{H^{1/2}(\Gamma)} \leqslant \|\widetilde{\psi}_k - \pi_h \widetilde{\psi}_k\|_{H^{1/2}(\Gamma)} + \|\pi_h \widetilde{\psi}_k - \dot{\psi}_{k,h}\|_{H^{1/2}(\Gamma)}. \tag{78}$$

The first term is the error due to interpolation. Proceeding as in the proof of Lemma 4.3 (but using π_h instead of Π_h), we obtain

$$\|\widetilde{\psi}_k - \pi_h \widetilde{\psi}_k\|_{H^{1/2}(\Gamma)} \lesssim C(\epsilon) h^{\widetilde{\gamma}-\epsilon} \|g\|_{H^{-1/2}(\Gamma)}. \tag{79}$$

For the second term, we first observe that

$$\|\pi_h \widetilde{\psi}_k - \dot{\psi}_{k,h}\|_{H^{1/2}(\Gamma)} \leqslant \frac{\sin(\pi s)}{\pi} k \sum_{l=-\mathtt{M}}^{\mathtt{N}} \eta_l^{1-s} \|\pi_h w(\eta_l) - w_h(\eta_l)\|_{H^{1/2}(\Gamma)}.$$

To simplify the notation, for any $\eta \in (0, \infty)$ we write $e(\eta) := \pi_h w(\eta) - w_h(\eta)$. The rest of the proof then consists in splitting the sum into three parts, namely consider

$$\eta_l \in (0,1), \quad \eta_l \in [1, h^{-\frac{2\alpha^*}{s-1/2}}), \quad \eta_l \in [h^{-\frac{2\alpha^*}{s-1/2}}, \infty),$$

and use different estimates for $\|e(\eta_l)\|_{H^{1/2}(\Gamma)}$.

[1] Case $\eta_l \in (0,1)$. We use (76) with $\delta = 1/2$ and $\tau = 0$. Since

$$\sigma_1 = \min\{1/2, \alpha_\Gamma\} + 1/2 \leqslant \widetilde{\gamma} \quad \text{and} \quad \sigma_2 = 2\min\{1/2, \alpha_\Gamma\} \lesssim \widetilde{\gamma}$$

we obtain

$$\|e(\eta_l)\|_{H^{1/2}(\Gamma)} \lesssim h^{\widetilde{\gamma}} \|g\|_{H^{-1/2}(\Gamma)} \lesssim h^{2\alpha^*} \|g\|_{H^{-1/2}(\Gamma)},$$

and thus

$$\begin{aligned}\frac{\sin(\pi s)}{\pi} k \sum_{0<\eta_l<1} \eta_l^{1-s} \|e(\eta_l)\|_{H^{1/2}(\Gamma)} &\lesssim k \sum_{0<\eta_l<1} \eta_l^{1-s} h^{2\alpha^*} \|g\|_{H^{-1/2}(\Gamma)} \\ &\lesssim h^{2\alpha^*} \|g\|_{H^{-1/2}(\Gamma)}.\end{aligned}$$

[2] Case $\eta_l \in [h^{-\frac{2\alpha^*}{s-1/2}}, \infty)$. Using triangle's inequality and the stability property (42) we have

$$\|e(\eta_l)\|_{H^{1/2}(\Gamma)} \leqslant \|w(\eta_l)\|_{H^{1/2}(\Gamma)} + \|w_h(\eta_l)\|_{H^{1/2}(\Gamma)}.$$

The first term can then be bounded proceeding as in the proof of Lemma 5.4 below, namely use (83) and (84) with $r = \beta = 1/2$ (with y replaced by $\eta_l = e^{y_l}$) together with the equivalence (24). The bound for the second term is similar upon replacing the role of $\mathcal{L}$ by $\mathcal{L}_h$ [9, 6]. Whence, we infer that

$$\|e(\eta_l)\|_{H^{1/2}(\Gamma)} \lesssim \eta_l^{-1/2}\|g\|_{H^{-1/2}(\Gamma)}.$$

Consequently, we find that

$$\begin{aligned}\frac{\sin(\pi s)}{\pi}k \sum_{\eta_l \geqslant h^{-\frac{2\alpha^*}{s-1/2}}} \eta_l^{1-s}\|e(\eta_l)\|_{H^\beta(\Gamma)} &\lesssim k \sum_{\eta_l \geqslant h^{-\frac{2\alpha^*}{s-1/2}}} \eta_l^{1/2-s}\|g\|_{H^{-1/2}(\Gamma)} \\ &\lesssim h^{2\alpha^*}\|g\|_{H^{-1/2}(\Gamma)};\end{aligned}$$

where the last inequality is derived as in the proof of Theorem 4.2 in [6].

3 Case $\eta_l \in [1, h^{-\frac{2\alpha^*}{s-1/2}})$. We decompose this case in three subcases and invoke (76) for different values of $\delta \in [1/2, 1)$ and $\tau \in [0, 1/2]$. In each case, we show that

$$\frac{\sin(\pi s)}{\pi}k \sum_{1 \leqslant \eta_l < h^{-\frac{2\alpha^*}{s-1/2}}} \eta_l^{1-s}\|e(\eta_l)\|_{H^{1/2}(\Gamma)} \lesssim \frac{1}{\epsilon}h^{2\alpha^*}\|g\|_{H^{-1/2}(\Gamma)}. \tag{80}$$

i) Case $\alpha_\Gamma \geqslant 1/2$. Since $1-\delta \in [0, 1/2)$ and $1/2 - \tau \in [0, 1/2]$ we have $\sigma_1 = \sigma_2 = 3/2 - \delta - \tau$. We then choose δ and τ so that

$$\delta + \tau = 3/2 + 1 - 2s + \epsilon \in [1/2, 3/2)$$

in which case (76) reads

$$\|e(\eta_l)\|_{H^{1/2}(\Gamma)} \lesssim \eta_l^{s-1-\epsilon/2}h^{2s-1-\epsilon}\|g\|_{H^{-1/2}(\Gamma)} \leqslant \eta_l^{s-1-\epsilon/2}h^{2\alpha^*}\|g\|_{H^{-1/2}(\Gamma)}.$$

Therefore, we get

$$\begin{aligned}&\frac{\sin(\pi s)}{\pi}k \sum_{1 \leqslant \eta_l < h^{-\frac{2\alpha^*}{s-1/2}}} \eta_l^{1-s}\|e(\eta_l)\|_{H^{1/2}(\Gamma)} \\ &\quad \lesssim k \sum_{1 \leqslant \eta_l < h^{-\frac{2\alpha^*}{s-1/2}}} \eta_l^{-\epsilon/2}h^{2\alpha^*}\|g\|_{H^{-1/2}(\Gamma)} \lesssim \frac{1}{\epsilon}h^{2\alpha^*}\|g\|_{H^{-1/2}(\Gamma)}.\end{aligned}$$

and (80) is proven in that subcase.

ii) Case $\alpha_\Gamma < 1/2$ and $\alpha_\Gamma \geqslant s - 1/2$. In this case, we choose

$$\delta = 3/2 - s + \epsilon/2 \in [1/2, 1) \qquad \text{and} \qquad \tau = 1 - s + \epsilon/2 \in [0, 1/2]$$

so that $\sigma_1 = \sigma_2 = 2s - 1 - \epsilon$. Whence, from (76) we get

$$\|e(\eta_l)\|_{H^{1/2}(\Gamma)} \lesssim \eta_l^{s-1-\epsilon/2}h^{2s-1-\epsilon}\|g\|_{H^{-1/2}(\Gamma)} \leqslant \eta_l^{s-1-\epsilon/2}h^{2\alpha^*}\|g\|_{H^{-1/2}(\Gamma)}$$

and we proceed as in the previous case to obtain (80).

iii) Case $\alpha_\Gamma < 1/2$ and $\alpha_\Gamma < s - 1/2$. This time we take

$$\delta = 1 - \alpha_\Gamma \in [1/2, 1) \qquad \text{and} \qquad \tau = 1/2 - \alpha_\Gamma \in [0, 1/2].$$

This gives $\sigma_1 = \sigma_2 = 2\alpha_\Gamma$ and thus (76) becomes

$$\|e(\eta_l)\|_{H^{1/2}(\Gamma)} \lesssim \eta_l^{\alpha_\Gamma - 1/2} h^{2\alpha_\Gamma} \|g\|_{H^{-1/2}(\Gamma)} \leqslant \eta_l^{s-1-\widetilde{\epsilon}/2} h^{2\alpha_\Gamma} \|g\|_{H^{-1/2}(\Gamma)},$$

where $\widetilde{\epsilon} := 2s - 1 - 2\alpha_\Gamma > 0$ because $\alpha_\Gamma - 1/2 < s - 1$. This implies that

$$\begin{aligned} \frac{\sin(\pi s)}{\pi} k \sum_{1 \leqslant \eta_l < h^{-\frac{2\alpha^*}{s-1/2}}} \eta_l^{1-s} \|e(\eta_l)\|_{H^{1/2}(\Gamma)} &\lesssim \widetilde{\epsilon}^{-1} h^{2\alpha_\Gamma} \|g\|_{H^{-1/2}(\Gamma)} \\ &\lesssim \frac{1}{\epsilon} h^{2\alpha^*} \|g\|_{H^{-1/2}(\Gamma)}, \end{aligned}$$

which is (80).

Gathering the above estimates for all three cases we have

$$\|\pi_h \widetilde{\psi}_k - \dot{\psi}_{k,h}\|_{H^{1/2}(\Gamma)} \lesssim \frac{1}{\epsilon} h^{2\alpha^*} \|g\|_{H^{-1/2}(\Gamma)} = \frac{1}{\epsilon} h^{\widetilde{\gamma} - \epsilon} \|g\|_{H^{-1/2}(\Gamma)}. \tag{81}$$

Estimates (79) and (81) in (78) yield the desired estimate. □

We now extend the stability provided by Lemma 4.1 in [6] to less regular right-hand side. The proof follows that in [6].

Lemma 5.4 *Suppose that $g \in H^{-r}(\Gamma)$ and let $v_k := Q_k^{-s}(\mathcal{L})g$. Then for any $\beta < 2s - r$ we have*

$$\|v_k\|_{\dot{H}^\beta(\Gamma)} \lesssim \max\left\{\frac{1}{1-s}, \frac{1}{2s - r - \beta}\right\} \|g\|_{\dot{H}^{-r}(\Gamma)}.$$

Consequently, there holds

$$\|v_k\|_{H^\beta(\Gamma)} \lesssim \max\left\{\frac{1}{1-s}, \frac{1}{2s - r - \beta}\right\} \|g\|_{H^{-r}(\Gamma)}$$

provided $\beta, r \in [-1 - \alpha_\Gamma, 1 + \alpha_\Gamma]$.

Proof. From the definition of v_k (recall that $\eta = e^y$), we deduce that

$$\|v_k\|_{\dot{H}^\beta(\Gamma)} \leqslant \frac{k \sin(\pi s)}{\pi} \sum_{l=-\mathtt{M}}^{\mathtt{N}} e^{(1-s)y_l} \|w(y_l)\|_{\dot{H}^\beta(\Gamma)}, \quad w(y_l) = (e^{y_l} I + \mathcal{L})^{-1} g. \tag{82}$$

We now estimate the $\dot{H}^\beta(\Gamma)$ norm of $w(y)$ for any $y \in \mathbb{R}$. Recall that $\{(\tau_j, e_j)\}_{j=1}^\infty$ denotes the eigenpairs of $\mathcal{L}$ and set $v := \mathcal{L}^{-r/2} g \in L^2(\Gamma)$ so that

$$\|w(y)\|^2_{\dot{H}^\beta(\Gamma)} = \|\mathcal{L}^{\frac{\beta}{2}}(e^{y_l}I+\mathcal{L})^{-1}\mathcal{L}^{\frac{r}{2}}v\|^2_{L^2(\Gamma)}$$
$$= \sum_{j=1}^{\infty} \tau_j^{\beta}(e^y+\tau_j)^{-2}\tau_j^r|v_j|^2 = \sum_{j=1}^{\infty}\left(\frac{\tau_j^{\frac{\beta+r}{2}}}{e^y+\tau_j}\right)^2|v_j|^2, \qquad (83)$$

where $v_j := \langle v, e_j\rangle_{L^2(\Gamma)}$ for $j \geqslant 1$. Since $\beta + r < 2s < 2$ and $\tau_j \geqslant 1$ for any $j \geqslant 1$, we have

$$\frac{\tau_j^{\frac{\beta+r}{2}}}{e^y+\tau_j} \leqslant \tau_j^{\frac{\beta+r}{2}-1} \leqslant 1.$$

Alternatively, Young's inequality $ab \leqslant a^p/p + b^q/q$ with $p = 1/(1-\theta)$, $q = 1/\theta$, and $\theta = (\beta+r)/2$, ensures that

$$\frac{\tau_j^{\frac{\beta+r}{2}}}{e^y+\tau_j} = e^{\left(\frac{\beta+r}{2}-1\right)y}\frac{e^{\left(1-\frac{\beta+r}{2}\right)y}\tau_j^{\frac{\beta+r}{2}}}{e^y+\tau_j} \leqslant e^{\left(\frac{\beta+r}{2}-1\right)y}. \qquad (84)$$

Whence, returning to (82) we obtain

$$\|v_k\|_{\dot{H}^\beta(\Gamma)} \leqslant \frac{k\sin(\pi s)}{\pi}\left[\sum_{l=-\mathtt{M}}^{0} e^{(1-s)y_l} + \sum_{l=1}^{\mathtt{N}} e^{\left(\frac{\beta+r}{2}-s\right)y_l}\right]\|v\|_{L^2(\Gamma)},$$

which is the first desired estimate in disguised upon noting that $\|v\|_{L^2(\Gamma)} = \|g\|_{\dot{H}^{-r}(\Gamma)}$.

The second estimate directly follows from the equivalence (24). □

References

1. A. V. Balakrishnan. Fractional powers of closed operators and the semigroups generated by them. *Pacific J. Math.*, 10:419–437, 1960.
2. P. Binev, A. Bonito, A. Cohen, W. Dahmen, R. DeVore, and G. Petrova. Solving PDEs with incomplete information. *SIAM J. Numer. Anal.*, 62(3):1278–1312, 2024.
3. A. Bonito, J. P. Borthagaray, R. H. Nochetto, E. Otárola, and A. J. Salgado. Numerical methods for fractional diffusion. *Comput. Vis. Sci.*, 19(5):19–46, 2018.
4. A. Bonito, A. Demlow, and R. H. Nochetto. Finite element methods for the Laplace–Beltrami operator. In *Handbook of Numerical Analysis*, volume 21, pages 1–103. Elsevier, 2020.
5. A. Bonito, D. Guignard, and W. Lei. Numerical approximation of Gaussian random fields on closed surfaces. *Comput. Methods Appl. Math.*, 2024.
6. A. Bonito and W. Lei. Approximation of the spectral fractional powers of the Laplace-Beltrami operator. *Numer. Math. Theory Methods Appl.*, 15(4):1193–1218, 2022.
7. A. Bonito, W. Lei, and J. E. Pasciak. On sinc quadrature approximations of fractional powers of regularly accretive operators. *J. Numer. Math.*, 27(2):57–68, 2019.
8. A. Bonito and J. E. Pasciak. Numerical approximation of fractional powers of elliptic operators. *Math. Comput.*, 84(295):2083–2110, 2015.
9. A. Bonito and J. E. Pasciak. Numerical approximation of fractional powers of regularly accretive operators. *IMA J. Numer. Anal.*, 37(3):1245–1273, 2017.

10. S. C. Brenner and L. R. Scott. *The mathematical theory of finite element methods*. Springer, 2008.
11. S. L. Brunton, B. R. Noack, and P. Koumoutsakos. Machine learning for fluid mechanics. *Annu. Rev. Fluid Mech.*, 52:477–508, 2020.
12. K. Duraisamy, G. Iaccarino, and H. Xiao. Turbulence modeling in the age of data. *Annu. Rev. Fluid Mech.*, 51:357–377, 2019.
13. M. Faustmann, J. M. Melenk, and M. Parvizi. On the stability of scott-zhang type operators and application to multilevel preconditioning in fractional diffusion. *ESAIM: M2AN*, 55(2):595–625, 2021.
14. L. Formaggia, J.-F. Gerbeau, F. Nobile, and A. Quarteroni. Numerical treatment of defective boundary conditions for the navier–stokes equations. *SIAM J. Numer. Anal.*, 40(1):376–401, 2002.
15. L. Grinberg and G. E. Karniadakis. Outflow boundary conditions for arterial networks with multiple outlets. *Ann. Biomed. Eng.*, 36(9):1496–1514, 2008.
16. P. Grisvard. *Elliptic Problems in Nonsmooth Domains*. Pitman, 1985.
17. A. Lischke, G. Pang, M. Gulian, F. Song, C. Glusa, X. Zheng, Z. Mao, W. Cai, M. M. Meerschaert, M. Ainsworth, and G. E. Karniadakis. What is the fractional laplacian? a comparative review with new results. *J. Comput. Phys.*, 404:109009, 2020.
18. J. Lund and K. L. Bowers. *Sinc methods for quadrature and differential equations*. SIAM, 1992.
19. C. A. Micchelli and T. J. Rivlin. Lectures on optimal recovery. In *Numerical Analysis Lancaster 1984*, pages 21–93. Springer Berlin Heidelberg, 1985.
20. M. Raissi, A. Yazdani, and G. E. Karniadakis. Hidden fluid mechanics: Learning velocity and pressure fields from flow visualizations. *Science*, 367(6481):1026–1030, 2020.
21. K. Rajagopal. On boundary conditions for fluids of the differential type. In *Navier-Stokes Equations and Related Nonlinear Problems*, pages 273–278. Springer, 1995.
22. P. J. Richards and S. E. Norris. Appropriate boundary conditions for computational wind engineering models revisited. *J. Wind. Eng. Ind. Aerodyn.*, 99(4):257–266, 2011.
23. P. J. Richards and S. E. Norris. Appropriate boundary conditions for computational wind engineering: Still an issue after 25 years. *J. Wind. Eng. Ind. Aerodyn.*, 190:245–255, 2019.
24. L. R. Scott and S. Zhang. Finite element interpolation of nonsmooth functions satisfying boundary conditions. *Math. Comput.*, 54(190):483–493, 1990.
25. H. Xu, W. Zhang, and Y. Wang. Explore missing flow dynamics by physics-informed deep learning: The parameterized governing systems. *Phys. Fluids*, 33(9):095116, 2021.

Constructions of Bounded Solutions of *div* u = f in Critical Spaces

Albert Cohen,
Ronald DeVore and
Eitan Tadmor

Abstract We construct uniformly bounded solutions of the equation $div\,\mathbf{u} = f$ for arbitrary data f in the critical spaces $L^d(\Omega)$, where Ω is a domain of $\mathbb{R}^d$. This question was addressed by Bourgain & Brezis, [BB2003], who proved that although the problem has a uniformly bounded solution, it is critical in the sense that there exists no linear solution operator for general L^d-data. We first discuss the validity of this existence result under weaker conditions than $f \in L^d(\Omega)$, and then focus our work on constructive processes for such uniformly bounded solutions. In the $d = 2$ case, we present a direct one-step explicit construction, which generalizes for $d > 2$ to a $(d-1)$-step construction based on induction. An explicit construction is also proposed for compactly supported data in $L^{d,\infty}(\Omega)$. We finally present constructive approaches based on optimization of a certain loss functional adapted to the problem. This approach provides a two-step construction in the $d = 2$ case. This optimization is used as the building block of a hierarchical multistep process introduced in [Tad2014] that converges to a solution in more general situations.

1 Introduction

Let Y denote a Banach space of functions defined on a d-dimensional domain $\Omega \subset \mathbb{R}^d$, where $d \geqslant 2$. We are concerned with the existence and construction of

---

Albert Cohen
Laboratoire Jacques-Louis Lions, UPMC Univ Paris 06, UMR 7598, Paris, France F-75005, e-mail: albert.cohen@sorbonne-universite.fr

Ronald DeVore
Department of Mathematics, Texas A&M University, College Station, TX 77840, USA, e-mail: rdevore@math.tamu.edu

Eitan Tadmor
Department of Mathematics, Institute for Physical sciences and Technology (IPST), University of Maryland, MD 20742-4015, USA, e-mail: tadmor@umd.edu

R. DeVore, A. Kunoth (eds.), *Multiscale, Nonlinear and Adaptive Approximation II*,
https://doi.org/10.1007/978-3-031-75802-7_9

uniformly bounded solutions u to the equation,

$$div\, \mathbf{u} = f, \tag{1}$$

whenever $f \in Y$. Namely, we ask whether there exists a $\gamma > 0$ such that for every $f \in Y$ there exists a solution $\mathbf{u} = (u_1, \dots, u_d) \in L^\infty(\Omega)$ to (1) such that

$$\|\mathbf{u}\|_{L^\infty} \leqslant \gamma \|f\|_Y. \tag{2}$$

Here for a vector valued function $\mathbf{v} = (v_1, \dots, v_d)$ such that each v_i belongs to a function space X, we use the simpler notation $\mathbf{v} \in X$ and $\|\mathbf{v}\|_X$ instead of X^d. We say the space Y is *admissible* if for all $f \in Y$, (1) admits a solution such that (2) holds.

There exists of course infinitely many solutions to (1) since as soon as one exists, we can add to it a null divergence function, for example a constant. The most natural candidate for a solution $\mathbf{u}$ when given f is to solve Laplace's equation with data f and then to take $\mathbf{u}$ as the gradient of the solution. More precisely, we introduce

$$\psi(x) = \nabla\phi(x) = \frac{C_d}{|x|^d}x, \quad x \in \mathbb{R}^d,$$

where ϕ is the fundamental solution of the Laplacian on $\mathbb{R}^d$, and define the so-called Helmholtz solution as

$$\mathbf{u}(x) = \mathbf{u}_{\mathrm{Hel}}(x) = \int_\Omega f(x)\psi(x-y)\mathrm{d}y = \tilde{f} * \psi(x),$$

where $\tilde{f}(x) = f(x)$ for $x \in \Omega$ and $\tilde{f}(x) = 0$ when $x \notin \Omega$. Note that $\mathbf{u}_{\mathrm{Hel}}$ depends linearly on f. When Ω is a bounded domain, it is readily seen that $\mathbf{u}_{\mathrm{Hel}}$ is a uniformly bounded solution of (1) for $f \in L^p(\Omega)$ whenever $p > d$, and therefore the spaces $Y = L^p$, $p > d$ are all admissible.

The question of whether $Y = L^d$ is admissible was addressed in the seminal work of Bourgain & Brezis [BB2003]. Their work studies the particular case where $\Omega = \mathbb{T}^d$ is the d-dimensional torus, which leads to assume in addition that $\int_{\mathbb{T}^d} f = 0$. They proved that the problem (1) is *critical* in the sense that it admits bounded solutions, but there is no *linear* solution operator from Y to L^∞. In particular, one cannot invoke the Helmholtz solution. We say a space Y is critical if it is admissible but there is no linear mapping taking $f \in Y$ into a solution $\mathbf{u} \in L^\infty$ of (2).

The main interest of the present paper is two-fold. We first ask which of the classical function spaces Y are admissible. Secondly, we are interested in explicit constructions of solutions to the Bourgain-Brezis problem. In particular, can we explicitly construct nonlinear mappings solving (2) when Y is critical. In section 2 we discuss theoretical aspects of the problem. We recall certain known results of Meyer which show that for $\Omega = \mathbb{R}^d$, the space G of all f that admit a solution $\mathbf{u} \in L^\infty$ to (2) is the dual space $W^{1,1}_{\mathrm{hom}}$ which is defined as the closure of the smooth test functions for the total variation. Therefore, any admissible space Y must be a subspace of G. In particular, Y is admissible if and only if $W^{1,1}_{\mathrm{hom}}$ embeds into Y^*. In particular, we show that not only is L^d admissible but also the larger space weak-L^d, i.e. $L^{d,\infty}(\Omega)$, is admissible, as well as even weaker Morrey spaces.

In section 3, we present explicit constructions of bounded solutions for certain critical admissible spaces Y. We give a one step formula in the $d = 2$ case with L^2-data, and we treat the case $d > 2$ with L^d-data by a $(d-1)$-step construction based on induction. [1] We end this section by a construction for $L^{d,\infty}$-data assuming in addition compact support. The reader may find these constructions interesting for their own sake. In section 4, we propose variational-based approaches for the constructions of bounded solutions. In the case $d = 2$, this approach delivers the solution in two steps. More generally, we use this optimization as the building block of a hierarchical decomposition that was used in [Tad2014] to construct solutions to the Bourgain-Brezis with L^d data by a limiting process. We use this multi-step hierarchical approach to construct solutions for more general data.

2 Theory

2.1 Existence of bounded solutions for L^d-data

Let $\Omega \subset \mathbb{R}^d$. The space

$$G = G(\Omega) := \{f = div\,(\mathbf{u}) \ : \ \mathbf{u} \in L^\infty(\Omega)\},$$

of distributions u whose divergence is uniformly bounded has been studied in various contexts, in particular image processing and nonlinear PDE's.

As noted in [Mey2002], for the case $\Omega = \mathbb{R}^d$, the space $G(\mathbb{R}^d)$ is the dual of the homogeneous space $W^{1,1}_{\text{hom}}(\mathbb{R}^d)$. The latter is defined as is the completion of the space of test functions $\mathcal{D}(\mathbb{R}^d)$ for the total variation, which defines a norm on this space.

Let us recall that the total variation of $v \in BV(\Omega)$ is defined as

$$|v|_{TV} := \sup_{\mathbf{w}} \int_\Omega v\, div\,\mathbf{w},$$

where the supremum is taken over all $\mathbf{w} = (w_1, \ldots, w_d) \in \mathcal{D}(\Omega)$ such that $\|\mathbf{w}\|_{L^\infty} = \sup_{x\in\Omega} |\mathbf{w}(x)|_2 \leqslant 1$. An equivalent quantity is defined in terms of finite difference:

$$|v|_{TV} \sim \sup_{h>0} h^{-1} \sup_{|y|\leqslant h} \|v - v(\cdot - y)\|_{L^1(\Omega_h)},$$

where $\Omega_h := \{x \in \Omega \ : \ \text{dist}(x, \partial\Omega) > h\}$. When $\nabla v \in L^1$, in particular when $v \in W^{1,1}_{\text{hom}}$, one simply has

[1] We recently learned that this type of construction was independently derived by D. Stolyarov [Sto2024], however unpublished

$$|v|_{TV} = \|\nabla v\|_{L^1}.$$

Let us stress that $W^{1,1}_{\text{hom}}(\mathbb{R}^d)$ is strictly smaller than $BV(\mathbb{R}^d)$.

Therefore, every f in a function space Y of locally integrable functions defined on $\mathbb{R}^d$ admits the representation $f = div\, \mathbf{u}$ with a uniformly bounded $\mathbf{u}$ satisfying the bound (2) if and only if for any test function $g \in \mathcal{D}(\mathbb{R}^d)$ one has

$$\left|\int_{\mathbb{R}^d} fg\right| \leqslant \gamma \|f\|_Y |g|_{TV}. \tag{3}$$

Note that this is equivalent to the condition that

$$\left|\int_E f\right| \leqslant \gamma \|f\|_Y \,\text{per}(E), \tag{4}$$

for all open sets E of finite perimeter. Indeed the above is obtained from (3) by taking $g = \varphi_\epsilon * \chi_E$ where φ_ϵ is a mollifier and letting $\epsilon \to 0$. But from the coarea formula

$$|g|_{TV} = \int_{-\infty}^{+\infty} \text{per}(E_t)\text{d}t, \quad E_t := \{x \ : \ g(x) > t\}, \tag{5}$$

see [EG1992], it also implies (3) for any $g \in \mathcal{D}(\mathbb{R}^d)$. Note that here, the perimeter per(E) coincides with the Hausdorff measure $\mathcal{H}^{d-1}(\partial E)$ only for sufficiently nice sets (for example with Jordan domains with rectifiable boundaries). More generally it should be be defined as $|\chi_E|_{TV}$ or equivalently as $\mathcal{H}^{d-1}(\partial E^*)$ where ∂E^* is the so-called reduced boundary as introduced by de Giorgi.

Remark 2.1 The co-area formula also shows that (4) actually implies the validity of (3) for any $g \in BV(\mathbb{R}^d)$. Therefore any $f \in G$ that is in addition locally integrable is also an element of the dual of BV. We give further in Remark 2.5 an example of a distribution that belongs to G but are not in the dual of BV.

For $Y = L^d(\mathbb{R}^d)$, the validity of (3) is ensured by the Sobolev embedding of $BV(\mathbb{R}^d)$ into $L^{d'}(\mathbb{R}^d)$ where $\frac{1}{d} + \frac{1}{d'} = 1$. Since for a general domain Ω, we can trivially extend any $f \in L^d(\Omega)$ by 0 to obtain a function of $L^d(\mathbb{R}^d)$ with the same norm, this implies that $Y = L^d(\Omega)$ is admissible.

In [BB2003], the same result is given in the periodic context, where $\Omega = \mathbb{T}^d$ is the d-dimensional torus. In this case, $Y = L^d(\Omega)$ is modified into

$$Y = L^d_\#(\mathbb{T}^d) = \left\{f \in L^d(\mathbb{T}^d) \ : \ \int_\Omega f = 0\right\}$$

As shown in Proposition 2 therein, there exists no linear solution operator $f \in L^d_\#(\mathbb{T}^d) \mapsto \mathbf{u} \in L^\infty(\mathbb{T}^d)$. Indeed, restricting attention to the simpler case of the two-dimensional torus, if $K : L^2_\#(\mathbb{T}^2) \mapsto L^\infty(\mathbb{T}^2)$ would be such a linear solution operator so that $div\, K = \mathbb{I}$ is the identity, then so is

$$\widetilde{K} := \int_{y\in\mathbb{T}^2} \tau_{-y} K \tau_y \mathrm{d}y,$$

which averages K over all 2D translations τ_y. Now $\widetilde{K}$ is translation invariant and it has a symbol, $\Lambda(n) := (\lambda_1(n), \lambda_2(n))$ such that $\widetilde{K}(e^{in\cdot x}) = \Lambda(n)e^{in\cdot x}$. Since $\widetilde{K}$ is assumed to boundedly map L^2 to L^∞, one should have $(\Lambda(n))_{n\in\mathbb{Z}^2} \in \ell^2(\mathbb{Z}^2)$. However, since $div\,\widetilde{K} = div\,K = \mathbb{I}$, that is $n \cdot \Lambda(n) = 1$, this implies $|\Lambda(n)|_2 \geqslant \dfrac{1}{|n|_2}$ which is a contradiction to $(\Lambda(n))_{n\in\mathbb{Z}^2} \in \ell^2(\mathbb{Z}^2)$.

The lack of linearity is attributed to the general fact that the problem of solving $\mathcal{L}\mathbf{u} = f$ with $\mathbf{u} \in X$ is critical if $\mathrm{Ker}(\mathcal{L})$ has no complement in X [BB2007, Aji2009]. This is the case of div in L^∞. One of the main themes in [BB2003] is the existence of solution with further $W^{1,d}$-regularity, similar to the Helmholtz solution $\mathbf{u}_{\mathrm{Hel}}$ that cannot be ensured to be uniformly bounded since it depends linearly on f.

2.2 Existence of solutions for $L^{d,\infty}$ data.

One first observation is that the Bourgain-Brezis problem has also a positive answer for the larger Lorentz space $Y = L^{d,\infty}(\Omega)$. Recall that a measurable function f is in $L^{d,\infty}(\Omega)$ if and only if $|\{x \in \Omega : |f(x)| > t\}| \leqslant C^d t^{-d}$, $t > 0$, and the smallest C for which this holds is its $L^{d,\infty}(\Omega)$ norm.

Theorem 2.1 *There exists a constant $\gamma = \gamma_d$ such that for any $f \in L^{d,\infty}(\Omega)$, there exists $\mathbf{u} \in L^\infty(\Omega)$ satisfying*

$$div\,\mathbf{u} = f, \quad \|\mathbf{u}\|_{L^\infty} \leqslant \gamma \|f\|_{L^{d,\infty}(\Omega)}. \tag{6}$$

Proof. By definition of $Y = L^{d,\infty}(\Omega)$, one has for any $f \in Y$ and measurable E,

$$\begin{aligned}\int_E |f| = \int_{t>0} |\{x \in E \,:\, |f(x)| > t\}|\mathrm{d}t &\leqslant \int_{t>0} \min\{|E|, \|f\|^d_{L^{d,\infty}} t^{-d}\}\mathrm{d}t \\ &\leqslant \frac{d}{d-1}\|f\|_{L^{d,\infty}}|E|^{\frac{d-1}{d}}\end{aligned}$$

Therefore (4) follows by application of the isoperimetric inequality with $\gamma_d = \frac{d}{d-1}K_d$ where K_d is the isoperimetry constant [Fed1969, 3.2.43]. □

Remark 2.2 An equivalent proof consists in establishing that $BV(\mathbb{R}^d)$ has continuous embedding in the dual Lorentz space $Y^* = L^{d',1}(\mathbb{R}^d)$, that is

$$\|g\|_{L^{d',1}(\mathbb{R}^d)} \leqslant \beta_d |g|_{TV}. \tag{7}$$

This readily follows by using the expression of the $L^{d,1}(\mathbb{R}^d)$ norm through the distribution function

$$\|g\|_{L^{d',1}(\mathbb{R}^d)} = d' \int_0^\infty |\{x \in \Omega \ : \ |g(x)| > t\}|^{\frac{d-1}{d}} \mathrm{d}t,$$

see e.g. [BC2011], and invoking the co-area formula and isoperimetric inequality to bound this quantity by the total variation of g.

Remark 2.3 The property assering that

$$\int_E |f| \leqslant C|E|^{\frac{d-1}{d}}$$

holds for all measurable sets E is actually equivalent to the membership of f in $L^{d,\infty}(\Omega)$. The smallest C for which this is valid gives an equivalent norm for $L^{d,\infty}$. We shall use this norm in going forward in this paper.

2.3 Beyond $L^{d,\infty}$

What is the largest Banach space Y of Borel measures μ which can be expressed as divergences of uniformly bounded $\mathbf{u}$? Placing ourselves in $\Omega = \mathbb{R}^d$, we know that such a Y should be embedded in $G(\mathbb{R}^d)$ the dual of $W^{1,1}_{\mathrm{hom}}(\mathbb{R}^d)$, that is, for all $\mu \in Y$ and $g \in W^{1,1}_{\mathrm{hom}}(\mathbb{R}^d)$ one has

$$\int_{\mathbb{R}^d} g\mathrm{d}\mu \leqslant \gamma\|\mathrm{d}\mu\|_Y |g|_{TV}. \tag{8}$$

Using the co-area formula, this is ensured in particular if

$$|\mu(E)| \leqslant \gamma\|\mu\|_Y \mathrm{per}(E) \tag{9}$$

for all sets $E \subset \mathbb{R}^d$ of finite perimeter.

Let us introduce the linear space of measures $S^d(\mathbb{R}^d)$ that satisfy the condition

$$|\mu|(B) \leqslant CR^{d-1}, \quad R > 0, \tag{10}$$

for all balls B or radius R, equipped with the norm

$$\|\mu\|_{S^d} := \sup R^{1-d}|\mu|(B), \tag{11}$$

where the supremum is taken over all balls B. For a general domain Ω, we define $S^d(\Omega)$ in a similar manner, replacing B by $B \cap \Omega$, and observe that any measure in this space has its extension by 0 contained in $S^d(\mathbb{R}^d)$ with a smaller or equal norm.

Remark 2.4 General conditions of the form $\mu(B) \leqslant CR^s$ were introduced by Otto Frostman in the study of fractional dimension of sets. In dimension $d = 2$, for positive measures, the specific condition $\mu(B) \leqslant CR$ was studied by Guy David for Dirac measures on a curve Γ. He proved that this condition is equivalent to the Ahlfors regularity of Γ and to the boundedness of the Cauchy integral operator acting on $L^2(\Gamma, \mu)$.

A distinction should be made between $S^d(\Omega)$ and the Morrey space $M^d(\Omega)$ that consists of all locally integrable f such that, for all ball B of radius R,

$$\int_{B\cap\Omega} |f| \leqslant CR^{d-1}, \tag{12}$$

with norm defined in a similar manner. This space is included in $S^d(\Omega)$ with equal norm when μ is of the form $f\,dx$, but the inclusion is strict: consider for example μ to be the Dirac measure on a segment of the plane in the case $d = 2$. In view of Remark 2.3, we have

$$L^{d,\infty} \subset M^d \subset S^d,$$

and these inclusions are strict. The following result that follows the arguments from [Mey2002] and [PT2017], shows that the Bourgain-Brezis problem has also a positive answer for $Y = M^d(\Omega)$ and $Y = S^d(\Omega)$.

Theorem 2.2 *There exists a constant $\gamma = \gamma_d$ such that the following holds. For any $f \in M^d(\Omega)$, there exists $\mathbf{u} \in L^\infty(\Omega)$ satisfying*

$$div\,\mathbf{u} = f, \quad \|\mathbf{u}\|_{L^\infty} \leqslant \gamma\|f\|_{M^d}. \tag{13}$$

For any $\mu \in S^d(\Omega)$, there exists $\mathbf{u} \in L^\infty(\Omega)$ satisfying

$$div\,\mathbf{u} = \mu, \quad \|\mathbf{u}\|_{L^\infty} \leqslant \gamma\|\mu\|_{S^d}. \tag{14}$$

Proof. Without loss of generality we work on $\Omega = \mathbb{R}^d$. As a first step, we use the boxing inequality [PT2008, Theorem 2.11] that states that any open set $E \subset \mathbb{R}^d$ of finite perimeter can be covered by balls B_j of radius R_j such that

$$\sum_j R_j^{d-1} \leqslant C\mathrm{per}(E), \tag{15}$$

where the constant C depends only on d. This shows that for any $f \in M^d(\mathbb{R}^d)$, we have

$$\int_E |f|\mathrm{d}x \leqslant C\|f\|_{M^d}\mathrm{per}(E).$$

which implies (4) and therefore proving (13).

Similarly, for any $\mu \in S^d(\mathbb{R}^d)$, we have

$$|\mu|(E) \leqslant C\|\mu\|_{S^d}\mathrm{per}(E).$$

For any test function $g \in \mathcal{D}(\mathbb{R}^d)$, we write $g = g_+ - g_-$ and

$$\left|\int_{\mathbb{R}^d} g\mathrm{d}\mu\right| \leqslant \int_{\mathbb{R}^d} g_+\mathrm{d}|\mu| + \int_{\mathbb{R}^d} g_-\mathrm{d}|\mu|.$$

For the first term, we have

$$\int_{\mathbb{R}^d} g_+ \mathrm{d}|\mu| = \int_0^\infty |\mu|(E_t)\mathrm{d}t \leqslant C\|\mu\|_{S^d} \int_0^\infty \mathrm{per}(E_t)$$

where $E_t := \{x \,:\, g(x) > t\}$. With a similar treatment of the second term and using the co-area formula, we reach

$$\left|\int_{\mathbb{R}^d} g\mathrm{d}\mu\right| \leqslant C\|\mu\|_{S^d}|g|_{TV}$$

which shows that μ belongs to the space $G(\mathbb{R}^d)$ and thereby proves (14). □

Remark 2.5 We stress that, in contrast to the functions of $M^d(\mathbb{R}^d)$, the measures of $S^d(\mathbb{R}^d)$ belong to $G(\mathbb{R}^d)$ but not to the dual of $BV(\mathbb{R}^d)$. This is due to the fact that the trace of a BV function on a $d-1$ dimensional surface could be meaningless. For example if μ is the Dirac measure on a segment of the $2d$ plane, we cannot apply it to the BV function $g = \chi_Q$ where Q is a square that admits this segment as one of its side.

Remark 2.6 As pointed out in [Mey2002] in the case $d = 2$, a positive measure belongs to $G(\mathbb{R}^d)$ if and only if it belongs to $S^d(\mathbb{R}^d)$. Indeed, on the one hand the above result shows that $S^d(\mathbb{R}^d)$ is contained in $G(\mathbb{R}^d)$. On the other hand, if μ is a positive measure that is contained in $G(\mathbb{R}^d)$ then to any ball $B = B(x_0, R)$ we associate the $W^{1,1}$ function

$$g(x) = \max\left\{0, 2 - \frac{|x - x_0|}{R}\right\}$$

Since μ is positive and belongs to $G(\mathbb{R}^d)$, and since g is positive and larger than 1 on B, we find that

$$\mu(B) \leqslant \int g\mathrm{d}\mu \leqslant C|g|_{TV}.$$

On the other hand, it is easily checked that $|g|_{TV} \leqslant C_d R^{d-1}$ where C_d only depend of d, therefore proving that $\mu \in S^d(\mathbb{R}^d)$.

3 Explicit constructions of bounded solutions

3.1 A one-step explicit construction for L^2-data

What follows is probably the simplest and most instructive construction of bounded solutions to the Bourgain-Brezis problem (1), at least in the d =-case with $Y = L^2(\Omega)$. Again, without loss of generality we will work on $\Omega = \mathbb{R}^2$.

For any $(x, y) \in \mathbb{R}^2$ and any fixed $f \in L^2(\mathbb{R}^2)$, we define

$$V^2(x) := \int_{-\infty}^{+\infty} |f(x,y)|^2\,dy, \quad H^2(y) := \int_{-\infty}^{+\infty} |f(x,y)|^2\,dx \tag{16}$$

and

$$\alpha(x,y) := \frac{V(x)}{H(y)+V(x)}, \quad \beta(x,y) := \frac{H(y)}{H(y)+V(x)}. \tag{17}$$

We then consider the splitting $f = f_1 + f_2$ where

$$f_1(x,y) := \alpha(x,y)f(x,y), \quad \text{and} \quad f_2(x,y) := \beta(x,y)f(x,y), \tag{18}$$

and we define

$$u_1(x,y) := \int_{-\infty}^{x} f_1(s,y)ds \quad \text{and} \quad u_2(x,y) := \int_{-\infty}^{y} f_2(x,t)dt \tag{19}$$

Therefore $\mathbf{u} = (u_1, u_2)$ satisfies $div\,\mathbf{u} = f_1 + f_2 = f$ and it remains to check that u is uniformly bounded. Let us bound $|u_1(x,y)|$ for any arbitrary but fixed $(x,y) \in \mathbb{R}^2$. If $H(y) = 0$, then obviously $u_1(x,y) = 0$ for all x so we assume $H(y) \neq 0$. Then, for any x we have

$$\begin{aligned}
|u_1(x,y)| &\leqslant \int_{-\infty}^{\infty} |f_1(s,y)|\,ds \leqslant \int_{-\infty}^{\infty} |f(s,y)|\frac{V(s)}{H(y)}\,ds \\
&\leqslant H(y)^{-1}\left(\int_{-\infty}^{\infty} |f(s,y)|^2\,ds\right)^{1/2}\left(\int_{-\infty}^{\infty} V(s)^2\,ds\right)^{1/2} \\
&= \left(\int_{-\infty}^{\infty} V(s)^2\,ds\right)^{1/2} = \|f\|_{L^2(\mathbb{R}^2)}.
\end{aligned}$$

In a similar way, we obtain the bound $\|u_2\|_{L^\infty} \leqslant \|f\|_{L^2(\mathbb{R}^2)}$, and so $\mathbf{u}$ is a bounded solution to the Bourgain-Brezis problem for f.

Remark 3.1 The above splitting of f into f_1 and f_2 is designed to ensure that the univariate primitive u_1 and u_2 are uniformly bounded. An interesting variant consists in taking

$$f_1(x,y) := f(x,y)\chi_{\{H(y)\leqslant V(x)\}} \quad \text{and} \quad f_2(x,y) := f(x,y)\chi_{\{V(x)<H(y)\}} \tag{20}$$

for which it is easily checked that the solution u also has each component u_1 and u_2 uniformly bounded by $\|f\|_{L^2}$. The extra feature of this choice is that f_1 and f_2 have disjoint supports. As we discuss next, in the more general d-dimensional case with $Y = L^d(\mathbb{R}^d)$, it is possible to explicitly construct a splitting $f = f_1 + \cdots + f_d$ also with disjoint supports and such that the univariate primitive u_j of f_j with respect to x_j are uniformly bounded.

3.2 A $(d-1)$-step explicit construction for L^d-data

We now consider the general d-dimensional case with data $f \in L^d(\mathbb{R}^d)$. Given any such f, we construct d pairwise disjoint sets $\Omega_j = \Omega_j(f)$ with $\mathbb{R}^d = \bigcup_{j=1}^d \Omega_j$, so that the functions $f_j := f\chi_{\Omega_j}$ satisfy $f = f_1 + \cdots + f_d$ as well as

$$\int_{\mathbb{R}} |f_j(x_1,\ldots,x_{j-1},s,x_{j+1},\ldots,x_d)|\,ds \leqslant \|f\|_{L^d(\mathbb{R}^d)}, \quad j=1,\ldots,d. \tag{21}$$

In turn the functions

$$u_j(x_1,\ldots,x_d) := \int_{-\infty}^{x_j} f_j(x_1,\ldots,x_{j-1},s,x_{j+1},\ldots,x_d)\,ds, \quad j=1,\ldots,d, \tag{22}$$

satisfy $\|u_j\|_{L^\infty(\mathbb{R}^d)} \leqslant \|f\|_{L^d(\mathbb{R}^d)}$. Hence, $\mathbf{u} = (u_1,\ldots,u_d)$ is a solution to (1) with $\|\mathbf{u}\|_{L^\infty(\mathbb{R}^d)} \leqslant \|f\|_{L^d(\mathbb{R}^d)}$.

The construction proceeds by induction on d. When $d=1$, given any $f \in L^1(\mathbb{R})$, we define $\Omega_1 = \mathbb{R}$ in which case the above claim is obvious. We assume we have shown how to construct such sets $\Omega_1(g).\ldots,\Omega_{d-1}(g)$ whenever $g \in L^{d-1}(\mathbb{R}^{d-1})$ and give the construction of $\Omega_1(f),\ldots,\Omega_d(f)$ whenever $f \in L^d(\mathbb{R}^d)$. Without loss of generality, we can assume $\|f\|_{L^d(\mathbb{R}^d)} = 1$.

We write any vector $x \in \mathbb{R}^d$ as (x_1, y) where $y = (x_2,\ldots,x_d) \in \mathbb{R}^{d-1}$ and define the thresholds

$$t(y)^{d-1} := \int_{\mathbb{R}} |f(x_1,y)|^d dx_1, \quad y \in \mathbb{R}^{d-1}.$$

Let

$$\Omega_1 := \Omega_1(f) := \{x = (x_1,y) \in \mathbb{R}^d : |f(x_1,y)| \geqslant \tau(y)\}, \tag{23}$$

and Ω_1' be its complement in $\mathbb{R}^d$. We define

$$f_1 := f\chi_{\Omega_1} \quad \text{and} \quad g := f - f_1 = f\chi_{\Omega_1'}. \tag{24}$$

This determines u_1 and for any $x = (x_1, y) \in \mathbb{R}^d$, we have

$$\int_{\mathbb{R}} |f_1(s,y)|ds \leqslant \int_{\mathbb{R}} |f(s,y)| \left(\frac{|f(s,y)|}{t(y)}\right)^{d-1} ds = t(y)^{1-d} \int_{\mathbb{R}} |f(s,y)|^d ds = 1.$$

This shows that (21) holds for f_1 and $\|u_1\|_{L^\infty(\mathbb{R})} \leqslant 1$ as desired.

We proceed to construct the set $\Omega_2,\ldots,\Omega_d$. For any fixed $x_1 \in \mathbb{R}$, we consider the function $g(x_1,y)$ as a function of $y \in \mathbb{R}^{d-1}$. We have

$$\int_{\mathbb{R}^{d-1}} |g(x_1,y)|^{d-1} dy \leqslant \int_{\mathbb{R}^{d-1}} t(y)^{d-1} dy = \|f\|_{L^d(\mathbb{R}^d)}^d = 1. \tag{25}$$

From the induction hypothesis, we can apply our construction to $g(x_1,\cdot)$ which is a function of $d-1$ variables $y = (y_2,\ldots,y_d)$. This gives $d-1$ disjoint sets $\Omega_j(x_1)$ for $j = 2,\ldots,d$ whose union is $\mathbb{R}^{d-1}$ and for which the above results are valid.

Therefore $g(x_1, \cdot)$. is split into functions $g_j(x_1, \cdot) = g(x_1, \cdot)\chi_{\Omega_j(x_1)}$ which satisfy

$$\int_{\mathbb{R}} |g_j(x_1, y_2, \ldots, y_{j-1}, s, y_{j+1}, \ldots, y_d)|\, ds \leqslant 1, \quad j = 2, \ldots, d. \tag{26}$$

We now define

$$\Omega_j = \Omega_j(f) := \{(x_1, y)) : x_1 \in \mathbb{R},\ y \in \Omega_j(x_1)\}, \quad j = 2, \ldots, d.$$

This completes the definition of the sets Ω_j and the functions f_j and u_j. We are left to check (21) for $j \neq 1$. Since $f_j(x_1, y) = g_j(x_1, y)$, it suffices to write

$$\int_{\mathbb{R}} |f_j(x_1, \ldots, x_{j-1}, s, x_{j+1}, \ldots, x_d)|\, ds$$
$$= \int_{\mathbb{R}} |g_j(x_1, y_2, \ldots, y_{j-1}, s, y_{j+1}, \ldots, y_{d-1})|\, ds \leqslant 1.$$

This establishes the properties we want of our construction for d dimensions.

3.3 A constructive decomposition for $L^{d,\infty}$

We know from the theoretical results of §2, that the space $Y = L^{d,\infty}(\Omega)$ is admissible whenever $\Omega \subset \mathbb{R}^d$ is measurable. In this section, for any $\tau > 1$ and any bounded measurable set $\Omega \subset \mathbb{R}^d$, we give an algorithm that takes any $f \in L^{d,\infty}(\Omega)$ and constructs a solution $\mathbf{u}$ to (1) such that $\|\mathbf{u}\|_{L^\infty} \leqslant \tau \|f\|_{L^{d,\infty}}$. For the sake of notational simplicity, we detail the construction in the case $d = 2$ and sketch its generalization to $d > 2$.

We fix Ω and f in going forward. We assume without loss of generality that $\Omega = [-R, R]^2$, for some $R > 0$ and $\|f\|_{L^{2,\infty}(\Omega)} = 1$. We define f to be zero outside of Ω. We construct a disjoint splitting $\Omega = \Omega_1 \cup \Omega_2$ and $f_j := f\chi_{\Omega_j}$, $j = 1, 2$, and take

$$u_1(x, y) := \int_{-\infty}^{x} f_1(s, y)\, ds, \quad u_2(x, y) := \int_{-\infty}^{y} f_2(x, s)\, ds. \tag{27}$$

Thus $div\,(\mathbf{u}) = f$ when $\mathbf{u} := (u_1, u_2)$, and the only issue will be to show that

$$\int_{-\infty}^{\infty} |f_1(s, y)|\, ds \leqslant \tau, \quad \int_{-\infty}^{\infty} |f_2(x, s)|\, ds \leqslant \tau, \tag{28}$$

for all $x, y \in \mathbb{R}$.

Let us first note that $f \in L^1(\mathbb{R}^2)$ and

$$M := \|f\|_{L^1(\mathbb{R}^2)} = \int_{\Omega} |f| \leqslant |\Omega|^{1/2}, \tag{29}$$

where we used Remark 2.3. We define the horizontal line $L_H(y) := \{(x,y) : x \in [-R,R]\}$ at level y and the vertical line $L_V(x) := \{(x,y) : y \in [-R,R]\}$ at level x. For any measurable function g that is supported on Ω, we define the energies

$$E_H(g,y) := \int_{L_H(y)} |g(x,y)|\, dx, \ y \in [-R,R], \qquad (30)$$

$$E_V(g,x) := \int_{L_V(x)} |g(x,y)|\, dy, \ x \in [-R,R],$$

which may be infinite.

Here is the first step of our construction. Let $A := \{y \in [-R,R] : E_H(f,y) \leqslant \tau\}$ and $A' := \{y \in [-R,R] : E_H(f,y) > \tau\}$ and

$$\Omega_H := \{(x,y) \in \mathbb{R}^2 : E_H(f,y) \leqslant \tau\} = \bigcup_{y \in A} L_H(y).$$

We define $f_1 := f$ on Ω_H and $f_2 := 0$ on Ω_H. Notice that we have

$$\int_{-\infty}^{x} |f_1(s,y)|\, ds \leqslant \tau, \quad y \in A. \qquad (31)$$

This means that (28) is satisfied for $y \in A$.

We proceed to the second step of our construction. Let $B := \{x \in [-R,R] : E_V(f,x) \leqslant \tau\}$ and $B' := \{x \in [-R,R] : E_V(f,x) > \tau\}$ and

$$\Omega_V := \{(x,y) \in \Omega : E_V(f,x) \leqslant \tau\} = \bigcup_{x \in B} L_V(x).$$

We define $f_2 = f$ on $\Omega_V \setminus \Omega_H$ and f_1 is defined to be zero on this set. We have

$$\int_{-\infty}^{y} |f_2(x,s)|\, ds \leqslant \tau, \quad x \in B. \qquad (32)$$

Thus far, we have defined f_1 and f_2 on $\Omega_H \cup \Omega_V$. Let $\Omega' := \Omega \setminus (\Omega_H \cup \Omega_V)$. The important thing to notice is that Ω' is gotten from Ω by removing horizontal and vertical strips. The following lemma shows that we have removed a significant portion of Ω in this construction.

Lemma 3.1 *The measure of Ω' satisfies*

$$|\Omega'| \leqslant \tau^{-2}|\Omega|. \qquad (33)$$

Proof: To prove this claim, we observe that

$$\Omega' := \{(x,y) \in \Omega : E_H(f,y) \geqslant \tau \text{ and } E_V(f,x) \geqslant \tau\} = A' \times B'.$$

Let $a := |A'|$ and $b := |B'|$ be the univariate Lebesgue measure of these sets. Then, we have

$$a\tau \leqslant \int_{\Omega} |f(x, y)| \, dxdy \leqslant |\Omega|^{1/2}. \tag{34}$$

A similar argument gives $b\tau \leqslant |\Omega|^{1/2}$. Hence, we have

$$|\Omega'| = ab \leqslant \tau^{-2}|\Omega|, \tag{35}$$

which proves the lemma. □

After applying the first step of our construction, we have defined f_1 and f_2 outside of Ω'. Let $\Omega^1 := \Omega'$ and let us repeat our construction for the set Ω^1 in place of Ω. This gives a new set $\Omega^2 := [\Omega^1]' \subset \Omega^1$ and thereby give the definitions of f_1, f_2 outside of Ω^2. The new residual set Ω^2 satisfies $|\Omega^2| \leqslant \tau^{-2}|\Omega^1| \leqslant \tau^{-4}|\Omega|$. Iterating this procedure gives in the limit a definition of f_1 and f_2 on all of Ω except for a set of measure zero. One easily checks that (28) holds. For example to check this for f_1, we note that if f_1 is defined to be nonzero on on a line $L_H(y)$ then at the (first) step k where it is defined to be nonzero it is completely defined on this line and (28) holds on this line.

The generalization of the construction to $d > 2$ with $\Omega = [-R, R]^d$ is based on a similar process assuming $\|f\|_{L^{d,\infty}(\Omega)} = 1$: introducing the energies

$$E_i(f, x_1, \ldots, x_{i-1}, x_{i+1}, \ldots, x_d) := \int_{-R}^{R} |f(x_1, \ldots, x_d)| dx_1$$

and comparing them with the level τ, we reccursively define $f_1, f_2, \ldots, f_d$ on $\Omega \setminus \Omega'$, where

$$\Omega' = \{(x_1, \ldots, x_d) \in \Omega \; : \; E_i(f, x_1, \ldots, x_{i-1}, x_{i+1}, \ldots, x_d) \geqslant \tau, \; i = 1, \ldots, d\}.$$

In order to conclude, we prove a contraction property similar to Lemma 3.1. For each $i = 1, \ldots, d$, we denote by $\Omega_i' \in [-R, R]^d$ the projection of Ω on the $d-1$-hyperplane orthogonal to the x_i axis, and invoke the Loomis-Whitney inequality

$$|\Omega'|^{d-1} \leqslant \prod_{i=1}^{d} |\Omega_i'|, \tag{36}$$

see [LW1949]. Obviously we have

$$\Omega_i' \subset \{(x_1, \ldots, x_{i-1}, x_{i+1}, \ldots, x_d) \in [-R, R]^{d-1} \; : \; E_i(f, x_1, \ldots, x_{i-1}, x_{i+1}, \ldots, x_d) \geqslant \tau\},$$

and therefore

$$\tau|\Omega_i'| \leqslant \int_{\Omega} |f(x_1, \ldots, x_d)| \leqslant |\Omega|^{\frac{d-1}{d}}.$$

Combining with (36), we obtain the contraction property $|\Omega'| \leqslant \tau^{-d}|\Omega|$ allowing us to conclude in a similar manner.

Remark 3.2 Note that the final result is independent of the support $[R, R]^d$, but the above construction does not seem to generalize in a straighforward manner to $\Omega = \mathbb{R}^d$ which is left as an open problem.

4 Variational-based constructions

4.1 Minimization problems

One natural way of approaching bounded solutions to (1) for data $f \in Y(\Omega)$ is to consider the minimization of functionals of the form

$$\mathcal{V}(\mathbf{u}) = \mathcal{V}_\lambda(\mathbf{u}) = \|\mathbf{u}\|_{L^\infty} + \lambda \|f - div\, \mathbf{u}\|_Y^p \tag{37}$$

for some fixed $p \geqslant 1$. Indeed, one intuition is that if a uniformly bounded solution to (1) exists, the minimizer of $\mathcal{V}_\lambda$ should tend to the solution $\mathbf{u}$ of (1) with minimal L^∞ norm as $\lambda \to \infty$.

We shall first see that in the case of $Y = L^2$ and $p = 2$, it is possible to avoid letting $\lambda \to \infty$ through a two-step constructive approach. We then discuss more general situations where we can construct a uniformly bounded solution $\mathbf{u}$ by a hierarchical decompositions based on iterated minimizations of the above functional.

Before going further let us observe that the existence of a minimizer for $\mathcal{V}$ can be derived by the elementary arguments under a mild assumption on the space Y.

Lemma 4.1 *Assume that $Y = Z'$ is a dual space of distribution, so that*

$$\|v\|_Y := \max\{\langle v, \varphi\rangle_{Y,Z} \ : \ \varphi \in \mathcal{D}(\Omega),\ \|\varphi\|_Z \leqslant 1\},$$

then there exists a minimizer $\mathbf{u}_1$ of $\mathcal{V}$

Proof. Consider a minimizing sequence $\mathbf{u}^n$, therefore such that $\|\mathbf{u}^n\|_{L^\infty}$ and $\|div\, \mathbf{u}^n\|_Y$ are are uniformly bounded. Then, up to a subsequence extraction, we have the following properties :

(i) Both $\|f - div\, \mathbf{u}^n\|_{L^2}$ and $\|\mathbf{u}^n\|_{L^\infty}$ have limits A and B such that $A + \lambda B$ is the infimum of $\mathcal{V}$.
(ii) $\mathbf{u}^n$ converges in the L^∞ weak-$*$ sense to some $\mathbf{u}_1 \in L^\infty$.
(iii) $div\, \mathbf{u}^n$ converges in the Y weak-$*$ sense to the (weak) divergence $div\, \mathbf{u}_1 \in L^2$.

From this and the properties of weak lower semi-continuity of norms, it readily follows that $\mathbf{u}_1$ is a minimizer of $\mathcal{V}$. □

Existence is therefore ensured for reflexive spaces Y such as $L^d(\Omega)$ in the d-dimensional case, but also for $Y = L^{d,\infty}$ which we have seen earlier to be an admissible choice for the existence of uniformly bounded property. We stress that uniqueness of minimizers, is in general not ensured, however in the case where Y is strictly convex, such as L^d, we find that $div\,(\mathbf{u}_1)$ is unique.

Note that the minimization of $\mathcal{V}$ may be computationally intensive, depending on the form of the Y norm. In the particular case of $Y = L^d$, it can be computed by solving relatively simple Euler-Lagrange equations, see [TT2011].

4.2 A two-step approach for L^2-data

Consider the case of a domain $\Omega \subset \mathbb{R}^2$ and $Y = L^2(\Omega)$. With the choise $p = 2$, the functional of interest is therefore

$$\mathcal{V}(\mathbf{u}) = \|\mathbf{u}\|_{L^\infty} + \lambda\|f - div\,\mathbf{u}\|_{L^2}^2 \tag{38}$$

An interesting property of the minimizers is given by the following.

Lemma 4.2 *Fix $\lambda > 0$ and let $r_\lambda = f - div\,\mathbf{u}_\lambda$ be the residual of the equation* (1) *for a minimizer $\mathbf{u}_\lambda$ of* (38). *Then r_λ belongs to $BV(\Omega)$ with*

$$|r_\lambda|_{TV} \leqslant \frac{1}{2\lambda}. \tag{39}$$

Proof. For any $\mathbf{z} \in \mathcal{D}(\Omega)$ and $\epsilon > 0$, we have

$$\begin{aligned}\mathcal{V}(\mathbf{u}_\lambda) &= \|\mathbf{u}_\lambda\|_{L^\infty} + \lambda\|f - div\,\mathbf{u}_\lambda\|_{L^2}^2 \leqslant \|\mathbf{u}_\lambda + \epsilon\mathbf{z}\|_{L^\infty} + \lambda\|f - div\,(\mathbf{u}_\lambda + \epsilon\mathbf{z})\|_{L^2}^2\\ &\leqslant \|\mathbf{u}_\lambda\|_{L^\infty} + \epsilon\|\mathbf{z}\|_{L^\infty} + \lambda\|r_\lambda\|_{L^2}^2 - 2\lambda\epsilon\int_\Omega r_\lambda div\,\mathbf{z} + o(\epsilon).\\ &= \mathcal{V}(\mathbf{u}_\lambda) + \epsilon\|\mathbf{z}\|_{L^\infty} - 2\lambda\epsilon\int_\Omega r_\lambda div\,\mathbf{z} + o(\epsilon)\end{aligned}$$

and by letting $\epsilon \downarrow 0$ we find that

$$\int_\Omega r_\lambda div\,\mathbf{z} \leqslant \frac{1}{2\lambda}\|\mathbf{z}\|_{L^\infty}, \tag{40}$$

for all $\mathbf{z} \in \mathcal{D}(\Omega)$, which shows that $r_\lambda \in BV(\Omega)$ with bound (39) for its total variation. □

Note that we also have the trivial bounds

$$\|r_\lambda\|_{L^2} \leqslant \|f\|_{L^2} \quad \text{and} \quad \|\mathbf{u}\|_{L^\infty} \leqslant \lambda\|f\|_{L^2}^2, \tag{41}$$

by comparing $\mathcal{V}(\mathbf{u}_\lambda)$ with $\mathcal{V}(0)$. However Lemma 4.2 shows a "regularization effect" $f \in L^2 \mapsto r_\lambda \in BV$. As noted in Remark 2.2, the space BV has a continuous embedding in the Lorentz space $L^{2,1}$ which is strictly smaller than L^2.

This effect leads us to a direct construction of a bounded solution. Without loss of generality, we again work on $\Omega = \mathbb{R}^2$, and denote by $\mathbf{u}_1$ the minimizer and $r_1 = f - div\,\mathbf{u}_1$ the residual, when taking the particular value $\lambda := \|f\|_{L^2}^{-1}$. Using both (39) and (7), we have on the one hand

$$\|r_1\|_{L^{2,1}} \leqslant \beta_2 |r_1|_{TV} \leqslant \frac{\beta_2}{2\lambda} = \frac{\beta_2}{2}\|f\|_{L^2}; \tag{42}$$

and on the other hand

$$\|\mathbf{u}_1\|_{L^\infty} \leqslant \|f\|_{L^2}, \tag{43}$$

in view of the second bound in (41). We then write $r_1 = div\, \mathbf{u}_2$, where

$$\mathbf{u}_2 := \psi * r_1 = \frac{1}{2\pi}\frac{x}{|x|^2} * r_1, \tag{44}$$

is the Helmholtz solution for the data r_1. Since $\psi \in L^{2,\infty}$ and $r_1 \in L^{2,1}$, it is readily seen that $\mathbf{u}_2$ is uniformly bounded by

$$\|\mathbf{u}_2\|_{L^\infty} \leqslant \|\psi\|_{L^{2,\infty}} \|r_1\|_{L^{2,1}} \leqslant C\|f\|_{L^2}, \quad C := \frac{\beta_2}{2}\|\psi\|_{L^{2,\infty}}. \tag{45}$$

Thus we end up with

$$\mathbf{u}_{2\text{step}} := \mathbf{u}_1 + \mathbf{u}_2, \tag{46}$$

as a two-step construction of a uniformly bounded solution to (1) which satisfies (2) with $\gamma = 1 + C$.

Remark 4.1 A similar regularization effect takes place in the $d > 2$ case for data $f \in L^d$

$$f \in L^d(\Omega) \mapsto r_1 \in L^{d,d-1}(\Omega).$$

However, since $\psi \in L^{d,\infty}$, this is not enough to derive a similar two-step construction by applying the Helmoltz solver to the residual. Instead, this will be addressed by the multi-step construction in the next section below.

Figure 1, quoted from [TT2011, Section 6], shows the two-step solution of the example due to L. Nirenberg, [BB2003, Remark 7], which demonstrates the unboundedness of $\|\mathbf{u}_{\text{Hel}}\|_{L^\infty}$ solved for $\mathbf{u} \in L^2_\#([-1,1]^2)$ with periodic boundary conditions, given by

$$f = \Delta v \qquad v(x_1, x_2) := x_1 |\log|x||^{1/3}\zeta(|x|), \qquad \zeta(r) = \chi_{(-1,1)} e^{-\frac{1}{1-r^2}}. \tag{47}$$

In this case, Helmholtz solution, $\mathbf{u}_{\text{Hel}} = \nabla V$, has a fractional logarithmic growth at the origin, which should be contrasted with the hounded two-step constructed solution shown in figure 1. Table 1 reports that the ratio between $N \times N$ grid discretization of $\|\mathbf{u}^N_{2\text{step}}\|_{L^\infty}$ and $\|f^N\|_{L^2}$ remains bounded when N is large, in contrast to the computed solution of Helmholtz.

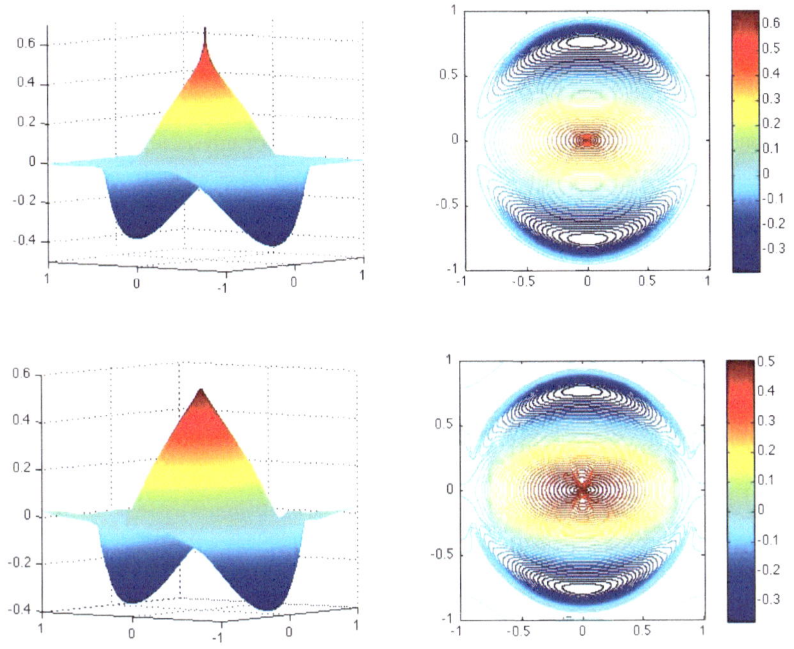

Fig. 1: Solution of Bourgain-Brezis problem with 2D data in (47).
Helmholtz solution, $\mathbf{u}_{\text{Hel}}$ (top), vs. two-step solution, $\mathbf{u}_{\text{2step}}$ (bottom).

The $N \times N$ grid	50×50	100×100	200×200	400×400	800×800
$\dfrac{\lVert \mathbf{u}^{1,N}_{\text{Hel}} \rVert_{L^\infty}}{\lVert f^N \rVert_{L^2}}$	0.2295	0.2422	0.2540	0.2650	0.2752
$\dfrac{\lVert \mathbf{u}^{1,N}_{\text{2step}} \rVert_{L^\infty}}{\lVert f^N \rVert_{L^2}}$	0.2096	0.2128	0.2144	0.2151	0.2154

Table 1: L^∞ norm of numerical solutions for different grids: Helmholtz vs. the two-step solution of (47) for different grids.

4.3 Hierarchical constructions for data in Fréchet differentiable spaces.

We now work with general data $f \in Y(\mathbb{R}^d)$. In this section, we use a hierarchical approach to construct uniformly bounded solution, under the assumption that the Y norm is Frechet differentiable.

Assumption (Fréchet differentiablity). *The Y-norm is* Fréchet differentiable, *namely — there exists $\phi : Y \to Y'$ such that*

$$\|v + \epsilon w\|_Y = \|v\|_Y + \epsilon\langle\phi(v), w\rangle + o(\epsilon) \quad \text{for all } v, w \in Y, \quad v \neq 0. \tag{48}$$

As an immediate consequence of this assumption, for any $p > 1$, the application $v \mapsto \|v\|_Y^p$ is also Fréchet differentiable and its derivative is given by

$$\phi_p(v) := p\|v\|_Y^{p-1}\phi(v).$$

As we discuss further, spaces admitting Fréchet differentiable norms are for example $Y = L^d$ as well as $Y = L^{d,q}$ when $1 < q < \infty$, but not $Y = L^{d,\infty}$.

Let us now consider for a $p > 1$ the general functional $\mathcal{V}_\lambda$ of (37), assuming as before that Y is a dual space. In the sequel we use the following two results which makes use of ϕ_p. The first is the generalization of Lemma 4.2 that shows a regularization effect, now on $\phi_p(r_\lambda)$ where $r_\lambda = f - div\, \mathbf{u}_\lambda$ is the residual.

Lemma 4.3 *If $\mathbf{u}_\lambda \in L^\infty$ is a minimizer of* (37) *with residual $r_\lambda = f - div\, \mathbf{u}_\lambda \in Y$ then*

$$|\phi_p(r_\lambda)|_{TV} \leqslant \frac{1}{\lambda}. \tag{49}$$

Proof. For any test function $\mathbf{z} \in \mathcal{D}(\Omega)$, we have

$$\begin{aligned} \mathcal{V}_\lambda(\mathbf{u}_\lambda) &= \|\mathbf{u}_\lambda\|_{L^\infty} + \lambda\|f - div\, \mathbf{u}_\lambda\|_Y^p \leqslant \|\mathbf{u}_\lambda + \epsilon\mathbf{z}\|_{L^\infty} + \lambda\|f - div\, (\mathbf{u}_\lambda + \epsilon\mathbf{z})\|_Y^p \\ &\leqslant \|\mathbf{u}_\lambda\|_{L^\infty} + |\epsilon| \cdot \|\mathbf{z}\|_{L^\infty} + \lambda\|r_\lambda\|_Y^p - \lambda\epsilon\big\langle\phi_p(r_\lambda), div\, \mathbf{z}\big\rangle + o(\epsilon). \\ &= \mathcal{V}_\lambda(\mathbf{u}_\lambda) + |\epsilon| \cdot \|\mathbf{z}\|_{L^\infty} - \lambda\epsilon\big\langle\phi_p(r_\lambda), div\, \mathbf{z}\big\rangle + o(\epsilon), \end{aligned}$$

and by letting $\epsilon \downarrow 0$ we find $|\phi_p(r_\lambda)|_{TV} = \sup\limits_{0 \neq \mathbf{z} \in \mathcal{D}(\Omega)} \dfrac{\int_\Omega \phi_p(r_\lambda)\, div\, \mathbf{z}}{\|\mathbf{z}\|_{L^\infty}} \leqslant \dfrac{1}{\lambda}$. □

Remark 4.2 Note that when $\lambda < \frac{1}{|\phi_p(f)|_{TV}}$, we then have a trivial minimizer $\mathbf{u}_\lambda = 0$ and $r_\lambda = f$. Lemma 4.3 is relevant when λ is large enough

$$\lambda > \frac{1}{|\phi_p(f)|_{TV}}. \tag{50}$$

Then $\mathbf{u}_\lambda \neq 0$, and (49) asserts the *BV* regularity of $\phi_p(r_\lambda)$. In fact, for large enough λ, one has the equality $|\phi_p(r_\lambda)|_{TV} = 1/\lambda$, and the minimizer $\mathbf{u}_\lambda$ with residual r_λ is characterized as an exteremal pair in the sense that

$$\int div\mathbf{u}_\lambda\phi_p(r_\lambda) = |\mathbf{u}_\lambda|_\infty|\phi_p(r_\lambda)|_{TV} = \frac{|\mathbf{u}_\lambda|_\infty}{\lambda}, \tag{51}$$

see [Mey2002, Theorem 3],[Tad2014, Lemma A.3].

We also need a second priori estimate which will be useful as a *closure bound* for the iterative procedure of hierarchical construction described below.

Lemma 4.4 *Assume that Y has a Fréchet differentiable norm and that BV is embedded in Y' in the sense that*

$$\|v\|_{Y'} \leqslant \beta|v|_{TV}, \quad v \in BV(\mathbb{R}^d). \tag{52}$$

Then, the following a priori bound holds

$$\|v\|_Y^{p-1} \leqslant \eta|\phi_p(v)|_{TV}, \qquad v \in Y, \tag{53}$$

with $\eta = \beta/p$.

Proof. Fixing $v \in Y$ and comparing the first order terms in

$$(1+\epsilon)^p\|v\|_Y^p = \|v+\epsilon v\|_Y^p = \|v\|_Y^p + \epsilon\langle\phi_p(v), v\rangle + o(\epsilon),$$

implies $p\|v\|_Y^p = \langle\phi_p(v), v\rangle$. Therefore

$$p\|v\|_Y^p \leqslant \|\phi_p(v)\|_{Y'}\|v\|_Y \leqslant \beta|\phi_p(v)|_{TV}\,\|v\|_Y,$$

which yields (53). □

Remark 4.3 Recalling that $\phi_p(v) := p\|v\|_Y^{p-1}\phi(v)$, (53) amounts to the lower bound, $|\phi(v)|_{TV} \geqslant \frac{1}{\beta}$, which indicates that the Fréchet differential must be oscillatory. Combining with lemma 4.3 we find

$$\frac{2}{\beta}\|r_\lambda\|_Y \leqslant 2\|r_\lambda\|_Y|\phi(r_\lambda)|_{TV} = |\phi_2(r_\lambda)|_{TV} \leqslant \frac{1}{\lambda}. \tag{54}$$

This means that given $f \in Y$ then minimization of the functional (38) yields a residual $r_\lambda \in Y$ with the closure bound $\|r_\lambda\|_Y \leqslant \eta/\lambda$. In section 4.5 below we show that in case of $Y = L^d$, the mapping $f \mapsto r$ is not only bounded but in fact, it is Hölder on L^d.

Following [Tad2014], we fix the value $p = 2$ and for a given sequence $(\lambda_j)_{j\geqslant 1}$, we define iteratively $\mathbf{u}_j$ as the minimizer of

$$\mathcal{V}_j(\mathbf{u}) = \|\mathbf{u}\|_{L^\infty} + \lambda_j\|r_{j-1} - div\,\mathbf{u}\|_Y^2, \tag{55}$$

and $r_j = r_{j-1} - div\,\mathbf{u}_j$. The following result shows that, with a proper choice of λ_j, the sum of the $\mathbf{u}_j$ admits a limit which is our desired uniformly bounded solution to (1).

Theorem 4.1 *Consider a Fréchet differentiable space Y such that* (52) *holds and set $\lambda_j = \lambda_1 2^{j-1}$, where $\lambda_1 := \frac{2\eta}{\|f\|_Y}$. Then, for any given $f \in Y$, the sum of the $\mathbf{u}_j$ converges in L^∞ to a limit $\mathbf{u} = \sum_{j=1}^{\infty}\mathbf{u}_j$ which is solution to* (1) *and satisfies*

$$\|\mathbf{u}\|_{L^\infty} \leqslant 2\beta\|f\|_Y. \tag{56}$$

Proof. Comparing $\mathbf{u}_j$ with the 0 solution, we find that

$$\|\mathbf{u}_j\|_{L^\infty} \leqslant \mathcal{V}_j(0) = \lambda_j \|r_{j-1}\|_Y^2,$$

and in particular

$$\|\mathbf{u}_1\|_{L^\infty} \leqslant \lambda_1 \|f\|_Y^2.$$

On the other hand, combining the closure bound (53) and the regulatization bound (49), we find that

$$\|r_{j-1}\|_Y \leqslant \eta|\phi_2(r_{j-1})|_{TV} \leqslant \frac{\eta}{\lambda_{j-1}}. \tag{57}$$

Therefore, for $j \geqslant 2$

$$\|\mathbf{u}_j\|_{L^\infty} \leqslant \eta^2 \frac{\lambda_j}{\lambda_{j-1}^2} = \frac{8\eta^2}{\lambda_1} 2^{-j}$$

and $\sum_{j=1}^{\infty} \mathbf{u}_j$ thus converges to a uniformly bounded limit with

$$\|\mathbf{u}\|_{L^\infty} \leqslant \lambda_1 \|f\|_Y^2 + \frac{4\eta^2}{\lambda_1} = 4\eta\|f\|_Y = 2\beta\|f\|_Y,$$

where we have used the chosen value of λ_1. In addition, a telescoping sum of $r_j = r_{j-1} - div\,\mathbf{u}_j$ yields $f = div\,(\sum_{k=1}^{j} \mathbf{u}_k) + r_j$ and the residual r_j tends to 0 in Y. This proves that $div\,(\mathbf{u}) = f$. □

Remark 4.4 Theorem 4.1 extends the hierarchical construction of Bourgain-Brezis problem in [Tad2014] for $Y = L^d$-data with $p = d$. In fact, the choice of the parameter $p > 1$ need not be tied to Y, which led to the simpler choice of $p = 2$ in (55).

Remark 4.5 The closure (53) implies that our choice of λ_1 is admissible in the sense that (50) holds,

$$\lambda_1 = \frac{2\eta}{\|f\|_Y} \geqslant \frac{2\eta}{\eta|\phi_2(f)|_{TV}} > \frac{1}{|\phi_2(f)|_{TV}}.$$

In other words — already the first variational iteration produces a non-trivial minimizer, $\mathbf{u}_1 \neq 0$. In fact, one can underestimate $\lambda_1 < \beta/\|f\|_Y$ in case β in (52) is not accessible, and yet the variational iterations become effective after the first $\log(\beta)$ iterations with zero minimizers.

Example 4.1 Inspired by [Mey2002], we demonstrate the hierarchical constriction of theorem 4.1 in the two-dimensional example of $f = \alpha\chi_R$, where α is a fixed constant and χ_R is the characteristic function of the ball of radius R. Of course, in this case of BV function we can simply solve $\mathbf{u} = \nabla\Delta^{-1} f = \alpha\frac{x}{2}\chi_R$. The minimization with $\lambda > 1/(4\pi\alpha)$ yields

$$f = div\mathbf{u}_\lambda + r_\lambda, \quad \mathbf{u}_\lambda = (\alpha-\beta)\nabla\Delta^{-1}\chi_R = (\alpha-\beta)\frac{x}{2}\chi_R \text{ and } r_\lambda = \beta\chi_R \text{ with } \beta := \frac{1}{4\pi\lambda}.$$

This is verified by checking that $|r_\lambda|_{TV} = \beta 2\pi R = \frac{1}{2\lambda}$, and the extremal property (see remark 4.2),

$$\int div\mathbf{u}_\lambda r_\lambda = (\alpha - \beta)\frac{R}{4\lambda} = |\mathbf{u}_\lambda|_\infty |r_\lambda|_{TV}, \quad |\mathbf{u}_\lambda|_\infty = (\alpha - \beta)\frac{R}{2}.$$

Iterating we find (for $\alpha > 8\pi$)

$$f = \sum_{j=1}^{\infty} \mathbf{u}_j, \qquad \mathbf{u}_j = \begin{cases} \left(\alpha - \dfrac{1}{8\pi}\right)\dfrac{x}{2}\chi_R \;\; j = 1, \\[2ex] \dfrac{1}{4\pi 2^j}\dfrac{x}{2}\chi_R, \quad j \geqslant 2. \end{cases}$$

The above theorem can be used to construct uniformly bounded solutions to (1) for data f in Lorentz spaces $Y = L^{d,q}(\mathbb{R}^d)$ when $1 < q < \infty$. Indeed, since $L^{d,q}$ is reflexive, they qualify as Asplund spaces with Fréchet differentiable norm, see [Asp1968] or [Phe1993, thm 2.12]. Except for the case $q = d$ corresponding to the space $Y = L^d$, the Fréchet derivative of the $L^{d,q}$ does not have a simple explicit form, however we note that this is not required for defining the hierarchical solution. Other applications of hierarchical constructions in inverse problems that arise in image processing can be found in [MNR2019].

In contrast, the space $Y = L^{d,\infty}(\mathbb{R}^d)$ does not have a Fréchet differentiable norm, as seen by the following counter-example due Luc Tatar, [Tar2011]. The purpose is to show that there exists f, g and $\alpha > 0$ such that

$$\|f + \epsilon g\|^p_{L^{p,\infty}} \geqslant \|f\|^p_{L^{p,\infty}} + \alpha|\epsilon| \tag{58}$$

proving that $\|\cdot\|^p_{L^{p,\infty}}$ is not Fréchet — not even Gateaux differentiable. To this end one restrict attention to the unit interval $(0, 1)$. Set $f(x) := x^{-\frac{1}{p}}$ and

$$g(x) := \sum_{k=0}^{\infty} (-1)^k 2^{\frac{k}{p}} \mathbb{1}_{\mathbb{I}_k}(x), \qquad \mathbb{I}_k = (2^{-(k+1)}, 2^{-k}).$$

The second rearrangement of f is given by $f^{**}(t) = \dfrac{p}{p-1} t^{-\frac{1}{p}}$, $0 < t < 1$. Since $|g(x)| \leqslant f(x)$ it follows that $F := f + \epsilon g \geqslant 0$ for $|\epsilon| < 1$, and hence

$$\begin{aligned} \|f + \epsilon g\|_{L^{p,\infty}} &\geqslant t^{\frac{1}{p}} F^{**}(t) = t^{-1+\frac{1}{p}} \int_0^t F^*(s)\mathrm{d}s \\ &\geqslant t^{-1+\frac{1}{p}} \int_0^t F(s)\mathrm{d}s = t^{-1+\frac{1}{p}} \int_0^t f^*(s)\mathrm{d}s + \epsilon t^{-1+\frac{1}{p}} \int_0^t g(s)\mathrm{d}s \\ &= \|f\|_{L^{p,\infty}} + \epsilon t^{-1+\frac{1}{p}} \int_0^t g(s)\mathrm{d}s \;\; \text{for all } \; t < 1. \end{aligned} \tag{59}$$

It remains to lower bound the term on the right. We compute

$$\int_{\mathbb{I}_k} g(s)\mathrm{d}s = (-1)^k 2^{\frac{k}{p}} |\mathbb{I}_k| = (-1)^k \rho^k, \qquad \rho := 2^{-1+\frac{1}{p}} < 1.$$

It follows that $\int_0^{2^{-k}} g(s)ds = \frac{(-1)^k \rho^k}{1+\rho}$ implying that $\epsilon t^{-1+\frac{1}{p}} \int_0^t g(s)ds_{\,|\,t=2^{-k}} = \frac{(-1)^k \epsilon}{1+\rho}$, and (58) follows from (59) at $t = 2^{-k}$ with $(-1)^k \epsilon > 0$ and $\alpha = {}^1/_{(1+\rho)}$.

4.4 A hierarchical construction for general data.

Finally, we work with general data $f \in Y(\Omega)$ for some $\Omega \in \mathbb{R}^d$, without making assumption on the Fréchet differentiability of the Y norm, but instead taking as a prior assumption that Y is a space such that uniformly bounded solution to (1) exist with the bound (2) for some $\gamma > 0$.

We also assuming that Y is a dual space so that there exists a minimizer to the functional F having the general form (37). Here we use the exponent $p = 1$ so that

$$\mathcal{V}(\mathbf{u}) = \mathcal{V}_\lambda(\mathbf{u}) = \|\mathbf{u}\|_{L^\infty} + \lambda \|div\, \mathbf{u} - f\|_Y \tag{60}$$

For a value of λ to be fixed later, we denote by $\mathbf{u}_1$ is the minimizer and $r_1 = f - div\, \mathbf{u}_1$ the residual. In addition to the trivial bound

$$\|r_1\|_Y \leqslant \|f\|_Y \quad \text{and} \quad \|\mathbf{u}_1\|_{L^\infty} \leqslant \lambda \|f\|_Y, \tag{61}$$

that are obtained by comparison of $\mathcal{V}(\mathbf{u}_1)$ and $\mathcal{V}(0)$, we can also compare $\mathcal{V}(\mathbf{u}_1)$ with $\mathcal{V}(\mathbf{u})$ where $\mathbf{u}$ is a solution to (1) satisfying the bound (2). It follows that

$$\lambda \|r_1\|_Y \leqslant \|\mathbf{u}_j\|_{L^\infty} \leqslant \gamma \|f\|_Y. \tag{62}$$

Therefore, taking $\lambda > \gamma$, we obtain a contraction property

$$\|r_1\|_Y \leqslant \rho \|f\|_Y, \tag{63}$$

with $\rho = \gamma/\lambda < 1$.

This suggests a hierarchical construction that was proposed in a more general context in [Tad2014]: with this fixed value of λ, we define iteratively $\mathbf{u}_j$ as the minimizer of

$$\mathcal{V}_j(\mathbf{u}) = \|\mathbf{u}\|_{L^\infty} + \lambda \|r_{j-1} - div\, \mathbf{u}\|_Y, \tag{64}$$

and $r_j = r_{j-1} - div\, \mathbf{u}_j$. By recursive application of the above contraction principle, it follows that

$$\|r_j\|_Y \leqslant \rho \|r_{j-1}\|_Y \leqslant \cdots \leqslant \rho^j \|f\|_Y. \tag{65}$$

as well as

$$\|\mathbf{u}_j\| \leqslant \gamma \|r_{j-1}\|_Y \leqslant \gamma \rho^{j-1} \|f\|_Y. \tag{66}$$

From this it follows that the hierarchical construction

$$\mathbf{u}_1 + \mathbf{u}_2 + \cdots + \mathbf{u}_j + \cdots, \tag{67}$$

converges to a uniformly bounded $\mathbf{u}$ that satisfies the equation (1) and the bound

$$\|\mathbf{u}\|_{L^\infty} \leqslant \frac{\gamma}{1-\rho}\|f\|_Y. \tag{68}$$

4.5 Hölder continuity

A main ingredient in the hierarchical decomposition is to secure that the mapping $T : f \mapsto r$ is bounded on Y. This is carried out in lemma 4.3 and 4.4 and clarified in remark 4.3. We show that the (nonlinear) mapping T is Hölder on $Y = L^d$. Indeed, given $f \in Y$ and $f' \in Y$ with the corresponding minimizing pairs $(\mathbf{u}_\lambda, r_\lambda)$ and $(\mathbf{u}'_\lambda, r'_\lambda)$. These pairs are characterized by the extremal property (51). In particular

$$\int div\,\mathbf{u}_\lambda \phi_p(r'_\lambda) \leqslant |\mathbf{u}_\lambda|_\infty |\phi_p(r'_\lambda)|_{TV} = \frac{|\mathbf{u}_\lambda|_\infty}{\lambda} = \int div\,\mathbf{u}_\lambda \phi_p(r_\lambda)$$

$$\int div\,\mathbf{u}'_\lambda \phi_p(r_\lambda) \leqslant |\mathbf{u}'_\lambda|_\infty |\phi_p(r_\lambda)|_{TV} = \frac{|\mathbf{u}'_\lambda|_\infty}{\lambda} = \int div\,\mathbf{u}'_\lambda \phi_p(r'_\lambda).$$

This implies the monotonicity

$$0 \leqslant \langle div\,\mathbf{u}_\lambda - div\,\mathbf{u}'_\lambda, \phi_p(r_\lambda) - \phi_p(r'_\lambda)\rangle,$$

or, given that $div\,\mathbf{u} = f - r_\lambda$ and $div\,\mathbf{u}' = f' - r'_\lambda$,

$$\langle r_\lambda - r'_\lambda, \phi_p(r_\lambda) - \phi_p(r'_\lambda)\rangle \leqslant \langle f - f', \phi_p(r_\lambda) - \phi_p(r'_\lambda)\rangle. \tag{69}$$

We now specify the case $Y = L^d$ with Fréchet differential (to simplify we switch from $p = 2$ to $p = d$) $\phi_d(r) = d\,\mathrm{sgn}(r)|r|^{d-1}$, for which

$$(r-s)\big(\phi_d(r) - \phi_d(s)\big) \geqslant d|r-s|^d, \qquad \phi_d(r) = d\,\mathrm{sgn}(r)|r|^{d-1}.$$

Then, the left of (69) admits the lower-bound

$$\langle r_\lambda - r'_\lambda, \phi_d(r_\lambda) - \phi_d(r'_\lambda)\rangle \geqslant d\|r_\lambda - r'_\lambda\|^d_{L^d}.$$

On the other hand, since $\|\phi_p(r_\lambda)\|_{L^{d'}} \leqslant \beta|\phi_p(r_\lambda)|_{TV} \leqslant \beta/\lambda$, the right of (69) admits the upper-bound

$$\langle f - f', \phi_p(r_\lambda) - \phi_p(r'_\lambda)\rangle \leqslant \|\phi_d(r_\lambda) - \phi_d(r'_\lambda)\|_{L^{d'}}\|f - f'\|_{L^d} \leqslant \frac{2\beta}{\lambda}\|f - f'\|_{L^d},$$

and we conclude $d\|r_\lambda - r'_\lambda\|^d_{L^d} \leqslant \frac{2\beta}{\lambda}\|f - f'\|_{L^d}$. Thus, the mapping $T : f \mapsto r_\lambda$ is Hölder of order $\theta = 1/d$:

$$\|Tf - Tf'\|_{L^d} \leqslant C_\lambda \|f - f'\|^\theta_{L^d}, \qquad C_\lambda = \left(\frac{2\beta}{d\lambda}\right)^\theta$$

clearly, by successive application of T with an increasing sequence of λ's, one recovers the solution of Bourgain-Brezis with L^d-data. Note that when Y is a Hilbert space, e.g., in the two-dimensional case, then T is in fact Lipschitz on L^2.

References

[Aji2009] S. Ajiev, On Chebyshev centres, retractions, metric projections and homogeneous inverses for Besov, Lizorkin-Triebel, Sobolev and other Banach spaces. East J. Approx. 15(3) (2009), 375-428.

[Asp1968] E. Asplund, Fréchet differentiability of convex functions, Acta Math. 121, (1968) 31-47.

[BC2011] H. Bahouri and A. Cohen, Refined Sobolev inequalities in Lorentz spaces, J Fourier Anal Appl (2011) 17:662-673.

[BS1988] C. Bennett and R. Sharpley, Interpolation of operators, Pure and Applied Mathematics, vol. 129, Academic Press, Orlando, 1988.

[BL1976] J. Bergh and J. Löfström, Interpolation Spaces, An Introduction, Springer, 1976.

[BB2003] J. Bourgain and H. Brezis, *On the equation* $\operatorname{div} Y = f$ *and application to control of phases*, J. Amer. Math. Soc. 16 (2003), no. 2, 393–426.

[BB2007] J. Bourgain and H. Brezis, *New estimates for elliptic equations and Hodge type systems* J. Eur. Math. Soc. (JEMS) 9 (2007), no. 2, 277–315.

[EG1992] L. Evans and R. Gariepy, Measure Theory and Fine Properties of Functions, CRC Press, 1992.

[LW1949] L. Loomis and H. Whitney An inequality related to the isoperimetric inequality, Bulletin of the American Mathematical Society 55 (10) (1949) 961–962.

[Fed1969] H. Federer, Geometric Measure Theory, Springer, 1969.

[Mey2002] Y. Meyer, *Oscillating Patterns in Image Processing and Nonlinear Evolution Equations*, University Lecture Series Volume 22, AMS 2002.

[MNR2019] K. Modin, A. Nachman and L. Rondic A multiscale theory for image registration and nonlinear inverse problems Advances in Math. 346 (2019) 1009-1066.

[PT2008] N. C. Phuc and M. Torres, Characterizations of the existence and removable singularities of divergence-measure vector fields, Indiana Univ. Math. J. 57 (2008), 1573-1597.

[PT2017] N. C. Phuc and M. Torres, Characterizations of signed measures in the dual of BV and related isometric isomorphisms. Annali della Scuola Normale Superiore di Pisa, Classe di Scienze (5). Vol. XVII (2017) 385-417

[ONe1963] R. O'Neil, *Convolution operators and* $L(p,q)$ *spaces*, Duke Math, J. 30, 129-143, 1963.

[Phe1993] R. Phelps, Convex Functions, Monotone Operators and Differentiability, Lecture Notes in Mathematics, Springer Verlag, 1993.

[Roc1974] R. T. Rockafeller, Conjugate Duality and Optimization, SIAM/CBMS Monograph Ser. 16, SIAM Pub. 1974.

[Sto2024] D. Stolyarov, https://math-cs.spbu.ru/wp-content/uploads/2017/10/DualGagliardoNirenberg.pdf, and private communication.

[TT2011] E. Tadmor and C. Tan, *Hierarchical construction of bounded solutions of* $divU = F$ *in critical regularity spaces*, "Nonlinear Partial Differential Equations", Proc. 2010 Abel Symposium (H. Holden & K. Karlsen eds.), Abel Symposia 7, Springer (2011), pp. 255-269.

[Tad2014] E. Tadmor, *Hierarchical construction of bounded solutions in critical regularity spaces*, Communications in Pure & Applied Mathematics 69(6) (2016) 1087-1109.

[Tar189] L. Tartar, *Lorentz spaces and applications*, Lecture notes, Carnegie Mellon University, 1989.

[Tar2011] L. Tartar, personal communication, 2011.

[Zie1989] W. Ziemer, Weakly Differentiable Functions, Graduate Texts in Math., Springer, vol. 120, 1989.

On the importance of the ε-regularization of the distribution-dependent Mumford–Shah model for hyperspectral image segmentation

Jan-Christopher Cohrs and
Benjamin Berkels

Abstract Recently, the distribution-dependent Mumford–Shah model for hyperspectral image segmentation was introduced. It approximates an image based on first and second order statistics using a data term, that is built of a Mahalanobis distance plus a covariance regularization, and the total variation as spatial regularization. Moreover, to achieve feasibility, the appearing matrices are restricted to symmetric positive-definite ones with eigenvalues exceeding a certain threshold. This threshold is chosen in advance as a data-independent parameter. In this article, we study theoretical properties of the model. In particular, we prove the existence of minimizers of the functional and show its Γ-convergence when the threshold regularizing the eigenvalues of the matrices tends to zero. It turns out that in the Γ-limit we lose the guaranteed existence of minimizers; and we give an example of an image where the Γ-limit indeed has no minimizer. Finally, we derive a formula for the minimum eigenvalues of the covariance matrices appearing in the functional that hints under which conditions the functional is able to handle the data without regularizing the eigenvalues. The results of this article demonstrate the significance and importance of the eigenvalue regularization to the model and that it cannot be dropped without substantial modifications.

1 Introduction

Image processing comprises several problems, such as image registration, deblurring, denoising, inpainting or segmentation. Image registration aims to align two or more

---

Jan-Christopher Cohrs
Aachen Institute for Advanced Study in Computational Engineering Science (AICES), RWTH Aachen University, Germany, e-mail: cohrs@aices.rwth-aachen.de

Benjamin Berkels
Aachen Institute for Advanced Study in Computational Engineering Science (AICES), RWTH Aachen University, Germany, e-mail: berkels@aices.rwth-aachen.de

R. DeVore, A. Kunoth (eds.), *Multiscale, Nonlinear and Adaptive Approximation II*,
https://doi.org/10.1007/978-3-031-75802-7_10

images of the same object. In image deblurring, one tries to recover the original image from a blurry version of it. Denoising deals with the reduction of noise in an image. The task of filling missing or damaged parts of an image is referred to as inpainting. Image segmentation is concerned with the task to partition an image into meaningful regions where the term *meaningful* depends on the respective setting, i.e., the image modality and which information one wants to extract from the image. A well-established model in image processing is the Mumford–Shah (MS) functional [16]. It is very popular in image segmentation, but also used for image restoration [3, 14]. For example, it has been adapted for inpainting [10, 12], but variants are also applied in deblurring [13]. The functional in its original form aims to find non-trivial piecewise smooth approximations to images and consists of the squared difference as a data term plus two regularizing terms [16]. For a fixed set of segment boundaries, it is shown in [8, Prop. 3] with a convexity argument that a unique minimizer exists. In [9, Th. 1.1] and for a slightly different version of the functional in [3, Th. 2], the authors prove that when the set of segment boundaries is replaced by the set of discontinuities of the approximating function, a minimizer of the functional exists. Rondi and Santosa [20] leverage the MS functional for electrical impedance tomography. They equip it with a data term measuring the difference of an operator applied to the approximating function and the original data in the norm of the space of bounded, linear operators that map from the space of traces of $L^2(\Omega)$ on $\partial\Omega$ with zero means to itself. Moreover, they prove existence results for a MS variant on two function classes for general data terms that satisfy a continuity condition. Fornasier et al. propose a structurally similar MS-based functional; however, their data term is the L^2-distance between a linear, non-invertible operator applied to the original image (in contrast to before where the operator was applied to the approximation) and the approximating function [11]. They provide counterexamples in which existence of minimizers cannot be guaranteed but prove two existence results for certain classes of linear operators. The authors of [6] introduce a variant of the MS model with the squared difference as data term including a blurring operator, but without the boundary regularization and prove the existence of a unique minimizer. In [5], this model is further modified with a data term that deals with images that are corrupted by Poisson noise; and also for this model the existence of a unique minimizer is proven. Mumford and Shah further propose to restrict their functional to piecewise constant functions and show for that restricted model the existence of minimizers for continuous input images [16, Th. 5.1].

Moreover, alternative data terms for the MS functional have been introduced, where existence results have not been proven yet. In [21], the authors generalize the original model by multiplying the data term with a location-dependent weighting factor to give less or no weight to noisy or corrupt pixels. Ramlau and Ring [19] build on the piecewise constant MS model to introduce a data term for X-ray tomography, which they define as the squared L^2-distance of the data and the Radon transform of an unknown density function.

In this article, we will consider the distribution-dependent MS model for hyperspectral image segmentation that was introduced in [7]. We will call it εAMS in the following, short for *ε-regularized anisotropic MS model*. It consists of the piecewise constant MS functional where the squared L^2-distance in the data term was replaced by a non-squared Mahalanobis distance [15] plus a regularizing term to penalize covariance matrix estimates with large eigenvalues. Covariance matrices are symmetric and positive semi-definite by definition. However, εAMS even requires them to be positive definite to ensure the invertibility and therefore the feasibility of the model. To this end, a regularizing parameter $\varepsilon > 0$ is introduced that defines a lower bound for the eigenvalues of the covariance matrices and therefore keeps them positive by setting eigenvalues of covariance matrices to ε, if they are smaller than this threshold. In terms of images one can think of ε as the minimum variance of the spectra possible in a segment. Consequences of this regularization are: it ensures invertibility of the covariance matrices and improves their conditions, the covariance matrices are optimized over a closed set and ε ensures a lower bound of the functional. However, ε is a data-independent parameter that has to be chosen before the application of the model. This choice is not trivial and it is also hard to interpret why a specific choice leads to good results and others do not. In particular, the choice of ε potentially excludes minima by restricting the admissible set. Nevertheless, the parameter ε plays a crucial role in the considered model and we will see more arguments why we need it in εAMS in the next section.

Since segmenting an image means finding minimizers of εAMS, the question arises if minimizers exist. We will state and prove in Section 2.2 an existence result, guaranteeing the existence of minimizers of εAMS. Also in the proof, a key factor is the ε-regularization, which emphasizes its importance. As a lower bound for the eigenvalues of the covariances matrices, ε is supposed to be chosen small. In Section 2.3, we will show the Γ-convergence of the functional to a Γ-limit when $\varepsilon \to 0$. It turns out that minimizers of the Γ-limit do not necessarily exist. In particular, we will see a counterexample for which the Γ-limit has no minimizer in Section 2.4. Finally, we derive a formula depending on the data that shows the minimum values for the eigenvalues of the covariance matrices in Section 2.5, which helps to understand what goes wrong in case of the Γ-limit.

2 The distribution-dependent Mumford–Shah model

The model that we will study is a variant of the MS segmentation functional [16], equipped with the indicator function proposed in [7]. We will see a detailed description of the model in Section 2.1 and show the existence of minimizers of the model afterwards (cf. Section 2.2). The model contains a (small) regularization parameter to ensure the feasibility of the model and which guarantees the existence of minimizers of the functional. We will show the Γ-convergence of the functional to a Γ-limit when this regularization parameter tends to 0 in Section 2.3. Under suitable conditions, not only the functional Γ-converges to a Γ-limit but also the minimizers, which would give us the existence of minimizers also for the limit functional. However, as we will see in Section 2.4, we do not have a guaranteed minimizer for the limit functional anymore, which emphasizes the importance of a regularization of the eigenvalues of the covariance estimates to bound them away from 0.

2.1 Model description of εAMS

Throughout this article, let $d \in \mathbb{N}$ and $\Omega \subseteq \mathbb{R}^d$ be a measurable, bounded domain with Lipschitz boundary. Furthermore, let $g \in L^\infty(\Omega, \mathbb{R}^L)$ be a hyperspectral image with $L \in \mathbb{N}$ channels, $\varepsilon, \eta > 0$ and $k \in \mathbb{N}$ the number of segments, which is chosen in advance. Since Ω is bounded, $g \in L^p(\Omega, \mathbb{R}^L)$ for every $p \in [1, \infty)$. For $A \subseteq \mathbb{R}^d$ measurable, let $|A|$ be the d-dimensional Lebesgue measure of A. Note that for such sets A we often equivalently consider the characteristic functions $\chi_A \colon \mathbb{R}^d \to \{0, 1\}$ and switch between these two representations without further notice. Let $O \subseteq \mathbb{R}^d$ be measurable. Since Ω is bounded, we get $|O \cap \Omega| < \infty$ and thus $\chi_O \in L^1(\Omega)$. Following [2, Def. 3.4, Def. 3.35], we define the *perimeter* $\mathrm{Per}\,(O, \Omega)$ *of* O *in* Ω as

$$\mathrm{Per}\,(O, \Omega) := \sup \left\{ \int_O \operatorname{div} \varphi \, \mathrm{d}x \;\middle|\; \varphi \in C_c^1(\Omega, \mathbb{R}^d), \, \| \|\varphi\|_2 \|_{L^\infty(\Omega)} \leqslant 1 \right\}.$$

Because $\chi_O \in L^1(\Omega)$, we have $\mathrm{Per}\,(O, \Omega) < \infty \Leftrightarrow \chi_O \in \mathrm{BV}(\Omega)$, cf. [2, Prop. 3.6]. In particular, $\mathrm{Per}\,(O, \Omega) = |\chi_O|_{\mathrm{BV}(\Omega)}$. Here, $|\cdot|_{\mathrm{BV}(\Omega)}$ denotes the Total Variation semi-norm and $\mathrm{BV}(\Omega)$ the space of functions of Bounded Variation. For a general introduction to $\mathrm{BV}(\Omega)$ and sets of finite perimeter, we refer to [2].

The set of all partitions of Ω into k measurable subsets with finite perimeter is

$$\begin{aligned} \mathcal{A} := \big\{ (O_1, O_2, \ldots, O_k) \;\big|\; \forall l \in \{1, \ldots, k\} : O_l \subseteq \Omega \text{ measurable}, \\ \mathrm{Per}\,(O_l, \Omega) < \infty, \; \forall m \neq l : O_l \cap O_m = \emptyset, \; \textstyle\bigcup_{l=1}^k O_l = \Omega \big\}. \end{aligned}$$

We want to point out here that we do not require the closure of the union to be equal to the closure of Ω but a slightly stronger condition. The reason is that we prove the partition property later using the characteristic functions of the involved sets which have to sum up to one at every element of Ω. This would not be the case for the condition with the closures. Then, the MS segmentation functional [16] is

$$J_{\mathrm{MS}} \colon \mathcal{A} \to \mathbb{R}, (O_1, O_2, \ldots, O_k) \mapsto \sum_{l=1}^k \left[\int_{O_l} f_l(x) \, \mathrm{d}x + \lambda \, \mathrm{Per}\,(O_l, \Omega) \right], \tag{1}$$

where $\lambda > 0$ is a parameter that balances the data term, i.e. the sum of the integrals over O_l, and the regularizing term, i.e. the sum over $\mathrm{Per}\,(O_l, \Omega)$. The so-called *indicator functions* $f_l \colon \Omega \to \mathbb{R}$ play a key role in the MS functional. They define the notion of homogeneity for segmentation and have to be chosen according to the task at hand. Let

$$\sigma \colon \left\{ M \in \mathbb{R}^{L \times L} \;\middle|\; M^T = M \right\} \to \mathbb{R}^L \tag{2}$$

be the mapping that returns the eigenvalues of a symmetric matrix in decreasing order, i.e., $\sigma(M)_1 \geqslant \sigma(M)_2 \geqslant \ldots \geqslant \sigma(M)_L$. Furthermore, we define the set of all symmetric and positive definite matrices as

$$\mathcal{P} := \left\{ M \in \mathbb{R}^{L\times L} \mid M^T = M,\ \forall z \in \mathbb{R}^L \setminus \{0\} : z^T M z > 0 \right\},$$

and based on that we define

$$\mathcal{P}_\varepsilon := \left\{ M \in \mathcal{P} \mid \forall i \in \{1,\ldots,L\} : \sigma(M)_i \geqslant \varepsilon^2 \right\},$$

which is the set of all symmetric and positive definite matrices with eigenvalues of at least ε^2. The indicator function for hyperspectral image segmentation from [7] is

$$f_l(x;\mu_l,\Sigma_l) = \sqrt{(g(x)-\mu_l)^T\,\Sigma_l^{-1}\,(g(x)-\mu_l)+\eta} + \log\det\Sigma_l,$$

where $\mu_l \in \mathbb{R}^L$ and $\Sigma_l \in \mathcal{P}_\varepsilon$. The variables μ_l and Σ_l are taking the roles of estimates of the mean and the covariance of the spectral distribution of the l-th segment, respectively. We will use the notation

$$\|z\|_{\Sigma_l^{-1},\eta} := \sqrt{z^T\Sigma_l^{-1}z+\eta}$$

for $z \in \mathbb{R}^L$, because the square root term can be seen as a regularized norm on $\mathbb{R}^L$, since $z \mapsto \sqrt{z^T M z}$ is a norm for any positive definite matrix M. Hence, we have

$$f_l(x;\mu_l,\Sigma_l) = \|g(x)-\mu_l\|_{\Sigma_l^{-1},\eta} + \log\det\Sigma_l.$$

The restriction of the eigenvalues of Σ_l to at least ε^2 was introduced to ensure the invertibility of Σ_l, which also implies the feasibility of $\log\det\Sigma_l$. As we will see in Section 2.4, this restriction cannot be dropped if we want to keep the feasibility and the guarantee that minimizers exist. Defining $X_\varepsilon := \mathcal{A} \times \left(\mathbb{R}^L\right)^k \times (\mathcal{P}_\varepsilon)^k$ and

$$\boldsymbol{O} := (O_1,\ldots,O_k) \in \mathcal{A},\ \boldsymbol{\mu} := (\mu_1,\ldots,\mu_k) \in \left(\mathbb{R}^L\right)^k, \boldsymbol{\Sigma} := (\Sigma_1,\ldots,\Sigma_k) \in (\mathcal{P}_\varepsilon)^k,$$

the objective function for hyperspectral image segmentation is

$$\begin{aligned} &J_\varepsilon\colon X_\varepsilon \to \mathbb{R},\\ &(\boldsymbol{O},\boldsymbol{\mu},\boldsymbol{\Sigma}) \mapsto \sum_{l=1}^k \left[\int_{O_l} \|g(x)-\mu_l\|_{\Sigma_l^{-1},\eta} + \log\det\Sigma_l \,\mathrm{d}x + \lambda\,\mathrm{Per}\,(O_l,\Omega)\right]. \end{aligned} \tag{3}$$

By optimizing J_ε over all parameters, we obtain a segmentation and estimates for the means μ_l and covariances Σ_l for $l \in \{1,\ldots,k\}$ from the data.

2.2 Existence of minimizers of εAMS

In this section, we use the *direct method in the calculus of variations* to show that minimizers of J_ε exist. The direct method consists of four steps:

1. finding a lower bound

$$\underline{J_\varepsilon} := \inf_{(O,\mu,\Sigma)\in X_\varepsilon} J_\varepsilon[O,\mu,\Sigma] > -\infty,$$

2. choosing a minimizing sequence $(O_n,\mu_n,\Sigma_n)_{n\in\mathbb{N}} \subseteq X_\varepsilon$, i.e., the sequence fulfills $J_\varepsilon[O_n,\mu_n,\Sigma_n] \overset{n\to\infty}{\longrightarrow} \underline{J_\varepsilon}$,
3. constructing a subsequence $(O_{n_m},\mu_{n_m},\Sigma_{n_m})_{m\in\mathbb{N}}$ and an element $(O^*,\mu^*,\Sigma^*) \in X_\varepsilon$ such that

$$J_\varepsilon[O^*,\mu^*,\Sigma^*] \leqslant \liminf_{m\to\infty} J_\varepsilon[O_{n_m},\mu_{n_m},\Sigma_{n_m}], \tag{4}$$

4. concluding that (O^*,μ^*,Σ^*) is a minimizer of J_ε due to

$$\begin{aligned}\underline{J_\varepsilon} &= \lim_{m\to\infty} J_\varepsilon[O_{n_m},\mu_{n_m},\Sigma_{n_m}] = \liminf_{m\to\infty} J_\varepsilon[O_{n_m},\mu_{n_m},\Sigma_{n_m}] \\ &\geqslant J_\varepsilon[O^*,\mu^*,\Sigma^*] \geqslant \underline{J_\varepsilon}.\end{aligned} \tag{5}$$

Before stating and proving the first main result, we state some definitions and lemmas. Note that we use *convergence in measure* (as defined below) when working with sequences of measurable, bounded sets.

Definition 2.1 Let $(A_n)_{n\in\mathbb{N}}$ be a sequence of measurable, bounded subsets of $\mathbb{R}^d$ and $A \subseteq \mathbb{R}^d$ a measurable, bounded set. We say that $(A_n)_{n\in\mathbb{N}}$ *converges to A in measure in Ω* if the Lebesgue measure of the symmetric difference in Ω converges to 0, i.e.,

$$\left|(A_n \bigtriangleup A)\cap\Omega\right| \to 0$$

for $n \to \infty$. We use the notation: $A_n \to A$ for $n \to \infty$.

This implies that also the sequence of Lebesgue measures of a convergent sequence of sets is convergent:

Lemma 2.1 *Let $(A_n)_{n\in\mathbb{N}}$ be a sequence of measurable, bounded subsets of $\mathbb{R}^d$ and $A \subseteq \mathbb{R}^d$ measurable and bounded. If $(A_n)_{n\in\mathbb{N}}$ converges to A in measure in Ω, then also the measures of $(A_n)_{n\in\mathbb{N}}$ restricted to Ω converge to the measure of A restricted to Ω in $\mathbb{R}$, i.e.,*

$$|A_n\cap\Omega| \to |A\cap\Omega| \text{ for } n\to\infty.$$

Proof. We have

$$\begin{aligned}\big|\,|A_n\cap\Omega| - |A\cap\Omega|\,\big| &= \left|\int_\Omega \chi_{A_n}(x)\,\mathrm{d}x - \int_\Omega \chi_A(x)\,\mathrm{d}x\right| \\ &\leqslant \int_\Omega |\chi_{A_n}(x)-\chi_A(x)|\,\mathrm{d}x = \|\chi_{A_n}-\chi_A\|_{L^1(\Omega)} = \left|(A_n\bigtriangleup A)\cap\Omega\right| \overset{n\to\infty}{\longrightarrow} 0.\end{aligned}$$

The last equality is true because the symmetric difference is equal to the L^1-difference of the corresponding characteristic functions. □

Lemma 2.2 *Let $\varepsilon > 0$. The set $\mathcal{P}_\varepsilon$ is a closed subset of $\mathbb{R}^{L\times L}$.*

Proof. The transpose depends continuously on the matrix as a linear function on a finite-dimensional space. Moreover, the eigenvalues of a matrix vary continuously with the matrix entries [17, 3.1.2]. Then, the closedness of $\mathcal{P}_\varepsilon$ follows since continuous equality and non-strict inequality conditions are preserved in the limit. □

Lemma 2.3 *Let $M \in \mathbb{R}^{L\times L}$ be a symmetric, positive semi-definite matrix with largest eigenvalue $\sigma(M)_1$ and smallest eigenvalue $\sigma(M)_L$. Then,*

$$\sigma(M)_L \|z\|_2^2 \leqslant z^T M z \leqslant \sigma(M)_1 \|z\|_2^2 \text{ for all } z \in \mathbb{R}^L.$$

The lemma follows from the principal axis theorem.

Theorem 2.1 *The functional J_ε, cf. (3), has a minimizer in X_ε.*

Proof. We use the *direct method in the calculus of variations.*

To show that J_ε has a lower bound, we rewrite the functional

$$J_\varepsilon[\boldsymbol{O}, \boldsymbol{\mu}, \boldsymbol{\Sigma}] = \sum_{l=1}^{k} \left[\int_{O_l} \|g(x) - \mu_l\|_{\Sigma_l^{-1},\eta} \, \mathrm{d}x + |O_l| \log\det \Sigma_l + \lambda \operatorname{Per}(O_l, \Omega) \right]. \quad (6)$$

For each $l \in \{1, \ldots, k\}$, we have $\|g(x) - \mu_l\|_{\Sigma_l^{-1},\eta} > 0$ because $\eta > 0$ and Σ_l is positive definite, implying that its inverse is positive definite, too. Moreover, it holds

$$|O_l| \log\det \Sigma_l = |O_l| \log \prod_{i=1}^{L} \sigma(\Sigma_l)_i \geqslant |O_l| \log \prod_{i=1}^{L} \varepsilon^2 = |O_l| \log \varepsilon^{2L} > -\infty, \quad (7)$$

because $\Sigma_l \in \mathcal{P}_\varepsilon$, the function log is monotonically increasing and for $|O_l|$ we have $\infty > |\Omega| \geqslant |O_l| \geqslant 0$ since $|\cdot|$ is a measure and $O_l \subseteq \Omega$. The last term in (6), namely $\lambda \operatorname{Per}(O_l, \Omega) = \lambda |\chi_{O_l}|_{\mathrm{BV}(\Omega)}$, is non-negative because $\lambda > 0$ and $|\chi_{O_l}|_{\mathrm{BV}(\Omega)} \geqslant 0$. Hence, we obtain

$$J_\varepsilon[\boldsymbol{O}, \boldsymbol{\mu}, \boldsymbol{\Sigma}] \geqslant \underline{c} \text{ for all } (\boldsymbol{O}, \boldsymbol{\mu}, \boldsymbol{\Sigma}) \in X_\varepsilon,$$

for some $\underline{c} \in \mathbb{R}$ with $\underline{c} > -\infty$. This implies the existence of a minimizing sequence

$$(\boldsymbol{O}_n, \boldsymbol{\mu}_n, \boldsymbol{\Sigma}_n)_{n\in\mathbb{N}} \subseteq X_\varepsilon.$$

In particular, the evaluations of the minimizing sequence are bounded as a convergent sequence of real numbers, i.e., there is a constant $\overline{C} < \infty$ such that

$$J_\varepsilon[\boldsymbol{O}_n, \boldsymbol{\mu}_n, \boldsymbol{\Sigma}_n] \leqslant \overline{C} \text{ for all } n \in \mathbb{N}. \quad (8)$$

Please note that we will need that $\overline{C}$ also bounds the summands inside the parentheses in (6) for every $l \in \{1, \ldots, k\}$ separately. However, if $\varepsilon < 1$, the term $\log\det \Sigma_l$ can be negative. In that case, $\overline{C}$ has to be chosen sufficiently large, e.g., by adding $|\Omega|\,|\log \varepsilon^{2L}|$, to fulfill that the summands are bounded above separately by $\overline{C}$.

We now construct a subsequence of the minimizing sequence and an element $(\boldsymbol{O}^*, \boldsymbol{\mu}^*, \boldsymbol{\Sigma}^*) \in X_\varepsilon$ that fulfill (4). Please note that every subsequence that we consider on our way to the final subsequence will be again denoted by $(\boldsymbol{O}_n, \boldsymbol{\mu}_n, \boldsymbol{\Sigma}_n)_{n\in\mathbb{N}}$ for ease of notation. The procedure to arrive at the sought subsequence of the minimizing sequence is to go through $(\boldsymbol{O}_n)_{n\in\mathbb{N}}$, $(\boldsymbol{\Sigma}_n)_{n\in\mathbb{N}}$ and $(\boldsymbol{\mu}_n)_{n\in\mathbb{N}}$ component-wise and show how to choose a subsequence based on the respective component satisfying the requirements. That is, we choose a subsequence of $(\boldsymbol{O}_n, \boldsymbol{\mu}_n, \boldsymbol{\Sigma}_n)_{n\in\mathbb{N}}$ such that $(O_{n,1})_{n\in\mathbb{N}}$ meets our requirements. From that new sequence $(\boldsymbol{O}_n, \boldsymbol{\mu}_n, \boldsymbol{\Sigma}_n)_{n\in\mathbb{N}}$, we choose a subsequence such that $(O_{n,2})_{n\in\mathbb{N}}$ behaves as desired. We continue choosing subsequences until $(O_{n,k})_{n\in\mathbb{N}}$ fulfills the requirements, meaning that we are done with the $\boldsymbol{O}_n$ part. In the same manner, we go through $(\boldsymbol{\Sigma}_n)_{n\in\mathbb{N}}$ and $(\boldsymbol{\mu}_n)_{n\in\mathbb{N}}$ and arrive at a subsequence $(\boldsymbol{O}_n, \boldsymbol{\mu}_n, \boldsymbol{\Sigma}_n)_{n\in\mathbb{N}}$ for which the requirements for all components are met.

The first step is to show that the sequence $(O_{n,l})_{n\in\mathbb{N}}$ as part of our minimizing sequence is bounded for each $l \in \{1, \ldots, k\}$. To this end, we consider the sequence of characteristic functions $(\chi_{O_{n,l}})_{n\in\mathbb{N}} \subseteq \mathrm{BV}(\Omega)$, which is in $\mathrm{BV}(\Omega)$ since $\boldsymbol{O}_n = (O_{n,1}, \ldots, O_{n,k}) \in \mathcal{A}$. First, $\partial\Omega$ is compact because it is bounded due to the boundedness of Ω and also closed since $\partial\Omega = \overline{\Omega} \cap \overline{\mathbb{R}^d \setminus \Omega}$. Since Ω also has Lipschitz boundary, Ω is an extension domain by [2, Prop. 3.21]. Furthermore, $(\chi_{O_{n,l}})_{n\in\mathbb{N}}$ is bounded for every $l \in \{1, \ldots, k\}$ because for every $n \in \mathbb{N}$, we find

$$\begin{aligned}\|\chi_{O_{n,l}}\|_{\mathrm{BV}(\Omega)} &= \|\chi_{O_{n,l}}\|_{L^1(\Omega)} + |\chi_{O_{n,l}}|_{\mathrm{BV}(\Omega)} \\ &= \int_\Omega |\chi_{O_{n,l}}(x)|\,\mathrm{d}x + \mathrm{Per}\left(O_{n,l}, \Omega\right) \leqslant |\Omega| + \frac{\overline{C}}{\lambda} < \infty,\end{aligned}$$

as $|\chi_{O_{n,l}}(x)| \leqslant 1$ and $\lambda\,\mathrm{Per}\left(O_{n,l}, \Omega\right) \leqslant \overline{C}$ since $O_{n,l}$ is part of the minimizing sequence. Due to the boundedness of $\|\chi_{O_{n,l}}\|_{\mathrm{BV}(\Omega)}$ and Ω being a bounded extension domain, [2, Th. 3.23] guarantees the existence of a subsequence $(\chi_{O_{n,l}})_{n\in\mathbb{N}}$ that converges weakly* in the sense of [2, Def. 3.11] with limit $u_l^* \in \mathrm{BV}(\Omega)$. In particular, this implies that this sequence converges strongly in $L^1(\Omega)$ to the same limit. Moreover, we choose a subsequence that converges even point-wise a.e. to u_l^*, cf. [1, Lemma 3.22 (1)]. This ensures that $u_l^*(x) \in \{0, 1\}$ for almost every $x \in \Omega$ because $\chi_{O_{n,l}}(x) \to u_l^*(x)$ in $\mathbb{R}$ and $\chi_{O_{n,l}}(x) \in \{0, 1\}$ for every $n \in \mathbb{N}$. Moreover, $O_l^* := (u_l^*)^{-1}(\{1\})$ is measurable because u_l^* is measurable as a function in $\mathrm{BV}(\Omega)$ and therefore the preimage of the measurable set $\{1\} \in \mathcal{B}(\mathbb{R})$ under u_l^* is also measurable. Furthermore, we have $\chi_{O_l^*} = u_l^*$.

To see that $\boldsymbol{O}^* = (O_1^*, \ldots, O_k^*)$ is a partition of Ω, note that

$$S := \left\{(a_1, \ldots, a_k) \in \mathbb{R}^k \;\middle|\; \textstyle\sum_{l=1}^k a_l = 1\right\}$$

is closed. Since we have chosen a subsequence $(\boldsymbol{O}_n, \boldsymbol{\mu}_n, \boldsymbol{\Sigma}_n)_{n\in\mathbb{N}}$ such that the sequence $(\chi_{O_{n,l}})_{n\in\mathbb{N}}$ converges point-wise a.e. for every $l \in \{1, \ldots, k\}$, we have for almost every $x \in \Omega$ a sequence $(\chi_{O_{n,1}}(x), \ldots, \chi_{O_{n,k}}(x))_{n\in\mathbb{N}}$ in S with limit

$(\chi_{O_1^*}(x), \ldots, \chi_{O_k^*}(x))$. It follows that $(\chi_{O_1^*}(x), \ldots, \chi_{O_k^*}(x)) \in S$ for almost every $x \in \Omega$ by the closedness of S. In this sense, the partition property of $\boldsymbol{O}^*$ holds almost everywhere. In other words, the limit $\boldsymbol{O}^*$ fulfills the required properties to belong to $\mathcal{A}$ only *almost everywhere* and resides therefore in the set

$$\mathcal{A}' := \big\{(O_1, O_2, \ldots, O_k) \;\big|\; \forall l \in \{1, \ldots, k\} : O_l \subseteq \Omega \text{ measurable}, \\ \operatorname{Per}(O_l, \Omega) < \infty,\ \forall m \neq l : |O_l \cap O_m| = 0, \big|\Omega \setminus \textstyle\bigcup_{l=1}^k O_l\big| = 0\big\},$$

which is a superset of $\mathcal{A}$. Those $x \in \Omega$ without guaranteed point-wise convergence are contained in a null set. Thus, we can change $\boldsymbol{O}^*$ such that the properties necessary to belong to $\mathcal{A}$ are fulfilled as this does not change the value of J_ε. Hence, we can assume that $\boldsymbol{O}^* \in \mathcal{A}$. Moreover, we have the equivalence of local convergence in measure and convergence in measure in Ω because Ω is bounded [2, Rem. 3.37]. Thus, $O \mapsto \operatorname{Per}(O, \Omega)$ is lower semi-continuous w.r.t. convergence in measure in Ω by [2, Prop. 3.38 (b)]. Consequently, the chosen $(\boldsymbol{O}_n)_{n\in\mathbb{N}}$ and $\boldsymbol{O}^*$ satisfy

$$\operatorname{Per}\big(O_l^*, \Omega\big) \leqslant \liminf_{n\to\infty} \operatorname{Per}\big(O_{n,l}, \Omega\big) \quad \text{for each } l \in \{1, \ldots, k\}. \tag{9}$$

In order to obtain subsequences $(\boldsymbol{\mu}_n)_{n\in\mathbb{N}} \subseteq (\mathbb{R}^L)^k$ and $(\boldsymbol{\Sigma}_n)_{n\in\mathbb{N}} \subseteq (\mathcal{P}_\varepsilon)^k$ that satisfy our requirements, we have to distinguish for every $l \in \{1, \ldots, k\}$ the cases (1) that in the limit, segment l has measure greater 0, i.e., $\big|O_l^*\big| > 0$; and (2) that in the limit, segment l has measure 0, i.e., $\big|O_l^*\big| = 0$.

Case (1), i.e. $\big|O_l^*\big| > 0$: Because we have a minimizing sequence, it holds

$$\big|O_{n,l}\big| \log\det \Sigma_{n,l} \leqslant \overline{C} \text{ for all } n \in \mathbb{N}.$$

Moreover, because of $\Sigma_{n,l} \in \mathcal{P}_\varepsilon$, we find for every $n \in \mathbb{N}$ the estimate

$$\begin{aligned} \log\det \Sigma_{n,l} &= \log \textstyle\prod_{i=1}^L \sigma(\Sigma_{n,l})_i \\ &\geqslant \log\Big(\sigma(\Sigma_{n,l})_1 \big[\sigma(\Sigma_{n,l})_L\big]^{L-1}\Big) \geqslant \log\sigma(\Sigma_{n,l})_1 + \log\varepsilon^{2(L-1)}. \end{aligned} \tag{10}$$

Since $(\chi_{O_{n,l}})_{n\in\mathbb{N}} \to \chi_{O_l^*}$ in $L^1(\Omega)$ and because the convergence of the characteristic functions in $L^1(\Omega)$ corresponds to the convergence of sets in measure which implies the convergence of the measures (cf. Lemma 2.1), we also have $\big|O_{n,l}\big| \to \big|O_l^*\big| =: c_l$. Thus, for fixed $\delta \in (0, c_l)$, there is $N_l \in \mathbb{N}$ such that for $n \geqslant N_l$, we have

$$0 < c_l - \delta \leqslant \big|O_{n,l}\big| \leqslant c_l + \delta. \tag{11}$$

By going to another subsequence (discarding the first $N_l - 1$ elements), we achieve that (11) holds for all $n \in \mathbb{N}$. If $\log\det \Sigma_{n,l} \geqslant 0$, we find with (10)

$$\overline{C} \geqslant |O_{n,l}| \log\det\Sigma_{n,l} \geqslant (c_l - \delta)\log\sigma(\Sigma_{n,l})_1 + (c_l - \delta)\log\varepsilon^{2(L-1)}$$
$$\Rightarrow \frac{\overline{C}}{c_l-\delta} - \log\varepsilon^{2(L-1)} \geqslant \log\sigma(\Sigma_{n,l})_1$$
$$\Rightarrow \infty > \exp\left(\frac{\overline{C}}{c_l-\delta} - \log\varepsilon^{2(L-1)}\right) \geqslant \sigma(\Sigma_{n,l})_1.$$

If $\log\det\Sigma_{n,l} < 0$, we start with

$$\overline{C} \geqslant |O_{n,l}| \log\det\Sigma_{n,l} \geqslant (c_l + \delta)\log\det\Sigma_{n,l},$$

and derive analogously

$$\frac{\overline{C}}{c_l+\delta} - \log\varepsilon^{2(L-1)} \geqslant \log\sigma(\Sigma_{n,l})_1 \;\Rightarrow\; \infty > \exp\left(\frac{\overline{C}}{c_l+\delta} - \log\varepsilon^{2(L-1)}\right) \geqslant \sigma(\Sigma_{n,l})_1.$$

Hence, $(\Sigma_{n,l})_{n\in\mathbb{N}}$ is bounded in the spectral norm $\|\cdot\|_{2\to2}$. In particular, this implies that there exists an $r > 0$ such that

$$(\Sigma_{n,l})_{n\in\mathbb{N}} \subseteq \overline{B_r^{\|\cdot\|_{2\to2}}(0)} \cap \mathcal{P}_\varepsilon \subseteq \mathbb{R}^{L\times L},$$

with $\overline{B_r^{\|\cdot\|_{2\to2}}(0)}$ being the closure of the ball with radius r around the origin with respect to the spectral norm. The intersection of the closed ball and $\mathcal{P}_\varepsilon$ (which is closed, cf. Lemma 2.2) is bounded and closed. By the Heine-Borel theorem, it is compact as a subset of a finite-dimensional space. Consequently, we can choose a convergent subsequence $(\Sigma_{n,l})_{n\in\mathbb{N}} \subseteq \mathcal{P}_\varepsilon$ that converges strongly to $\Sigma_l^* \in \mathcal{P}_\varepsilon$.

Now, we consider $(\mu_{n,l})_{n\in\mathbb{N}} \subseteq \mathbb{R}^L$. By the reverse triangle inequality, we get

$$\|\mu_{n,l} - g(x)\|_2 \geqslant |\|\mu_{n,l}\|_2 - \|g(x)\|_2| \geqslant \|\mu_{n,l}\|_2 - \|g(x)\|_2,$$

for all $x \in \Omega$. Moreover, the eigenvalues of $\Sigma_{n,l}^{-1}$ are the reciprocals of the eigenvalues of $\Sigma_{n,l}$. Since $(\mu_{n,l})_{n\in\mathbb{N}}$ is part of the minimizing sequence, we have

$$\begin{aligned}
\overline{C} &\geqslant \int_{O_{n,l}} \|g(x) - \mu_{n,l}\|_{\Sigma_{n,l}^{-1},\eta}\,\mathrm{d}x \\
&\geqslant \int_{O_{n,l}} \sqrt{(g(x) - \mu_{n,l})^T\,\Sigma_{n,l}^{-1}\,(g(x) - \mu_{n,l})}\,\mathrm{d}x \\
&\overset{\text{Lemma 2.3}}{\geqslant} \int_{O_{n,l}} \sqrt{\frac{1}{\sigma(\Sigma_{n,l})_1}}\,\|g(x) - \mu_{n,l}\|_2\,\mathrm{d}x \\
&\geqslant \sqrt{\frac{1}{\sigma(\Sigma_{n,l})_1}}\left(\int_{O_{n,l}} \|\mu_{n,l}\|_2\,\mathrm{d}x - \int_{O_{n,l}} \|g(x)\|_2\,\mathrm{d}x\right) \\
&\geqslant \sqrt{\frac{1}{\sigma(\Sigma_{n,l})_1}}\left(|O_{n,l}|\;\|\mu_{n,l}\|_2 - |O_{n,l}|\;\|\|g\|_2\|_{L^\infty(O_{n,l})}\right) \\
&\geqslant \sqrt{\frac{1}{\sigma(\Sigma_{n,l})_1}}\left(|O_{n,l}|\;\|\mu_{n,l}\|_2 - |\Omega|\;\|\|g\|_2\|_{L^\infty(\Omega)}\right).
\end{aligned}$$

As a consequence, we find

$$\sqrt{\sigma(\Sigma_{n,l})_1}\,\overline{C} + |\Omega|\;\| \|g\|_2 \|_{L^\infty(\Omega)} \geqslant |O_{n,l}|\;\|\mu_{n,l}\|_2. \tag{12}$$

The left-hand side is bounded from above because Ω is bounded, $g \in L^\infty(\Omega, \mathbb{R}^L)$ and $\|\Sigma_{n,l}\|_{2\to 2} \leqslant \tilde{C} < \infty$ for all $n \in \mathbb{N}$ and a constant $\tilde{C} \in \mathbb{R}$, as shown above. Combining (11) and (12), we get

$$\sqrt{\tilde{C}}\,\overline{C} + |\Omega|\;\| \|g\|_2 \|_{L^\infty(\Omega)} \geqslant |O_{n,l}|\;\|\mu_{n,l}\|_2 \geqslant (c_l - \delta)\,\|\mu_{n,l}\|_2,$$

and the boundedness of $(\mu_{n,l})_{n\in\mathbb{N}}$ follows from division by $c_l - \delta$. With the same reasoning as before, using the Heine-Borel theorem, we can choose a strongly convergent subsequence with limit $\mu_l^* \in \mathbb{R}^L$.

In order to show the inequality (4) for the l-th summand, we first note that the transpose and inverse of a matrix vary continuously with their entries and that square root, determinant and logarithm are continuous functions. Furthermore, we use (9) and that the sum and the product of functions of point-wise a.e. convergent sequences converge point-wise a.e., too. Since $(O_{n,l})_{n\in\mathbb{N}}$, $(\mu_{n,l})_{n\in\mathbb{N}} \subseteq \mathbb{R}^L$ and $(\Sigma_{n,l})_{n\in\mathbb{N}} \subseteq \mathcal{P}_\varepsilon$ converge to O_l^* (in measure and point-wise a.e. as characteristic functions), $\mu_l^* \in \mathbb{R}^L$ and $\Sigma_l^* \in \mathcal{P}_\varepsilon$, respectively, we get for the l-th summand of the outer sum of J_ε (3)

$$\begin{aligned}
&\int_{O_l^*} \|g(x) - \mu_l^*\|_{(\Sigma_l^*)^{-1},\eta} + \log\det\Sigma_l^*\,\mathrm{d}x + \lambda\,\mathrm{Per}\left(O_l^*, \Omega\right)\\
&= \int_\Omega \chi_{O_l^*}(x)\left(\|g(x) - \mu_l^*\|_{(\Sigma_l^*)^{-1},\eta} + \log\det\Sigma_l^*\right)\,\mathrm{d}x + \lambda\,\mathrm{Per}\left(O_l^*, \Omega\right)\\
&= \int_\Omega \lim_{n\to\infty} \chi_{O_{n,l}}(x)\left(\|g(x) - \mu_{n,l}\|_{\Sigma_{n,l}^{-1},\eta} + \log\det\Sigma_{n,l}\right)\,\mathrm{d}x + \lambda\,\mathrm{Per}\left(O_l^*, \Omega\right)\\
&\leqslant \liminf_{n\to\infty} \int_\Omega \chi_{O_{n,l}}(x)\left(\|g(x) - \mu_{n,l}\|_{\Sigma_{n,l}^{-1},\eta} + \log\det\Sigma_{n,l}\right)\,\mathrm{d}x\\
&\quad + \liminf_{n\to\infty} \lambda\,\mathrm{Per}\left(O_{n,l}, \Omega\right)\\
&\leqslant \liminf_{n\to\infty} \int_{O_{n,l}} \|g(x) - \mu_{n,l}\|_{\Sigma_{n,l}^{-1},\eta} + \log\det\Sigma_{n,l}\,\mathrm{d}x + \lambda\,\mathrm{Per}\left(O_{n,l}, \Omega\right).
\end{aligned} \tag{13}$$

The second equality uses that the integrand converges point-wise a.e. due the convergence properties shown before. The penultimate inequality follows from (9) and Fatou's lemma [2, Th. 1.20], which holds, in particular, for measurable and possibly negative functions that are bounded below by a function in $L^1(\Omega)$. Such an integrable lower bound exists due to the boundedness of Ω and $\log\det\Sigma_{n,l} \geqslant \log\varepsilon^{2L}$, cf. (7).

Summing up the above, for $l \in \{1, \ldots, k\}$ such that $|O_l^*| > 0$, we have constructed a convergent subsequence of the minimizing sequence and obtained elements O_l^*, μ_l^* and Σ_l^* as the limit of the sequence that fulfill the inequality stated in (4).

We consider now case (2): First, for any $n \in \mathbb{N}$ with $|O_{n,l}| = 0$, we set $\mu_{n,l} := 0 \in \mathbb{R}^L$ and $\Sigma_{n,l} := I_L$, where I_L denotes the identity matrix in $\mathbb{R}^{L\times L}$. Note that this does not change the value of J_ε and the sequence keeps all properties shown before. We start by considering the second summand in (6). Since $|O_{n,l}| \to 0$ and $\log\det\Sigma_{n,l} \geqslant 2L\log\varepsilon$ for all $n \in \mathbb{N}$, cf. (7), two cases have to be considered:

1. If $\log\det\Sigma_{n,l}$ is bounded from above, i.e., $\log\det\Sigma_{n,l} \leqslant \alpha$ for some $\alpha < \infty$, it follows with (10)

$$\alpha \geqslant \log\det\Sigma_{n,l} \geqslant \log\sigma(\Sigma_{n,l})_1 + \log\varepsilon^{2(L-1)}$$
$$\Rightarrow \infty > \exp\left[\alpha - \log\varepsilon^{2(L-1)}\right] \geqslant \sigma(\Sigma_{n,l})_1.$$

Hence, in this case $(\Sigma_{n,l})_{n\in\mathbb{N}}$ is bounded in the spectral norm and therefore, by the Heine-Borel theorem, we can choose a convergent subsequence with limit $\Sigma_l^* \in \mathcal{P}_\varepsilon$. As a consequence, we obtain

$$\left|O_l^*\right| \log\det\Sigma_l^* = \int_\Omega \chi_{O_l^*}(x) \log\det\Sigma_l^* \,\mathrm{d}x$$
$$= \int_\Omega \lim_{n\to\infty} \chi_{O_{n,l}}(x) \log\det\Sigma_{n,l} \,\mathrm{d}x \leqslant \liminf_{n\to\infty} \int_\Omega \chi_{O_{n,l}}(x) \log\det\Sigma_{n,l} \,\mathrm{d}x,$$

again using point-wise a.e. convergence of the integrand and Fatou's lemma [2, Th. 1.20] for a sequence of functions with an integrable lower bound.

2. If $\log\det\Sigma_{n,l}$ is not bounded from above, we do not have control over $\Sigma_{n,l}$. Therefore, different from the typical use of the direct method, we do not try to find a convergent subsequence but just a (possibly non-convergent) subsequence $(\Sigma_{n,l})_{n\in\mathbb{N}}$ and an element $\Sigma_l^* \in \mathcal{P}_\varepsilon$ such that (4) is fulfilled. Hence, if $\log\det\Sigma_{n,l}$ is not bounded from above, we choose a subsequence of $(\Sigma_{n,l})_{n\in\mathbb{N}}$ such that $\log\det\Sigma_{n,l}$ is monotonically increasing and non-negative (and diverging to ∞). Setting $\Sigma_l^* := \varepsilon^2 I_L \in \mathcal{P}_\varepsilon$, we obtain

$$\underbrace{\left|O_l^*\right|}_{=0} \log\det\Sigma_l^* = 0 \leqslant \liminf_{n\to\infty} \int_\Omega \underbrace{\chi_{O_{n,l}}(x)}_{\geqslant 0} \underbrace{\log\det\Sigma_{n,l}}_{\geqslant 0} \,\mathrm{d}x.$$

We will now consider the integral in (6). Also for $\left(\mu_{n,l}\right)_{n\in\mathbb{N}}$, we cannot hope to find a convergent subsequence in this case and will therefore only aim to satisfy (4) with $\left(\mu_{n,l}\right)_{n\in\mathbb{N}}$ and an element $\mu_l^* \in \mathbb{R}^L$. Since the integrand is always positive by the positive definiteness of $\Sigma_{n,l}^{-1}$ and the positivity of η, the integral is obviously non-negative for every $n \in \mathbb{N}$. Combined with choosing $\mu_l^* := 0 \in \mathbb{R}^L$ and again using $|O_l^*| = 0$, we obtain

$$\int_{O_l^*} \|g(x) - \mu_l^*\|_{(\Sigma_l^*)^{-1},\eta} \,\mathrm{d}x = 0 \leqslant \liminf_{n\to\infty} \int_{O_{n,l}} \|g(x) - \mu_{n,l}\|_{\Sigma_{n,l}^{-1},\eta} \,\mathrm{d}x.$$

Combining all considerations for the case $\left|O_l^*\right| = 0$ and using again (9), we get

$$\int_{O_l^*} \|g(x) - \mu_l^*\|_{(\Sigma_l^*)^{-1},\eta} \,\mathrm{d}x + \left|O_l^*\right| \log\det\Sigma_l^* + \lambda \operatorname{Per}\left(O_l^*, \Omega\right)$$

$$\leqslant \liminf_{n\to\infty} \int_{O_{n,l}} \|g(x) - \mu_{n,l}\|_{\Sigma_{n,l}^{-1},\eta} \, dx + \int_{\Omega} \chi_{O_{n,l}}(x) \log\det\Sigma_{n,l} \, dx$$
$$+ \lambda \operatorname{Per}\left(O_{n,l}, \Omega\right).$$

In total, we got a minimizing sequence $(\boldsymbol{O}_n, \boldsymbol{\mu}_n, \boldsymbol{\Sigma}_n)_{n\in\mathbb{N}} \subseteq X_\varepsilon$ and $(\boldsymbol{O}^*, \boldsymbol{\mu}^*, \boldsymbol{\Sigma}^*) \in X_\varepsilon$ with $\boldsymbol{O}^* = (O_1^*, \ldots, O_k^*)$, $\boldsymbol{\mu}^* = (\mu_1^*, \ldots, \mu_k^*)$ and $\boldsymbol{\Sigma}^* = (\Sigma_1^*, \ldots, \Sigma_k^*)$ such that

$$\int_{O_l^*} \|g(x) - \mu_l^*\|_{(\Sigma_l^*)^{-1},\eta} \, dx + \left|O_l^*\right| \log\det\Sigma_l^* + \lambda \operatorname{Per}\left(O_l^*, \Omega\right)$$
$$\leqslant \liminf_{n\to\infty} \int_{O_{n,l}} \|g(x) - \mu_{n,l}\|_{\Sigma_{n,l}^{-1},\eta} \, dx + \left|O_{n,l}\right| \log\det\Sigma_{n,l} + \lambda \operatorname{Per}\left(O_{n,l}, \Omega\right)$$

is true for every $l \in \{1, \ldots, k\}$. Using again that the sum of the limit inferiors of two sequences is smaller or equal to the limit inferior of the sum of the sequences, the inequality (4) is fulfilled. This implies (5), i.e., $(\boldsymbol{O}^*, \boldsymbol{\mu}^*, \boldsymbol{\Sigma}^*)$ is a minimizer of J_ε. □

2.3 Gamma convergence of J_ε when $\varepsilon \to 0$

In Section 2.2, we have seen that ε is crucial in several parts of the proof of Theorem 2.1; namely, to show that a lower bound of J_ε exists, to choose subsequences of the minimizing sequence that fulfill (4), and, most importantly, to achieve that we optimize w.r.t. Σ_l over a closed set $\mathcal{P}_\varepsilon$. However, introduced to ensure that the inverse of Σ_l exists, the parameter ε is supposed to be chosen close to 0 and the question arises what happens if $\varepsilon \to 0$. The notion of Γ-convergence helps answering this question. In particular, we will show the Γ-convergence of J_ε to a Γ-limit.

Let $X := \mathcal{A} \times \left(\mathbb{R}^L\right)^k \times \mathcal{P}^k$. We use the definition of Γ-convergence given in [4, Def. 1.5, Def. 1.45] and also refer to [4] for general information about Γ-convergence. It requires a notion of convergence for sequences in X, which we choose as follows.

Definition 2.2 We say that a sequence $(\boldsymbol{O}_n, \boldsymbol{\mu}_n, \boldsymbol{\Sigma}_n)_{n\in\mathbb{N}} \subseteq X$ *converges weakly* to* $(\boldsymbol{O}, \boldsymbol{\mu}, \boldsymbol{\Sigma}) \in X$ if for every $l \in \{1, \ldots, k\}$ it holds

1. $\chi_{O_{n,l}} \overset{*}{\rightharpoonup} \chi_{O_l}$ in $BV(\Omega)$ in the sense of [2, Def. 3.11],
2. $\mu_{n,l} \to \mu_l$ in $\mathbb{R}^L$,
3. $\Sigma_{n,l} \to \Sigma_l$ in $\mathbb{R}^{L\times L}$,

when $n \to \infty$. We write: $(\boldsymbol{O}_n, \boldsymbol{\mu}_n, \boldsymbol{\Sigma}_n) \overset{*}{\rightharpoonup} (\boldsymbol{O}, \boldsymbol{\mu}, \boldsymbol{\Sigma})$.

Please note that if Ω is sufficiently regular, weak* convergence in $BV(\Omega)$ in the sense of [2, Def. 3.11] corresponds to weak* convergence in the usual sense [2, Rem. 3.12]. In that case, Definition 2.2 describes, in fact, the usual weak* convergence in X. Obviously, we have $X_\varepsilon \subseteq X$. It is necessary to extend J_ε for $\varepsilon > 0$ from X_ε to X as the domain and from $\mathbb{R}$ to $\overline{\mathbb{R}} := \mathbb{R} \cup \{+\infty\}$ as the image. From now on, J_ε accepts any symmetric and positive definite matrix as input and penalizes it with a functional value of ∞ if this matrix is not in $\mathcal{P}_\varepsilon$. Hence, it is given as

$$\begin{aligned} &J_\varepsilon \colon X \to \overline{\mathbb{R}}, \\ &(\boldsymbol{O}, \boldsymbol{\mu}, \boldsymbol{\Sigma}) \mapsto \sum_{l=1}^{k} \left[\int_{O_l} \|g(x) - \mu_l\|_{\Sigma_l^{-1},\eta} + \log\det \Sigma_l \,\mathrm{d}x + \lambda \operatorname{Per}(O_l, \Omega) + I_{\mathcal{P}_\varepsilon}(\Sigma_l) \right]. \end{aligned} \tag{14}$$

The function $I_{\mathcal{P}_\varepsilon}(\Sigma_l)$ for the set $\mathcal{P}_\varepsilon$ is defined as

$$I_{\mathcal{P}_\varepsilon}(\Sigma_l) := \begin{cases} 0 & \text{if } \Sigma_l \in \mathcal{P}_\varepsilon, \\ \infty & \text{if } \Sigma_l \notin \mathcal{P}_\varepsilon. \end{cases}$$

Apparently, Theorem 2.1 also ensures the existence of minimizers for the extended J_ε.

To establish the Γ-convergence of $(J_\varepsilon)_{\varepsilon>0}$ to a Γ-limit $J_0 \colon X \to \overline{\mathbb{R}}$, two inequalities have to be shown for each sequence $(\varepsilon_n)_{n\in\mathbb{N}}$ with $\varepsilon_n > 0$ for all $n \in \mathbb{N}$ and $\varepsilon_n \to 0$ when $n \to \infty$ and each $(\boldsymbol{O}, \boldsymbol{\mu}, \boldsymbol{\Sigma}) \in X$:

(a) *lim inf inequality*: For every weakly* convergent sequence $(\boldsymbol{O}_n, \boldsymbol{\mu}_n, \boldsymbol{\Sigma}_n)_{n\in\mathbb{N}} \subseteq X$ with limit $(\boldsymbol{O}, \boldsymbol{\mu}, \boldsymbol{\Sigma})$ it holds

$$J_0[\boldsymbol{O}, \boldsymbol{\mu}, \boldsymbol{\Sigma}] \leqslant \liminf_{n\to\infty} J_{\varepsilon_n}[\boldsymbol{O}_n, \boldsymbol{\mu}_n, \boldsymbol{\Sigma}_n],$$

(b) *lim sup inequality*: There is a weakly* convergent sequence $(\boldsymbol{O}_n, \boldsymbol{\mu}_n, \boldsymbol{\Sigma}_n)_{n\in\mathbb{N}} \subseteq X$ with limit $(\boldsymbol{O}, \boldsymbol{\mu}, \boldsymbol{\Sigma})$ such that

$$J_0[\boldsymbol{O}, \boldsymbol{\mu}, \boldsymbol{\Sigma}] \geqslant \limsup_{n\to\infty} J_{\varepsilon_n}[\boldsymbol{O}_n, \boldsymbol{\mu}_n, \boldsymbol{\Sigma}_n].$$

If these two conditions are satisfied, then $(J_\varepsilon)_{\varepsilon>0}$ Γ-converges to J_0, which will be shown next, including the specification of the limit J_0.

Theorem 2.2 *Consider the family of functionals $(J_\varepsilon)_{\varepsilon>0}$, as defined in* (14) *with $J_\varepsilon \colon X \to \overline{\mathbb{R}}$. Define the functional J_0 as*

$$\begin{aligned} &J_0 \colon X \to \overline{\mathbb{R}}, \\ &(\boldsymbol{O}, \boldsymbol{\mu}, \boldsymbol{\Sigma}) \mapsto \sum_{l=1}^{k} \left[\int_{O_l} \|g(x) - \mu_l\|_{\Sigma_l^{-1},\eta} + \log\det \Sigma_l \,\mathrm{d}x + \lambda \operatorname{Per}(O_l, \Omega) \right]. \end{aligned} \tag{15}$$

Then, $(J_\varepsilon)_{\varepsilon>0}$ Γ-converges to J_0 when $\varepsilon \to 0$ with respect to the weak convergence in X (cf. Definition 2.2).*

Proof. Let $(\boldsymbol{O}, \boldsymbol{\mu}, \boldsymbol{\Sigma}) \in X$ and $(\varepsilon_n)_{n\in\mathbb{N}}$ such that $\varepsilon_n > 0$ and $\varepsilon_n \overset{n\to\infty}{\longrightarrow} 0$. We start by showing the lim inf inequality. Let $(\boldsymbol{O}_n, \boldsymbol{\mu}_n, \boldsymbol{\Sigma}_n)_{n\in\mathbb{N}} \subseteq X$ be a sequence such that $(\boldsymbol{O}_n, \boldsymbol{\mu}_n, \boldsymbol{\Sigma}_n) \overset{*}{\rightharpoonup} (\boldsymbol{O}, \boldsymbol{\mu}, \boldsymbol{\Sigma}) \in X$. Then, $(J_{\varepsilon_n}[\boldsymbol{O}_n, \boldsymbol{\mu}_n, \boldsymbol{\Sigma}_n])_{n\in\mathbb{N}} \subseteq \overline{\mathbb{R}}$ is the sequence of the corresponding functional values. Without loss of generality, we can assume

$$\lim_{n\to\infty} J_{\varepsilon_n}[O_n, \mu_n, \Sigma_n] = \liminf_{n\to\infty} J_{\varepsilon_n}[O_n, \mu_n, \Sigma_n],$$

otherwise we select a subsequence of $(O_n, \mu_n, \Sigma_n)_{n\in\mathbb{N}}$ such that the corresponding functional values converge to $\liminf_{n\to\infty} J_{\varepsilon_n}[O_n, \mu_n, \Sigma_n]$. First, consider the term $I_{\mathcal{P}_\varepsilon}(\Sigma_l)$ of (14). For any $\varepsilon > 0$ and $\Sigma \in \mathcal{P}$, we have $I_{\mathcal{P}_\varepsilon}(\Sigma) \geqslant 0$ and $I_{\mathcal{P}}(\Sigma) = 0$, i.e.,

$$I_{\mathcal{P}}(\Sigma_l) = 0 \leqslant \liminf_{n\to\infty} I_{\mathcal{P}_{\varepsilon_n}}(\Sigma_{n,l}) \text{ for each } l \in \{1, \dots, k\}. \tag{16}$$

We now focus on the remaining terms of (14). Since $(O_n, \mu_n, \Sigma_n) \overset{*}{\rightharpoonup} (O, \mu, \Sigma)$, we have that $\chi_{O_{n,l}} \overset{*}{\rightharpoonup} \chi_{O_l}$ in $\mathrm{BV}(\Omega)$ and therefore $\chi_{O_{n,l}} \to \chi_{O_l}$ in $L^1(\Omega)$ for each $l \in \{1, \dots, k\}$, guaranteed by [2, Prop. 3.13]. In particular, we choose a subsequence of $(O_n, \mu_n, \Sigma_n)_{n\in\mathbb{N}}$ such that $(\chi_{O_{n,l}})_{n\in\mathbb{N}}$ for $l = 1$ converges point-wise a.e. to χ_{O_l}, from which we select another subsequence such that $(\chi_{O_{n,l}})_{n\in\mathbb{N}}$ for $l = 2$ converges point-wise a.e. to χ_{O_l}, and so on until $l = k$. The existence of such subsequences follows from [1, Lemma 3.22 (1)]. The resulting subsequence is again denoted with $(O_n, \mu_n, \Sigma_n)_{n\in\mathbb{N}}$. Analogously to the derivation in (13), we get

$$\begin{aligned} &\int_{O_l} \|g(x) - \mu_l\|_{\Sigma_l^{-1},\eta} + \log\det\Sigma_l \,\mathrm{d}x + \lambda \operatorname{Per}(O_l, \Omega) \\ \leqslant &\liminf_{n\to\infty} \int_{O_{n,l}} \|g(x) - \mu_{n,l}\|_{\Sigma_{n,l}^{-1},\eta} + \log\det\Sigma_{n,l} \,\mathrm{d}x + \lambda \operatorname{Per}\left(O_{n,l}, \Omega\right) \end{aligned}$$

for each $l \in \{1, \dots, k\}$. Combined with (16), we conclude

$$J_0[O, \mu, \Sigma] \leqslant \liminf_{n\to\infty} J_{\varepsilon_n}[O_n, \mu_n, \Sigma_n].$$

This shows the lim inf inequality.

To show the lim sup inequality, let $(O, \mu, \Sigma) \in X$ and $(\varepsilon_n)_{n\in\mathbb{N}} \subseteq \mathbb{R}$ such that $\varepsilon_n > 0$ and $\varepsilon_n \to 0$ when $n \to \infty$. We choose the constant sequence $(O_n, \mu_n, \Sigma_n)_{n\in\mathbb{N}}$ with $(O_n, \mu_n, \Sigma_n) = (O, \mu, \Sigma)$ for every $n \in \mathbb{N}$. This sequence obviously converges weakly* to (O, μ, Σ). Moreover, note that if $\varepsilon'' \geqslant \varepsilon' > 0$, then it follows $\mathcal{P}_{\varepsilon''} \subseteq \mathcal{P}_{\varepsilon'}$. Combined with $\varepsilon_n \to 0$ and $\Sigma_l \in \mathcal{P}$, there exists an $N \in \mathbb{N}$ such that $\Sigma_{n,l} = \Sigma_l \in \mathcal{P}_{\varepsilon_n}$ for every $l \in \{1, \dots, k\}$ and every $n \geqslant N$. As a consequence, we obtain that only for finitely many $n \in \mathbb{N}$, namely for at most $N - 1$, it can hold

$$J_{\varepsilon_n}[O_n, \mu_n, \Sigma_n] = J_{\varepsilon_n}[O, \mu, \Sigma] = \infty.$$

For all $n \geqslant N$, it must hold $I_{\mathcal{P}_{\varepsilon_n}}(\Sigma_{n,l}) = I_{\mathcal{P}_{\varepsilon_n}}(\Sigma_l) = 0$, and therefore

$$J_{\varepsilon_n}[O_n, \mu_n, \Sigma_n] = J_{\varepsilon_n}[O, \mu, \Sigma] = J_0[O, \mu, \Sigma].$$

We can conclude that the sequence of evaluations $(J_{\varepsilon_n}[O_n, \mu_n, \Sigma_n])_{n\in\mathbb{N}} \subseteq \overline{\mathbb{R}}$ is convergent with limit $J_0[O, \mu, \Sigma]$. This results in

$$J_0[O, \mu, \Sigma] = \lim_{n\to\infty} J_{\varepsilon_n}[O_n, \mu_n, \Sigma_n] = \limsup_{n\to\infty} J_{\varepsilon_n}[O_n, \mu_n, \Sigma_n]$$

and shows the lim sup inequality. □

Please note that J_0 is weakly* lower semi-continuous as a Γ-limit [4].

2.4 Loss of closedness of $\mathcal{P}_\varepsilon$ when $\varepsilon \to 0$

The Γ-convergence of a family of functionals to a Γ-limit implies under certain conditions also the convergence of minimizers of the family of functionals to a minimizer of the Γ-limit (cf. [4, Def. 1.19, Th. 1.21]). However, these conditions are not fulfilled in our case. In particular, we will now see an explicit example where a minimizer does not exist if we do not use the regularization with $\varepsilon > 0$, i.e., if we allow $\Sigma_l \in \mathcal{P}$ instead of restricting it to $\Sigma_l \in \mathcal{P}_\varepsilon$.

While $\mathbf{O} \in \mathcal{A}$ and $\boldsymbol{\mu} \in \left(\mathbb{R}^L\right)^k$ do not cause any trouble as J_0 is bounded below with respect to these parameters, the data term of J_0, which is the square root in (15), and the term $\log\det\Sigma_l$ do not always balance each other. For certain images g, the latter can make the functional arbitrarily small because certain terms of the former vanish. To see this, we consider for $l \in \{1, \dots, k\}$ the corresponding parts of J_0 from (15) and let $\Sigma_l = V_l D_l V_l^T$ be the eigendecomposition of Σ_l with an orthogonal matrix $V_l \in \mathbb{R}^{L\times L}$ and a diagonal matrix $D_l \in \mathbb{R}^{L\times L}$ with the eigenvalues of Σ_l on its diagonal. For Σ_l^{-1}, it holds $\Sigma_l^{-1} = V_l D_l^{-1} V_l^T$. We obtain

$$
\begin{aligned}
&\int_{O_l} \|g(x) - \mu_l\|_{\Sigma_l^{-1},\eta} + \log\det\Sigma_l \,\mathrm{d}x \\
&= \int_{O_l} \sqrt{\left[V_l^T\left(g(x)-\mu_l\right)\right]^T D_l^{-1}\left[V_l^T\left(g(x)-\mu_l\right)\right] + \eta}\,\mathrm{d}x + |O_l|\log\left(\prod_{i=1}^L \sigma(\Sigma_l)_i\right) \\
&= \int_{O_l} \sqrt{\sum_{i=1}^L \frac{\left[V_l^T\left(g(x)-\mu_l\right)\right]_i^2}{\sigma(\Sigma_l)_i} + \eta}\,\mathrm{d}x + |O_l|\sum_{i=1}^L \log\sigma(\Sigma_l)_i.
\end{aligned} \tag{17}
$$

The data term benefits from $\sigma(\Sigma_l)_i$ for $i \in \{1, \dots, L\}$ being very large as this makes the fraction small and approach 0. On the contrary, the $\log\det\Sigma_l$ term benefits from bringing $\sigma(\Sigma_l)_i$ as close to 0 as possible. In this sense, the two terms are antagonists.

An example in which they do not balance each other is as follows: Let $g \in L^\infty(\Omega, \mathbb{R}^L)$ such that $g(\Omega)$ is $(L-m)$-dimensional for a fixed $m \in \{1, \dots, L-1\}$. Since Ω is bounded, we get $g \in L^2(\Omega, \mathbb{R}^L)$. Fix $O_l \subseteq \Omega$ and $\mu_l \in \mathbb{R}^L$. We take an orthonormal basis of $\operatorname{span}(g(\Omega))$ and complete it to an orthonormal basis of $\mathbb{R}^L$ that we write as columns into a matrix $V_l \in \mathbb{R}^{L\times L}$. This basis consists of vectors $v_1, \dots, v_{L-m} \in \mathbb{R}^L$ that are a basis of $\operatorname{span}(g(\Omega))$ and vectors $v_{L-m+1}, \dots, v_L \in \mathbb{R}^L$ that are a basis of $\operatorname{span}(g(\Omega))^\perp$. We choose a diagonal matrix $D_l \in \mathbb{R}^{L\times L}$ with entries $\sigma_1, \dots, \sigma_L > 0$ on its diagonal and set $\Sigma_l := V_l D_l V_l^T$. It holds $\Sigma_l \in \mathcal{P}$. Then, $\langle\cdot,\cdot\rangle\colon \mathbb{R}^L \times \mathbb{R}^L \to \mathbb{R}, (y, z) \mapsto y^T\Sigma_l^{-1}z$ defines an inner product on $\mathbb{R}^L$. We have $\langle g(x), v_{L-m+i}\rangle = 0$ for almost every $x \in \Omega$ and every $i \in \{1, \dots, m\}$.

Let $P\colon \mathbb{R}^L \to \mathbb{R}^L$ denote the orthogonal projection with respect to $\langle\cdot,\cdot\rangle$ onto the subspace spanned by $v_{L-m+1},\ldots,v_L$ and I the identity operator. We find

$$\begin{aligned}
&\|g(x)-\mu_l\|^2_{\Sigma_l^{-1},\eta} = \|g(x)-(I-P)\mu_l - P\mu_l\|^2_{\Sigma_l^{-1},\eta}\\
&= \langle g(x)-(I-P)\mu_l, g(x)-(I-P)\mu_l\rangle + \langle P\mu_l, P\mu_l\rangle + \eta\\
&= \|g(x)-(I-P)\mu_l\|^2_{\Sigma_l^{-1},\eta} + \langle P\mu_l, P\mu_l\rangle \geqslant \|g(x)-(I-P)\mu_l\|^2_{\Sigma_l^{-1},\eta}.
\end{aligned}$$

The third equality holds by the linearity of the inner product and the orthogonality of $g(\Omega)$ and $v_{L-m+1},\ldots,v_L$. The inequality is true because of the positive definiteness of an inner product. That means replacing μ_l with $(I-P)\mu_l$ does not increase (17), i.e., we can assume $\mu_l \in \operatorname{span}(g(\Omega))$ when searching for a minimizer. Using this, we obtain analogously to the derivation of (17)

$$\begin{aligned}
&\int_{O_l} \|g(x)-\mu_l\|_{\Sigma_l^{-1},\eta} + \log\det\Sigma_l \,\mathrm{d}x\\
&= \int_{O_l} \sqrt{\sum_{i=1}^{L} \frac{\left[V_l^T(g(x)-\mu_l)\right]_i^2}{\sigma_i} + \eta}\,\mathrm{d}x + |O_l|\sum_{i=1}^{L}\log\sigma_i\\
&= \int_{O_l} \sqrt{\sum_{i=1}^{L-m} \frac{\left[v_i^T(g(x)-\mu_l)\right]^2}{\sigma_i} + \eta}\,\mathrm{d}x + |O_l|\sum_{i=1}^{L}\log\sigma_i.
\end{aligned}$$

The last equality holds by orthogonality. We observe that in this example the eigenvalues $\sigma_{L-m+1},\ldots,\sigma_L$ are not present in the data term. As a consequence, during minimization the functional benefits from bringing these values closer to 0 since the data term does not change but $\log\det\Sigma_l$ decreases. In this particular case, we even see that J_0 is unbounded from below because having the eigenvalues tending to 0 will let the value of J_0 diverge to $-\infty$. One could circumvent this divergence, e.g., by replacing log with $\log(\cdot+1)$, which is bounded below for positive arguments resulting in a lower bound for J_0. Nevertheless, the problem that J_0 benefits from bringing the eigenvalues closer to 0 remains. When applying J_ε in its original version defined on X_ε (cf. (3)) to such an image g, the functional will push $\sigma_{L-m+1},\ldots,\sigma_L$ to be equal to ε^2 in the minimum. Since $\mathcal{P}_\varepsilon$ is closed, such minimizers are still in $\mathcal{P}_\varepsilon$. The extension of J_ε to the open set $\mathcal{P}$ (cf. (14)) stayed coercive with respect to Σ_l due to the addition of $I_{\mathcal{P}_\varepsilon}$ since it prevents J_ε from making the eigenvalues smaller than ε^2 and thus avoids the problem that $\mathcal{P}$ is open and the eigenvalues could approach 0. Note that here the term *coercivity* has to be defined in a more general way than usual, describing that the functional shows a rapid growth at the extremes of the domain. However, in the Γ-limit when $\varepsilon \to 0$, we lose the closedness of the admissible set without compensating for it. Hence, J_0 does not satisfy a condition like coercivity (in the general sense explained above) in the eigenvalues of Σ_l and we also lose the existence of a minimizer in general.

2.5 Bound for eigenvalues of covariance estimates

We have seen in Section 2.4 that, for certain images g, we lose the boundedness from below when not regularizing with ε and allowing $\Sigma_l \in \mathcal{P}$ for $l \in \{1, \ldots, k\}$. However, it was already hinted before that the data term $\|g(x) - \mu_l\|_{\Sigma_l^{-1},\eta}$ and the $\log\det\Sigma_l$ term in (15) balance each other when the image g is sufficiently benign and $|O_l| > 0$. We first note that we can consider each summand of the outer sum over $l \in \{1, \ldots, k\}$ in (15) separately and that we only have to consider the integral because only here the eigenvalues of Σ_l play a role. Furthermore, we will use that for a real-valued concave function f with domain $D \subseteq \mathbb{R}$ the inequality $f\left(\sum_{i=1}^n a_i x_i\right) \geqslant \sum_{i=1}^n a_i f(x_i)$ holds for coefficients $a_1, \ldots, a_n \geqslant 0$ with $\sum_{i=1}^n a_i = 1$ and $x_1, \ldots, x_n \in D$. Let $\Sigma_l = V_l D_l V_l^T$ be the eigendecomposition of Σ_l with an orthogonal matrix $V_l \in \mathbb{R}^{L\times L}$ with columns $v_{l,i}$ for $i \in \{1, \ldots, L\}$ and a diagonal matrix $D_l \in \mathbb{R}^{L\times L}$ with the eigenvalues of Σ_l on its diagonal. It follows with (17)

$$
\begin{aligned}
&\int_{O_l} \|g(x) - \mu_l\|_{\Sigma_l^{-1},\eta} + \log\det\Sigma_l \,\mathrm{d}x \\
&\geqslant \int_{O_l} \sqrt{\sum_{i=1}^{L} \frac{\left[V_l^T (g(x) - \mu_l)\right]_i^2}{\sigma(\Sigma_l)_i}} \,\mathrm{d}x + |O_l| \sum_{i=1}^{L} \log\sigma(\Sigma_l)_i \\
&= \int_{O_l} \sqrt{L} \sqrt{\sum_{i=1}^{L} \frac{1}{L} \left[v_{l,i}^T (g(x) - \mu_l)\right]^2 \frac{1}{\sigma(\Sigma_l)_i}} \,\mathrm{d}x + |O_l| \sum_{i=1}^{L} \log\sigma(\Sigma_l)_i \\
&\geqslant \int_{O_l} \frac{\sqrt{L}}{L} \sum_{i=1}^{L} \sqrt{\left[v_{l,i}^T (g(x) - \mu_l)\right]^2 \frac{1}{\sigma(\Sigma_l)_i}} \,\mathrm{d}x + |O_l| \sum_{i=1}^{L} \log\sigma(\Sigma_l)_i \\
&= \int_{O_l} \frac{1}{\sqrt{L}} \sum_{i=1}^{L} \left|v_{l,i}^T (g(x) - \mu_l)\right| \frac{1}{\sqrt{\sigma(\Sigma_l)_i}} \,\mathrm{d}x + |O_l| \sum_{i=1}^{L} \log\sigma(\Sigma_l)_i \\
&= \sum_{i=1}^{L} \left(\frac{1}{\sqrt{L}} \int_{O_l} \left|v_{l,i}^T (g(x) - \mu_l)\right| \,\mathrm{d}x \,\frac{1}{\sqrt{\sigma(\Sigma_l)_i}} + |O_l| \log\sigma(\Sigma_l)_i \right).
\end{aligned}
$$

The summands of the outer sum are uncoupled with respect to $\sigma(\Sigma_l)_i$, thus they can be considered separately. For each $i \in \{1, \ldots, L\}$, the summand is a function of $\sigma(\Sigma_l)_i$ of the form

$$a/\sqrt{x} + b\log x, \tag{18}$$

for coefficients $a \geqslant 0$ and $b > 0$. This function is defined for $x > 0$ and has a minimum at $x^* = \left(\frac{a}{2b}\right)^2$ if $a > 0$. For $x < x^*$, it is strictly monotonically decreasing and, for $x > x^*$, strictly monotonically increasing. Hence, the minimum at x^* is global. Consequently, the following $\sigma(\Sigma_l)_i$ minimizes the i-th summand

$$\sigma(\Sigma_l)_i^* = \left(\int_{O_l} \left|v_{l,i}^T (g(x) - \mu_l)\right| \,\mathrm{d}x \Big/ \left(2\sqrt{L}\,|O_l|\right) \right)^2. \tag{19}$$

While the denominator is always greater 0 under the assumption of O_l having positive measure, the integral in the numerator can be 0. This is the case when $v_{l,i}$ and $g(x) - \mu_l$ are perpendicular for almost every $x \in O_l$. Geometrically speaking, we require that $g(x)$ deviates from μ_l in the direction of $v_{l,i}$ for almost every $x \in O_l$ or, in terms of spectra of images, that there is variation in the spectra of segment l along the direction of $v_{l,i}$. A visualization is given in Fig. 1. A heuristic for practical

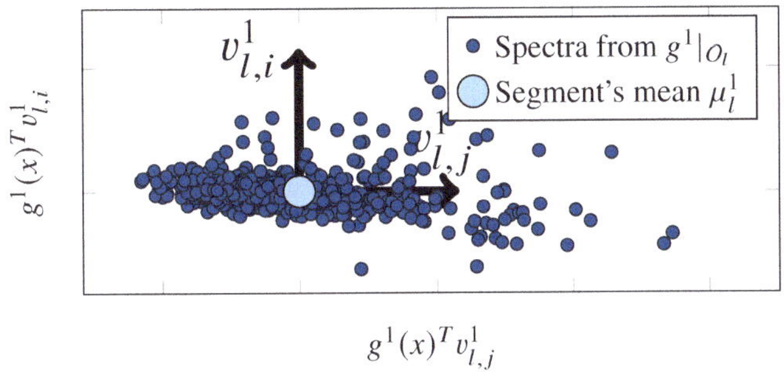

a) Samples from spectral distribution with variation in both depicted directions.

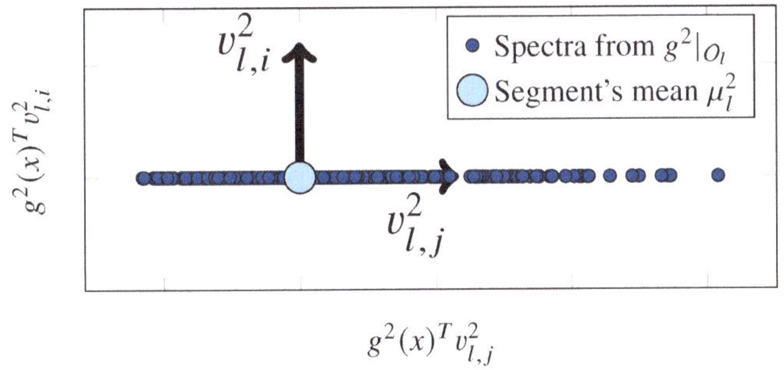

b) Samples from spectral distribution with variation in only one direction.

Fig. 1: The two figures show samples from two exemplary spectral distributions of $g^m|_{O_l}$ for a segment $O_l \subseteq \Omega$ and images $g^m \in L^\infty(\Omega, \mathbb{R}^L)$, $m \in \{1, 2\}$. The arrows $v^m_{l,i} \in \mathbb{R}^L$ and $v^m_{l,j} \in \mathbb{R}^L$ are the i-th and j-th eigenvector of the symmetric and positive-semidefinite covariance matrix $\Sigma^m_l \in \mathbb{R}^{L \times L}$. The projections of the sampled spectra onto those vectors are plotted. In the top figure, there is variation of the spectra in both directions $v^1_{l,i}$ and $v^1_{l,j}$, which results in estimates for the i-th and j-th eigenvalue of Σ^1_l being bounded from below with a positive bound (cf. Eq. (19)) and hence $\Sigma^1_l \in \mathcal{P}$ (provided that we also have variation in the directions of all other eigenvectors $v^1_{l,t}$ for $t \in \{1, \ldots, L\} \setminus \{i, j\}$). In the bottom figure, there is only variation of the spectra in direction $v^2_{l,j}$, not in direction $v^2_{l,i}$. This results in an estimate for the i-th eigenvalue of Σ^2_l equal to 0, which is not admissible anymore since then $\Sigma^2_l \notin \mathcal{P}$. Technically, we have $a = 0$ in (18) in this situation, making the function unbounded from below and Eq. (19) infeasible.

applications to achieve that $\sigma(\Sigma_l)_i^* > 0$ for all $i \in \{1, \ldots, L\}$ is to apply the principal components analysis (PCA) [18] to the data before processing it and keep only the components needed to describe 99.9% of the variance.

3 Conclusion

We analyzed the distribution-dependent Mumford–Shah model for hyperspectral image segmentation and its covariance regularization with a parameter ε. We showed the existence of minimizers and the Γ-convergence of the functional for $\varepsilon \rightarrow 0$. Moreover, we gave an example of an image where the Γ-limit does not have a minimizer. It turned out that the regularization with ε plays a very important role in the segmentation model since it ensures the feasibility of the model, allows to find bounds for the minimizing sequence such that convergent subsequences can be chosen and most importantly makes the set of admissible covariance matrices closed. In the Γ-limit, the set of admissible covariance matrices is not closed anymore. Together with the expression for the minimum values of the eigenvalues that was derived, we saw that this allows to push the eigenvalues of the covariance matrices arbitrarily close to 0 and therefore the covariance matrices out of the admissible set when there is no variation in the data in direction of the corresponding eigenvectors. Consequently, the functional in its current form needs the ε-regularization.

We plan to follow two future directions: we will investigate conditions on the data that guarantee the existence of minimizers in the Γ-limit, and without altering the functional. The formula for the minimum eigenvalues can serve as good starting point since we need the numerator to be always greater than a positive constant. Moreover, as already mentioned, the set of admissible covariance matrices is not closed anymore in the Γ-limit. Therefore, we will secondly aim to find a modification of the functional that guarantees the existence of minimizers without regularizing the covariance estimates or any condition on the data. An idea how to achieve that is to alter the model such that it fulfills a coercivity condition that makes the functional values explode when an eigenvalue of a covariance matrix is approaching 0 and running out of the admissible set. The point at which the functional explodes, when the eigenvalues fall below it, should, however, depend on the data. Also for this future direction the expression for the minimum eigenvalues is a good starting point.

Acknowledgements Jan-Christopher Cohrs was funded by the Deutsche Forschungsgemeinschaft (DFG, German Research Foundation) - 333849990/GRK2379 (IRTG Modern Inverse Problems).

Competing Interests The authors have no conflicts of interest to declare that are relevant to the content of this chapter.

References

1. Alt, H.W.: Linear Functional Analysis: An Application-Oriented Introduction. Universitext. Springer London, London (2016)
2. Ambrosio, L., Fusco, N., Pallara, D.: Functions of Bounded Variation and Free Discontinuity Problems. Oxford Mathematical Monographs. Oxford University Press (2000)
3. Bar, L., Chan, T.F., Chung, G., Jung, M., Kiryati, N., Mohieddine, R., Sochen, N., Vese, L.A.: Mumford and Shah Model and its Applications to Image Segmentation and Image Restoration. In: O. Scherzer (ed.) Handbook of Mathematical Methods in Imaging, chap. 25, pp. 1097–1157. Springer, New York, NY (2011). DOI 10.1007/978-0-387-92920-0
4. Braides, A.: Gamma-convergence for Beginners. Oxford University Press (2002). DOI 10.1093/acprof:oso/9780198507840.001.0001
5. Cai, X., Chan, R., Nikolova, M., Zeng, T.: A Three-Stage Approach for Segmenting Degraded Color Images: Smoothing, Lifting and Thresholding (SLaT). Journal of Scientific Computing **72**, 1313–1332 (2017). DOI 10.1007/s10915-017-0402-2
6. Cai, X., Chan, R., Zeng, T.: A Two-Stage Image Segmentation Method Using a Convex Variant of the Mumford–Shah Model and Thresholding. SIAM Journal on Imaging Sciences **6**(1), 368–390 (2013). DOI 10.1137/120867068
7. Cohrs, J.C., Bajaj, C., Berkels, B.: A Distribution-Dependent Mumford–Shah Model for Unsupervised Hyperspectral Image Segmentation. IEEE Transactions on Geoscience and Remote Sensing **60**, 1–21 (2022). DOI 10.1109/TGRS.2022.3227061
8. David, G.: Singular Sets of Minimizers for the Mumford-Shah Functional. Progress in Mathematics. Birkhäuser Basel (2005). DOI 10.1007/b137039
9. De Giorgi, E., Carriero, M., Leaci, A.: Existence theorem for a minimum problem with free discontinuity set. Archive for Rational Mechanics and Analysis **108**, 195–218 (1989). DOI 10.1007/BF01052971
10. Esedoglu, S., Shen, J.: Digital inpainting based on the Mumford–Shah–Euler image model. European Journal of Applied Mathematics **13**(4), 353–370 (2002). DOI 10.1017/S0956792502004904
11. Fornasier, M., March, R., Solombrino, F.: Existence of minimizers of the Mumford–Shah functional with singular operators and unbounded data. Annali di Matematica Pura ed Applicata **192**(3), 361–391 (2011). DOI 10.1007/s10231-011-0228-8
12. Halim, A., Rohim, A., Kumar, B.V.R., Saha, R.: A Fractional Image Inpainting Model Using a Variant of Mumford-Shah Model (2022). DOI 10.48550/arXiv.2208.01887
13. Hammer, A., Dumoulin, J., Vozel, B., Chehdi, K.: Deblurring of UAV aerial images for civil structures inspections using Mumford–Shah / Total variation regularisation. In: 2007 5th International Symposium on Image and Signal Processing and Analysis, pp. 262–267 (2007). DOI 10.1109/ISPA.2007.4383702
14. Jung, M., Bresson, X., Chan, T.F., Vese, L.A.: Nonlocal Mumford–Shah Regularizers for Color Image Restoration. IEEE Transactions on Image Processing **20**(6), 1583–1598 (2011). DOI 10.1109/TIP.2010.2092433
15. Mahalanobis, P.C.: On the generalized distance in statistics. Proceedings of the National Institute of Sciences (Calcutta) **2**, 49–55 (1936)
16. Mumford, D., Shah, J.: Optimal approximations by piecewise smooth functions and associated variational problems. Communications on Pure and Applied Mathematics **42**(5), 577–685 (1989). DOI 10.1002/cpa.3160420503
17. Ortega, J.M.: Numerical analysis: a second course. Classics in applied mathematics; 3. Society for Industrial and Applied Mathematics, Philadelphia, Pa (1990). DOI 10.1137/1.9781611971323
18. Pearson, K.: LIII. on lines and planes of closest fit to systems of points in space. The London, Edinburgh, and Dublin Philosophical Magazine and Journal of Science **2**(11), 559–572 (1901). DOI 10.1080/14786440109462720
19. Ramlau, R., Ring, W.: A Mumford–Shah level-set approach for the inversion and segmentation of X-ray tomography data. Journal of Computational Physics **221**(2), 539–557 (2007). DOI 10.1016/j.jcp.2006.06.041

20. Rondi, L., Santosa, F.: Enhanced electrical impedance tomography via the Mumford–Shah functional. ESAIM: Control, Optimisation and Calculus of Variations **6**, 517–538 (2001)
21. Tsai, A., Yezzi, A., Willsky, A.S.: Curve evolution implementation of the Mumford–Shah functional for image segmentation, denoising, interpolation, and magnification. IEEE Transactions on Image Processing **10**(8), 1169–1186 (2001). DOI 10.1109/83.935033

A Note on Best n-term Approximation for Generalized Wiener Classes

Ronald DeVore, Guergana Petrova and
Przemysław Wojtaszczyk

Abstract We determine the best n-term approximation of generalized Wiener model classes in a Hilbert space H. This theory is then applied to several special cases.

1 Introduction

One of the main themes in approximation theory is to prove theorems on how well functions can be approximated in a Banach space norm $\|\cdot\|_X$ by methods of linear or nonlinear approximation. The present paper is exclusively concerned with approximation in a separable Hilbert space H equipped with norm $\|\cdot\|$, induced by a scalar product $\langle\cdot,\cdot\rangle$. Let $\mathcal{D} := \{\phi_i, i \in \mathbb{N}\}$ be an orthonormal basis for H. This means that any function $f \in H$ has the unique representation

$$f = \sum_{j=1}^{\infty} f_j \phi_j, \quad \text{where} \quad \|f\|^2 = \sum_{j=1}^{\infty} |f_j|^2.$$

We are concerned with n term approximation of the elements $f \in H$. We denote by $\Sigma_n := \{S = \sum_{j\in\Lambda} c_j e_j : \Lambda \subset \mathbb{N}, |\Lambda| = n\}$ and let

$$\sigma_n(f) := \inf_{S\in\Sigma_n} \|f - S\|, \quad n \geqslant 1,$$

Ronald DeVore
Department of Mathematics, Texas A& M University, College Station, TX 77843, USA, e-mail: rdevore@tamu.edu

Guergana Petrova
Department of Mathematics, Texas A& M University, College Station, TX 77843, USA, e-mail: gpetrova@tamu.edu

Przemysław Wojtaszczyk
Institut of Mathematics, Polish Academy of Sciences, ul. Sniadeckich 8, 00-656, Warsawa, Poland, e-mail: Wojtaszczyk@impan.pl

R. DeVore, A. Kunoth (eds.), *Multiscale, Nonlinear and Adaptive Approximation II*,
https://doi.org/10.1007/978-3-031-75802-7_11

be the *error of n-term approximation* of f. Given any $f \in H$, a best n-term approximation S_n of f is given by

$$S_n = S_n(f) := \sum_{j \in \Lambda_n} f_j \phi_j,$$

where $\Lambda_n := \Lambda_n(f)$ is a set of n indices j for which $|f_j| \geqslant |f_i|$ whenever $j \in \Lambda$ and $i \notin \Lambda$. Even though the set $\Lambda_n(f)$ is not uniquely defined because of possible ties in terms of the size of the absolute values of the coefficients f_j, the error of approximation $\sigma_n(f)$ is uniquely defined.

We are interested in model classes $K \subset H$ that are given by imposing conditions on the coefficients (f_j) of f. For such sets K, we define

$$\sigma_n(K) := \sup_{f \in K} \sigma_n(f) \tag{1}$$

and we are interested in the asymptotic decay of $\sigma_n(K) \to 0$ as $n \to \infty$.

Since we are only considering the approximation to take place in H, in going further it is sufficient to consider only the case

$$H = \ell_2 := \left\{ \mathbf{x} = (x_1, x_2, \ldots,) : \|x\|_{\ell_2}^2 := \sum_{j=1}^{\infty} |x_j|^2 < \infty \right\}.$$

A classical result in this case is the following. Let $0 < p < 2$, and consider the unit ball in ℓ_p,

$$K = U(\ell_p) := \left\{ \mathbf{x} = (x_1, x_2, \ldots,) : \|x\|_{\ell_p}^p := \sum_{j=1}^{\infty} |x_j|^p \leqslant 1 \right\} \subset \ell_2.$$

It is known in this case that

$$\sigma_n(U(\ell_p)) \asymp n^{-1/p+1/2}, \quad n \to \infty, \tag{2}$$

with absolute constants in the equivalency[1] . This result is attributed to Stechkin [9].

Other results of the above type have been frequently obtained in the literature. To describe a general setting, let

$$\mathbf{w} := (w_j)_{j \in \mathbb{N}}, \quad 1 \leqslant w_1 \leqslant w_2 \leqslant \ldots,$$

be a monotonically nondecreasing sequence of positive weights. We consider the weighted ℓ_p space $\ell_p(\mathbf{w})$ defined as the set of all real valued sequences $\mathbf{x} \in \ell_2$ such that

$$\ell_p(\mathbf{w}) := \{\mathbf{x} \in \ell_2 : \|\mathbf{x}\|_{\ell_p(\mathbf{w})} < \infty\}, \quad 0 < p \leqslant \infty,$$

---

[1] We use the notation $A \asymp B$ when there are absolute constants$C_1, C_2 > 0$such that we have $C_1 B \leqslant A \leqslant C_2 B$

where

$$\|\mathbf{x}\|_{\ell_p(\mathbf{w})} := \begin{cases} \left[\sum_{j=1}^{\infty} |w_j x_j|^p\right]^{1/p}, & 0 < p < \infty, \\ \sup_j |w_j x_j|, & p = \infty. \end{cases} \tag{3}$$

We denote by $U(\ell_p(\mathbf{w}))$ the unit ball of this class

$$U(\ell_p(\mathbf{w}))) := \{\mathbf{x} \in \ell_p(\mathbf{w}) : \|\mathbf{x}\|_{\ell_p(\mathbf{w})} \leqslant 1\}$$

and derive the rate of the error $\sigma_n(U_p(\mathbf{w}))$, see Theorem 2.1. It can happen that $\sigma_n(U_p(\mathbf{w}))$ is infinite for all n. Also, note that the class $\ell_p(\mathbf{w})$ is different from the classical weighted sequence spaces since the weights in the $\ell_p(\mathbf{w})$ norm are raised to a power. The problem considered here has been investigated in the context of best n-term approximation of diagonal operators in the general case of approximation in ℓ_q. The works of Stepanets, see [6, 7, 8], determine the rate of $\sigma_n(U(\ell_p(\mathbf{w}))))_{\ell_q}$ for general $0 < q \leqslant \infty$ under the restriction $\lim_{k\to\infty} w_k = \infty$. Recently, in [4], based on results from [1], the condition on the sequence $\mathbf{w}$ in the case $0 < p < q$ has been removed, see Theorem 2.1(i). The case $q < p < \infty$ has also been considered, but under additional assumptions on the weight sequence $\mathbf{w}$, see Theorem 2.1(ii). In this paper, we consider only the case $q = 2$ and provide a simple, different unified approach for finding the rate of $\sigma_n(U(\ell_p(\mathbf{w}))))_{\ell_2}$ for all $0 < p \leqslant \infty$ with no restriction on the weight $\mathbf{w}$.

We then go on to apply our result in the case of several special weights

$$w_j = j^{\alpha}(1 + \log j)^{\beta}, \quad j \in \mathbb{N},$$

provided $\alpha > 0$, $\beta \in \mathbb{R}$, or $\alpha = 0$, $\beta \geqslant 0$. We show in Corollary 3.1 that when $\alpha = 0$, $\beta \geqslant 0$,

$$\sigma_n(U(\ell_p(\mathbf{w})) \asymp n^{-(1/p-1/2)}[\log n]^{-\beta}, \quad n > 1, \quad 0 < p < 2.$$

In the case $\alpha = \beta = 0$, this recovers Stechkin's result (2). In Corollary 3.2, we consider the case $\alpha > 0$, $\beta \in \mathbb{R}$, $0 < p \leqslant \infty$ and show that

$$U(\ell_p(\mathbf{w}))) \asymp n^{-(\alpha+1/p-1/2)}[\log n]^{-\beta}, \quad n > 1, \quad \alpha + 1/p - 1/2 > 0.$$

We call $\ell_p(\mathbf{w})$ a *generalized Wiener class* in analogy with the definition of Wiener spaces in Fourier analysis when $H = L_2([0, 1])$ and ϕ_j is the Fourier basis. Our results have some overlap with the study of Wiener classes in the Fourier setting, that is, when $\mathcal{D}$ is the Fourier basis $\mathcal{F}$. When considering the specific case of Fourier basis, our results, which are restricted to approximation in Hilbert spaces, are valid for L_2. Several results in the literature consider the approximation of Wiener classes in L_q when the basis is the Fourier basis $\mathcal{F}$. For example, the case $\alpha > 1/2$, $\beta = 0$, and $p = 1$ has been analyzed in [2] and upper bounds for the error in L_∞, have been obtained for the Wiener spaces with $\mathcal{D}$ being the d-dimensional Fourier basis $\mathcal{F}^d$, see Lemma 4.3(i). Recently, these results have been improved in [3], where matching up to logarithm lower and upper bounds for multidimensional Wiener spaces with $\mathcal{D} = \mathcal{F}^d$ are given for the case $\beta = 0$, $\alpha > 0$, $0 < p \leqslant 1$, see Corollary 4.3 in [3], and $\beta = 0$, $\alpha > 1 - 1/p$, $1 < p \leqslant q$, all when the error is measured in L_q, $2 \leqslant q \leqslant \infty$, see Theorem 4.5 in [3]. In particular, when the dimension $d = 1$ and the error is measured in the Hilbert space norm (i.e. $q = 2$), the results from [3] give the rate $\sigma_n(U_p(\mathbf{w}, \mathcal{F}))_{L_2} \asymp n^{-(\alpha+1/p-1/2)}$, provided

$$\alpha > 0, \quad 0 < p \leqslant 1, \quad \text{or} \quad \alpha > 1 - 1/p, \quad 1 < p \leqslant 2.$$

This latter result is a special case of our analysis.

2 Best n term approximation for $U(\ell_p(\mathbf{w}))$

In going forward, we assume $H = \ell_2$ with its canonical basis e_j, $j \in \mathbb{N}$. Before presenting our main theorem, let us introduce the decreasing rearrangement

$$\mathbf{x}^* = (x_j^*)_{j \in \mathbb{N}}$$

of the absolute values of the coordinates x_j of a sequence $\mathbf{x} = (x_j)_{j \in \mathbb{N}}$ that is an element of the sequence space $\mathbf{c_0}$ (consisting of all sequences whose elements converge to 0). Namely, we have that x_1^* is the largest of the numbers $|x_j|$, $j \in \mathbb{N}$, then x_2^* is the next largest, and so on. It follows that

$$x_1^* \geqslant x_2^* \geqslant \dots,$$

and $\|\mathbf{x}\|_{\ell_p} = \|\mathbf{x}^*\|_{\ell_p}$ for all $0 < p \leqslant \infty$. For each $n \geqslant 1$ and $\mathbf{x} \in \ell_2$, we have that

$$\sigma_n(\mathbf{x}) = \left[\sum_{j>n} [x_j^*]^2 \right]^{1/2}, \tag{4}$$

where $\sigma_n(\mathbf{x})$ is the error of n-term approximation of $\mathbf{x}$ in the ℓ_2 norm. In order to prove our main result, we will need the following lemma.

Lemma 2.1 *If* $\mathbf{x} \in \ell_p(\mathbf{w})$, *then* $\mathbf{x}^* \in \ell_p(\mathbf{w})$, $\sigma_n(\mathbf{x}) = \sigma_n(\mathbf{x}^*)$, *and*

$$\|\mathbf{x}^*\|_{\ell_p(\mathbf{w})} \leqslant \|\mathbf{x}\|_{\ell_p(\mathbf{w})}. \tag{5}$$

Proof. It follows directly from the definitions of σ_n and $\mathbf{x}^*$ that $\sigma_n(\mathbf{x}) = \sigma_n(\mathbf{x}^*)$. Let the sequence $\mathbf{x} = (x_1, x_2, \dots)$ be in $\ell_p(\mathbf{w})$. We can assume that all x_j are non-negative since changing the signs of its entries does not effect neither its rearrangement nor its $\ell_p(\mathbf{w})$ norm. We shall first construct sequences $\mathbf{y}^{(m)} = (y_1^{(m)}, y_2^{(m)}, \dots)$, $m = 0, 1, \dots$, such that each $\mathbf{y}^{(m+1)}$ is gotten by swapping the positions of two of the entries in $\mathbf{y}^{(m)}$ with the indices of these entries each larger than m. Also, the first m

entries of $\mathbf{y}^{(m)}$ satisfy

$$y_j^{(m)} = x_j^*, \quad j = 1, \ldots, m.$$

Indeed, we start with $\mathbf{y}^0 = \mathbf{x}$. Assuming that $\mathbf{y}^{(m)}$ has been defined, we let $j > m$ be the smallest index larger than m such that $y_j^{(m)}$ is the largest of the entries $y_k^{(m)}$, $k > m$. We swap the entries with positions $m+1$ and j to create the sequence $\mathbf{y}^{(m+1)}$ from $\mathbf{y}^{(m)}$. We have for $m = 0, 1, 2, \ldots,$

$$\|\mathbf{y}^{(m+1)}\|_{\ell_p(\mathbf{w})} \leqslant \|\mathbf{y}^{(m)}\|_{\ell_p(\mathbf{w})} \leqslant \ldots \leqslant \|\mathbf{y}^{(0)}\|_{\ell_p(\mathbf{w})} = \|\mathbf{x}\|_{\ell_p(\mathbf{w})},$$

because the weights in $\mathbf{w}$ are non-decreasing. Note that for every m

$$\sum_{j=1}^{m} [w_j x_j^*]^p = \sum_{j=1}^{m} [w_j y_j^{(m)}]^p \leqslant \|\mathbf{y}^{(m)}\|_{\ell_p(\mathbf{w})}^p \leqslant \|\mathbf{x}\|_{\ell_p(\mathbf{w})}^p,$$

and therefore

$$\|\mathbf{x}^*\|_{\ell_p(\mathbf{w})} \leqslant \|\mathbf{x}\|_{\ell_p(\mathbf{w})},$$

which completes the proof. □

Now we are ready to determine the rate of $\sigma_n(U(\ell_p(\mathbf{w}))$ for all n and all weight sequences $\mathbf{w}$. We fix $\mathbf{w}$, n, and $0 < p < \infty$. From the sequence $\mathbf{w}$, we define the numbers

$$W_m := [w_1^p + w_2^p + \ldots + w_m^p]^{1/p}, \quad m \geqslant 1. \tag{6}$$

Then the following theorem holds

Theorem 2.1 *For any* $0 < p < \infty$ *and* $\mathbf{w}$, *we have*

$$\max_{m \geqslant n} (m-n)[W_m]^{-2} \leqslant \sigma_n^2(U(\ell_p(\mathbf{w}))) \leqslant \max_{m \geqslant n} (m-n+1)[W_m]^{-2}, \tag{7}$$

and for $p = \infty$ *we have that*

$$\sigma_n^2(U(\ell_\infty(\mathbf{w}))) \asymp \sum_{j=n+1}^{\infty} w_j^{-2}.$$

Proof. We fix, $n, p, \mathbf{w}$. For every sequence $\mathbf{x} \in U(\ell_p(\mathbf{w}))$, we consider its decreasing rearrangement $\mathbf{x}^*$, which according to Lemma 2.1 is also an element of the unit ball and has the same n-term approximation. We next construct a new sequence $\tilde{\mathbf{x}}$ from $\mathbf{x}^*$ by making its first n entries equal to x_n^* and not touching the rest of the sequence, that is,

$$\tilde{x}_j = \begin{cases} x_n^*, & j = 1, \ldots, n, \\ x_j^*, & j > n. \end{cases}$$

Note that because the sequence $\mathbf{x}^*$ is nonincreasing and the weights are nondecreasing, we have

$$\|\tilde{\mathbf{x}}\|_{\ell_p(\mathbf{w})} \leqslant \|\mathbf{x}^*\|_{\ell_p(\mathbf{w})} \leqslant 1, \quad \text{and} \quad \sigma_n(\tilde{\mathbf{x}}) = \sigma_n(\mathbf{x}^*) = \sigma_n(\mathbf{x}).$$

Let us denote by $b := x_n^*$ and notice that $\sum_{j=1}^n w_j^p \tilde{x}_j^p = W_n^p b^p$. We have

$$\sigma_n(\mathbf{x}) = \sigma_n(\tilde{\mathbf{x}}) = \sum_{j=n+1}^{\infty} [\tilde{x}_j]^2 = \sum_{j=n+1}^{\infty} [x_j^*]^2.$$

We are now interested in how to change the tail of $\tilde{\mathbf{x}}$, that is, how to change the x_j^*'s with $j > n$ to new quantities y_j, $j > n$ so that we maximize the n-term approximation $\sigma_n^2(\mathbf{y}) = \sum_{j=n+1}^{\infty} [y_j]^2$, under the restrictions $b = x_n^* \geqslant y_{n+1} \geqslant \cdots$ and

$$\sum_{j=n+1}^{\infty} w_j^p y_j^p \leqslant 1 - W_n^p b^p =: S. \tag{8}$$

Notice that an investment of $w_j^p y_j^p$ towards S gives a return y_j^2 at coordinate j. Since the w_j are non-decreasing, to maximize $\sigma_n(\mathbf{y})$, it is best to invest as much as we can for $j = n+1$, $j = n+2$, and so on. So, the sequence which will maximize $\sigma_n^2(\mathbf{y})$ has to have $y_j = b$, $j = n+1, \ldots$ until we have used up our capital S. In other words, given that our sequence $\mathbf{y}$ has first n coordinates b, then to maximize $\sigma_n(\mathbf{y})$ we should take $\mathbf{y} = (b, b, \ldots, b, c, 0, 0, \ldots) \in U(\ell_p(\mathbf{w}))$, where $0 \leqslant c < b$. The membership in the unit ball of $\ell_p(\mathbf{w})$ requires that

$$b^p \sum_{j=1}^{m} w_j^p + c^p w_{m+1}^p \leqslant 1 \quad \Rightarrow \quad b^p [W_m]^p \leqslant 1$$

and

$$\sigma_n^2(\mathbf{y}) = (m-n)b^2 + c^2 < (m-n+1)b^2 \leqslant (m-n+1)[W_m]^{-2}.$$

Therefore, we have

$$\sigma_n^2(U(\ell_p(\mathbf{w}))) \leqslant \sup_{m \geqslant n} (m-n+1) W_m^{-2}.$$

Next, consider the special sequence $\mathbf{s}^{(m)}$ with entries $\mathbf{s}_j^{(m)}$ given by

$$\mathbf{s}_j^{(m)} = \begin{cases} W_m^{-1}, & j = 1, \ldots, m, \\ 0, & j > m. \end{cases}$$

Clearly $\mathbf{s}^{(m)} \in U(\ell_p(\mathbf{w}))$ for all m and

$$\sigma_n^2(\mathbf{s}^{(m)}) = (m-n)[W_m]^{-2}, \quad m > n.$$

Such sequences provides the lower bound in the case $1 < p < \infty$, and thus (7) is proven.

In the case $p = \infty$, we have that the entries x_j of a sequence $\mathbf{x} \in \ell_\infty(\mathbf{w})$ satisfy

$$|x_j| \leqslant \|\mathbf{x}\|_{\ell_\infty(\mathbf{w})} w_j^{-1} \quad \Rightarrow \quad \sigma_n^2(\mathbf{x}) \leqslant \left[\sum_{j=n+1}^{\infty} w_j^{-2}\right] \|\mathbf{x}\|_{\ell_\infty(\mathbf{w})}.$$

On the other hand, for the sequence

$$\mathbf{w}^{-1} := (w_1^{-1}, w_2^{-1}, \ldots) \in U(\ell_\infty(\mathbf{w})),$$

we have that $\sigma_n^2(\mathbf{w}^{-1}) = \sum_{j=n+1}^{\infty} w_j^{-2}$, and the proof is completed. □

3 Special cases of sequence spaces

In this section, we discuss several special cases of sequences $\mathbf{w}$ that are used in the definition of the classical Wiener spaces, see [3, 5] and the references therein.

Corollary 3.1 *Consider the classes $\ell_p(\mathbf{w})$ with $w_j := (1+\log j)^\beta$, $\beta \geqslant 0$, $0 < p < 2$. Then we have*

$$\sigma_n(U(\ell_p(\mathbf{w}))) \asymp m^{-(1/p-1/2)} [\log m]^{-\beta}.$$

Proof. Let us start by calculating the W_m,

$$W_m^p = \sum_{j=1}^{m} w_j^p = \sum_{j=1}^{m} (1 + \log j)^{\beta p} \geqslant \frac{m}{2}(1 + \log(m/2))^{\beta p}.$$

But we also have $W_m^p \leqslant m(1 + \log m)^{\beta p}$, so for m sufficiently big we get

$$W_m \asymp m^{1/p} [\log m]^\beta. \tag{9}$$

Using this in Theorem 2.1 gives that

$$\sigma_n(U(\ell_p(\mathbf{w}))) \asymp n^{1/2-1/p} [\log n]^{-\beta},$$

provided $0 < p < 2$. □

Corollary 3.2 *Consider the classes $\ell_p(\mathbf{w})$, $0 < p \leqslant \infty$, with*

$$w_j := \max_{1 \leqslant i \leqslant j} i^\alpha \log(i+1)^\beta, \quad j \in \mathbb{N}.$$

with $\alpha > 0$, $\beta \in \mathbb{R}$. Then we have

$$\sigma_n(U(\ell_p(\mathbf{w}))) \asymp n^{-(\alpha+1/p-1/2)} [\log(n+1)]^{-\beta},$$

provided $\alpha + 1/p - 1/2 > 0$.

Proof. **Case** $0 < p < \infty$: Let us observe that for any $\delta > 0$ and $\gamma \in \mathbb{R}$, the function $\varphi(x) := x^\delta \log(x+1)^\gamma$, is an increasing function on $[c, \infty)$, if $c = c(\delta, \gamma)$ is sufficiently large. Therefore, we have that

$$W_m^p = \sum_{j=1}^{m} j^{\alpha p} \log(j+1)^{\beta p} \asymp m^{\alpha p+1} \log(m+1)^{\beta p}, \quad m \geqslant M,$$

provided M is sufficiently large. This gives

$$W_m \asymp m^{\alpha+1/p} \log(m+1)^{\beta}, \quad m \geqslant M. \tag{10}$$

Theorem 2.1 now gives that

$$\sigma_n(U(\ell_p(\mathbf{w}))) \asymp n^{1/2-\alpha-1/p}[\log(n+1)]^{-\beta},$$

for n sufficiently large as desired.

Case $p = \infty$**:** In this case we have the restriction that $2\alpha > 1$. According to Theorem 2.1, we have that

$$\sigma_n^2(U(\ell_\infty(\mathbf{w}))) \asymp \sum_{j=n+1}^{\infty} j^{-2\alpha}[\log(j+1)]^{-2\beta} \asymp n^{-2\alpha+1} \log(n+1)^{-2\beta}.$$

□

Remark 3.1 Note that when $\beta = 0$, the ranges of α and p in Corollary 3.2 are

$$\alpha > 0, \quad 0 < p \leqslant 2 \quad \text{or} \quad \alpha > 1/2 - 1/p > 0, \quad 2 < p \leqslant \infty.$$

Acknowledgements This work was supported by the NSF Grant DMS 2134077 and the ONR Contract N00014-20-1-278. We would like to thank V. K. Nguyen for pointing out to us several valuable references.

References

1. Gao F., : Exact value of the n-term approximation of a diagonal operator, J. Approx. Theory, **162** (2010), 646–652.
2. Jahn T., Ullrich T., and Voigtlaender F., : Sampling numbers of smoothness classes via ℓ^1-minimization. Journal of Complexity, **79** (2023).
3. Moeller M., Gelfand numbers and best m-term trigonometric approximation for weighted mixed Wiener classes in L_2, Master Thesis, `https://www-user.tu-chemnitz.de/$~$momoe/MA.pdf`.
4. Nguyen V.D., Nguyen V.K., :Best n-term approximation of the diagonal operators and application to function spaces with mixed smoothness, Analysis Math., **48** (2022) 1127–1152.
5. Nguyen V.D., Nguyen V.K., W. Sickel, W., : s-numbers of embeddings of weighted Wiener algebras. Journal of Approximation Theory, **279** (2022).
6. Stepanets A. I.,: Approximation characteristics of the space S_φ^p in different metrics, Ukrainian Math. J., **53** (2001), 1340–1374.
7. Stepanets A. I.,: Methods of Approximation Theory, VSP (Leiden, Boston, 2005).
8. Stepanets A. I.,: Problems of approximation theory in linear spaces, Ukrainian Math. J., **58** (2006), 54–102.
9. Temlyakov V., : Approximation of functions with bounded mixed derivatives. Trudy Mat. Inst. Steklov, **178**, 3–113 (1986).

Stable Truncation and Root-Independent Normalization of Tree Tensor Networks

Lars Grasedyck,
Sebastian Krämer and
Dieter Moser

Abstract A ubiquitous tool in numerical linear algebra is the singular value decomposition, which can be lifted to higher order tensors and provides a useful mechanic for quasi-optimal approximations. The stable computation of a therefor required hierarchical singular value decomposition is non-trivial if it is to be efficiently achieved only via data-sparse, hierarchical low rank representations. Even if the initial representation is stable, the standard approach for the decomposition based on Gram matrices involves a squaring of singular values which will typically reduce the attainable accuracy. We provide an efficient and stable variant of the hierarchical singular value decomposition related to (root-independent) normal forms of tensors that can be computed in linear complexity of the dimension or order of the tensor and is accurate up to machine precision. Numerical tests highlight the higher accuracy of the introduced approach.

1 Introduction

A fundamental tool in matrix computations is the famous singular value decomposition. On the one hand, it reveals the rank of the matrix and the subspace of the image or kernel, and on the other hand, it allows to directly access best approximations of lower rank. Such a type of useful normal form can also be obtained in higher dimension [7], where it is based on a family of matricizations (reshapings) of the tensor, corresponding to an underlying tree structure of the modes. However, the efficient computation of such a hierarchical SVD within data sparse tree formats is non-trivial. In the matrix case, if $A \in \mathbb{R}^{n \times m}$ is given in factorized form

Lars Grasedyck, Sebastian Krämer and Dieter Moser
Institut für Geometrie und Praktische Mathematik, RWTH Aachen University, Templergraben 55, 52056 Aachen, Germany, e-mail: `lgr@igpm.rwth-aachen.de`, e-mail: `kraemer@igpm.rwth-aachen.de`, e-mail: `moser@igpm.rwth-aachen.de`

R. DeVore, A. Kunoth (eds.), *Multiscale, Nonlinear and Adaptive Approximation II*,
https://doi.org/10.1007/978-3-031-75802-7_12

$$A = BC^T, \quad B \in \mathbb{R}^{n\times r}, \; C \in \mathbb{R}^{m\times r}, \quad r \ll n, m,$$

then a compact SVD omitting the zero singular values and corresponding singular vectors can be computed in $O((n+m)r^2)$. In [7], a generalized computation in $O(dnk^3)$ is introduced for a tensor in $\mathbb{R}^{n^d}$ of initial representation rank k, based on Gram matrices. However, in the calculation of such, one essentially squares the corresponding singular values, which will typically reduce the obtainable, relative accuracy to the square root of machine precision.

This problem is well solved for linear trees yielding tensor trains [14] or matrix product states (MPS) [16], as well as the (non-hierarchical) Tucker format or HOSVD [4], where a normal form can be computed efficiently and close to machine precision by local SVDs so that a subsequent quasi-optimal truncation to lower ranks is possible without loss of accuracy — provided the initial representation is stable [1].

In a general tree format, the hierarchical Tucker (HT) format [11, 8], we will provide normal forms, including a root-independent one, as well as a corresponding SVD-based quasi-best truncation to lower rank that is accurate up to machine precision, related to approaches discussed for instance in [3, 2, 10] – again provided the initial representation is stable.

In Section 2, we first introduce a compact notation based on a consistent labeling of modes of tensors needed to conveniently handle subsequent topics, while also staying close to an actual implementation. In Section 3, we develop a stable normalization procedure that provides the hierarchical singular value decomposition of a tensor that is initially given in HT representation. A crucial detail here is the possible occurrence of relatively small singular values that can lead to instability problems with respect to normalization constraints. In Section 4, we therefore introduce a slightly weaker variant, so-called decoupled networks and derive a root-independent normalization. We show how a tensor with prescribed singular values, as it is often desired for numerical tests, can be obtained by factor-wise independent calculations. Finally, in Section 5, we provide numerical tests that highlight the differences between the classical Gram matrix based approach and the introduced stable normalization.

2 Label-Based Tensor Arithmetic

The concepts presented in this section follow a principle, similar to [5, 13], of not describing tensor networks by means of static maps, but of letting them emerge from the properties of the factors themselves. In a certain sense, one realizes the Einstein notation but with consistently predefined indices instead of ones repeatedly written down in sequence. With some initial work in Section 2.1, the formal description of hierarchical Tucker decompositions in Section 2.4 is all the more compact, representing just one particular instance of a well-defined, more general product elaborated throughout Sections 2.2 and 2.3.

Instead of conventional indexing of tensors via tuples of natural numbers, we do so via *labeled tuples* as defined below. To that end, we assign an abstract set of symbolic labels to each tensor, where the set of all such labels is assumed to be ordered (in some unspecified way). For a set of labels $\Lambda = \{\lambda_1, \dots, \lambda_K\}$, with $\lambda_1 < \dots < \lambda_K$, and any tuple $t = (t_1, \dots, t_K)$, we treat a labeled tuple I both as set and the function defined through it,

$$I = \Lambda[t] := \{(\lambda_k, t_k)\}_{k=1}^{K}, \quad I_{\lambda_k} := t_k, \quad k = 1, \dots, K.$$

For a singleton label set $\{\gamma\}$ and any s, we abbreviate $\gamma[s] := \{\gamma\}[(s)] = \{(\gamma, s)\}$.

2.1 Labeled Tensors

We regard a *labeled tensor* $\mathbf{T}$ with labels $\Lambda = \Lambda(\mathbf{T})$ as a function that maps labeled tuples $\Lambda[i] \in \mathbb{N}^{\Lambda}$, $i \in \mathbb{N}^{|\Lambda|}$, that is, functions from Λ to $\mathbb{N}$, to single entries,

$$\mathbf{T} : \mathbb{N}^{\Lambda} \to \mathbb{R}.$$

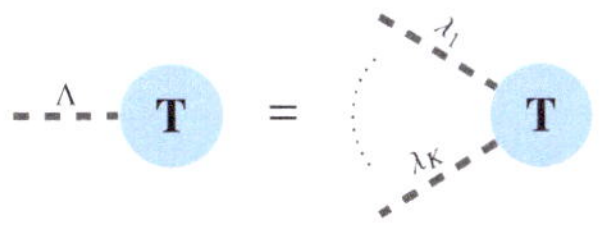

We assign *mode domain sizes* through a (not necessarily minimal) map $\mathcal{D} = \mathcal{D}(\mathbf{T}) \in (\mathbb{N} \cup \{\infty\})^{\Lambda}$ that is required to hold

$$\operatorname{supp}(\mathbf{T}) \subseteq \{I \in \mathbb{N}^{\Lambda} \mid I_{\lambda} \leqslant \mathcal{D}_{\lambda},\ \lambda \in \Lambda\}.$$

These sizes work in an analogous way to explicit size limits, but are preferred for notational convenience. Instead of $\mathbf{T} \in \mathbb{R}^{3\times4\times5}$, we may for instance state $\Lambda(\mathbf{T}) = \{\lambda_1, \lambda_2, \lambda_3\}$ and $\mathcal{D}(\mathbf{T}) = \{(\lambda_1, 3), (\lambda_2, 4), (\lambda_3, 5)\}$, where within the corresponding range, the labeled tensor then holds the same entries as an unlabeled one. As generalization of fibers or slices of a tensor, we extend $\mathbf{T}$ to map $S \in \mathbb{N}^{\Gamma}$, for any $\Gamma \subsetneq \Lambda(\mathbf{T})$, to the tensor $\mathbf{T}(S)$, defined via

$$\Lambda(\mathbf{T}(S)) := \Lambda(\mathbf{T}) \setminus \Gamma, \quad \mathbf{T}(S)(\cdot) = \mathbf{T}(S \cup \cdot), \quad \mathcal{D}(\mathbf{T}(S)) = \mathcal{D}(\mathbf{T})|_{\Lambda(\mathbf{T}(S))}.$$

For instance, for a labeled tensor $\mathbf{T}$ with $\Lambda(\mathbf{T}) = \{\lambda_1, \lambda_2, \lambda_3\}$ and a labeled tuple $S = \{(\lambda_1, 4), (\lambda_3, 5)\} \in \mathbb{N}^{\{\lambda_1, \lambda_3\}}$, the restricted tensor is the map $\mathbf{T}(S) : \mathbb{N}^{\{\lambda_2\}} \to \mathbb{R}$ with $\mathbf{T}(S)(\{(\lambda_2, i)\}) = \mathbf{T}(\{(\lambda_1, 4), (\lambda_2, i), (\lambda_3, 5)\})$ for all $i \in \mathbb{N}$. As analogue to the conventional tensor product, we define the product $\mathbf{T}_1 \otimes \mathbf{T}_2$ of two tensors with disjoint labels via

$$\Lambda(\mathbf{T}_1 \otimes \mathbf{T}_2) := \Lambda(\mathbf{T}_1) \cup \Lambda(\mathbf{T}_2), \quad (\mathbf{T}_1 \otimes \mathbf{T}_2)(S_1 \cup S_2) := \mathbf{T}_1(S_1) \cdot \mathbf{T}_2(S_2),$$

for all $S_1 \in \mathbb{N}^{\Lambda(\mathbf{T}_1)}$ and $S_2 \in \mathbb{N}^{\Lambda(\mathbf{T}_2)}$. Its mode domain sizes are consequently set through the accumulated labeled tuple $\mathcal{D}(\mathbf{T}_1 \otimes \mathbf{T}_2) = \mathcal{D}(\mathbf{T}_1) \cup \mathcal{D}(\mathbf{T}_2)$. For the general case, we define the *labeled tensor product*

$$\mathbf{T}_1 \boxtimes \mathbf{T}_2 := \sum_{S \in \mathbb{N}^\Gamma} \mathbf{T}_1(S) \otimes \mathbf{T}_2(S), \qquad \Gamma := \Lambda(\mathbf{T}_1) \cap \Lambda(\mathbf{T}_2).$$

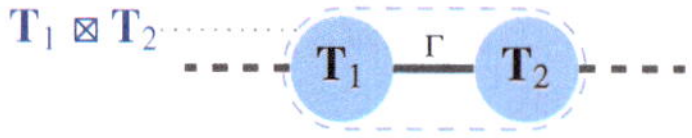

For a collection of labeled tensors $\{\mathbf{T}_k\}_{k=1}^K$, in which every label appears at most twice, we also abbreviate the product

$$\boxtimes(\{\mathbf{T}_k\}_{k=1}^K) := \mathbf{T}_1 \boxtimes \ldots \boxtimes \mathbf{T}_K,$$

being well-defined due to the commutativity and associativity of the right-hand side.

Example 2.1 (Labeled Tensor Product) Let $\mathbf{T}_1 : \mathbb{N}^{\Lambda_1} \to \mathbb{R}$ and $\mathbf{T}_2 : \mathbb{N}^{\Lambda_2} \to \mathbb{R}$ be two tensors with labels $\Lambda_1 := \Lambda(\mathbf{T}_1) = \{\lambda_1, \lambda_2, \lambda_4\}$ and $\Lambda_2 := \Lambda(\mathbf{T}_2) = \{\lambda_2, \lambda_3, \lambda_4, \lambda_5\}$, where $\lambda_1 < \ldots < \lambda_5$. Then for $\Gamma := \Lambda_1 \cap \Lambda_2 = \{\lambda_2, \lambda_4\}$ and each $S \in \{\Gamma[(i,j)] \mid i,j \in \mathbb{N}\}$, the restricted tensors $\mathbf{T}_1(S) : \mathbb{N}^{\{\lambda_1\}} \to \mathbb{R}$ and $\mathbf{T}_2(S) : \mathbb{N}^{\{\lambda_3,\lambda_5\}} \to \mathbb{R}$ have disjoint labels. Their product $\mathbf{H} := \mathbf{T}_1 \boxtimes \mathbf{T}_2$ is hence the tensor $\mathbf{H} : \mathbb{N}^{\{\lambda_1,\lambda_3,\lambda_5\}} \to \mathbb{R}$, with

$$\begin{aligned} &\mathbf{H}(\{(\lambda_1, s), (\lambda_3, t), (\lambda_5, u)\}) \\ &= \sum_{i,j \in \mathbb{N}} \mathbf{T}_1(\underbrace{\{(\lambda_2, i), (\lambda_4, j)\}}_{\Gamma[(i,j)]} \cup \{(\lambda_1, s)\}) \cdot \mathbf{T}_2(\underbrace{\{(\lambda_2, i), (\lambda_4, j)\}}_{\Gamma[(i,j)]} \cup \{(\lambda_3, t), (\lambda_5, u)\}) \end{aligned}$$

for all $s, t, u \in \mathbb{N}$. This corresponds to the following graphical visualization.

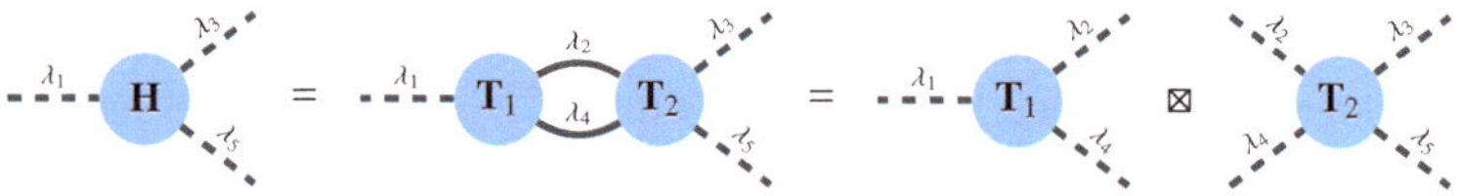

The mode domains $\mathcal{D}(\mathbf{H}) : \{\lambda_1, \lambda_3, \lambda_5\} \to \mathbb{N} \cup \{\infty\}$ of $\mathbf{H}$ are given via $\mathcal{D}(\mathbf{H}) = \{(\lambda_1, \mathcal{D}(\mathbf{T}_1)_{\lambda_1}), (\lambda_3, \mathcal{D}(\mathbf{T}_2)_{\lambda_3}), (\lambda_5, \mathcal{D}(\mathbf{T}_2)_{\lambda_5})\}$.

2.2 Labeled Matrices

We regard a *labeled matrix* $\mathbf{A}$, with labels $\Lambda = \Lambda(\mathbf{A})$, as a function

$$\mathbf{A} : \mathbb{N}^\Lambda \times \mathbb{N}^\Lambda \to \mathbb{R}.$$

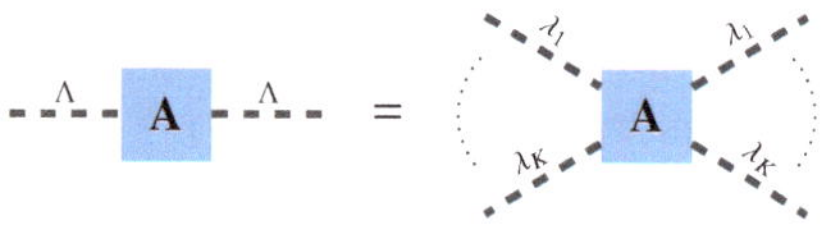

Analogously to tensors, the (mode domain) sizes of a matrix $\mathbf{A}$ are assigned through two maps $\mathcal{D}_i = \mathcal{D}_i(\mathbf{A}) : \Lambda \to \mathbb{N} \cup \{\infty\}$, $i \in \{1, 2\}$, that are required to hold

$$\operatorname{supp}(\mathbf{A}) \subseteq \{(I, J) \in \mathbb{N}^\Lambda \times \mathbb{N}^\Lambda \mid I_\lambda \leqslant \mathcal{D}_{1\lambda},\ J_\lambda \leqslant \mathcal{D}_{2\lambda},\ \lambda \in \Lambda\}.$$

We define the transpose $\mathbf{A}^T$ as matrix defined via $\mathbf{A}^T(S_1, S_2) := \mathbf{A}(S_2, S_1)$, with accordingly switched sizes. For notational convenience, we set $\mathbf{T}^T$ to do nothing in case $\mathbf{T}$ is a tensor[1]. We further generalize *rows* $\mathbf{A}(S_1, \cdot)$ and *columns* $\mathbf{A}(\cdot, S_2)$ to be each the tensor with identical labels that fulfills

$$\mathbf{A}(S_1, \cdot)(S_2) = \mathbf{A}(S_1, S_2) = \mathbf{A}(\cdot, S_2)(S_1), \quad S_1, S_2 \in \mathbb{N}^\Lambda.$$

For $\Gamma := \Lambda(\mathbf{A}) \subseteq \Lambda(\mathbf{T})$, we define the tensors $\mathbf{T} \boxtimes \mathbf{A}$ and $\mathbf{A} \boxtimes \mathbf{T}$ via

$$(\mathbf{T} \boxtimes \mathbf{A})(S) := \mathbf{T} \boxtimes \mathbf{A}(\cdot, S), \quad (\mathbf{A} \boxtimes \mathbf{T})(S) := \mathbf{A}(S, \cdot) \boxtimes \mathbf{T}, \quad S \in \mathbb{N}^\Gamma.$$

For two labeled matrices $\mathbf{A_1}$ and $\mathbf{A_2}$, we further define $\mathbf{A_1} \boxtimes \mathbf{A_2}$ as the labeled matrix given via $\Lambda(\mathbf{A_1} \boxtimes \mathbf{A_2}) = \Lambda(\mathbf{A_1}) \cup \Lambda(\mathbf{A_2})$,

$$(\mathbf{A_1} \boxtimes \mathbf{A_2})(S_1 \cup C_2, C_1 \cup S_2) := \mathbf{A_1}(S_1, C_1) \boxtimes \mathbf{A_2}(C_2, S_2),$$

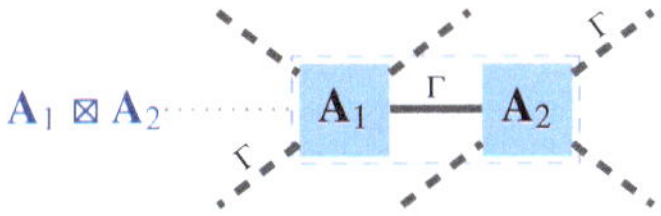

for $S_1 \in \mathbb{N}^{\Lambda(\mathbf{A_1})}$, $C_1 \in \mathbb{N}^{\Lambda(\mathbf{A_1})\setminus\Gamma}$, $C_2 \in \mathbb{N}^{\Lambda(\mathbf{A_2})\setminus\Gamma}$, $S_2 \in \mathbb{N}^{\Lambda(\mathbf{A_2})}$, where the contraction is again over $\Gamma := \Lambda(\mathbf{A}_1) \cap \Lambda(\mathbf{A}_2)$. For matrices with identical labels, in particular for singleton sets, the prior is merely a designated reformulation of ordinary matrix products.

2.3 Orthogonality

In the following, we translate orthogonality constraints and decompositions to labeled objects. For a labeled tensor $\mathbf{T}$ and $\Gamma \subseteq \Lambda(\mathbf{T})$, we say $\mathbf{T}$ is Γ-orthogonal if

$$\|\mathbf{T} \boxtimes \mathbf{X}\|_F = \|\mathbf{X}\|_F \text{ holds for all } \mathbf{X} \text{ s.t. } \Lambda(\mathbf{X}) = \Gamma,\ \mathcal{D}(\mathbf{X}) = \mathcal{D}(\mathbf{T})|_\Gamma.$$

For singletons $\Gamma = \{\gamma\}$, we skip set brackets. To indicate orthogonality, we use an arrow in the corresponding visualization.

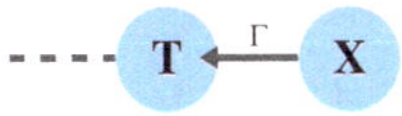

[1] In principle, one may still think of the transposition as reversing the order of indices. Since however the index is a set anyway, this does not change the labeled tensor.

For a matrix, we say $\mathbf{A}$ is (column-)orthogonal if

$$\|\mathbf{A} \boxtimes \mathbf{X}\|_F = \|\mathbf{X}\|_F \text{ holds for all } \mathbf{X} \text{ s.t. } \Lambda(\mathbf{X}) = \Lambda(\mathbf{A}),\ \mathcal{D}(\mathbf{X}) = \mathcal{D}_2(\mathbf{A}).$$

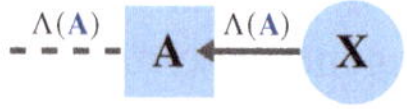

We call $\mathbf{A}$ row-orthogonal if $\mathbf{A}^T$ is (column-)orthogonal. By linearity, both norm equalities then also hold true for tensors as well as matrices $\mathbf{X}$ with suitable support with $\Lambda(\mathbf{X}) \cap \Lambda(\mathbf{T}) = \Gamma$ or $\Lambda(\mathbf{A}) \subseteq \Lambda(\mathbf{X})$, respectively. The following Lemma 2.1 describes how far orthogonality is retained in products. Although phrased more generally, the two cases required are for either $\Gamma_2 = \emptyset$, as it appears in a path of tensors, or $\Gamma_{12} = \emptyset$, when different branches of a tree are combined.

Lemma 2.1 *Let* $\mathbf{T}_1$ *and* $\mathbf{T}_2$ *be two tensors and let* $\Gamma_1 \subset \Lambda(\mathbf{T}_1)$ *and* $\Gamma_2 \subset \Lambda(\mathbf{T}_2)$ *such that* $\Gamma_1 \cap \Lambda(\mathbf{T}_2) = \emptyset$ *and* $\Gamma_2 \cap \Lambda(\mathbf{T}_1) = \emptyset$. *Let further* $\Gamma_{12} = \Lambda(\mathbf{T}_1) \cap \Lambda(\mathbf{T}_2)$ *as well as* $\mathbf{T}_2$ *be* $\Gamma_2 \dot\cup \Gamma_{12}$*-orthogonal. Then the following two properties are equivalent:*

(a) $\mathbf{T}_2 \boxtimes \mathbf{T}_1$ *is* $\Gamma_1 \dot\cup \Gamma_2$*-orthogonal*
(b) $\mathbf{T}_1$ *is* Γ_1*-orthogonal*

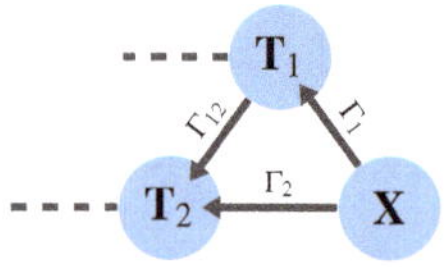

Proof. Let $\mathbf{X}$ be a tensor with $\Lambda(\mathbf{X}) = \Gamma_1 \dot\cup \Gamma_2$ and according sizes $\mathcal{D}(\mathbf{X}) = \mathcal{D}(\mathbf{T}_1)|_{\Gamma_1} \dot\cup \mathcal{D}(\mathbf{T}_2)|_{\Gamma_2}$. Then $\Lambda(\mathbf{X}) \cap \Lambda(\mathbf{T}_1) = \Gamma_1$ and $\Lambda(\mathbf{T}_1 \boxtimes \mathbf{X}) \cap \Lambda(\mathbf{T}_2) = \Gamma_2 \dot\cup \Gamma_{12}$. Consequently, the two properties imply

$$\|\mathbf{X}\|_F \overset{(a)}{=} \|(\mathbf{T}_2 \boxtimes \mathbf{T}_1) \boxtimes \mathbf{X}\|_F = \|\mathbf{T}_1 \boxtimes \mathbf{X}\|_F \overset{(b)}{=} \|\mathbf{X}\|_F$$

Thus, either both or neither of the equalities are valid for all such $\mathbf{X}$. □

We say $(\mathbf{Q}, \mathbf{R})$ is a QR-decomposition of a tensor $\mathbf{T}$ (with finite Frobenius norm) if for some $\Gamma \subseteq \Lambda(\mathbf{T})$,

$$\mathbf{Q} \boxtimes \mathbf{R} = \mathbf{T}, \quad \Lambda(\mathbf{Q}) = \Lambda(\mathbf{T}), \quad \Lambda(\mathbf{R}) = \Gamma,$$

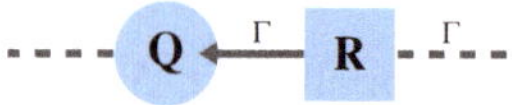

where $\mathbf{Q}$ is Γ-orthogonal and $\mathbf{R}$ is an upper triangular matrix; that is, it holds $\mathbf{R}(\Gamma[i], \Gamma[j]) = 0$ for all $i > j$ in colexicographic ordering. We say $(\mathbf{U}, \mathbf{\Sigma}, \mathbf{V})$ is an SVD of a tensor $\mathbf{T}$ (with finite Frobenius norm) if

$$\mathbf{U} \boxtimes \mathbf{\Sigma} \boxtimes \mathbf{V}^T = \mathbf{T}, \quad \Lambda(\mathbf{U}) \cap \Lambda(\mathbf{V}) = \Gamma = \Lambda(\mathbf{\Sigma}),$$

where $\mathbf{U}$ as well as $\mathbf{V}$ are Γ-orthogonal and $\mathbf{\Sigma}$ is a diagonal matrix with non-negative, non-increasing diagonal; that is, it holds $\mathbf{\Sigma}(\Gamma[i], \Gamma[j]) = 0$ for $i \neq j$. We in particular allow $\mathbf{V}$ to be either a labeled tensor (in which case $\mathbf{V}^T = \mathbf{V}$) or a matrix (in which case $\mathbf{V}$ is column-orthogonal). For a singleton $\Gamma = \{\gamma\}$, we as usual call the decomposition thin, if $\mathbf{\Sigma}$ is a square matrix with size $\mathcal{D}_i(\mathbf{\Sigma})_\gamma = r$, $i \in \{1, 2\}$, for some $r \in \mathbb{N}$. We call it compact if r equals the rank of $\mathbf{\Sigma}$. In all cases, we assume that the sizes of $\mathbf{U}$ and $\mathbf{V}$ in Γ equal the corresponding ones of $\mathbf{\Sigma}$.

2.4 Hierarchical Tucker Format / Tree Tensor Networks

We base the herein described hierarchical Tucker format, also referred to as tree tensor networks, each on an (undirected) tree graph

$$\mathcal{G} = (\mathcal{V}, \mathcal{E}, \mathcal{L}), \quad \mathcal{E} \subseteq \{\{v, w\} \mid v \neq w \in \mathcal{V}\},$$

augmented with a subset of *legs* $\mathcal{L} \subseteq \mathcal{V}$ (which do not necessarily need to be leaves). To such a graph, we assign abstract labels

$$\Lambda(\mathcal{G})_v = \{\beta_e\}_{e \in \mathcal{E}: v \in e} \cup \begin{cases} \{\alpha_v\} & \text{if } v \in \mathcal{L}, \\ \emptyset & \text{otherwise}, \end{cases} \qquad v \in \mathcal{V}.$$

The corresponding tensor format is then set as the accumulation of all instances

$$\{\mathbf{N}_v\}_{v \in \mathcal{V}}, \quad \Lambda(\mathbf{N}_v) = \Lambda(\mathcal{G})_v, \quad v \in \mathcal{V}.$$

Consequently, the tensor $\mathbf{T}$ represented by the network (cf. Fig. 1) is $d = |\mathcal{L}|$ dimensional (or of order d) with mode sizes contained in n, with

$$\mathbf{T} := \boxtimes(\{\mathbf{N}_v\}_{v \in \mathcal{V}}), \quad \Lambda(\mathbf{T}) = \{\alpha_\ell\}_{\ell \in \mathcal{L}}, \quad n_v := \begin{cases} \mathcal{D}(\mathbf{T})_{\alpha_v}, & v \in \mathcal{L}, \\ 1, & v \in \mathcal{V} \setminus \mathcal{L}. \end{cases}$$

We note here that we assign singleton label sets $\{\alpha_v\}$ for all legs $v \in \mathcal{L}$ for simplicity, whereas there is no substantial difference to the general case in which one node $\mathbf{N}_v$ may contain multiple labels shared with the represented tensor.

For the following, we specify some further, common graph theoretical notation. The neighbors of $v \in \mathcal{V}$ are abbreviated $\operatorname{neigh}(v) := \{w \in \mathcal{V} \mid \{v, w\} \in \mathcal{E}\}$. For $b \in \mathcal{V}$, we denote the branches relative to a root $c \in \mathcal{V}$ as

$$\operatorname{branch}_c(b) := \{b\} \cup \{h \in \mathcal{V} \mid \text{there exists a path } (c, \ldots, b, \ldots, h)\},$$

such that thereby

$$\operatorname{branch}_c(c) = \mathcal{V} = \{c\} \,\dot\cup\, \dot{\bigcup}_{b \in \operatorname{neigh}(c)} \operatorname{branch}_c(b).$$

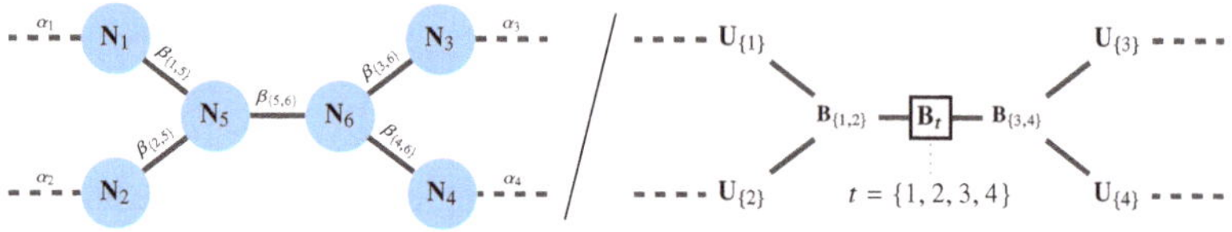

Fig. 1: Left: network corresponding to $\mathcal{G} = (\mathcal{V}, \mathcal{E}, \mathcal{L})$ for $\mathcal{V} = \{1, \ldots, 6\}$, $\mathcal{E} = \{\{1,5\}, \{2,5\}, \{5,6\}, \{3,6\}, \{4,6\}\}$ and $\mathcal{L} = \{1, \ldots, 4\}$. We for instance have $\mathrm{branch}_5(6) = \{3,4,5,6\}$, $\mathrm{neigh}(6) = \{3,4,5\}$, $\mathrm{desc}_2(5) = \{1,6\}$ and $\mathrm{pred}_4(5) = 6$. The set of leaves, for a root $c \in \{5,6\}$, is $\{1,2,3,4\}$. Right: corresponding graph with distinguished root with nodes as commonly named for the hierarchical Tucker decomposition [7].

Further, we denote the descendants of a vertex $b \in \mathcal{V}$ as

$$\mathrm{desc}_c(b) := \mathrm{neigh}(b) \cap \mathrm{branch}_c(b).$$

All vertices with an empty set of descendants are called leaves. For all vertices $b \in \mathcal{V}$ but the root, we set $p = \mathrm{pred}_c(b)$ via $\{p\} := \mathrm{neigh}(b) \setminus \mathrm{desc}_c(b)$.

3 Normalization

It is well known that hierarchical tensor formats can be truncated quasi-optimally [7]. That is, among all representations of accordingly reduced size, one finds a decomposition of a tensor that is, up to a (small) constant, as close to the initial one as possible. In order to do so, it is sufficient (though not necessary to that extent, cf. Section 4) to first calculate a certain normalized form, related to the singular vectors of matricizations of the represented tensor. We describe this in more detail in Section 3.2.

For completeness and subsequent sections, we first reiterate the commonly known [7] orthogonalization process Algorithm 5 in the herein used notation. Subsequently, we elaborate on properties of root-normal forms and derive the stable normalization Algorithm 6.

3.1 Orthogonalization

Definition 3.1 We say a network $\{\mathbf{O}_v\}_{v \in \mathcal{V}}$ corresponding to the tree $\mathcal{G} = (\mathcal{V}, \mathcal{E}, \mathcal{L})$ is orthogonalized with respect to $c \in \mathcal{V}$ (or short, c-orthogonalized) if each $\mathbf{O}_b$ is $\{\beta_{\{p,b\}}\}$-orthogonal, $p = \mathrm{pred}_c(b)$, $b \in \mathcal{V} \setminus \{c\}$.

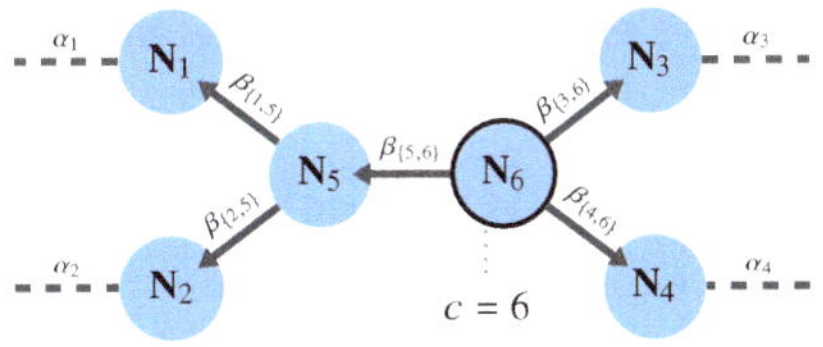

Fig. 2: Visualization of the network as in Fig. 1 when orthogonalized with respect to $c = 6$. The drawn circle around $\mathbf{N}_6$ has no further meaning but to indicate the chosen root.

For a c-orthogonalized network, we can under inductive use of Lemma 2.1 for instance conclude that $\|\boxtimes(\{\mathbf{O}_v\}_{v\in\mathcal{V}})\|_F = \|\mathbf{O}_c\|_F$, which yields a stable calculation of the norm of a decomposed tensor.

In the remainder of this section, we describe how to orthogonalize a network $\{\mathbf{N}_v\}_{v\in\mathcal{V}}$ corresponding to the tree $\mathcal{G} = (\mathcal{V}, \mathcal{E}, \mathcal{L})$ (assuming that its factors have finite norms). For a fixed root $c \in \mathcal{V}$, we first set $\mathbf{M}_\ell := \mathbf{N}_\ell$ for all leaves $\ell \in \mathcal{V}$. In leaves-to-root direction, we then recursively define (cf. Algorithm 5) each the tensors $\mathbf{Q}_b$ and matrices $\mathbf{R}_e$ as factors of QR-decompositions

$$\mathbf{Q}_b \boxtimes \mathbf{R}_{\{p,b\}} := \mathbf{M}_b, \quad \Lambda(\mathbf{R}_e) = \{\beta_e\}, \quad b \in \mathrm{desc}_c(p), \quad p \in \mathcal{V}. \tag{1}$$

The tensors $\mathbf{M}_b$ are in turn defined by

$$\mathbf{M}_b := \boxtimes(\{\mathbf{R}_{\{h,b\}}\}_{h\in\mathrm{desc}_c(b)}) \boxtimes \mathbf{N}_b, \quad b \in \mathcal{V}. \tag{2}$$

Each node (except the leaves) thus receives matrices from its descendants,

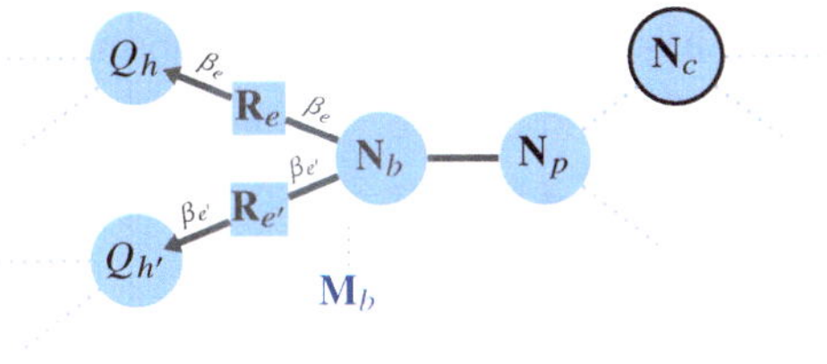

and is then factorized itself (except for the root).

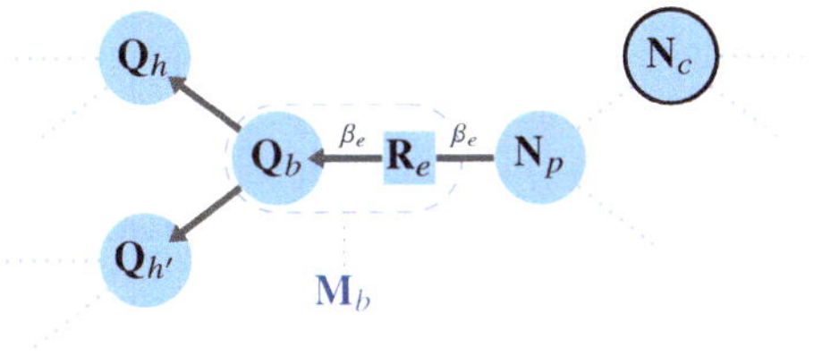

Lemma 3.1 *For any $S \subsetneq \mathcal{V}$, for which $B := \dot{\bigcup}_{s \in S} \mathrm{branch}_c(s)$ is a disjoint union, the network $\mathbf{O}^{(S)}$ with nodes*

$$\mathbf{O}_v^{(S)} = \begin{cases} \mathbf{Q}_v, & v \in B \setminus S, \\ \mathbf{M}_v, & v \in S, \\ \mathbf{N}_v, & v \in \mathcal{V} \setminus B, \end{cases}$$

corresponds to the same graph and represents the same tensor as $\{\mathbf{N}_v\}_{v \in \mathcal{V}}$. Further, $\{\mathbf{O}_v^{(\{c\})}\}_{v \in \mathcal{V}}$ is orthogonalized with respect to the root c.

Proof. We show the statement inductively. For the set of leaves S, it holds true trivially. Let the statement now hold true for some S and let $b \notin S$ be a vertex with $\mathrm{desc}_c(b) \subseteq S$. Then for $S_b := \{b\} \cup S \setminus \mathrm{desc}_c(b)$, the networks $\mathbf{O}^{(S_b)}$ and $\mathbf{O}^{(S)}$ differ only in the nodes at vertices $\{b\} \cup \mathrm{desc}_c(b)$. For the product of such, we have

$$\begin{aligned} &\boxtimes\left(\{\mathbf{O}_h^{(S_b)}\}_{h \in \mathrm{desc}_c(b)}\right) \boxtimes \mathbf{O}_b^{(S_b)} = \boxtimes(\{\mathbf{Q}_h\}_{h \in \mathrm{desc}_c(b)}) \boxtimes \mathbf{M}_b \\ = &\boxtimes\left(\{\mathbf{Q}_h \boxtimes \mathbf{R}_{\{h,b\}}\}_{h \in \mathrm{desc}_c(b)}\right) \boxtimes \mathbf{N}_b = \boxtimes(\{\mathbf{M}_h\}_{h \in \mathrm{desc}_c(b)}) \boxtimes \mathbf{N}_b \\ = &\boxtimes\left(\{\mathbf{O}_h^{(S)}\}_{h \in \mathrm{desc}_c(b)}\right) \boxtimes \mathbf{O}_b^{(S)}. \end{aligned}$$

As thus also $\boxtimes(\{\mathbf{O}_v^{(S_b)}\}_{v \in \mathcal{V}}) = \boxtimes(\{\mathbf{O}_v^{(S)}\}_{v \in \mathcal{V}})$, the statement follows inductively for all according S. □

3.2 Root-Normal Form

Definition 3.2 (Root-Normal Form) We say a network $\{\boldsymbol{\Upsilon}_v\}_{v \in \mathcal{V}}$ corresponding to the tree $\mathcal{G} = (\mathcal{V}, \mathcal{E}, \mathcal{L})$ has (compact) root-normal form, if for each $e = \{p, b\}$, $b \in \mathrm{desc}_c(p)$, $p \in \mathcal{V}$, there is a (compact) SVD

$$\hat{\mathbf{U}}_e \boxtimes \boldsymbol{\Sigma}_e \boxtimes \hat{\mathbf{V}}_e^T = \mathbf{T} = \boxtimes(\{\boldsymbol{\Upsilon}_v\}_{v \in \mathcal{V}}), \tag{3}$$

such that

$$\hat{\mathbf{U}}_e \boxtimes \boldsymbol{\Sigma}_e = \boxtimes(\{\boldsymbol{\Upsilon}_v\}_{v \in \mathrm{branch}_b(p)}), \quad \hat{\mathbf{V}}_e^T = \boxtimes(\{\boldsymbol{\Upsilon}_v\}_{v \in \mathrm{branch}_p(b)}).$$

One of the key factors of a representation in root-normal form (using the notation in Definition 3.2) is that for each edge $e \in \mathcal{E}$, the projection to the span of the first $\ell_e \in \mathbb{N}$ singular vectors can be done in an instant. That is, the tensor defined by

$$\pi_e^{(\ell)}(\mathbf{T}) := (\mathbf{T} \boxtimes \hat{\mathbf{V}}_e^{(\ell)}) \boxtimes \hat{\mathbf{V}}_e^{(\ell)T}, \quad \hat{\mathbf{V}}_e^{(\ell)}(\beta_e[i]) = \hat{\mathbf{V}}_e(\beta_e[i]),\ i = 1, \ldots, \ell_e,$$

with $\mathcal{D}(\hat{\mathbf{V}}_e^{(\ell)})_{\beta_e} = \ell_e$, is directly obtained by accordingly restricting $\boldsymbol{\Upsilon}_v$, $v \in e$. Further, restricting with respect to all edges to $\ell \in \mathbb{N}^{\mathcal{E}}$ (cf. Section 4.2), one obtains

Algorithm 5: Network Orthogonalization

Input: network $\{\mathbf{O}_v\}_{v\in\mathcal{V}}$ corresponding to the tree $\mathcal{G} = (\mathcal{V}, \mathcal{E}, \mathcal{L})$ with root $c \in \mathcal{V}$
Output: network $\{\mathbf{O}_v\}_{v\in\mathcal{V}}$ orthogonalized with respect to root c

```
function ortho({O_v}_{v∈V}, c)
    declare {O_v}_{v∈V} global variable                 // variables will be overwritten

    orthorec(c, neigh(c))
end

function orthorec(b, H)                                  // H = desc_c(b)
    for h ∈ H do in parallel
        if D := neigh(h) \ {b} ≠ ∅ then                  // D = desc_c(h)
            orthorec(h, D)
        end

        compute (compact) QR-decomposition               // cf. Eq. (1)

            O_h ⊠ R_{b,h} := O_h,    Λ(R_{b,h}) = {β_{b,h}}
    end

    for h ∈ H do                                         // cf. Eq. (2)
        O_b := R_{b,h} ⊠ O_b
    end
end
```

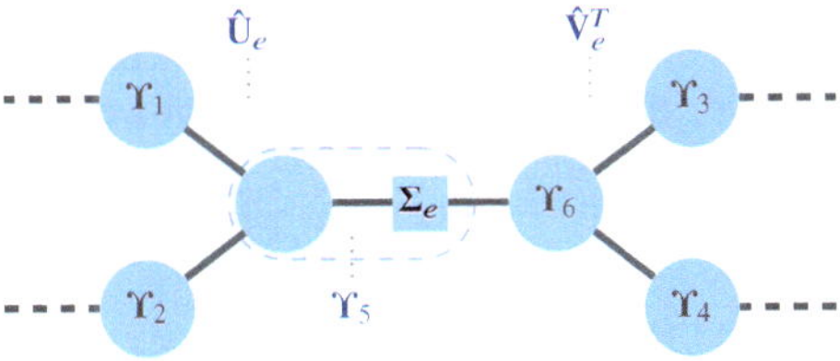

Fig. 3: Visualization of the network as in Fig. 1 in root-normal form for the root $c = 5$ together with the large-scale SVD with respect to the edge $e = \{5, 6\}$ of the represented tensor $\mathbf{T}$. The partition of nodes follows $\text{branch}_5(6) = \{3, 4, 6\}$ and $\text{branch}_6(5) = \{1, 2, 5\}$. Orthogonality constraints are not displayed here.

$$\boxtimes(\{\Upsilon_v^{(\ell)}\}_{v\in\mathcal{V}}) = \pi_{e_{|\mathcal{E}|}}^{(\ell)} \circ \ldots \circ \pi_{e_1}^{(\ell)}(\mathbf{T}),$$

for any sequence of edges $\{e_i\}_{i=1}^{|\mathcal{E}|} = \mathcal{E}$ with $e_i \subseteq \text{branch}_c(v_{i-1})$, $v_{i-1} \in e_{i-1}$, for all $i = 2, \ldots, |\mathcal{E}|$. In particular, these hierarchical projections yield a quasi-best approximation to the original tensor, in the sense that

$$\| \boxtimes (\{\Upsilon_v^{(\ell)}\}_{v\in\mathcal{V}}) - \mathbf{T}\|_F \leqslant \sqrt{|\mathcal{E}|} \min_{\{\tilde{\mathbf{N}}_v^{(\ell)}\}_{v\in\mathcal{V}}} \| \boxtimes (\{\tilde{\mathbf{N}}_v^{(\ell)}\}_{v\in\mathcal{V}}) - \mathbf{T}\|_F,$$

where the minimization is over all equally sized networks. This well-known bound has originally been introduced and proven by [7].

The priorly described orthogonalization is not only a first step towards normalization, but c-orthogonality is in particular retained in the root-normal form.

Lemma 3.2 *A root-normal network is orthogonalized with respect to its root.*

Proof. Follows directly by inductive use of Lemma 2.1. □

Through the close relation to matricizations (reshapings) of the represented tensor, analogous uniqueness constraints as for the ordinary matrix SVD as well as HOSVD [4] also carry over to this general case.

Theorem 3.1 (Essential Uniqueness) *Let $\{\Upsilon_v^{(1)}\}_{v\in\mathcal{V}}$ and $\{\Upsilon_v^{(2)}\}_{v\in\mathcal{V}}$ be both in compact root-normal form corresponding to $\mathcal{G} = (\mathcal{V}, \mathcal{E}, \mathcal{L})$ that represent the same tensor. Then there exist orthogonal, square matrices $\{\mathbf{W}_e\}_{e\in\mathcal{E}}$ with*

$$\mathbf{W}_e \boxtimes \mathbf{\Sigma}_e = \mathbf{\Sigma}_e \boxtimes \mathbf{W}_e, \quad \Lambda(\mathbf{W}_e) = \Lambda(\mathbf{\Sigma}_e), \quad e \in \mathcal{E}, \tag{4}$$

such that

$$\Upsilon_v^{(1)} = \boxtimes(\{\mathbf{W}_e\}_{e\in\mathcal{E}: v\in e}) \boxtimes \Upsilon_v^{(2)}. \tag{5}$$

Proof. Let $\hat{\mathbf{V}}_e^{(\cdot)T} := \boxtimes(\{\Upsilon_v^{(\cdot)}\}_{v\in\mathrm{branch}_b(h)})$ for $e = \{b, h\}$, $h \in \mathrm{desc}_c(b)$, $b \in \mathcal{V}$. By definition of normalization, each $\hat{\mathbf{V}}_e^{(1)T}$ and $\hat{\mathbf{V}}_e^{(2)T}$ are corresponding components of an SVD of $\mathbf{T}$. Thus, based on the essential uniqueness of the usual matrix SVD, for each $e \in \mathcal{E}$, there exists $\mathbf{W}_e$ that fulfills $Eq.$ (4) and for which

$$\hat{\mathbf{V}}_e^{(1)T} = \mathbf{W}_e \boxtimes \hat{\mathbf{V}}_e^{(2)T}. \tag{6}$$

For every fixed $v \in \mathcal{V}$, we thereby have

$$\mathbf{B}_v := \Upsilon_v^{(1)} \boxtimes (\{\hat{\mathbf{V}}_{\{v,h\}}^{(1)T}\}_{h\in\mathrm{desc}_c(v)}) = \Upsilon_v^{(1)} \boxtimes (\{\mathbf{W}_{\{v,h\}} \boxtimes \hat{\mathbf{V}}_{\{v,h\}}^{(2)T}\}_{h\in\mathrm{desc}_c(v)})$$

For $v = c$, with $\mathrm{desc}_c(c) = \mathrm{neigh}(c)$, it then follows

$$\Upsilon_c^{(2)} \boxtimes (\{\hat{\mathbf{V}}_{\{c,h\}}^{(2)T}\}_{h\in\mathrm{desc}_c(c)}) = \mathbf{T} = \mathbf{B}_c.$$

Due to orthogonality of each $\hat{\mathbf{V}}_e^{(\cdot)T}$, we have $\Upsilon_c^{(2)} = \Upsilon_c^{(1)} \boxtimes (\{\mathbf{W}_{\{c,h\}}\}_{h\in\mathrm{neigh}(c)})$, which is equivalent to Eq. (5) due to orthogonality of each $\mathbf{W}_e$. Assume in turn that $v \neq c$. Then for $\{p\} := \mathrm{neigh}(v) \setminus \mathrm{desc}_c(v)$, we obtain

$$\mathbf{W}_{\{p,v\}} \boxtimes \Upsilon_v^{(2)} \boxtimes (\{\hat{\mathbf{V}}_{\{v,h\}}^{(2)T}\}_{h\in\mathrm{desc}_c(v)}) = \mathbf{W}_{\{p,v\}} \boxtimes \hat{\mathbf{V}}_{\{p,v\}}^{(2)T} = \hat{\mathbf{V}}_{\{p,v\}}^{(1)T} = \mathbf{B}_v.$$

Analogously to the first case, this yields Eq. (5) for all remaining vertices. □

3.3 Normalization via Orthogonal Exchange-Matrices

We in the following carry out how to construct a root-normal form (cf. Definition 3.2). Let therefor $\{\mathbf{Q}_v\}_{v\in\mathcal{V}\setminus\{c\}} \cup \{\mathbf{M}_c\}$ be a c-orthogonalized network (cf. Section 3.1). We firstly set $\mathbf{C}_c := \mathbf{M}_c$. In paths from root-to-leaves, we then recursively define the tensors $\mathbf{U}_e$ with $\Lambda(\mathbf{U}_e) = \Lambda(\mathbf{C}_b) = \Lambda(\mathbf{N}_b)$, the square matrices $\mathbf{\Sigma}_e$ as well as (not necessarily square) matrices $\mathbf{V}_e$ with $\Lambda(\mathbf{\Sigma}_e) = \Lambda(\mathbf{V}_e) = \{\beta_e\}$ as factors of SVDs

$$\mathbf{U}_e \boxtimes \mathbf{\Sigma}_e \boxtimes \mathbf{V}_e^T := \mathbf{C}_p, \quad e = \{p, b\}, \quad b \in \mathrm{desc}_c(p), \quad p \in \mathcal{V}. \tag{7}$$

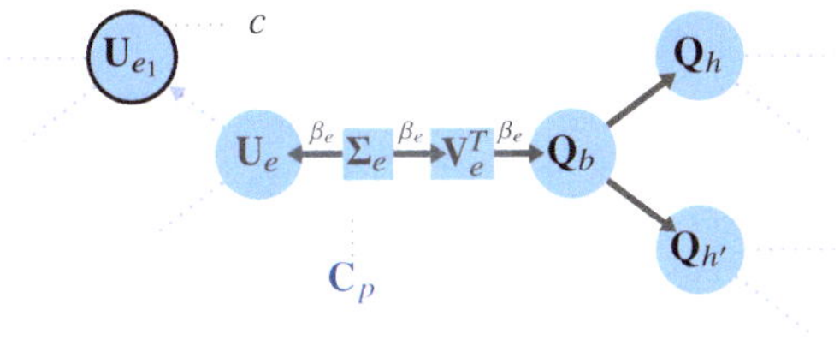

The tensors $\mathbf{C}_b$ in turn are defined via

$$\mathbf{C}_b := \mathbf{\Sigma}_e \boxtimes \mathbf{V}_e^T \boxtimes \mathbf{Q}_b, \quad e = \{p, b\}, \quad b \in \mathrm{desc}_c(p), \quad p \in \mathcal{V}. \tag{8}$$

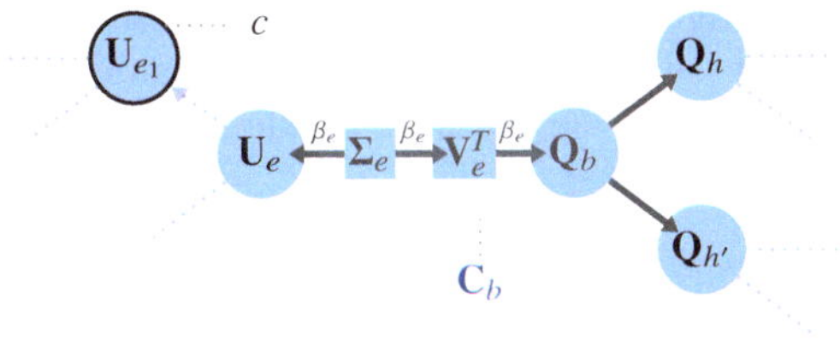

Lemma 3.3 *For each $p \in \mathcal{V}$, where $P = (c, \ldots, p)$ is the path from c to p, the network $\mathbf{Y}^{(p)}$ with nodes*

$$\mathbf{Y}_v^{(p)} = \begin{cases} \mathbf{U}_{\{v\}\cup \mathrm{desc}_c(v)\cap P}, & v \in P \setminus \{p\}, \\ \mathbf{C}_v, & v = p, \\ \mathbf{Q}_v, & v \in \mathcal{V} \setminus P, \end{cases}$$

corresponds to the same graph and represents the same tensor as $\{\mathbf{N}_v\}_{v\in\mathcal{V}}$. Further, each $\mathbf{Y}_h^{(p)}$ is orthogonalized with respect to p.

Proof. We show the statement inductively in $p \in \mathcal{V}$. For $p = c$ (where $P = (c)$), it holds true trivially. Let the statement now hold true for some $p \in \mathcal{V}$ and let $b \in \mathrm{desc}_c(p)$ (for which $P = (c, \ldots, p, b)$). Then the networks $\mathbf{Y}^{(p)}$ and $\mathbf{Y}^{(b)}$ differ only in the nodes at the edge $e = \{p, b\}$. For the product of such, we have

$$\mathbf{Y}_p^{(p)} \boxtimes \mathbf{Y}_b^{(p)} = \mathbf{C}_p \boxtimes \mathbf{Q}_b = \mathbf{U}_e \boxtimes \mathbf{\Sigma}_e \boxtimes \mathbf{V}_e^T \boxtimes \mathbf{Q}_b = \mathbf{U}_e \boxtimes \mathbf{C}_b = \mathbf{Y}_p^{(b)} \boxtimes \mathbf{Y}_b^{(b)}.$$

As thus also $\boxtimes(\{\mathbf{Y}_v^{(p)}\}_{v\in\mathcal{V}}) = \boxtimes(\{\mathbf{Y}_v^{(b)}\}_{v\in\mathcal{V}})$, the statement follows inductively for all according $p \in \mathcal{V}$. Orthogonality again follows directly by construction. □

For the initial, c-orthogonalized network $\{\mathbf{Y}_v^{(c)}\}_{v\in\mathcal{V}} = \{\mathbf{C}_c\} \cup \{\mathbf{Q}_v\}_{v\in\mathcal{V}\setminus\{c\}}$ (cf. Lemma 3.3) with $\mathbf{C}_c = \mathbf{M}_c$, and the exchange matrices $\{\mathbf{V}_e\}_{e\in\mathcal{E}}$ determined by Eq. (7), we now define the network

$$\Upsilon_v := \mathbf{Y}_v^{(c)} \boxtimes (\{\mathbf{V}_{\{v,w\}}\}_{w\in\text{neigh}(v)}), \quad \mathbf{Y}_v^{(c)} = \begin{cases} \mathbf{C}_v, & v = c, \\ \mathbf{Q}_v, & \text{otherwise}, \end{cases}, \quad v \in \mathcal{V}. \quad (9)$$

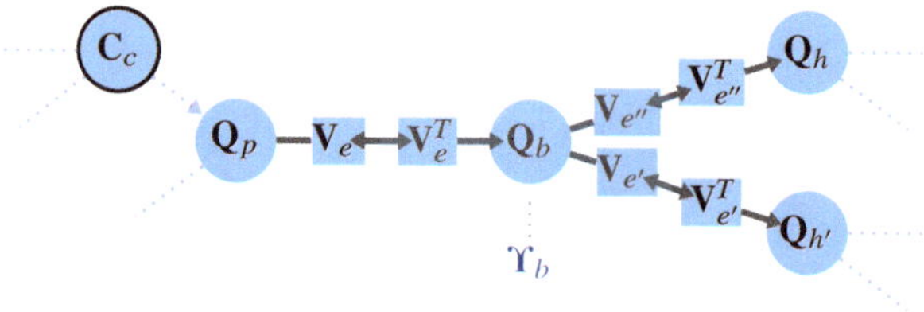

We summarize the prior construction in Algorithm 6. Note that, within the degrees of freedom of each SVD (cf. Theorem 3.1), one may choose the exchange matrices $\{\mathbf{V}_e\}_{e\in\mathcal{E}}$ to be as close to an identity matrix as possible. This has the advantage that in particular an already root-normalized network remains the same. The step Eq. (9) is not explicitly required (see Algorithm 7), though it can serve the purpose of parallelization (see Algorithm 6) or implicit normalization. Firstly, the following Lemma 3.4 shows that the application of exchange matrices does not change the represented tensor even when the exchange matrices are not square, in which case $\mathbf{V}_e \boxtimes \mathbf{V}_e^T$ is not the identity.

Lemma 3.4 *For each path $P = (c, \ldots, p)$ and $e = \{p, b\}$, $b \in \text{desc}_c(p)$, it is*

$$\boxtimes(\{\mathbf{Y}_v^{(c)}\}_{v\in P}) \boxtimes \mathbf{V}_e \boxtimes \mathbf{V}_e^T = \boxtimes(\{\mathbf{Y}_v^{(c)}\}_{v\in P}).$$

Proof. Firstly, we have

$$\boxtimes(\{\mathbf{Y}_v^{(c)}\}_{v\in P}) = \boxtimes(\{\mathbf{Y}_v^{(p)}\}_{v\in P}) = \boxtimes(\{\mathbf{Y}_v^{(p)}\}_{v\in P\setminus\{p\}}) \boxtimes \mathbf{C}_p.$$

With $\mathbf{C}_p = \mathbf{U}_e \boxtimes \mathbf{\Sigma}_e \boxtimes \mathbf{V}_e^T$ and the β_e-orthogonality of $\mathbf{V}_e$ then follows $\mathbf{C}_p \boxtimes \mathbf{V}_e \boxtimes \mathbf{V}_e^T = \mathbf{C}_p$ and thus the to be shown statement. □

The subsequent Lemma 3.5 based on Lemma 3.4 in other words states that certain subproducts of the network are each within the tensor product of the spans of corresponding singular vectors, as it for instance is similarly covered in form of minimal subspaces in [10, 6].

Lemma 3.5 *For any connected vertex subset $S \subsetneq \mathcal{V}$ that contains c, it is*

$$\boxtimes(\{\boldsymbol{\Upsilon}_v\}_{v\in S}) = \boxtimes(\{\mathbf{Y}_v^{(c)}\}_{v\in S}) \boxtimes (\{\mathbf{V}_{\{\ell,h\}}\}_{h\in\mathrm{desc}_c(\ell),\ell\in L(S)}),$$

for subleaves $L(S) := \{\ell \in S \mid \mathrm{desc}_c(\ell) \cap S = \emptyset\}$. *In particular, the network* $\boxtimes(\{\boldsymbol{\Upsilon}_v\}_{v\in\mathcal{V}})$ *represents the same tensor as* $\boxtimes(\{\mathbf{Y}_v^{(c)}\}_{v\in\mathcal{V}})$.

Proof. We prove by induction over S, where the statement holds true by definition for $S = \{c\}$. Let now $S_b = S \cup \{b\}$ for $b \in \mathrm{desc}_c(p)$, $p \in L(S)$. Then for $e = \{p, b\}$ and since S contains the path $P = (c, \ldots, p)$, we can conclude with Lemma 3.4 that

$$\boxtimes(\{\mathbf{Y}_v^{(c)}\}_{v\in S_b}) = \boxtimes(\{\mathbf{Y}_v^{(c)}\}_{v\in S}) \boxtimes \mathbf{V}_e \boxtimes \mathbf{V}_e^T \boxtimes \mathbf{Y}_p^{(c)}.$$

Thus, by induction hypothesis,

$$\begin{aligned}
&\boxtimes(\{\mathbf{Y}_v^{(c)}\}_{v\in S_b}) \boxtimes (\{\mathbf{V}_{\{\ell,h\}}\}_{h\in\mathrm{desc}_c(\ell),\ell\in L(S_b)})\\
&= \boxtimes(\{\mathbf{Y}_v^{(c)}\}_{v\in S}) \boxtimes (\{\mathbf{V}_{\{\ell,h\}}\}_{h\in\mathrm{desc}_c(\ell),\ell\in L(S)}) \boxtimes \mathbf{V}_e^T \boxtimes \mathbf{Y}_b^{(c)} \boxtimes (\{\mathbf{V}_{\{b,h\}}\}_{h\in\mathrm{desc}_c(b)})\\
&= \boxtimes(\{\boldsymbol{\Upsilon}_v^{(c)}\}_{v\in S}) \boxtimes \boldsymbol{\Upsilon}_b^{(c)} = \boxtimes(\{\boldsymbol{\Upsilon}_v^{(c)}\}_{v\in S_b}).
\end{aligned}$$

The general case then follows by induction. □

Algorithm 6: Parallelizable Exchange Matrix Calculation

Input: c-orthogonalized network $\{\mathbf{O}_v\}_{v\in\mathcal{V}}$ corresponding to tree $\mathcal{G} = (\mathcal{V}, \mathcal{E}, \mathcal{L})$
Output: exchange matrices $\{\mathbf{V}_e\}_{e\in\mathcal{E}}$ and singular value matrices $\{\boldsymbol{\Sigma}_e\}_{e\in\mathcal{E}}$

function exchange($\{\mathbf{O}_v\}_{v\in\mathcal{V}}$, c)
  declare $\{\mathbf{O}_v\}_{v\in\mathcal{V}}$, $\{\mathbf{V}_e\}_{e\in\mathcal{E}}$ *and* $\{\boldsymbol{\Sigma}_e\}_{e\in\mathcal{E}}$ *global variables*
  exchangerec($\mathbf{O}_c$, c, neigh(c))
end

function exchangerec($\mathbf{C}_b$, b, H) // $H = \mathrm{desc}_c(b)$
  for $h \in H$ **do in parallel**
    $e := \{b, h\}$
    compute (compact) SVD // cf. Eq. (7)

$$\mathbf{U}_e \boxtimes \boldsymbol{\Sigma}_e \boxtimes \mathbf{V}_e^T := \mathbf{C}_b, \quad \Lambda(\mathbf{U}_e) = \Lambda(\mathbf{O}_b), \quad \Lambda(\boldsymbol{\Sigma}_e) = \{\beta_e\}$$

    if $D := \mathrm{neigh}(h) \setminus \{b\} \neq \emptyset$ **then** // $D = \mathrm{desc}_c(h)$
      exchangerec($\boldsymbol{\Sigma}_e \boxtimes \mathbf{V}_e^T \boxtimes \mathbf{O}_h$, h, D) // cf. Eq. (8)
    end
  end
end

In Theorem 3.2, we prove that the priorly constructed network $\{\Upsilon_v\}_{v\in\mathcal{V}}$ is indeed in root-normal form, provided that all involved SVDs (cf. Eq. (7)) are either compact or lead to (finite) square exchange matrices $\{\mathbf{V}_e\}_{e\in\mathcal{E}}$, which hence are also *row-orthogonal*. It may however not be possible to distinguish zero singular values, and thus compact SVDs, in the presence of round-off errors, such that one can not rely on the first case. Square matrices in turn may be enforced (in the finite case, as it for example also results from the Gram matrix approach Section 3.4). We however take a different approach in the subsequent Section 4 in order to establish an equivalence between root-normal forms and certain node-wise constraints.

Theorem 3.2 (Root-Normal Form) *The network $\{\Upsilon_v\}_{v\in\mathcal{V}}$ derived via Eq.* (9) *either for finite, square $\{\mathbf{V}_e\}_{e\in\mathcal{E}}$ or via compact SVDs, has root-normal form and represents the same tensor $\mathbf{T}$ as $\{\mathbf{N}_v\}_{v\in\mathcal{V}}$.*

Proof. For fixed $e=\{p,b\}$, $b\in \mathrm{desc}_c(p)$, $p\in\mathcal{V}$, let

$$\hat{\mathbf{U}}_e = \boxtimes(\{\mathbf{Y}_v^{(p)}\}_{v\in\mathrm{branch}_b(p)\setminus\{p\}})\boxtimes\mathbf{U}_e,\qquad \hat{\mathbf{V}}_e^T=\mathbf{V}_e^T\boxtimes(\{\mathbf{Y}_v^{(p)}\}_{v\in\mathrm{branch}_p(b)}).$$

Then $\hat{\mathbf{U}}_e$ and $\hat{\mathbf{V}}_e$ are both β_e-orthogonal and

$$\hat{\mathbf{U}}_e\boxtimes\mathbf{\Sigma}_e\boxtimes\hat{\mathbf{V}}_e^T=\boxtimes(\{\mathbf{Y}_v^{(p)}\}_{v\in\mathrm{branch}_b(p)\setminus\{p\}})\boxtimes\mathbf{C}_p\boxtimes(\{\mathbf{Y}_v^{(p)}\}_{v\in\mathrm{branch}_p(b)})=\mathbf{T}.$$

Thus, the triplet $(\hat{\mathbf{U}}_e,\mathbf{\Sigma}_e,\hat{\mathbf{V}}_e)$ indeed forms an SVD of $\mathbf{T}$. In turn, we know by Lemma 3.5 that

$$\begin{aligned}\boxtimes(\{\Upsilon_v\}_{v\in\mathrm{branch}_b(p)})&=\boxtimes(\{\mathbf{Y}_v^{(c)}\}_{v\in\mathrm{branch}_b(p)})\boxtimes\mathbf{V}_e=\boxtimes(\{\mathbf{Y}_v^{(p)}\}_{v\in\mathrm{branch}_b(p)})\boxtimes\mathbf{V}_e\\&=\boxtimes(\{\mathbf{Y}_v^{(p)}\}_{v\in\mathrm{branch}_b(p)\setminus\{p\}})\boxtimes\mathbf{U}_e\boxtimes\mathbf{\Sigma}_e=\hat{\mathbf{U}}_e\boxtimes\mathbf{\Sigma}_e.\end{aligned}$$

Further, $\mathbf{T}=\boxtimes(\{\Upsilon_v\}_{v\in\mathcal{V}})$. Using compact SVDs, we necessarily have

$$\hat{\mathbf{V}}_e^T=\boxtimes(\{\Upsilon_v\}_{v\in\mathrm{branch}_p(b)}).$$

In the other case, not only each $\mathbf{V}_{e'}$ but also each $\mathbf{V}_{e'}^T$, $e'\in\mathcal{E}$, is $\{\beta_{e'}\}$-orthogonal. Thus, $\tilde{\mathbf{V}}_e=\boxtimes(\{\Upsilon_v\}_{v\in\mathrm{branch}_p(b)})$ is β_e-orthogonal and

$$\mathbf{T}=\hat{\mathbf{U}}_e\boxtimes\mathbf{\Sigma}_e\boxtimes\tilde{\mathbf{V}}_e^T$$

is likewise an SVD, which remained to be shown. □

The prior normalization algorithm offers a direct approach to prove the usual tensor decomposition in the first place. While this particular construction is not necessarily recommendable in practice, we include it here for the sake of completeness given its brevity.

Corollary 3.1 *For every tree graph $\mathcal{G}=(\mathcal{V},\mathcal{E},\mathcal{L})$, root $c\in\mathcal{V}$ and tensor $\mathbf{T}$ with $\Lambda(\mathbf{T})=\{\alpha_\ell\}_{\ell\in\mathcal{L}}$, there exists an essentially unique, compact, root-normalized decomposition $\mathbf{T}=\boxtimes(\{\Upsilon_v\}_{v\in\mathcal{V}})$.*

Proof. We define a network implicitly, where $\mathbf{O}_c$ is set to contain the same set of entries as $\mathbf{T}$ and each $\mathbf{O}_h$, $h \in \mathrm{desc}_c(b)$, $b \in \mathcal{V}$, is a $\{\beta_{\{b,h\}}\}$-orthogonal permutation tensor (with whichever support is sufficient) such that $\mathbf{T} = \boxtimes(\{\mathbf{O}_v\}_{v\in\mathcal{V}})$. The remainder then follows with the priorly described procedure as the network is orthogonalized with respect to its root. □

3.4 Gram-Matrix Approach

Theoretically identical exchange matrices can also be calculated with the same order of computational complexity via recursively accumulated Gram-matrices, as introduced in [7]. While we do not lay out the corresponding recursion, each acquired $\mathbf{V}_e$, $e = \{p, b\} \in \mathcal{E}$, $b \in \mathrm{desc}_c(p)$, $p \in \mathcal{V}$, then stems from an eigenvalue decomposition or SVD

$$\mathbf{V}_e \boxtimes \mathbf{\Sigma}_e^2 \boxtimes \mathbf{V}_e^T := \mathbf{G}_e, \quad \Lambda(\mathbf{G}_e) = \{\beta_e\},$$

of the explicitly determined, symmetric matrix[2] (cf. Lemma 4.1)

$$\mathbf{G}_e(\beta_e[i], \beta_e[j]) := \mathbf{B}(\beta_e[i]) \boxtimes \mathbf{B}(\beta_e[j]), \; i, j \in \mathbb{N}, \quad \mathbf{B} := \boxtimes(\{\mathbf{O}_v\}_{v\in P}),$$

where each $P = (c, \ldots, p)$ is the path from the root c to the vertex p. However, as expected, the squaring of singular values can lead to instabilities in the normalization and a consequent loss of accuracy as investigated in Section 5.

4 Decoupled Networks and Root-Independent Normal Forms

In the following, we first show that $\{\mathbf{\Upsilon}_v\}_{v\in\mathcal{V}}$ as priorly constructed fulfills certain *decoupling* constraints (Definition 4.1, similar to [10, comments after Theorem 1]) also without further conditions with respect to the exchange matrices as in Theorem 3.2 (compactness or row-orthogonal, square exchange matrices). *Decoupled* networks are nearly in root-normal form, and allow for quasi-optimal truncations exactly as for root-normal forms[3] (cf. Theorem 4.2), but its factors are generally not orthogonal. Apart from a different perspective, the subsequent line of reasoning can be taken over for decoupled networks constructed by other means such as from root-independent normal forms (cf. Section 4.3) and to generally relate the global SVD constraints of root-normal forms to simple, node-wise constraints (cf. Theorem 4.1).

Apart from the corresponding Proposition 4.1, we in that regard only assume that $\{\mathbf{\Upsilon}_v\}_{v\in\mathcal{V}}$ corresponds to the tree $\mathcal{G}$ and is indeed decoupled. We likewise

[2] With the notation introduced in Eq. (10), we may also write $\mathbf{G}_e = \mathbf{B} \boxtimes_{\setminus\beta_e} \mathbf{B}$.

[3] The statement becomes trivial in the case that the network resulted from a truncation of only zero singular values in the first place.

do not postulate that the corresponding matrices $\{\mathbf{\Sigma}_e\}_{e\in\mathcal{E}}$ are singular values of some represented tensor by prior construction, but allow them to be any matrices of such sort. That is, we only assume them to be square, diagonal matrices with non-negative, non-increasing diagonal (with corresponding labels, conforming sizes and finite Frobenius-norm).

4.1 Decoupled Networks

For the following, we require a modified contraction. For two tensors $\mathbf{T}_1$, $\mathbf{T}_2$ with $\Gamma \subseteq \Lambda := \Lambda(\mathbf{T}_1) = \Lambda(\mathbf{T}_2)$, we define the labeled matrix $\mathbf{A} = \mathbf{T}_1 \boxtimes_{\setminus\Gamma} \mathbf{T}_2$ via

$$\Lambda(\mathbf{A}) := \Gamma, \quad \mathbf{A}(\Gamma[i],\Gamma[j]) := \mathbf{T}_1(\Gamma[i]) \boxtimes \mathbf{T}_2(\Gamma[j]), \quad i,j \in \mathbb{N}^{|\Gamma|}. \tag{10}$$

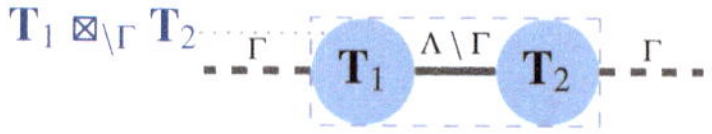

The domain sizes are correspondingly set as $\mathcal{D}_i(\mathbf{A}) := \mathcal{D}(\mathbf{T}_i)$, $i = 1, 2$.

Definition 4.1 We say a network $\{\mathbf{\Upsilon}_v\}_{v\in\mathcal{V}}$ together with $\{\mathbf{\Sigma}_e\}_{e\in\mathcal{E}}$ is decoupled if for each $e = \{b,t\}$, $t \in \mathrm{neigh}(b)$, $b \in \mathcal{V}$, it holds

$$\mathbf{\Phi}_b \boxtimes_{\setminus\beta_e} \mathbf{\Phi}_b = \mathbf{\Sigma}_e^2, \quad \mathbf{\Phi}_b := \begin{cases} \mathbf{\Upsilon}_c, & b = c, \\ \mathbf{\Sigma}_{\{\mathrm{pred}_c(b),b\}} \boxtimes \mathbf{\Upsilon}_b, & b \neq c. \end{cases}$$

For all factors except the root, the decoupling constraints in Definition 4.1 do not depend on entries that meet zero singular values in each $\mathbf{\Sigma}_{\{\mathrm{pred}_c(b),b\}}$. Depending on the construction of the network, it is hence not surprising that in particular orthogonality does not matter for such parts of the nodes. However, proving that this intuition does indeed hold true for any decoupled network, without prior information, involves some technicalities. After Proposition 4.1, we hence take careful steps towards the main Theorem 4.2.

Proposition 4.1 *The network $\{\mathbf{\Upsilon}_v\}_{v\in\mathcal{V}}$ derived via Eq. (9) is decoupled.*

Proof. Let $b \in \mathrm{desc}_c(p)$, $p \in \mathcal{V}$. For $b \neq c$, we have by construction that

$$\mathbf{\Phi}_b = \mathbf{\Sigma}_{\{p,b\}} \boxtimes \mathbf{V}^T_{\{p,b\}} \boxtimes \mathbf{Y}_b^{(c)} \boxtimes (\{\mathbf{V}_{\{b,h\}}\}_{h\in\mathrm{desc}_c(b)}) = \mathbf{C}_b \boxtimes (\{\mathbf{V}_{\{b,h\}}\}_{h\in\mathrm{desc}_c(b)})$$

Let now $e = \{b,t\}$, $t \in \mathrm{desc}_c(b)$. Then

$$\begin{aligned} \mathbf{\Phi}_b \boxtimes (\{\mathbf{V}^T_{\{b,h\}}\}_{h\in\mathrm{desc}_c(b)\setminus\{t\}}) &= \mathbf{C}_b \boxtimes (\{\mathbf{V}_{\{b,h\}} \boxtimes \mathbf{V}^T_{\{b,h\}}\}_{h\in\mathrm{desc}_c(b)\setminus\{t\}}) \boxtimes \mathbf{V}_e \\ &= \mathbf{C}_b \boxtimes \mathbf{V}_e = \mathbf{U}_e \boxtimes \mathbf{\Sigma}_e. \end{aligned}$$

Thereby, since $\mathbf{U}_e$ is β_e-orthogonal,

$$
\begin{aligned}
&\boldsymbol{\Phi}_b \boxtimes_{\setminus\beta_e} \boldsymbol{\Phi}_b \\
&= (\boldsymbol{\Phi}_b \boxtimes (\{\mathbf{V}^T_{\{b,h\}}\}_{h\in\mathrm{desc}_c(b)\setminus\{t\}})) \boxtimes_{\setminus\beta_e} (\boldsymbol{\Phi}_b \boxtimes (\{\mathbf{V}^T_{\{b,h\}}\}_{h\in\mathrm{desc}_c(b)\setminus\{t\}})) \\
&\quad (\mathbf{U}_e \boxtimes \boldsymbol{\Sigma}_e) \boxtimes_{\setminus\beta_e} (\mathbf{U}_e \boxtimes \boldsymbol{\Sigma}_e) = \boldsymbol{\Sigma}_e^2.
\end{aligned}
$$

The same holds true verbatim for $b = c$ by omitting $\boldsymbol{\Sigma}_{\{p,b\}}$. In turn, for $e = \{b, p\}$ (with $b \neq c$), we have

$$
\boldsymbol{\Phi}_b \boxtimes (\{\mathbf{V}^T_{\{b,h\}}\}_{h\in\mathrm{desc}_c(b)}) = \mathbf{C}_b = \boldsymbol{\Sigma}_e \boxtimes \mathbf{V}_e^T \boxtimes \mathbf{Y}_b^{(c)}.
$$

Since both $\mathbf{V}_e^T$ and $\mathbf{Y}_b^{(c)}$ are β_e-orthogonal, we also obtain $\boldsymbol{\Phi}_b \boxtimes_{\setminus\beta_e} \boldsymbol{\Phi}_b = \boldsymbol{\Sigma}_e^2$ in this case. □

Throughout this section, we abbreviate certain domain sizes and, as before, assume these to be consistent. For all $e = \{b, h\}$, $h \in \mathrm{desc}_c(b)$, $b \in \mathcal{V}$, let $k, r \in \mathbb{N}^{\mathcal{E}}$ be defined via

$$
k_e := \mathcal{D}(\boldsymbol{\Upsilon}_b)_{\beta_e} = \mathcal{D}(\boldsymbol{\Upsilon}_b)_{\beta_e} = \mathcal{D}_1(\boldsymbol{\Sigma}_e)_{\beta_e} = \mathcal{D}_2(\boldsymbol{\Sigma}_e)_{\beta_e}, \quad r_e := \mathrm{rank}(\boldsymbol{\Sigma}_e).
$$

Lemma 4.1 *Let $\{\boldsymbol{\Upsilon}_v\}_{v\in\mathcal{V}}$ be decoupled. Then for each path $P = (c, \ldots, b)$ and $e = \{b, h\}$, $h \in \mathrm{desc}_c(b)$, $b \in \mathcal{V}$, it is*

$$
(\boxtimes_{v\in P}\boldsymbol{\Upsilon}_v) \boxtimes_{\setminus\beta_e} (\boxtimes_{v\in P}\boldsymbol{\Upsilon}_v) = \boldsymbol{\Sigma}_e^2.
$$

In particular, it follows that $r_e \leqslant \prod_{v\in\mathrm{branch}_h(b)} n_v$.

Proof. Firstly, the case $c = b$ follows by definition. Let otherwise $P = (v_1, \ldots, v_k)$ be the path P and let $e_1, \ldots, e_{k-1} \in \mathcal{E}$ be the edges along that path. Then

$$
\begin{aligned}
&(\boldsymbol{\Upsilon}_{v_k} \boxtimes \ldots \boxtimes \boldsymbol{\Upsilon}_{v_1}) \boxtimes_{\setminus\beta_e} (\boldsymbol{\Upsilon}_{v_1} \boxtimes \ldots \boxtimes \boldsymbol{\Upsilon}_{v_k}) \\
&= (\boldsymbol{\Upsilon}_{v_k} \boxtimes \ldots \boxtimes \boldsymbol{\Upsilon}_{v_2}) \boxtimes_{\setminus\beta_e} ((\boldsymbol{\Phi}_c \boxtimes_{\setminus\beta_{e_1}} \boldsymbol{\Phi}_c) \boxtimes \boldsymbol{\Upsilon}_{v_2} \boxtimes \ldots \boxtimes \boldsymbol{\Upsilon}_{v_k}) \\
&= (\boldsymbol{\Upsilon}_{v_k} \boxtimes \ldots \boxtimes \boldsymbol{\Upsilon}_{v_2}) \boxtimes_{\setminus\beta_e} (\boldsymbol{\Sigma}_{e_1}^2 \boxtimes \boldsymbol{\Upsilon}_{v_k} \boxtimes \ldots \boxtimes \boldsymbol{\Upsilon}_{v_2}) \\
&= (\boldsymbol{\Upsilon}_{v_k} \boxtimes \ldots \boxtimes \boldsymbol{\Upsilon}_{v_3}) \boxtimes_{\setminus\beta_e} ((\boldsymbol{\Phi}_{v_2} \boxtimes_{\setminus\beta_{e_2}} \boldsymbol{\Phi}_{v_2}) \boxtimes \boldsymbol{\Upsilon}_{v_3} \boxtimes \ldots \boxtimes \boldsymbol{\Upsilon}_{v_k}) = \boldsymbol{\Sigma}_e^2.
\end{aligned}
$$

□

Theorem 4.1 *A network $\{\boldsymbol{\Upsilon}_v\}_{v\in\mathcal{V}}$ is in root-normal form if and only if it is c-orthogonalized as well as decoupled (and $k_e \leqslant \prod_{v\in\mathrm{branch}_h(b)} n_v$ holds true).*

Proof. ⇒: Let $\{\boldsymbol{\Upsilon}_v\}_{v\in\mathcal{V}}$ be in root-normal form and let $\hat{\mathbf{U}}_e$ as well as $\hat{\mathbf{V}}_e$ each be the β_e-orthogonal tensors postulated by Definition 3.2, $e \in \mathcal{E}$. For dimensionality reasons, we have

$$
k_e = \mathcal{D}(\hat{\mathbf{U}}_e)_{\beta_e} \leqslant \prod_{\lambda\in\Lambda(\hat{\mathbf{U}}_e):\lambda\neq\beta_e} \mathcal{D}(\hat{\mathbf{U}}_e)_\lambda = \prod_{v\in\mathrm{branch}_h(b)} n_v.
$$

Further, via inductive application of Lemma 2.1, we can conclude that the network is orthogonal with respect to its root c. Thirdly, let $e = \{b, h\}$, $h \in \mathrm{desc}_c(b)$, $b \neq c$ and $p = \mathrm{pred}_c(b)$. Then due to orthogonality,

$$\mathbf{\Phi}_b \boxtimes_{\backslash\beta_e} \mathbf{\Phi}_b = \mathbf{B} \boxtimes_{\backslash\beta_e} \mathbf{B}, \quad \mathbf{B} := \hat{\mathbf{U}}_{\{p,b\}} \boxtimes \mathbf{\Phi}_b \boxtimes (\{\hat{\mathbf{V}}^T_{\{b,w\}}\}_{w\in\mathrm{desc}_c(b):w\neq h}).$$

Since further $\mathbf{B} = \boxtimes(\{\mathbf{\Upsilon}_v\}_{v\in\mathrm{branch}_h(b)}) = \hat{\mathbf{U}}_{\{b,h\}} \boxtimes \mathbf{\Sigma}_{\{b,h\}}$, it directly follows

$$\mathbf{B} \boxtimes_{\backslash\beta_e} \mathbf{B} = \mathbf{\Sigma}^2_{\{b,h\}}.$$

The case $b = c$ follows verbatim by removing the factor $\hat{\mathbf{U}}_{\{p,b\}}$.
$\Leftarrow$: Let $e = \{b, h\}$, $h \in \mathrm{desc}_c(b)$, $b \in \mathcal{V}$. Let further $(v_1, \ldots, v_k) = (c, \ldots, b)$ be the path from c to b with edges $e_1, \ldots, e_{k-1} \in \mathcal{E}$ along that path. Then for $\mathbf{B} := \boxtimes(\{\mathbf{\Upsilon}_v\}_{v\in\mathrm{branch}_h(b)})$, using Lemma 4.1 and c-orthogonality, we have

$$\mathbf{B} \boxtimes_{\backslash\beta_e} \mathbf{B} = (\mathbf{\Upsilon}_{v_k} \boxtimes \ldots \boxtimes \mathbf{\Upsilon}_{v_1}) \boxtimes_{\backslash\beta_e} (\mathbf{\Upsilon}_{v_1} \boxtimes \ldots \boxtimes \mathbf{\Upsilon}_{v_k}) = \mathbf{\Sigma}^2_e.$$

Now, considering the size constraint towards k_e, there exists a β_e-orthogonal $\hat{\mathbf{U}}_e$ with $\hat{\mathbf{U}}_e \boxtimes \mathbf{\Sigma}_e = \mathbf{B}$. Since further $\hat{\mathbf{V}}^T_e := \boxtimes(\{\mathbf{\Upsilon}_v\}_{v\in\mathrm{branch}_b(h)})$ is orthogonal due to the c-orthogonality of the network, we have constructed an SVD as required by the root-normal form. □

Corollary 4.1 *A decoupled, compact network ($k = r$) is in root-normal form.*

Proof. With $k = r$ follows the size constraint $k_e \leqslant \prod_{v\in\mathrm{branch}_h(b)} n_v$ via Lemma 4.1. Further, due to the regularity of each $\mathbf{\Sigma}_e$, the network must be c-orthogonal (cf. Definition 4.1). Root-normality then follows via Theorem 4.1. □

4.2 Truncation and Normalization of Decoupled Networks

For $\ell \in \mathbb{N}^{\mathcal{E}}$ with $\ell_e \leqslant k_e$, $e \in \mathcal{E}$, we define the following *lift* matrices $\{\mathbf{E}^{(\ell)}_e\}_{e\in\mathcal{E}}$ via

$$\Lambda(\mathbf{E}^{(\ell)}_e) := \beta_e, \quad \mathcal{D}_1(\mathbf{E}^{(\ell)}_e)_{\beta_e} := k_e, \quad \mathcal{D}_2(\mathbf{E}^{(\ell)}_e)_{\beta_e} := \ell_e,$$

$$\mathbf{E}^{(\ell)}_e(\beta_e[i], \beta_e[j]) := \begin{cases} 1, & i = j \in \{1, \ldots, \ell_e\}, \\ 0, & \text{otherwise}. \end{cases}$$

Thereby, each $\mathbf{E}^{(\ell)}_e$ is column-orthogonal. These matrices mainly serve the purpose to lift unequally sized matrices or tensors to a common size, without changing their product. We also use them to define restricted nodes

$$\mathbf{\Upsilon}^{(\ell)}_v := \mathbf{\Upsilon}_v \boxtimes (\{(\mathbf{E}^{(\ell)}_{\{v,w\}})^T\}_{w\in\mathrm{neigh}(v)}).$$

Hence, these tensors each contain the same entries within the smaller domain size $\mathcal{D}(\mathbf{\Upsilon}^{(\ell)}_v)_{\beta_{\{v,w\}}} = \ell_{\{v,w\}} \leqslant k_{\{v,w\}}$, $w \in \mathrm{neigh}(v)$.

Definition 4.2 Let $\{\Upsilon_v\}_{v\in\mathcal{V}}$ be a decoupled network. In direction of leaves to root, for each $e = \{p, b\}$, $b \in \mathcal{V} \setminus \{c\}$, $p = \mathrm{pred}_c(b)$, we recursively set sizes k^*_e with

$$r_e \leqslant k^*_e \leqslant \min(k_e, n_b \prod_{h\in\mathrm{desc}_c(b)} k^*_{\{b,h\}}, \prod_{v\in\mathrm{branch}_b(p)} n_v),$$

and define the β_e-orthogonal nodes

$$\mathbf{\Upsilon}^*_b := \tilde{\mathbf{U}}_e \boxtimes \mathbf{W}^T_e,$$

for factors of a thin SVD

$$\tilde{\mathbf{U}}_e \boxtimes \mathbf{\Sigma}^{(k^*)}_e \boxtimes \mathbf{W}^T_e = \mathbf{\Phi}^{(k^*)}_b, \tag{11}$$

such that $\mathbf{W}_e$ is a square matrix. For the root c, we subsequently set $\mathbf{\Upsilon}^*_c = \mathbf{\Phi}^{(k^*)}_c$.

Note here that in Eq. (11), $\mathbf{\Phi}^{(k^*)}_b = (\mathbf{\Sigma}_e \boxtimes \mathbf{\Upsilon}_b)^{(k^*)} = \mathbf{\Sigma}^{(k^*)}_e \boxtimes \mathbf{\Upsilon}^{(k^*)}_b$, whereas multiplying with the singular values from left or right yields the same result. Further, when $k = k^*$ is a valid choice, then the calculations for each vertex are independent of one another.

Lemma 4.2 *The construction of* $\{\mathbf{\Upsilon}^*_v\}_{v\in\mathcal{V}}$ *is well-defined. Further, for each* $e = \{p, b\}$, $b \in \mathcal{V} \setminus \{c\}$, $p = \mathrm{pred}_c(b)$, *it is* $k^*_e \geqslant r_e$ *and*

$$\mathbf{E}^{(k^*)}_e \boxtimes \mathbf{\Sigma}^{(k^*)}_e \boxtimes \mathbf{\Upsilon}^*_b \boxtimes (\{\mathbf{E}^{(k^*)T}_{\{b,h\}}\}) = \mathbf{\Phi}_b$$

In particular, the network is likewise decoupled, but also c-orthogonal.

Proof. Due to decoupling properties (cf. Definition 4.1) and since by recursive construction, $k^*_{\{b,h\}} \geqslant r_{\{b,h\}}$ holds true for $h \in \mathrm{desc}_c(b)$, it also follows both $k^*_e \geqslant r_e$ as well as

$$\mathbf{\Phi}_b = \mathbf{E}^{(k^*)} \boxtimes \mathbf{\Phi}^{(k^*)}_b \boxtimes (\{\mathbf{E}^{(k^*)T}_{\{b,h\}}\}). \tag{12}$$

Thereby,

$$\mathbf{\Sigma}^2_e = \mathbf{\Phi}_b \boxtimes_{\setminus\beta_e} \mathbf{\Phi}_b = \mathbf{E}^{(k^*)} \boxtimes (\mathbf{\Phi}^{(k^*)}_b \boxtimes_{\setminus\beta_e} \mathbf{\Phi}^{(k^*)}_b) \boxtimes \mathbf{E}^{(k^*)T}.$$

Consequently, $\mathbf{\Phi}^{(k^*)}_b \boxtimes_{\setminus\beta_e} \mathbf{\Phi}^{(k^*)}_b = (\mathbf{\Sigma}^{(k^*)}_e)^2$ and the singular values $\mathbf{\Sigma}^{(k^*)}_e$ indeed appear in the SVD Eq. (11) (which hence is well-defined). Further, $\mathbf{W}_e$ as well as $\mathbf{W}^T_e$ commute with $\mathbf{\Sigma}^{(k^*)}_e$. This then yields

$$\mathbf{\Sigma}^{(k^*)}_e \boxtimes \mathbf{\Upsilon}^*_b = \tilde{\mathbf{U}}_e \boxtimes \mathbf{W}^T_e \boxtimes \mathbf{\Sigma}^{(k^*)}_e = \mathbf{\Phi}^{(k^*)}_b.$$

Combined with Eq. (12), we obtain the to be shown equality. □

Theorem 4.2 *Let* $\{\mathbf{\Upsilon}_v\}_{v\in\mathcal{V}}$ *be decoupled and* $\ell \in \mathbb{N}^{\mathcal{E}}$ *with* $\ell_e \geqslant k_e$, $e \in \mathcal{E}$. *Then*

$$\boxtimes(\{\Upsilon_v^{(\ell)}\}_{v\in\mathcal{V}}) = \boxtimes(\{\Upsilon_v^{*(\ell^*)}\}_{v\in\mathcal{V}}), \quad \ell_e^* := \min(\ell_e, k_e^*), e \in \mathcal{E}.$$

If $\ell_e \geqslant r_e$, $e \in \mathcal{E}$, then both $\{\Upsilon_v^{(\ell)}\}_{v\in\mathcal{V}}$ and $\{\Upsilon_v^{*(\ell^*)}\}_{v\in\mathcal{V}}$ remain decoupled.

Proof. Let $E_\infty := \{e \in \mathcal{E} \mid \ell_e = r_e = \infty\}$ be the not truncated edges corresponding to infinite representation ranks and let $V_{c,\infty} := \{c\} \cup \{b \in \mathcal{V} \setminus \{c\} \mid \{\mathrm{pred}_c(b), b\} \in E_\infty\}$. Firstly, for all $b \in V_\infty$, Lemma 4.2 yields that $\Upsilon_b^* = \Upsilon_b^{(k^*)}$. Further, we have for each $e = \{b, h\}$, $h \in \mathrm{desc}_c(b)$, $b \in \mathrm{desc}_c(p)$, $p \in \mathcal{V}$, the corresponding path $(v_1, \ldots, v_k)$ from $v_1 = c$ to $v_k = b$, as well as $\mathbf{P}_b^{(\ell)} := \Upsilon_{v_1}^{(\ell)} \boxtimes \ldots \boxtimes \Upsilon_{v_k}^{(\ell)}$, that (cf. Definition 4.1 and Lemma 4.1)

$$\mathcal{N}_e(\mathbf{P}_b^{(\ell)}) \supseteq \mathcal{N}_e(\boldsymbol{\Sigma}_e^{(\ell)}), \quad \mathcal{N}_e(\mathbf{Y}) := \mathrm{null}(\mathbf{X}_e \mapsto \mathbf{Y} \boxtimes \mathbf{X}_e), \ \Lambda(\mathbf{X}_e) = \{\beta_e\}. \tag{13}$$

This is inductively proven through (once again by omitting $\boldsymbol{\Sigma}_e$ for $b = c$)

$$\begin{aligned}\mathcal{N}_e(\boldsymbol{\Sigma}_e^{(\ell)}) &= \mathcal{N}_e(\boldsymbol{\Sigma}_{\{p,b\}} \boxtimes \Upsilon_b \boxtimes \mathbf{E}_e^{(\ell)T}) \subseteq \mathcal{N}_e((\boldsymbol{\Sigma}_{\{p,b\}} \boxtimes \Upsilon_b)^{(\ell)}) \\ &= \mathcal{N}_e(\boldsymbol{\Sigma}_{\{p,b\}}^{(\ell)} \boxtimes \Upsilon_b^{(\ell)}) \subseteq \mathcal{N}_e(\mathbf{P}_p^{(\ell)} \boxtimes \Upsilon_b^{(\ell)}) = \mathcal{N}_e(\mathbf{P}_b^{(\ell)}).\end{aligned}$$

We can now introduce restricted (iff $r_e < \ell_e$) but by Eq. (13) redundant identities via

$$\boxtimes(\{\Upsilon_v^{(\ell)}\}_{v\in\mathcal{V}}) = \boxtimes(\{\mathbf{E}_e^{(\ell)T} \boxtimes \boldsymbol{\Sigma}_e^\dagger \boxtimes \boldsymbol{\Sigma}_e \boxtimes \mathbf{E}_e^{(\ell)} \boxtimes \Upsilon_b^{(\ell)}\}_{b\in\mathcal{V}\setminus V_{c,\infty}}) \boxtimes (\{\Upsilon_b^{(\ell)}\}_{b\in V_{c,\infty}}),$$

where each $e = \{\mathrm{pred}_c(b), b\}$. Let in the following $\mathbf{I}_e^{(\ell)} := \mathbf{E}_e^{(\ell)} \boxtimes \mathbf{E}_e^{(\ell)T}$, $e \in \mathcal{E}$. In each of these factors for $e \notin E_\infty$, we can then use Lemma 4.2 and replace

$$\begin{aligned}&\boldsymbol{\Sigma}_e \boxtimes \mathbf{E}_e^{(\ell)} \boxtimes \Upsilon_b^{(\ell)} \boxtimes (\{\mathbf{E}_{\{b,h\}}^{(\ell)T}\}_{h\in\mathrm{desc}_c(b)}) = \mathbf{I}_e^{(\ell)} \boxtimes \boldsymbol{\Phi}_b \boxtimes (\{\mathbf{I}_{\{b,h\}}^{(\ell)}\}_{h\in\mathrm{desc}_c(b)}) \\ &= \mathbf{I}_e^{(\ell)} \boxtimes \mathbf{E}_e^{(k^*)} \boxtimes \boldsymbol{\Sigma}_e^{(k^*)} \boxtimes \Upsilon_b^* \boxtimes (\{\mathbf{E}_{\{b,h\}}^{(k^*)T} \boxtimes \mathbf{I}_{\{b,h\}}^{(\ell)}\}_{h\in\mathrm{desc}_c(b)}) \\ &= \mathbf{E}_e^{(\ell^*)} \boxtimes \boldsymbol{\Sigma}_e^{(\ell^*)} \boxtimes \Upsilon_b^{*(\ell^*)} \boxtimes (\{\mathbf{E}_{\{b,h\}}^{(\ell^*)T}\}).\end{aligned}$$

Further, it is $\boldsymbol{\Sigma}_e^\dagger \boxtimes \mathbf{E}_e^{(\ell^*)} \boxtimes \boldsymbol{\Sigma}_e^{(\ell^*)} \boxtimes \Upsilon_b^{*(\ell^*)} = \mathbf{E}_e^{(\ell^*)} \boxtimes \Upsilon_b^{*(\ell^*)}$. The case $e \in E_\infty$ in turn follows analogously by omitting $\boldsymbol{\Sigma}_e$ in all equations. By cancellation of orthogonal lift matrices, we then obtain $\boxtimes(\{\Upsilon_v^{(\ell)}\}_{v\in\mathcal{V}}) = \boxtimes(\{\Upsilon_v^{*(\ell^*)}\}_{v\in\mathcal{V}})$. In turn, for $\ell_e \geqslant r_e$, $e \in \mathcal{E}$, the networks and decoupling constraints Definition 4.1 remain valid (though for *smaller* singular values matrices). □

Corollary 4.2 *In the situation of Theorem 4.2, it follows in particular that*

$$\boxtimes(\{\Upsilon_v\}_{v\in\mathcal{V}}) = \boxtimes(\{\Upsilon_v^*\}_{v\in\mathcal{V}}).$$

Further, $\{\Upsilon_v^*\}_{v\in\mathcal{V}}$ is in root-normal form and $\{\boldsymbol{\Sigma}_e\}_{e\in\mathcal{E}}$ indeed correspond to the singular values of the represented tensor.

Proof. The choice $\ell = k$ and thus $\ell^* = k^*$ directly shows the equality of both decoupled representations. Further, by construction, $\{\Upsilon_v^*\}_{v\in\mathcal{V}}$ is c-orthogonalized, decoupled and fulfills the necessary size constraints as required by Theorem 4.1. It thus is in root-normal form. □

4.3 Root-Independent Normal Forms

In Definition 4.1, the node-wise decoupling constraints suggest a root-independent description using the factors $\mathbf{\Phi}_v$, $v \in \mathcal{V}$. For any root $c \in \mathcal{V}$, a corresponding root-normal form $\{\mathbf{\Upsilon}_v^*\}_{v\in\mathcal{V}}$ (of a by $\{\mathbf{\Phi}_v\}_{v\in\mathcal{V}}$ indirectly, but uniquely defined tensor) can directly be obtained by node-wise SVDs as described in Definition 4.2. In particular, a different root may be chosen for this construction than it may have initially been applied to determine the root-independent terms. We summarize a more efficient albeit less parallelizable method (compared to Algorithm 6) how to construct such nodes in Algorithm 7. To avoid redundant calculations, Algorithm 7 however skips steps for the set of leaves $L \subset \mathcal{V}$, as for such, we have the simple relations $\mathbf{\Phi}_\ell = \mathbf{\Sigma}_{\{\mathrm{pred}_c(\ell),\ell\}} \boxtimes \mathbf{\Upsilon}_\ell$, $\ell \in L$, with already orthogonal $\mathbf{\Upsilon}_\ell$.

More from a theoretical perspective, one may also turn such a representation to the outside, so to say, provided it is compact as discussed. For a linear tree, which yields the tensor train or MPS format, this format is described in [16] as canonical MPS. Further discussion for general trees has appeared in [13].

Definition 4.3 (Root-Independent Normal Form) We say $(\{\mathbf{\Xi}_v\}_{v\in\mathcal{V}}, \{\mathbf{\Sigma}_e\}_{e\in\mathcal{E}})$ has compact root-independent normal form, if for every $c \in \mathcal{V}$, the network defined via

$$\mathbf{\Upsilon}_b^{(c)} := \mathbf{\Xi}_b \boxtimes (\{\mathbf{\Sigma}_{\{b,h\}}\}_{h\in\mathrm{desc}_c(b)}), \quad b \in \mathcal{V},$$

has compact root-normal form (with respect to the root c).

Note here that the notation applied in Definition 4.3 is consistent in the sense that $\mathbf{T} := \boxtimes(\{\mathbf{\Upsilon}_v^{(c)}\}_{v\in\mathcal{V}})$ indeed yields the singular value matrices $\{\mathbf{\Sigma}_e\}_{e\in\mathcal{E}}$. Further, a root-independent normal form can directly be constructed from a root-normal form as laid out by Corollary 4.3.

Corollary 4.3 *Let $\{\mathbf{\Upsilon}_v^{(c)}\}_{v\in\mathcal{V}}$ be a compact root-normal form with respect to the root $c \in \mathcal{V}$. Then*

$$\mathbf{\Xi}_b := \{\mathbf{\Upsilon}_b^{(c)}\}_{b\in\mathcal{V}} \boxtimes (\{\mathbf{\Sigma}^\dagger_{\{b,h\}}\}_{h\in\mathrm{desc}_c(b)}), \quad b \in \mathcal{V},$$

together with the corresponding $\{\mathbf{\Sigma}_e\}_{e\in\mathcal{E}}$ has compact root-independent normal form.

Proof. By construction, we have $\mathbf{\Xi}_v \boxtimes (\{\mathbf{\Sigma}_{\{v,w\}}\}_{w\in\mathrm{neigh}(v)}) = \mathbf{\Phi}_v$, $v \in \mathcal{V}$ (cf. Definition 4.1) as well as $\mathbf{\Upsilon}_b^{(c)} = \mathbf{\Sigma}^\dagger_{\{\mathrm{pred}_c(b),b\}} \boxtimes \mathbf{\Xi}_b$, $b \in \mathcal{V}\setminus\{c\}$, for any root $c \in \mathcal{V}$. Thus, $\{\mathbf{\Upsilon}_v^{(c)}\}_{v\in\mathcal{V}}$ is compact and decoupled. Corollary 4.1 then yields that it is hence in root-normal form. □

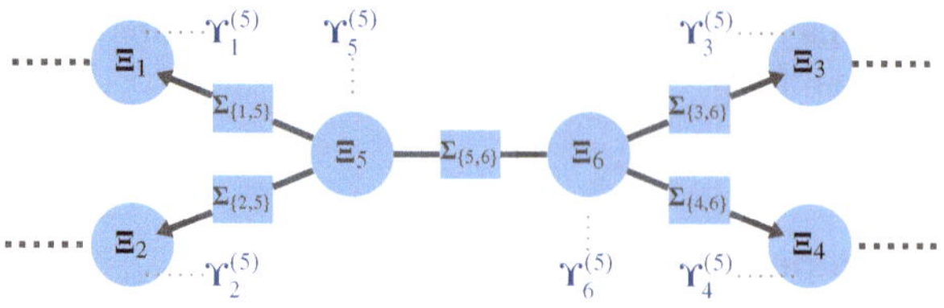

Fig. 4: Root-independent normal form for the network in Fig. 1. The different root-normal forms (cf. Fig. 5) are obtained by different assignments of the singular values matrices on the edges. Additionally to the indicated orthogonality constraints, $\boldsymbol{\Upsilon}_6^{(5)}$ is $\{\beta_{\{5,6\}}\}$-orthogonal. If $\boldsymbol{\Sigma}_{\{5,6\}}$ is set as distinguished root instead of a part of $\boldsymbol{\Upsilon}_5^{(5)}$, then one obtains the same structure and conditions as for the conventional hierarchical Tucker format depicted in Fig. 1. Moreover, the then appearing tensor $\boldsymbol{\Upsilon}_5^{(6)}$ would be $\{\beta_{\{5,6\}}\}$-orthogonal as well.

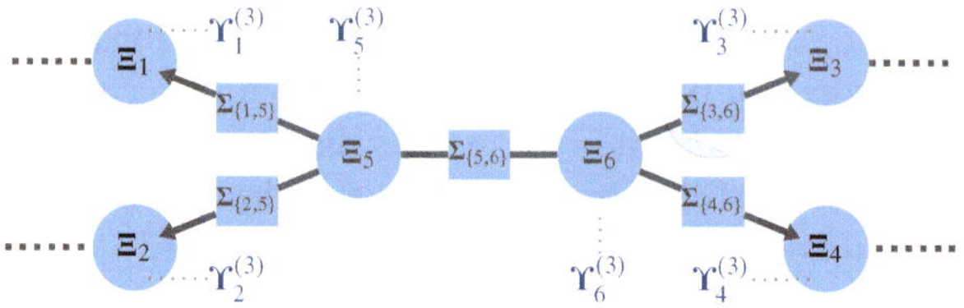

Fig. 5: Same situation as in Fig. 4, but with different assignment of singular value matrices. Additionally to the indicated orthogonality constraints, $\boldsymbol{\Upsilon}_5^{(3)}$ is $\{\beta_{\{5,6\}}\}$-orthogonal and $\boldsymbol{\Upsilon}_6^{(3)}$ is $\{\beta_{\{3,6\}}\}$-orthogonal.

Except for the leaves, each $\boldsymbol{\Xi}_b$, $b \in \mathcal{V}$, has finite Frobenius norm if and only if every neighboring $\boldsymbol{\Sigma}_e^{\dagger}$ has finite Frobenius norm. Still, the multiplication with any neighboring $\boldsymbol{\Sigma}_e$, for a different edge e, converges and has yet finite Frobenius norm by construction in that it equals a different root-normal form. The essential uniqueness of the root-normal form also transfers to this case.

Corollary 4.4 *Let $\{\boldsymbol{\Xi}_v^{(1)}\}_{v\in\mathcal{V}}$ and $\{\boldsymbol{\Xi}_v^{(2)}\}_{v\in\mathcal{V}}$ together with $\{\boldsymbol{\Sigma}_e\}_{e\in\mathcal{E}}$ be two root-independent normal forms of the same tensor. Then there exist matrices as in Eq.* (4) *such that*

$$\boldsymbol{\Xi}_v^{(1)} = \boxtimes(\{\mathbf{W}_e\}_{e\in\mathcal{E}:v\in e}) \boxtimes \boldsymbol{\Xi}_v^{(2)}. \tag{14}$$

Proof. Let $c \in \mathcal{V}$ take the role of the root. Then by Definition 4.3, the two root-independent normal forms yield root-normal forms $\{\boldsymbol{\Upsilon}_v^{(1,c)}\}_{v\in\mathcal{V}}$ and $\{\boldsymbol{\Upsilon}_v^{(2,c)}\}_{v\in\mathcal{V}}$. By Theorem 3.1, we then obtain matrices $\{\mathbf{W}_e\}_{e\in\mathcal{E}}$ that satisfy Eq. (4) such that for each $v \in \mathcal{V}$ it holds

$$\Xi_v^{(1)} \boxtimes (\{\boldsymbol{\Sigma}_{\{v,h\}}\}_{h \in \mathrm{desc}_c(v)}) = \boldsymbol{\Upsilon}_v^{(1,c)} = \boxtimes(\{\mathbf{W}_e\}_{e \in \mathcal{E}: v \in e}) \boxtimes \boldsymbol{\Upsilon}_v^{(2,c)}$$
$$= \boxtimes (\{\mathbf{W}_e\}_{e \in \mathcal{E}: v \in e}) \boxtimes \left(\Xi_v^{(2)} \boxtimes (\{\boldsymbol{\Sigma}_{\{v,h\}}\}_{h \in \mathrm{desc}_c(v)})\right)$$
$$= \left(\boxtimes (\{\mathbf{W}_e\}_{e \in \mathcal{E}: v \in e}) \boxtimes \Xi_v^{(2)}\right) \boxtimes (\{\boldsymbol{\Sigma}_{\{v,h\}}\}_{h \in \mathrm{desc}_c(v)}).$$

Note here that the last step requires $\mathbf{W}_e$ to commute with $\boldsymbol{\Sigma}_e$ as $\Xi_v^{(2)}$ is not a matrix. Due to the postulated compactness for root-normal forms, we obtain $\Xi_v^{(1)} = \boxtimes(\{\mathbf{W}_e\}_{e \in \mathcal{E}: v \in e}) \boxtimes \Xi_v^{(2)}$, which concludes the proof. □

As indicated, the node-wise decoupling constraints Definition 4.1 can likewise be translated into this case (cf. Corollary 4.3).

Corollary 4.5 *The pair* $(\{\Xi_v\}_{v \in \mathcal{V}}, \{\boldsymbol{\Sigma}_e\}_{e \in \mathcal{E}})$ *is in root-independent normal form if and only if for all* $v \in \hat{e}$, $\hat{e} \in \mathcal{E}$, *the tensor*

$$\Xi_v \boxtimes (\{\boldsymbol{\Sigma}_e\}_{e \in \mathcal{E} \setminus \{\hat{e}\}: v \in e}) \tag{15}$$

is $\{\beta_{\hat{e}}\}$*-orthogonal.*

4.4 Assignment and Feasibility of Singular Values

Corollary 4.5 implies that the conditions for a root-independent normal form are only local orthogonality constraints (cf. Eq. (15)), pairwise indirectly coupled via the supposed, common singular values on the edges. In particular, this decouples the condition for feasibility of singular value tuples within in single tensor.

Corollary 4.6 *Let* $\{\boldsymbol{\Sigma}_e\}_{e \in \mathcal{E}}$ *be a family of square, regular matrices. Then there exists a tensor* $\mathbf{T}$ *with* $\Lambda(\mathbf{T}) = \{\alpha_\ell\}_{\ell \in \mathcal{L}}$ *that yields corresponding singular values if and only if there exists* $\{\Xi_v\}_{v \in \mathcal{V}}$ *that fulfills the orthogonality constraints Eq.* (15).

Proof. From the existence of a corresponding tensor $\mathbf{T}$ follows the existence of $\{\Xi_v\}_{v \in \mathcal{V}}$ by Corollary 3.1. Vice versa, by Corollaries 4.3 and 4.5, $\{\Xi_v\}_{v \in \mathcal{V}}$ provides a root-normal form representing $\mathbf{T}$ that has singular values $\{\boldsymbol{\Sigma}_e\}_{e \in \mathcal{E}}$. □

The reduction by Corollary 4.6 to node-wise orthogonality constraints is used by Algorithm 8 to construct a representation of a tensor with prescribed (feasible) singular values. The nodes returned by it need only be transferred to either a root-independent and rooted normal form through rescaling. An analogous version of that algorithm has been introduced and analyzed in [12] for the tensor train format.

Corollary 4.7 *Let* $\{\boldsymbol{\Sigma}_e\}_{e \in \mathcal{E}}$ *be equal (modulo their labels). Then there exists a tensor* $\mathbf{T}$ *with* $\Lambda(\mathbf{T}) = \{\alpha_\ell\}_{\ell \in \mathcal{L}}$ *that yields these corresponding singular values.*

Algorithm 7: Calculation of Decoupled Factors

Input: c-orthogonalized network $\{\mathbf{O}_v\}_{v\in\mathcal{V}}$ corresponding to tree $\mathcal{G} = (\mathcal{V}, \mathcal{E}, \mathcal{L})$
Output: for leaves $L \subset \mathcal{V}$: decoupled $\{\boldsymbol{\Phi}_v\}_{v\in\mathcal{V}\setminus L}$, matching orthogonal $\{\boldsymbol{\Upsilon}_\ell\}_{\ell\in L}$ and singular value matrices $\{\boldsymbol{\Sigma}_e\}_{e\in\mathcal{E}}$

function normal($\{\mathbf{O}_v\}_{v\in\mathcal{V}}, c$)
  declare $\{\mathbf{O}_v\}_{v\in\mathcal{V}}$, $\{\boldsymbol{\Phi}_v\}_{v\in\mathcal{V}\setminus L}$, $\{\boldsymbol{\Upsilon}_v\}_{v\in L}$ *and* $\{\boldsymbol{\Sigma}_e\}_{e\in\mathcal{E}}$ *global variables*

  normalrec($\mathbf{O}_c, c$, neigh(c))
end

function normalrec($\mathbf{C}_b, b, H$) // $H = \mathrm{desc}_c(b)$
  for $h \in H$ **do**
    $e := \{b, h\}$

    compute (compact) SVD // cf. Eq. (7)

$$\mathbf{U}_e \boxtimes \boldsymbol{\Sigma}_e \boxtimes \mathbf{V}_e^T := \mathbf{C}_b, \quad \Lambda(\mathbf{U}_e) = \Lambda(\mathbf{C}_b), \quad \Lambda(\boldsymbol{\Sigma}_e) = \{\beta_e\}$$

    $\mathbf{C}_b := \mathbf{U}_e \boxtimes \boldsymbol{\Sigma}_e$ // can leave out $\mathbf{V}_e$ for subsequent SVDs on $\mathbf{C}_b$

    if $D :=$ neigh$(h) \setminus \{b\} \neq \emptyset$ **then** // $D = \mathrm{desc}_c(h)$
      normalrec($\boldsymbol{\Sigma}_e \boxtimes \mathbf{V}_e^T \boxtimes \mathbf{O}_h, h, D$) // cf. Eq. (8)
    else
      $\boldsymbol{\Upsilon}_h := \mathbf{V}_e^T \boxtimes \mathbf{O}_h$ // $\boldsymbol{\Phi}_h$ skipped for leaves h, $\boldsymbol{\Upsilon}_h$ is orthogonal
    end
  end

  $\boldsymbol{\Phi}_b := \mathbf{C}_b$
end

Proof. Let σ be the $r \in \mathbb{N} \cup \{\infty\}$ positive singular values stated to be equal for all edges. Then the hyperdiagonal tensors $\boldsymbol{\Xi}_v$ with

$$\boldsymbol{\Xi}_v(\Lambda(\boldsymbol{\Xi}_v)[(i_1, i_2, \ldots)]) = \begin{cases} \sigma_{i_1}^{-(K_v-1)}, & r \geqslant i_1 = i_2 = \ldots, \\ 0, & \text{otherwise}, \end{cases}$$

where K_v is the number of edges at v, assigned with common sizes $\mathcal{D}(\boldsymbol{\Xi}_v)_\gamma = r$, $\gamma \in \Lambda(\boldsymbol{\Xi}_v)$, $v \in \mathcal{V}$, fulfill the orthogonality constraints in Corollary 4.5. □

Algorithm 8: Assignment of Feasible Singular Values

Input: feasible constellation $\{\boldsymbol{\Sigma}_e\}_{e\in\mathcal{E}}$
Output: nodes $\boldsymbol{\Phi}_v = \Xi_v \boxtimes (\{\boldsymbol{\Sigma}_{\{v,w\}}\}_{w\in\mathrm{neigh}(v)}),\ v \in \mathcal{V}$, for a root-independent normal form $(\{\Xi_v\}_{v\in\mathcal{V}}, \{\boldsymbol{\Sigma}_e\}_{e\in\mathcal{E}})$

```
function assign {Σ_e}_{e∈E}
    for v ∈ V do in parallel
        randomly initialize Φ_v
        while not convergent do
            for w ∈ neigh(v) do
                e := {v, w}
                compute SVD
                    U_e ⊠ Θ_e ⊠ V_e^T := Φ_v,   Λ(U_e) = Λ(O_v),   Λ(Θ_e) = {β_e}
                Φ_v := U_e ⊠ Σ_e
            end
        end
    end
end
```

5 Numerical Experiments

In the following, we compare the accuracy of the truncation via the priorly described node-wise singular value decompositions in Algorithm 6, with the one obtained through the Gram-matrix approach introduced by [7] (cf. Section 3.4). This is to empirically verify that the squaring of singular values involved in the calculation of the Gram-matrices indeed leads to the expected stability issues, but also to problems for high accuracy truncations. We further compare with the more efficient but less parallelizable Algorithm 7 and investigate the normalization accuracies of each calculated representation.

5.1 Setting and Presented Results

In all following experiments, a given representation $\{\mathbf{N}_v\}_{v\in\mathcal{V}}$ is first orthogonalized via Algorithm 5.

Then, it is root-normalized to $\{\boldsymbol{\Upsilon}_v\}_{v\in\mathcal{V}}$ as described by Eq. (9) via square exchange matrices $\{\mathbf{V}_e\}_{e\in\mathcal{E}}$. The latter are either determined through thin SVDs in Algorithm 6 or through the Gram-matrix approach summarized in Section 3.4.

As an alternative, we also determine decoupled nodes $\{\mathbf{\Phi}_v\}_v$ via Algorithm 7 to derive a root-normal form via Definition 4.2. There, we once use thin and once compact SVDs (with respect to relative machine accuracy).

In all cases, the representations are subsequently truncated (if not already smaller) to the same sizes $\ell_e \equiv \overline{\ell}$, $e \in \mathcal{E}$ (cf. Section 4.2). While this may usually not be adequate, the tensors considered here have very similar or equal singular values $\{\sigma_e\}_{e\in\mathcal{E}}$, allowing for a simpler comparison.

After these truncations, we also apply a *post-normalization* as by Definition 4.2 to reestablish root-normal constraints in case of $\ell_e \geqslant r_e$, $e \in \mathcal{E}$ (with respect to relative machine accuracy). We also apply the procedure as a heuristic for truncations with loss, in particular as no definite distinguishment of zero singular values is possible. In that case however (except for the trivial $\overline{\ell} = 1$), there is (for now) no theoretical guarantee for when the normalization constraints are fulfilled at least approximately. We consider three different aspects:

1. Singular Values

We compare the differently calculated singular values σ_e for one edge $e \in \mathcal{E}$, where we use the in that regard most precise SVD call that MATLAB 2023B allows for (`s = svd(A)`). We select the edge e as each a certain worst case among all available edges.

2. Truncation

We compare the relative truncation accuracies

$$\frac{\| \boxtimes (\{\mathbf{\Upsilon}_v^{(\ell)}\}_{v\in\mathcal{V}}) - \boxtimes(\{\mathbf{N}_v\}_{v\in\mathcal{V}})\|_F}{\| \boxtimes (\{\mathbf{N}_v\}_{v\in\mathcal{V}})\|_F}$$

for different values $\overline{\ell} \in \mathbb{N}$. This distance between two tensors is each calculated via the norm of the root factor of a subsequently orthogonalized, combined representation of the difference of the two tensors (cf. Definition 3.1, [8, 15]).

3. Normalization

We compare the normalization concerning node-wise decoupling constraints (cf. Definition 4.1) as well as c-orthogonality. Each accuracy is set as maximum over all following values. From root to leaves, for each $e = \{p, b\}$, $b \in \mathrm{desc}_c(p)$, $p \in \mathcal{V}$, we determine the accuracy of decoupling in node p by

$$\frac{\|\tilde{\mathbf{V}}_e \boxtimes \tilde{\mathbf{\Sigma}}_e^{(\ell)} \boxtimes \tilde{\mathbf{V}}_e^T - \tilde{\mathbf{\Sigma}}_e^{(\ell)}\|_2}{\|\tilde{\mathbf{\Sigma}}_e^{(\ell)}\|_2}, \quad \tilde{\mathbf{U}}_e \boxtimes \tilde{\mathbf{\Sigma}}_e^{(\ell)} \boxtimes \tilde{\mathbf{V}}_e := \begin{cases} \mathbf{\Upsilon}_c^{(\ell)}, & b = c, \\ \tilde{\mathbf{\Sigma}}_{\{\mathrm{pred}_c(b),b\}}^{(\ell)} \boxtimes \mathbf{\Upsilon}_b^{(\ell)}, & b \neq c. \end{cases}$$

given the thin SVD with $\Lambda(\tilde{\mathbf{U}}_e) = \Lambda(\boldsymbol{\Phi}_p^{(\ell)})$ and $\Lambda(\tilde{\boldsymbol{\Sigma}}_e^{(\ell)}) = \{\beta_e\}$. Further, we determine the accuracy of orthogonalization in node b by

$$\frac{\|\tilde{\mathbf{V}}_e \boxtimes \boldsymbol{\Theta}_e \boxtimes \tilde{\mathbf{V}}_e^T - \mathbf{I}_e\|_2}{\|\mathbf{I}_e\|_2}, \quad \tilde{\mathbf{U}}_e \boxtimes \boldsymbol{\Theta}_e \boxtimes \tilde{\mathbf{V}}_e := \boldsymbol{\Upsilon}_b^{(\ell)},$$

for a corresponding identity matrix $\mathbf{I}_e$, given the thin SVD with $\Lambda(\tilde{\mathbf{U}}_e) = \Lambda(\boldsymbol{\Upsilon}_b^{(\ell)})$, $\Lambda(\boldsymbol{\Theta}_e) = \{\beta_e\}$.

5.2 Perturbation of Representations

In the following, by *perturbating a representation* $\{\mathbf{N}_v\}_{v \in \mathcal{V}}$, we mean the following. For random, orthogonal, square matrices $\{\mathbf{W}_e\}_{e \in \mathcal{E}}$ of according sizes, we replace $\mathbf{N}_v$ by $\mathbf{N}_v \boxtimes (\{\mathbf{W}_{\{v,w\}}\}_{w \in \mathrm{neigh}(v)})$, $v \in \mathcal{V}$. This perturbation does not change the represented tensor, but the result will (with reliable likelihood) not be in root-normal form. In principle, one may also apply non-orthogonal matrices that cancel out each other. However, even for well-conditioned ones, their potentially large number can cause the representation to become ill-conditioned in itself and lead to an a-priori approximation error already orders of magnitude larger than machine accuracy (see Fig. 12). For further reading on stability of representations, we refer to [1].

5.3 Truncation Accuracy of Exponential Sums

For a dimension $d \in \mathbb{N}$ and uniform mode sizes $n_\ell := \overline{n} \in \mathbb{N}$, $\ell \in \mathcal{L}$, we consider the tensor $\mathbf{T}$ with

$$\mathbf{T}(\{\alpha_v\}_{v \in \mathcal{L}}[i]) := \|i\|_2^{-1}, \quad i \in \{1, \ldots, \overline{n}\}^d.$$

We first formulate a reasonably accurate CP-decomposition of $\mathbf{T}$ based on exponential sums $x^{-1/2} \approx \sum_{k=1}^{K} a_k \exp(-b_k x)$ for $x \in [1, R]$, with $K = 35$ summands [9] as balanced, binary, hierarchical Tucker format $\{\mathbf{N}_v\}_{v \in \mathcal{V}}$. For the leaves $\mu \in \mathcal{L}$ with $\Lambda(\mathbf{N}_\mu) = \{\alpha_\mu, \beta_e\}$, we explicitly set

$$\mathbf{N}_\mu(\alpha_\mu[i_\mu], \beta_e[j]) := a_\ell^{1/d} \exp(-b_k i_\mu^2), \quad j = 1, \ldots, K, \quad i_\mu = 1, \ldots, \overline{n}.$$

All other nodes, $v \in \mathcal{V} \setminus \mathcal{L}$, are set as hyperdiagonal tensors

$$\mathbf{N}_v(\{\beta_{\{v,w\}}\}_{w \in \mathrm{neigh}(v)}[(i_1, i_2, i_3)]) := \begin{cases} 1, & i_1 = i_2 = i_3 \leqslant K, \\ 0, & \text{otherwise}. \end{cases}$$

All sizes corresponding to edges are set as $k_e = K$, $e \in \mathcal{E}$. Note here that such respresentations are already close to their normal forms, whereby a calculation of singular values beyond machine precision is possible. Further, given certain circumstances, even the Gram-matrix approach can yield a sufficient accuracy. We therefore also perform the same experiment but for the perturbed representation (cf. Section 5.2). For different parameter choices, Figs. 6 and 7 display the results described in Section 5.1.

5.4 Truncation Accuracy for Assigned Singular Values

For a dimension $d \in \mathbb{N}$ and uniform mode size $n_\ell := \overline{n} \in \mathbb{N}$, $\ell \in \mathcal{L}$, we consider randomly initialized decompositions with assigned, shared singular values in all edges. We generate such representations via Algorithm 8 with singular value matrices

$$\boldsymbol{\Sigma}_e(\beta_e[i], \beta_e[j]) = \begin{cases} \overline{\sigma}_i, & i = j \in \{1, \dots, \overline{n}\}, \\ 0, & \text{otherwise}, \end{cases} \qquad e \in \mathcal{E}.$$

All sizes corresponding to edges are set as $k_e = \overline{n}$, $e \in \mathcal{E}$. Assigning the same singular values for all edges here ensures that these are indeed feasible (cf. Corollary 4.7). The thereby constructed root-normal form $\{\boldsymbol{\Upsilon}_v\}_{v \in \mathcal{V}}$ is then perturbed (cf. Section 5.2) into a decomposition $\{\mathbf{N}_v\}_{v \in \mathcal{V}}$ that still represents the same tensor, but is not in normal form. We further run the same experiments each with single instead of double precision. Since the recovered singular values as well as truncation accuracies barely vary for different random initializations, we show just a single instance.

5.5 Conclusions on Numerical Experiments

We in the following first comment on results specific to the two tested kinds of representations. The sections thereafter contain general observations.

5.5.1 Exponential Sums

In the unperturbed case for an exponential sum, Fig. 6, Algorithm 6 indeed manages to calculate singular values with an accuracy exceeding relative machine precision. In contrast, it seems that the more direct calculation in Algorithm 7 remedies the benefit of a format almost in root-normal form, also concerning the truncation accuracy. In the perturbed case, Fig. 7, there is almost no difference between the presented methods. In contrast, in both Figs. 6 and 7, we observe the loss of accuracy caused by the Gram matrix approach within the expected magnitude.

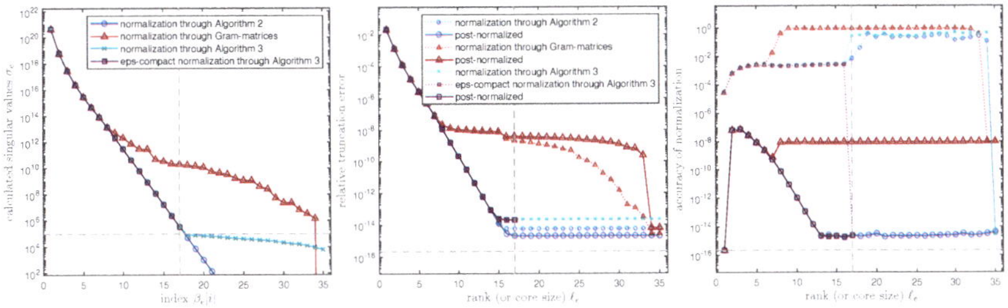

Fig. 6: Results for a (near-normalized, unperturbed) exponential sum decomposition for $d = 16$ and $\overline{n} = 1000$. See Section 5.5.1 for conclusions.

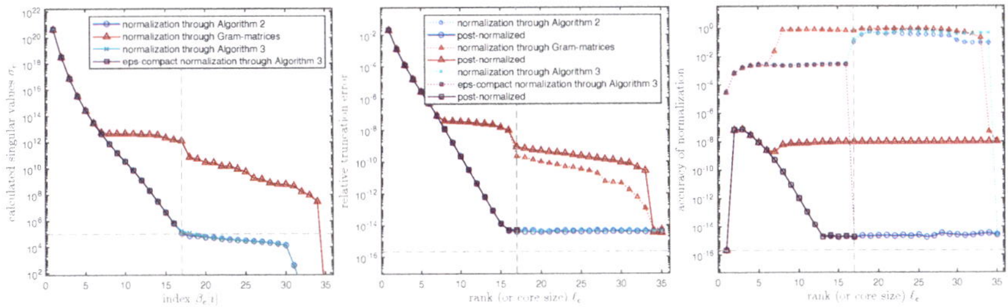

Fig. 7: Results for a perturbed exponential sum decomposition for $d = 16$ and $\overline{n} = 1000$. See Section 5.5.1 for conclusions.

5.5.2 Assigned Singular Values

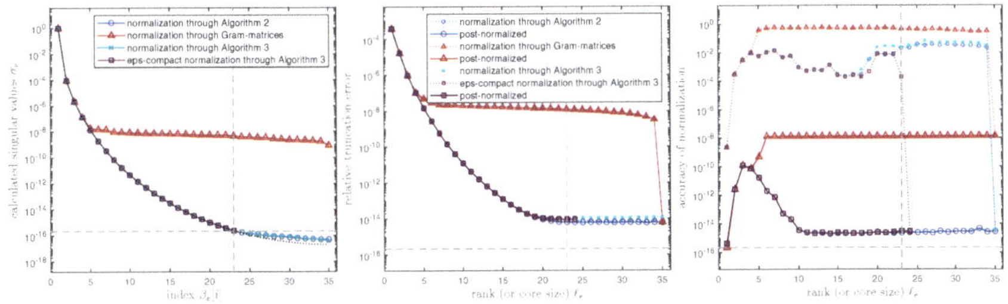

Fig. 8: Results for $\overline{\sigma}_k := (0.09\text{EPS})^{\sqrt{1-(1-t_k).^2}}$, where t is a uniform grid from 0 to 1 of length 35 (such that $\overline{\sigma}$ is a quarter of a circle in logarithmic scale), for $d = 32$ and $\overline{n} = 35$. See Sections 5.5.2 and 5.5.3 for conclusions.

The same observations as for the perturbed exponential sums hold true also for assigned singular values. In some cases, Algorithm 6 exhibits slightly more accurate results than Algorithm 7, likely because the multiplication of orthogonal exchange matrices results in almost no error. Note that here we can compare to the exact singular values, revealing only minor errors for singular values near machine precision. The compact normalization does however not reduce the accuracy in comparison to a full normalization by Algorithm 7. The post-normalization (after truncation) in all cases does not worsen the truncation accuracy, as expected for a truncation of only (near) zero singular values.

5.5.3 Normalization Accuracy

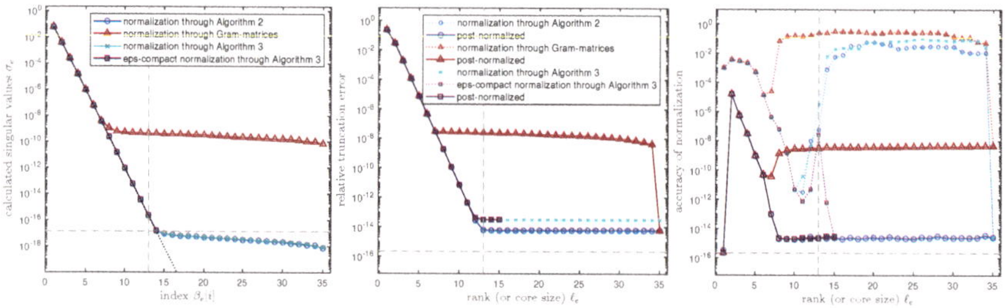

Fig. 9: Results for $\overline{\sigma}_k := 16^{-k}$, for $d = 32$ and $\overline{n} = 35$. See Sections 5.5.2 and 5.5.3 for conclusions.

The accuracy of the root-normalization does in turn depend on the experiment. Generally, as expected even for (near) loss-free truncation, the formats do not remain normalized, whereas post-normalization corrects this. However, there are some exceptions to this. For compact-normalization by Algorithm 7, when no singular values remain to be truncated, the format simply stays in normal form as it is not changed. This likewise holds true for the other algorithms, when no truncation is actually performed, except for the Gram matrix approach. The highest accuracy of normalization in most remaining cases, after truncation but before post-normalization, is reached for a slightly lower value than the theoretically postulated rank at which exactly all (near) zero singular values are truncated. However, while clearly visible in Fig. 9, not even such a peak is reliable in practice, as indicated by very slowly declining singular values in Fig. 8. In turn, when truncating non-zero singular values, the accuracy of the heuristic post-normalization representation depends heavily on the studied case, though it can be surprisingly good.

5.5.4 Single Precision

Comparing Fig. 11 to Fig. 9 (representing the general behavior), we can observe that the singular values are indeed calculated up to each the respective machine accuracy (again except for the Gram matrix approach). Whereas the best truncation accuracy for double precision is already roughly two orders of magnitude higher than the according machine accuracy, this error stays the same for single precision and is thus

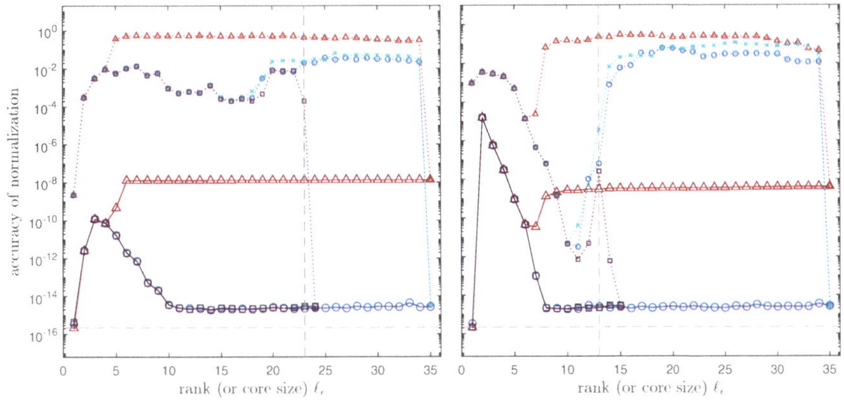

Fig. 10: Direct comparison of normalization accuracies for (left) $\overline{\sigma}_k := (0.09\text{EPS})^{\sqrt{1-(1-t_k).^2}}$ as in Fig. 8 and (right) $\overline{\sigma}_k := 16^{-k}$ as in Fig. 9 .

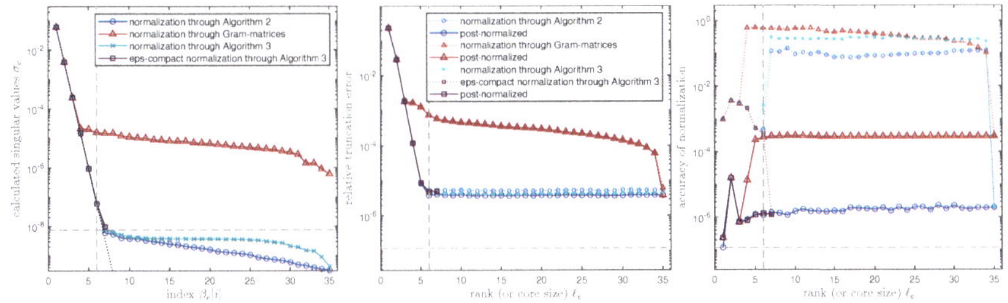

Fig. 11: Results for experiment as in Fig. 9 but in single precision. See Section 5.5.4 for conclusions.

much larger in relative terms. Single precision may however still be viable for fast and low accuracy calculations.

5.5.5 Perturbation by Non-Orthogonal Matrices

While the accuracy attained through the normalization of representations perturbed by orthogonal matrices is near machine accuracy, this does no longer hold true for theoretically equalizing pairs of non-orthogonal matrices and their inverses. It is thus likely that the perturbed representations are themselves ill-conditioned (cf. [1]). This is also confirmed by the measured, already large euclidean distance between the original representation and its perturbed version.

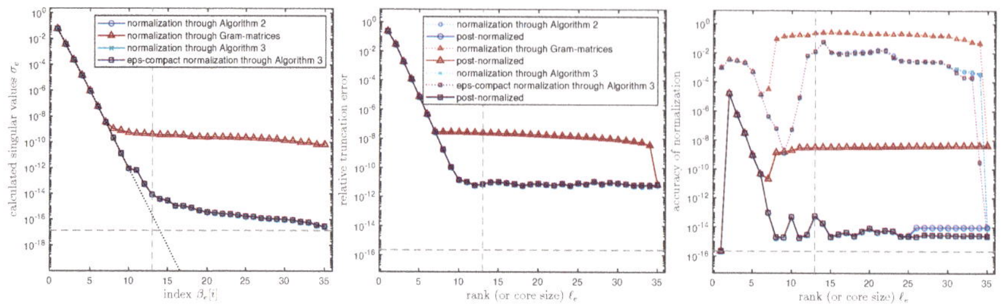

Fig. 12: Results for experiment as in Fig. 9 but for random (generated via RANDN calls and inverted via INV), non-orthogonal perturbation matrices. See Section 5.5.5 for conclusions on results.

References

1. Bachmayr, M., Kazeev, V.: Stability of low-rank tensor representations and structured multilevel preconditioning for elliptic pdes. Foundations of Computational Mathematics **20**(5), 1175–1236 (2020). DOI 10.1007/s10208-020-09446-z
2. Bachmayr, M., Schneider, R.: Iterative methods based on soft thresholding of hierarchical tensors. Foundations of Computational Mathematics **17**(4), 1037–1083 (2016). DOI 10.1007/s10208-016-9314-z
3. Ceruti, G., Lubich, C., Sulz, D.: Rank-adaptive time integration of tree tensor networks. SIAM Journal on Numerical Analysis **61**(1), 194–222 (2023). DOI 10.1137/22m1473790
4. De Lathauwer, L., De Moor, B., Vandewalle, J.: A multilinear singular value decomposition. SIAM Journal on Matrix Analysis and Applications **21**(4), 1253–1278 (2000). DOI 10.1137/S0895479896305696
5. Etter, S.: Parallel als algorithm for solving linear systems in the hierarchical tucker representation. SIAM Journal on Scientific Computing **38**(4), A2585–A2609 (2016). DOI 10.1137/15M1038852
6. Falcó, A., Hackbusch, W.: On minimal subspaces in tensor representations. Foundations of Computational Mathematics **12**(6), 765–803 (2012). DOI 10.1007/s10208-012-9136-6
7. Grasedyck, L.: Hierarchical singular value decomposition of tensors. SIAM Journal on Matrix Analysis and Applications **31**(4), 2029–2054 (2010). DOI 10.1137/090764189
8. Hackbusch, W.: Numerical tensor calculus. Acta Numerica **23**, 651–742 (2014). DOI 10.1017/S0962492914000087
9. Hackbusch, W.: Computation of best l^∞ exponential sums for 1/x by remez' algorithm. Computing and Visualization in Science **20**(1-2), 1–11 (2019). DOI 10.1007/s00791-018-00308-4
10. Hackbusch, W.: Truncation of tensors in the hierarchical format. SeMA Journal **78**(2), 175–192 (2019). DOI 10.1007/s40324-018-00184-5
11. Hackbusch, W., Kühn, S.: A new scheme for the tensor representation. Journal of Fourier Analysis and Applications **15**(5), 706–722 (2009). DOI 10.1007/s00041-009-9094-9
12. Krämer, S.: A geometric description of feasible singular values in the tensor train format. SIAM Journal on Matrix Analysis and Applications **40**(3), 1153–1178 (2019). DOI 10.1137/18M1192408
13. Krämer, S.: Tree tensor networks, associated singular values and high-dimensional approximation. Dissertation, RWTH Aachen University, Aachen (2020). DOI 10.18154/RWTH-2020-05412. Veröffentlicht auf dem Publikationsserver der RWTH Aachen University; Dissertation, RWTH Aachen University, 2020
14. Oseledets, I.V.: Tensor-train decomposition. SIAM Journal on Scientific Computing **33**(5), 2295–2317 (2011). DOI 10.1137/090752286

15. Tobler, C.: Low-rank tensor methods for linear systems and eigenvalue problems. Ph.D. thesis, ETH Zürich (2012)
16. Vidal, G.: Efficient classical simulation of slightly entangled quantum computations. Phys. Rev. Lett. **91**, 147902 (2003). DOI 10.1103/PhysRevLett.91.147902

Tree-Based Nonlinear Reduced Modeling

Diane Guignard and Olga Mula

Abstract This paper is concerned with model order reduction of parametric Partial Differential Equations (PDEs) using tree-based library approximations. Classical approaches are formulated for PDEs on Hilbert spaces and involve one single linear space to approximate the set of PDE solutions. Here, we develop reduced models relying on a collection of linear or nonlinear approximation spaces called a library, and which can also be formulated on general metric spaces. To build the spaces of the library, we rely on greedy algorithms involving different splitting strategies which lead to a hierarchical tree-based representation. We illustrate through numerical examples that the proposed strategies have a much wider range of applicability in terms of the parametric PDEs that can successfully be addressed. While the classical approach is very efficient for elliptic problems with strong coercivity, we show that the tree-based library approaches can deal with diffusion problems with weak coercivity, convection-diffusion problems, and with transport-dominated PDEs posed on general metric spaces such as the L^2-Wasserstein space.

1 Introduction

1.1 Context

This article, dedicated to Prof. Wolfgang Dahmen, develops tree-based nonlinear model order reduction of parametric Partial Differential Equations (PDEs). Let us start by drawing the general landscape. Parametric PDEs are of common use to model

Diane Guignard
Department of Mathematics and Statistics, University of Ottawa, Ottawa, ON K1N 6N5, Canada, e-mail: dguignar@uottawa.ca

Olga Mula
Eindhoven University of Technology, Den Dolech 2, P.O. Box 513, 5600 Eindhoven, NL, e-mail: o.mula@tue.nl

R. DeVore, A. Kunoth (eds.), *Multiscale, Nonlinear and Adaptive Approximation II*,
https://doi.org/10.1007/978-3-031-75802-7_13

complex physical systems, and are routinely involved in design and decision-making processes. Such equations can generally be written in abstract form as

$$\mathcal{P}(u, y) = 0, \tag{1}$$

where $\mathcal{P}$ is a partial differential operator, and $y = (y_1, \ldots, y_p)$ is a vector of scalar parameters ranging in some domain $\mathrm{Y} \subset \mathbb{R}^p$. The parameters represent physical quantities such as diffusivity, viscosity, velocity, source terms, or the geometry of the physical domain in which the PDE is posed. In several relevant instances, y may be high or even countably infinite dimensional, that is, $p \gg 1$ or $p = \infty$ (for simplicity, we consider $p < \infty$ in this work).

We assume well-posedness, that is, for any $y \in \mathrm{Y}$ the problem admits a unique solution $u = u(y)$ in a metric space (V, dist) whose elements depend on a physical variable x ranging in a domain $\Omega \subset \mathbb{R}^q$. The variable x usually refers to space but it is not limited to that meaning, and it may also refer to more elaborate sets of variables such as space and time. We may thus regard u as a function $(x, y) \mapsto u(x, y)$ from $\Omega \times \mathrm{Y}$ to $\mathbb{R}$, or we may also consider the *parameter-to-solution map*

$$y \in \mathrm{Y} \mapsto u(y) \in V \tag{2}$$

from Y to V. This map is typically nonlinear, and the image set

$$\mathcal{M} := u(\mathrm{Y}) = \{u(y) \ : \ y \in Y\} \subset V \tag{3}$$

is usually called the *solution manifold*. Throughout the paper, we assume that Y is compact in $\mathbb{R}^p$ and that the map (2) is continuous. Therefore, $\mathcal{M}$ is a compact set of V.

In numerous design and decision-making tasks, one is often confronted to optimize over the solution manifold $\mathcal{M}$, and this requires evaluating solutions $u(y)$ on a large number of dynamically updated parameters $y \in \mathrm{Y}$. Computations cannot be addressed rapidly unless the overall complexity has been appropriately reduced. This motivates the search for accurate methods to approximate the family of solutions very quickly at a reduced computational cost. This task, usually known as *reduced modelling* or *model order reduction*, has classically been considered in Hilbert and Banach spaces $(V, \|\cdot\|_V)$ by approximating $\mathcal{M}$ with well-chosen linear subspaces of V. In other words, one finds an appropriate n-dimensional subspace $V_n^{\mathrm{lin}} \subset V$, and builds a map

$$\hat{u}_n : \mathrm{Y} \to V_n^{\mathrm{lin}}, \quad y \mapsto \hat{u}_n(y),$$

in which evaluations of $\hat{u}_n$ scale only with n in complexity. This approach is suitable if the approximation error

$$e(\mathcal{M}, \hat{u}_n, V) := \sup_{y \in \mathrm{Y}} \mathrm{dist}(u(y), \hat{u}_n(y))$$

decays fast as $n \to \infty$. This is typically the case if $\hat{u}_n$ has a near-optimal approximation property in the sense that there exists $C(n) > 0$ such that

$$e(\mathcal{M}, \hat{u}_n, V) \leqslant C(n) \operatorname{dist}(\mathcal{M}, V_n^{\text{lin}}),$$

where

$$\operatorname{dist}(\mathcal{M}, V_n^{\text{lin}}) := \sup_{u \in \mathcal{M}} \inf_{v \in V_n^{\text{lin}}} \operatorname{dist}(u, v) \tag{4}$$

is the error of best approximation of $\mathcal{M}$ in V_n^{lin}. If the product $C(n) \operatorname{dist}(\mathcal{M}, V_n^{\text{lin}})$ decays fast as $n \to \infty$, the approximation of the parameter-to-solution map (2) by $\hat{u}_n$ is computationally very efficient. However, we can expect this to happen only when $C(n)$ mildly increases with n, and the Kolmogorov n-width of $\mathcal{M}$

$$\kappa(\mathcal{M}, \mathcal{V}_n^{\text{lin}}, V) := \inf_{V_n^{\text{lin}} \in \mathcal{V}_n^{\text{lin}}} \operatorname{dist}(\mathcal{M}, V_n^{\text{lin}}) \tag{5}$$

decays fast with n. In the above definition, V_n^{lin} runs over the set $\mathcal{V}_n^{\text{lin}}$ of all linear subspaces of dimension at most n in the Banach space V. While this is the case for certain families of parabolic or elliptic problems (see [1]), most transport-dominated problems are expected to present a slow decaying width and require to study nonlinear approximation methods (see, e.g., [2, Chapter 3]).

1.2 Library approximation

There is a tremendous activity in overcoming the limitations of working with linear subspaces for model reduction, and this paper is a contribution in this direction. We formulate a library approach which can be expressed in general metric spaces, and for which one can recover more classical formulations when instantiating it for PDEs on Hilbert and Banach spaces.

The idea is as follows. Given a metric space V, one considers an approximation family $\mathcal{V}_n$ which is not necessarily restricted to linear subspaces $\mathcal{V}_n^{\text{lin}}$ as in the above discussion. Each member $V_n \in \mathcal{V}_n$ can in general denote a linear or nonlinear approximation space parametrized by at most n degrees of freedom, and we define

$$\operatorname{dist}(\mathcal{M}, V_n) := \sup_{u \in \mathcal{M}} \inf_{v \in V_n} \operatorname{dist}(u, v) \tag{6}$$

and

$$\kappa(\mathcal{M}, \mathcal{V}_n, V) := \inf_{V_n \in \mathcal{V}_n} \operatorname{dist}(\mathcal{M}, V_n)$$

for such general spaces V_n exactly in the same spirit as (4) and (5). The quantity $\kappa(\mathcal{M}, \mathcal{V}_n, V)$ becomes a generalized Kolmogorov n-width which boils down to the classical definition when V is a Banach space and $\mathcal{V}_n^{\text{lin}}$ is the family of n-dimensional linear subspaces. Among the many possible examples of nonlinear $\mathcal{V}_n$, we may mention a family $\mathcal{V}_n^{NN}$ generated by neural network functions with different architectures. Each $V_n \in \mathcal{V}_n^{NN}$ would be the set of neural network functions with a given architecture and having at most n parameters varying in a certain range. Here, the variability of the architectures generates $\mathcal{V}_n$ in a Hilbert or a Banach space.

Another example, which we will develop later on, is when V is the L^2-Wasserstein space of measures. In this case, we can consider $\mathcal{V}_n = \mathcal{V}_n^{\mathrm{Bar}}$ as the set of all possible barycenters generated by at most n measures. This set will be defined more precisely in equation (23).

We next go beyond this idea with a library approximation. Instead of finding only one space $V_n \in \mathcal{V}_n$ for approximating $\mathcal{M}$, we consider a collection of spaces, which we call a library $\mathcal{L} = \{V^{(\alpha)}\}_{\alpha \in A}$ of cardinality $m := |\mathcal{L}| = |A| \in \mathbb{N}^*$, and for which each $V^{(\alpha)}$ is a linear or nonlinear space of dimension $n(\alpha)$. In our work, we rely on greedy algorithms to build spaces with different splitting strategies which will eventually lead to a representation of $\mathcal{L}$ in terms of a tree of subspaces. This is interesting because the representation of the subspaces $V^{(\alpha)} \in \mathcal{L}$ is thus hierarchical. The goal is then to explore how much we can benefit from this compression depending on different tree constructions.

To keep track of the tree structure, α will denote a multi-index set of the form $\alpha = (a_1, \ldots, a_k) \in \mathbb{N}_0^k$ for some $k \in \mathbb{N}^*$, and A is the set of multi-indices that are needed to define $\mathcal{L}$. Once $\mathcal{L}$ is obtained, we build an approximation

$$\hat{u}_{\mathcal{L}} : \mathrm{Y} \to \mathcal{L}, \quad \hat{u}_{\mathcal{L}}(y) \in V^{(\iota(y))}, \tag{7}$$

where $\iota : \mathrm{Y} \to A$ is a mapping from the parameters to the subspace index.

In the same spirit as the notions $\mathrm{dist}(\mathcal{M}, V_n)$ and $\kappa(\mathcal{M}, \mathcal{V}_n, V)$, we can now define the error of best approximation of $\mathcal{M}$ with a given library $\mathcal{L}$ as

$$\mathrm{dist}(\mathcal{M}, \mathcal{L}) := \sup_{u \in \mathcal{M}} \inf_{V^{(\alpha)} \in \mathcal{L}} \inf_{v \in V^{(\alpha)}} \mathrm{dist}(u, v)$$

and a notion of library width as

$$\kappa(\mathcal{M}, \mathcal{L}_{m,n}, V) := \inf_{\mathcal{L} \in \mathcal{L}_{m,n}} \mathrm{dist}(\mathcal{M}, \mathcal{L}) \tag{8}$$

with

$$\mathcal{L}_{m,n} := \{\mathcal{L} = (V^{(\alpha)})_{\alpha \in A} \ : \ |\mathcal{L}| \leqslant m, \text{ and } n(\alpha) \leqslant n \text{ for all } \alpha \in A\} \tag{9}$$

the set of libraries of cardinality at most m and containing subspaces $V^{(\alpha)}$ of dimension $n(\alpha) \leqslant n$. This approach is expected to give better accuracy than the one without a library since we have

$$\kappa(\mathcal{M}, \mathcal{L}_{m,n}, V) \leqslant \kappa(\mathcal{M}, \mathcal{L}_{1,n}, V) = \kappa(\mathcal{M}, \mathcal{V}_n, V). \tag{10}$$

1.3 Contribution and connection with earlier works

The idea of library approximation is not new in model order reduction. However, the common denominator of most approaches is that they rely on the assumption that the parameter space Y is a cartesian product or has a very simple shape (we refer to [3, 4, 5, 6, 7, 8, 9, 10] for works in this direction). The construction of the library then relies on identifying appropriate splittings of $\mathcal{M}$ induced by rectangular splittings of Y (or splittings involving a very explicit expression in Y). While this type of splitting is very practical because the definition of the index mapping ι is straightforward, this point of view cannot be easily extended for general Y, and there is no real exploitation of a potential tree structure in the representation of the library. With respect to this body of works, one of the main novelties in this paper is therefore the introduction of a splitting algorithm that is valid for any shape of the parameter domain and the compressed tree representation. The algorithm that we explore actually emerged in a discussion between Prof. Ron DeVore and Diane Guignard (one the authors of this paper).

The second main contribution is the phrasing of our algorithms in the general setting of nonlinear metric spaces. While the lack of linearity of V may pose numerical challenges, working in nonlinear metric spaces comes with the significant advantage of potentially capturing physical effects intrinsically in the geometry of the space. This gives room to reduce complexity even further, preserve certain physical properties by construction (e.g. mass conservation, symplecticity,...), and to treat different PDE classes under the same functional framework. A salient example is the family of Wasserstein gradient flows for which it is proven that some of the most common PDEs such as the heat equation or the Fokker-Planck equation can be seen as gradient flows in the Wasserstein space $(\mathcal{P}_2(\Omega), W_2)$ (which we will define more rigorously later on).

Probably [11] is one of the first works exploring model reduction on metric spaces where a strategy for the L^2–Wasserstein space was introduced. In that work, a parametric family of Wasserstein barycenters plays the role of V_n, and one can define a barycentric n-width exactly in the same manner as (5) but where V_n is infimized over all barycenter approximation sets $\mathcal{V}_n$ generated by at most n measures. We refer to [12, 13, 14, 15] for further works and applications of the idea. The work [15] is of particular interest for this paper since it develops a fully adaptive library approach. The main difference with the present approach is that it is not tree-based, and it involves a much larger library $\mathcal{V}_n$. This makes the definition of the operator ι much more costly. We discuss more in detail the differences between both approaches later on in Section 2.2.2.

1.4 Plan of the paper

The paper is organized as follows. Since our tree-based library construction heavily relies on greedy algorithms, we recall in Section 2 the vanilla versions of this algorithm for $\mathcal{M}$ in Banach spaces, and in the L^2-Wasserstein space. In Section 3 we explain our tree-based library approaches. Section 4 contains numerical experiments illustrating the performances of the algorithms.

2 The plain greedy algorithm for model reduction

In this section, we recall the greedy algorithms for model reduction that we use for our library approximation. The first is the classical greedy algorithm in Hilbert and Banach spaces whose convergence properties were analyzed in [16, 17]. The second greedy algorithm is less standard, and was originally introduced in [11] for model reduction in the L^2-Wasserstein space.

2.1 Formulation in Hilbert and Banach spaces

Let $(V, \|\cdot\|_V)$ be a Banach space, and consider $\mathcal{V}_n = \mathcal{V}_n^{\text{lin}}$. The greedy algorithm builds a space $V_n^{\text{lin}} \in \mathcal{V}_n^{\text{lin}}$ in a nested, iterative way. For $n = 1$, we find

$$u_1 \in \arg\max_{u \in \mathcal{M}} \|u\|_V \tag{11}$$

and define

$$\text{U}_1 := \{u_1\}, \quad V_1^{\text{lin}} := \text{span}\{u_1\}.$$

For $n > 1$, given U_{n-1} and V_{n-1}^{lin}, we find

$$u_n \in \arg\max_{u \in \mathcal{M}} \min_{v \in V_{n-1}^{\text{lin}}} \|u - v\|_V \tag{12}$$

and set

$$\text{U}_n = \text{U}_{n-1} \cup \{u_n\}, \quad V_n^{\text{lin}} = \text{span}\{\text{U}_n\}.$$

In [16, 17], it was proven that this greedy procedure generates a sequence of spaces $\{V_n^{\text{lin}}\}_n$ for which $\text{dist}(\mathcal{M}, V_n^{\text{lin}})$ decays at a comparable rate as the Kolmogorov n-width $\kappa(\mathcal{M}, \mathcal{V}_n^{\text{lin}}, V)$ as $n \to \infty$.

Before going further, several comments on the practical implementation of the algorithm are in order. Note that we have assumed that the above optimization problems over $\mathcal{M}$ admit a maximum. In general, this is not true and, in addition, one may not have a perfect optimization procedure to find the exact maximum. We can overcome both issues by formulating a weak greedy version of (11)-(12) where, for a parameter $\gamma \in (0, 1]$, we search for

$$u_1 \in \{u \in \mathcal{M} \ : \ \|u\|_V \geqslant \gamma \sup_{v \in \mathcal{M}} \|v\|_V\} \tag{13}$$

and for $n > 1$,

$$u_n \in \{u \in \mathcal{M} \ : \ \inf_{v \in V_{n-1}^{\text{lin}}} \|u - v\|_V \geqslant \gamma \sup_{z \in \mathcal{M}} \inf_{v \in V_{n-1}^{\text{lin}}} \|z - v\|_V\}. \tag{14}$$

When $\gamma = 1$, we recover the original greedy algorithm.

The implementation of the weak greedy algorithm is still computationally expensive in its current form because it requires having access to all $u \in \mathcal{M}$, and this means that we need to compute the solution $u(y)$ for all parameters $y \in \mathrm{Y}$. The complexity can be greatly reduced when working with PDEs defined in Hilbert spaces via weak formulations. In this case, we can replace the computation of full PDE solutions and distance evaluations by evaluation of residuals. The justification of this is as follows. In the Hilbert space setting, (1) can be expressed as an operator equation of the form $\mathcal{P}(u, y) = A(y)u(y) - f(y) = 0$. Here the solution $u(y)$ lives in the solution Hilbert space V, we have identified a test Hilbert space Z such that $f(y)$ belongs to the dual Z' of Z, and the operators are boundedly invertible mappings from V to the dual Z'. We assume that the invertibility is uniform in $y \in \mathrm{Y}$, that is, there exists $0 < r \leqslant R < \infty$ such that

$$R(y) := \|A(y)\|_{V \to Z'} \leqslant R, \quad \text{and} \quad r(y) := \|A(y)^{-1}\|_{Z' \to V} \leqslant r^{-1}, \quad \forall y \in \mathrm{Y}. \tag{15}$$

This setting covers a wide range of PDEs, such as classical elliptic problems with $Z = V$, as well as saddle-point problems, indefinite problems and unsymmetric problems such as convection-diffusion PDEs, or space-time formulations of parabolic problems. For some of these PDEs, Z sometimes needs to be chosen different from V to ensure a moderate value of the condition number $\nu := R/r \geqslant 1$. For example, for a simple elliptic operator of the form $A(y)u(y) = -\mathrm{div}(a(y)\nabla u(y))$ where the parametric diffusion $a(y)$ is uniformly bounded away from 0 and ∞, then the constants r and R directly relate to $\min_{y \in \mathrm{Y}} a(y)$ and $\max_{y \in \mathrm{Y}} a(y)$. It follows from (15) that one has the equivalence

$$r\|u(y) - v\|_V \leqslant \mathcal{R}(v, y) \leqslant R\|u(y) - v\|_V, \tag{16}$$

where

$$\mathcal{R}(v, y) := \|A(y)v - f(y)\|_{Z'}$$

is the residual of the PDE for a state $v \in V$ and a parameter $y \in \mathrm{Y}$. As a consequence, one can further modify the weak greedy algorithm into the search for $n > 1$ of

$$y_n \in \{y \in \mathrm{Y} \ : \ \Delta_{n-1}(y) \geqslant \gamma \sup_{y \in \mathrm{Y}} \Delta_{n-1}(y)\}, \tag{17}$$

where

$$\Delta_{n-1}(y) := \inf_{v \in V_{n-1}^{\mathrm{lin}}} \mathcal{R}(v, y)$$

is a quantity that one can estimate without computing exact solutions of the PDE.

Once y_n is obtained from (17), we set $u_n = u(y_n)$. Thanks to the equivalence relation (16), u_n satisfies a weak greedy property with respect to the ambient error norm,

$$\inf_{v \in V_{n-1}^{\mathrm{lin}}} \|u_n - v\|_V \geqslant \widetilde{\gamma} \sup_{z \in \mathcal{M}} \inf_{v \in V_{n-1}^{\mathrm{lin}}} \|z - v\|_V,$$

for a parameter

$$\widetilde{\gamma} := \frac{\gamma}{\nu} \leqslant \gamma.$$

This means that working with the equivalent residual degrades the tightness of the weak greedy algorithm by a factor ν.

Step (17) only requires computing residuals that do not directly involve PDE solutions. However, we still have to face the fact that we need to make a search over the whole parameter set Y. Depending on the type of residual, one directly deals with the optimization task with classical optimization methods. This is the case when the residual depends affinely on the parameters. For more complex dependencies, the search over Y is replaced by a search over a large, finite training set $\mathrm{Y}^{(N)}$ of parameters (which generates a sampled version $\mathcal{M}^{(N)}$ of the manifold). This is a change that affects the global approximation properties of the algorithm, and quantifying how many samples are enough to preserve the same convergence rate as the non-sampled version is a difficult question. Some elements of answer can be found in [18].

Remark 2.1 When an explicit formula for the y-dependent constant $r(y)$ in (15) is available, then it is customary to use the *a posteriori* error estimator $\Delta_n(y)/r(y)$ rather than $\Delta_n(y)$ in the weak greedy procedure, see for instance [19, 20]. Indeed, we have

$$\frac{1}{R(y)}\Delta_n(y) \leqslant e_n(y) \leqslant \frac{1}{r(y)}\Delta_n(y), \quad e_n(y) := \inf_{v \in V_n^{\mathrm{lin}}} \|u(y) - v\|_V,$$

and thus this estimator provides a certified control of the error with constant one, and its effectivity index belongs to $[1, R(y)/r(y)]$ (the effectivity index of $\Delta_n(y)$ lies in $[r(y), R(y)]$). In practice, including the constant $r(y)$ or not can impact the selection of the snapshots, in particular when $r(y)$ varies significantly with y. However, this sensitivity will not be investigated in this work.

For subsequent developments, for a given Banach space V, a given target accuracy $\varepsilon > 0$ and a given maximal amount of iterations $n_{\max}$, it will be convenient to define the routine

$$[\mathrm{U}_n, V_n^{\mathrm{lin}}, e] = \mathtt{greedy}[\mathcal{M}, V, \varepsilon, n_{\max}] \tag{18}$$

which iterates over the steps (11)-(12) of the greedy algorithm until

$$n = \min\{n_{\max}, n_\varepsilon\}$$

with

$$n_\varepsilon = \min\{n \in \mathbb{N}^* \ : \ \mathrm{dist}(\mathcal{M}, V_n) \leqslant \varepsilon\}.$$

This routine yields the space $V_n^{\mathrm{lin}} = \mathrm{span}\{\mathrm{U}_n\}$ which approximates $\mathcal{M}$ with accuracy $e = \mathrm{dist}(\mathcal{M}, V_n) \leqslant \varepsilon$ provided that reaching this accuracy takes less than $n_{\max}$ steps. If it takes more than $n_{\max}$ iterations to reach ε accuracy, then $e = \mathrm{dist}(\mathcal{M}, V_n) > \varepsilon$ and the approximation error does not meet the target accuracy.

It will also be necessary to introduce a restarted version of the greedy algorithm where we resume iterations from a given space V_k^{lin} with $k \leqslant n_{\max}$, and we compute the iterations (12) starting at iteration $k+1$. This routine is denoted

$$[\mathrm{U}_n, V_n^{\mathrm{lin}}, e] = \texttt{restarted\_greedy}[\mathcal{M}, V, V_k^{\mathrm{lin}}, \varepsilon, n_{\max}]. \quad (19)$$

and the output is of dimension $n \in [k, n_{\max}]$.

2.2 A greedy barycentric algorithm in the L^2-Wasserstein space

The above weak greedy algorithm can be formulated in an analogous manner in the L^2-Wasserstein space of measures using barycenters to approximate measures. To explain this, we first recall the definition of this space and of barycenters.

2.2.1 The L^2-Wasserstein space and barycenters

We assume in the following that Ω is a subset in the normed vector space $(\mathbb{R}^q, |\cdot|)$, with $|\cdot|$ the Euclidean norm on $\mathbb{R}^q$. We denote $\mathcal{M}(\Omega)$ the space of Borel regular measures on Ω with finite total mass and

$$\mathcal{M}^+(\Omega) := \{\nu \in \mathcal{M}(\Omega) \ : \ \nu \geqslant 0\}, \quad \mathcal{P}(\Omega) := \{\nu \in \mathcal{M}^+(\Omega) \ : \ \nu(\Omega) = 1\}.$$

The Wasserstein space $\mathcal{P}_2(\Omega)$ is defined as the set of probability measures $\nu \in \mathcal{P}(\Omega)$ with finite second order moments, namely

$$\mathrm{Pr} := \{\nu \in \mathcal{P}(\Omega) \ : \ \int_\Omega |x|^2 \, \nu(\mathrm{d}x) \ < +\infty\}.$$

We endow the output space $\mathcal{P}_2(\Omega)$ with the metric that is induced by optimal transport theory which is defined as follows. Let μ and ν be two probability measures in $\mathcal{P}_2(\Omega)$. We define $\Pi(\mu,\nu) \subset \mathcal{P}_2(\Omega\times\Omega)$ as the subset of probability distributions π on $\Omega\times\Omega$ with marginal distributions equal to μ and ν. The Wasserstein distance between μ and ν is defined as:

$$W_2(\mu,\nu) := \inf_{\pi\in\Pi(\mu,\nu)} \left(\int_{\Omega\times\Omega} |x-y|^2 \, \pi(dx,dy) \right)^{1/2}, \quad \forall(\mu,\nu) \in \mathcal{P}_2(\Omega)\times\mathcal{P}_2(\Omega). \quad (20)$$

As detailed in [21], the space $\mathcal{P}_2(\Omega)$ endowed with the distance W_2 is a metric space that is usually called the L^2-Wasserstein space.

In this space, Wasserstein barycenters arise as natural objects to define a reasonable, and computationally tractable approximation family $\mathcal{V}_n$ in $V = (\mathrm{Pr}, W_2)$. Their definition and properties are relatively well understood, especially since the works of [22]. Let $n \in \mathbb{N}^*$ denote the number of observations in our dataset and consider the simplex in $\mathbb{R}^n$:

$$\Sigma_n := \left\{ \Lambda_n = (\lambda_1, \ldots, \lambda_n)^T \in \mathbb{R}^n \ : \ \lambda_i \geqslant 0, \ \sum_{i=1}^n \lambda_i = 1 \right\}.$$

Given a set of weights $\Lambda_n = (\lambda_i)_{1 \leqslant i \leqslant n} \in \Sigma_n$, and given a set $\mathrm{U}_n = \{u_i\}_{1 \leqslant i \leqslant n}$ of n probability measures from $\mathcal{P}_2(\Omega)$, we say that $\mathrm{Bar}(\Lambda_n, \mathrm{U}_n) \in \mathcal{P}_2(\Omega)$ is a barycenter associated to Λ_n and U_n if and only if

$$\mathrm{Bar}(\Lambda_n, \mathrm{U}_n) \in \operatorname*{arg\,min}_{\nu \in \mathcal{P}_2(\Omega)} \sum_{i=1}^n \lambda_i W_2^2(\nu, u_i) . \tag{21}$$

We can thus define an approximation space V_n^{Bar} in Pr by varying the weights $\Lambda_n \in \Sigma_n$,

$$V_n^{\mathrm{Bar}} := \mathrm{Bar}(\Sigma_n, \mathrm{U}_n) := \{\mathrm{Bar}(\Lambda_n, \mathrm{U}_n) \ : \ \Lambda_n \in \Sigma_n\} \subset \mathcal{P}_2(\Omega) , \tag{22}$$

and this space belongs to the approximation class

$$\mathcal{V}_n^{\mathrm{Bar}} := \{\mathrm{Bar}(\Sigma_n, \mathrm{U}_n) \ : \ \mathrm{U}_n \in \overset{n}{\mathrm{Pr}}\}. \tag{23}$$

2.2.2 The barycentric greedy algorithm

If $\mathcal{M} \subset V = (\mathrm{Pr}, W_2)$, we can build a barycentric space $V_n = \mathrm{Bar}(\Sigma_n, \mathrm{U}_n)$ to approximate $\mathcal{M}$ by building U_n in a nested manner with a greedy algorithm. This is done exactly in the same spirit as the construction for the Hilbert/Banach case.

The algorithm is initialized with the two elements of $\mathcal{M}$ that lie at furthest distance, namely

$$(u_1, u_2) \in \operatorname*{arg\,max}_{(\nu_1, \nu_2) \in \mathcal{M} \times \mathcal{M}} W_2(\nu_1, \nu_2),$$

and we set

$$\mathrm{U}_2 := \{u_1, u_2\}, \quad V_2^{\mathrm{Bar}} := \mathrm{Bar}(\Sigma_2, \mathrm{U}_2).$$

Next, for $n \geqslant 3$, given U_{n-1} and the associated $V_{n-1}^{\mathrm{Bar}} = \mathrm{Bar}(\Sigma_{n-1}, \mathrm{U}_{n-1})$, we search for

$$u_n \in \operatorname*{arg\,max}_{\nu \in \mathcal{M}} \inf_{\zeta \in V_{n-1}^{\mathrm{Bar}}} W_2(\nu, \zeta),$$

and we set

$$\mathrm{U}_n = \mathrm{U}_{n-1} \cup \{u_n\}, \quad V_n^{\mathrm{Bar}} := \mathrm{Bar}(\Sigma_n, \mathrm{U}_n). \tag{24}$$

The convergence properties of this algorithm are mathematically not well understood yet but there is ample numerical evidence that, for many PDE classes, the algorithm builds spaces V_n^{Bar} that efficiently approximate $\mathcal{M}$ in the sense of presenting a fast convergence rate $\mathrm{dist}(\mathcal{M}, V_n^{\mathrm{Bar}})$ (see equation (6)). Like in the Banach space case, the practical implementation of the algorithm requires replacing $\mathcal{M}$ by a finite set of $N \gg 1$ training samples $\mathrm{U}^{(N)} = \{u(y) \ : \ y \in \mathrm{Y}^{(N)}\}$ computed from a finite set of N parameters $\mathrm{Y}^{(N)}$. Computations present an additional challenge compared to the Banach space setting which is connected to the complexity of evaluating Wasserstein distances and barycenters. In one space dimension ($q = 1$), one can leverage closed form characterizations involving inverse cumulative distributions. This is no longer possible for $q > 1$ and, in this setting, a popular approach is to resort to entropy regularized versions of the distance and the barycenter (see [23, 15]).

We can now easily adapt the definition of the `greedy` and `restarted_greedy` routines introduced in (18) and (19) for $V = (\mathrm{Pr}, W_2)$. We just need to replace V_n^{lin} by the barycentric space V_n^{Bar} from (24).

A library version of this algorithm was recently introduced in [15]. We briefly recall it next in order to compare it to the tree-based library approach that we introduce in the next section. The starting point in that work is to observe that, since in practice we use a training set $\mathrm{U}^{(N)}$ for the greedy algorithm, the V_n that is obtained in the plain greedy algorithm belongs to the approximation class

$$\mathcal{V}_{n,N}^{\mathrm{Bar}} := \mathrm{Bar}(\Sigma_N^n, \mathrm{U}^{(N)}) = \{\mathrm{Bar}(\Lambda_N, \mathrm{U}^{(N)}) : \Lambda_N \in \Sigma_N^n\} \subset \mathcal{V}_n^{\mathrm{Bar}}, \tag{25}$$

where

$$\Sigma_N^n := \{\Lambda_N \in \Sigma_N : |\mathrm{spt}(\Lambda_N)| \leqslant n\}$$

is the set of n-sparse vectors from the simplex Σ_N. Here, $|\mathrm{spt}(\Lambda_N)|$ denotes the cardinality of the support of Λ_N, namely

$$\mathrm{spt}(\Lambda_N) := \{i \in \{1, \dots, N\} : \lambda_i \neq 0\}.$$

It is interesting to remark that Σ_N^n is the union of the n-simplices embedded in $\mathbb{R}^N$, and there are $\binom{N}{n}$ of these simplices. This means that $|\mathcal{V}_{n,N}^{\mathrm{Bar}}| = \binom{N}{n}$. In [15], the authors build a model reduction method which can be summarized as the construction of a mapping

$$\hat{u}_{\mathcal{V}_{n,N}^{\mathrm{Bar}}} : \mathrm{Y} \to \mathcal{V}_{n,N}^{\mathrm{Bar}}, \quad u_{\mathcal{V}_{n,N}^{\mathrm{Bar}}} \in V_n^{\mathrm{Bar}}(y) \in \mathcal{V}_{n,N}^{\mathrm{Bar}}.$$

The mapping involves a procedure to adapt the approximation space $V_n^{\mathrm{Bar}}(y)$ for each $y \in \mathrm{Y}$. This is done via solving a challenging optimization problem involving sparsity constraints (we refer to [15] for further details). As a consequence, this is a library approach of the form (7) where $\mathcal{L} := \mathcal{V}_{n,N}^{\mathrm{Bar}}$. One of the main differences compared to the strategy that we develop next is that, for each $y \in \mathrm{Y}$, the search of $V_n^{\mathrm{Bar}}(y)$ is done over the whole $\binom{N}{n}$ members of the family which is very large if $N \gg 1$. In this context, our tree-based approach can be understood as a way of narrowing-down the amount of spaces in which one needs to make the search.

3 Tree-based library approaches

We next present two tree-based library approaches that build on the previous greedy algorithms. The strategy of Section 3.1 requires Y to have a tensor product structure. The second approach, which is one of the main novelties of the paper, can be applied to any shape of parameter domain Y. We present it in Section 3.2.

3.1 Cartesian splittings of Y

One can build on the plain greedy algorithm to derive a library approach where $\mathcal{L} = \{V^{(\alpha)}\}_{\alpha \in A}$ is composed of spaces $V^{(\alpha)}$ that are built in order to approximate a subset $\mathcal{M}^{(\alpha)}$ of the manifold in such a way that

$$\operatorname{dist}(\mathcal{M}^{(\alpha)}, V^{(\alpha)}) \leqslant \varepsilon, \quad \forall \alpha \in A.$$

The subsets $\mathcal{M}^{(\alpha)}$ must form a covering of $\mathcal{M}$ in the sense that

$$\mathcal{M} = \cup_{\alpha \in A} \mathcal{M}^{(\alpha)}.$$

Since $\mathcal{M}$ has a natural parametrization through Y, the most natural approach is to build a splitting of Y,

$$\mathrm{Y} = \cup_{\alpha \in A} \mathrm{Y}^{(\alpha)},$$

which we then use to build

$$\mathcal{M}^{(\alpha)} = \{u(y) \ : \ y \in \mathrm{Y}^{(\alpha)}\}.$$

Splitting Y is particularly easy when it is a cartesian product since we can build an approach based on simple dyadic partitions. To explain the algorithm, we assume without loss of generality that $\mathrm{Y} = [0, 1]^p$, and we define indices

$$\alpha = \{(k_m, i_m)\}_{m=1}^{p}, \quad \text{for } k_m \in \mathbb{N}_0,\ 0 \leqslant i_m < 2^{k_m}. \tag{26}$$

We associate to α the parameter subdomain

$$\mathrm{Y}^{(\alpha)} := \times_{m=1}^{p} \left[\frac{i_m}{2^{k_m}}, \frac{i_m + 1}{2^{k_m}}\right] \subseteq \mathrm{Y}.$$

Note that with this notation

$$\mathrm{Y} = \mathrm{Y}^{(\alpha)},\ \mathcal{M} = \mathcal{M}^{(\alpha)} \quad \text{for } \alpha = \{(0, 0)\}_{m=1}^{p},$$

and for a given $\alpha = \{(k_m, i_m)\}_{m=1}^{p}$ we can express partitions of $\mathrm{Y}^{(\alpha)}$ along a certain parameter $\tilde{p} \in \{1, \dots, p\}$ as

$$\mathrm{Y}^{(\alpha)} = \mathrm{Y}^{(\alpha(0,\tilde{p}))} \cup \mathrm{Y}^{(\alpha(1,\tilde{p}))}$$

with

$$\alpha(0,\tilde{p}) = \{(k_1,i_1),\dots,(k_{\tilde{p}-1},i_{\tilde{p}-1}),(k_{\tilde{p}}+1,2i_{\tilde{p}}),(k_{\tilde{p}+1},i_{\tilde{p}+1}),\dots,(k_p,i_p)\},$$
$$\alpha(1,\tilde{p}) = \{(k_1,i_1),\dots,(k_{\tilde{p}-1},i_{\tilde{p}-1}),(k_{\tilde{p}}+1,2i_{\tilde{p}}+1),(k_{\tilde{p}+1},i_{\tilde{p}+1}),\dots,(k_p,i_p)\}.$$

Note that for $m \in \{1,2,\dots,p\}$, k_m controls the length of the subinterval in the direction y_m while i_m controls its location in $[0,1]$.

We can now formulate the following algorithm to generate a tree-based library approximation. We set the target tolerance $\varepsilon > 0$ and the maximal dimension $n_{\max}$ of the elements of the library. We then proceed iteratively as follows:

- Initialization ($k = 1$): We define

$$\alpha_k = \{(0,0)\}_{m=1}^{p}$$
$$A_k = \{\alpha_k\}$$
$$\mathrm{Y}_k = \cup_{\alpha\in A_k}\mathrm{Y}^{(\alpha)} = \{\mathrm{Y}\}$$

  and we run

$$[\mathrm{U}^{(\alpha_k)}, V^{(\alpha_k)}, e^{(\alpha_k)}] = \texttt{greedy}(\mathcal{M}^{(\alpha_k)}, V, \varepsilon, n_{\max}),$$

  where, by construction, $\mathcal{M}^{(\alpha_k)} = \mathcal{M}$. We then set

$$\mathrm{L}_k = \{V^{(\alpha)}\}$$

  which is a library containing one space of dimension at most $n_{\max}$. Therefore, $\mathrm{L}_k \in \mathcal{L}_{1,n_{\max}}$ according to Definition (9).
  If $e^{(\alpha_k)} \leqslant \varepsilon$, the algorithm stops and we set $K = 1$. Otherwise we set $k = 2$ and the algorithm goes into the inductive step.
- Inductive step $k > 1$: From the previous step, we receive A_{k-1}, Y_{k-1}, and $\mathrm{L}_{k-1} = \{V^{(\alpha)}\}_{\alpha\in A_{k-1}}$. We start by setting $A_k = A_{k-1}$, and we are going to update the definition of A_k through the following procedure. For every $\alpha \in A_{k-1}$:
  - If $e^{(\alpha)} = \mathrm{dist}(\mathcal{M}^{(\alpha)}, V^{(\alpha)}) \leqslant \varepsilon$, the algorithm has reached accuracy ε on $\mathcal{M}^{(\alpha)}$ so we do not need to split this subset.
  - If $e^{(\alpha)} = \mathrm{dist}(\mathcal{M}^{(\alpha)}, V^{(\alpha)}) > \varepsilon$, we split $\mathcal{M}^{(\alpha)}$ along one parameter direction. To discover the best parameter to choose, we run a greedy algorithm for every possible dyadic split. This means that for every parameter $\tilde{p} \in \{1,\dots,p\}$, we compute

$$\begin{cases} [\mathrm{U}^{(\alpha(0,\tilde{p}))}, V^{(\alpha(0,\tilde{p}))}, e^{(\alpha(0,\tilde{p}))}] & = \texttt{greedy}[\mathcal{M}^{((\alpha(0,\tilde{p}))}, V, \varepsilon, n_{\max}] \\ [\mathrm{U}^{(\alpha(1,\tilde{p}))}, V^{(\alpha(1,\tilde{p}))}, e^{(\alpha(1,\tilde{p}))}] & = \texttt{greedy}[\mathcal{M}^{(\alpha(1,\tilde{p}))}, V, \varepsilon, n_{\max}] \end{cases}$$

    and we define

$$e_{\tilde{p}}^{(\alpha)} := \max(e^{(\alpha(0,\tilde{p}))}, e^{(\alpha(1,\tilde{p}))}). \tag{27}$$

We next choose the parameter along which we do the splitting by finding

$$p_{\text{split}} := \underset{\tilde{p} \in \{1,\dots,p\}}{\arg\min}\ e_{\tilde{p}}^{(\alpha)}. \tag{28}$$

Then we decompose $\mathrm{Y}^{(\alpha)} \in \mathrm{Y}_{k-1}$ as

$$\mathrm{Y}^{(\alpha)} = \mathrm{Y}^{(\alpha(0,p_{\text{split}}))} \cup \mathrm{Y}^{(\alpha(1,p_{\text{split}}))}$$

and we update the set of indices

$$A_k \leftarrow (A_k \setminus \{\alpha\}) \cup \{\alpha(0, p_{\text{split}}), \alpha(1, p_{\text{split}})\},$$

i.e., we replace the splitted index α by the refined sets $\alpha(0, p_{\text{split}})$ and $\alpha(1, p_{\text{split}})$.

Once all the indices in A_{k-1} have been visited (and splitted if needed), we finally set

$$\begin{aligned} \mathrm{Y}_k &= \cup_{\alpha \in A_k} \mathrm{Y}^{(\alpha)} \\ \mathrm{L}_k &= \{V^{(\alpha)}\}_{\alpha \in A_k}. \end{aligned}$$

If

$$e_k := \max_{\alpha \in A_k} e^{(\alpha)} \leqslant \varepsilon,$$

the algorithm stops and we set $K = k$, otherwise we increase the index k and repeat the inductive step.

This algorithm, referred to as `Y-cart-split` in what follows, produces a library $\mathcal{L} := \mathrm{L}_K \in \mathcal{L}_{2^{K-1}, n_{\max}}$. That is, the library $\mathcal{L}$ has cardinality $|\mathcal{L}| \leqslant 2^{K-1}$ and contains spaces of dimension at most $n_{\max}$. Moreover, the total number of snapshots generating the library is $\sum_{\alpha \in A_K} \dim(V^{(\alpha)}) \leqslant n_{\max} 2^{K-1}$.

Once $\mathcal{L}$ has been computed, we need to define the final approximation $\hat{u}_{\mathcal{L}}$ introduced in equation (7), namely

$$\hat{u}_{\mathcal{L}} : \mathrm{Y} \to \mathcal{L}, \quad \hat{u}_{\mathcal{L}}(y) \in V^{(\iota(y))}.$$

Thanks to the dyadic splittings in the parameter domain, the mapping ι has a very simple characterization, and it reads

$$\iota(y) = \{\alpha \in A_k \ :\ y \in \mathrm{Y}^{(\alpha)}\}.$$

3.2 An approach which does not depend on the shape of Y

The previous strategy heavily relies on the assumption that the parameter space Y has a tensor product structure. We introduce in this subsection an alternative algorithm which can be used to generate a library $\mathcal{L} = \{V^{(\alpha)}\}_{\alpha\in A}$ for a generic Y. The main difference lies in the definition of the subdomain $\mathrm{Y}^{(\alpha)} \subseteq \mathrm{Y}$ and the space $V^{(\alpha)}$ associated to an index $\alpha \in A$. However, once defined, the subdomain $\mathrm{Y}^{(\alpha)} \subseteq \mathrm{Y}$ generates again the subset

$$\mathcal{M}^{(\alpha)} = \{u(y)\ :\ y \in \mathrm{Y}^{(\alpha)}\} \subseteq \mathcal{M}.$$

Here, the indices α will be binary, namely of the form $\alpha = (a_1, \dots, a_k) \in \{0,1\}^k$. The root of the tree is encoded with $\alpha = (1)$ and we work with the convention

$$\mathcal{M}^{(\alpha)} = \mathcal{M},\ \mathrm{Y}^{(\alpha)} = \mathrm{Y}, \quad \text{for } \alpha = (1).$$

Moreover, for $u \in \mathcal{M}$, we will use the notation

$$\mathrm{dist}(u, V^{(\alpha)}) := \inf_{v\in V^{(\alpha)}} \mathrm{dist}(u, v).$$

The second algorithm proposed to generate a tree-based library approximation can be formulated as follows. We set the target tolerance $\varepsilon > 0$. We then proceed iteratively as follows:

- Initialization ($k = 1$): We define

$$\alpha_k = (1), \qquad A_k = \{\alpha_k\}, \qquad \mathrm{Y}_k = \cup_{\alpha\in A_k} \mathrm{Y}^{(\alpha)} = \{\mathrm{Y}\}.$$

  When V is a Hilbert or a Banach space, we select a first snapshot applying one step of the plain greedy algorithm, that is,

$$[\mathrm{U}^{(\alpha_k)}, V^{(\alpha_k)}, e^{(\alpha_k)}] = \mathtt{greedy}(\mathcal{M}^{(\alpha_k)}, V, \varepsilon, n_{\max} = 1), \tag{29}$$

  where $\mathrm{U}^{(\alpha_k)} = \{u_{(\alpha_k)}\}$ and where $\mathcal{M}^{(\alpha_k)} = \mathcal{M}$ by construction. We then set

$$\mathrm{L}_k = \{V^{(\alpha_k)}\}.$$

  When $V = (\mathrm{Pr}, W_2)$, we apply the same strategy but we need to select two snapshots to initialize the greedy algorithm so we should use $n_{\max} = 2$ in (29). If $e^{(\alpha_k)} \leqslant \varepsilon$, the algorithm stops, otherwise we set $k = 2$ and the algorithm goes into the inductive step.
- Inductive step $k > 1$: From the previous step, we receive A_{k-1}, Y_{k-1}, and $\mathrm{L}_{k-1} = \{V^{(\alpha)}\}_{\alpha\in A_{k-1}}$. We set $A_k = A_{k-1}$, and we are going to update A_k with the procedure that follows. For every $\alpha \in A_{k-1}$ for which $e^{(\alpha)} > \varepsilon$, we split $\mathcal{M}^{(\alpha)}$ into two subsets as follows. We make two steps of a restarted greedy algorithm that takes $V^{(\alpha)}$ as a starting approximation space,

$$[\widetilde{\mathrm{U}}^{(\alpha)}, \widetilde{\mathrm{V}}^{(\alpha)}, e] = \mathtt{restarted\_greedy}[\mathcal{M}^{(\alpha)}, V, V^{(\alpha)}, \varepsilon, n_{\max}]$$

  with $n_{\max} = \dim(V^{(\alpha)}) + 2$. By construction,

$$\widetilde{\mathrm{U}}^{(\alpha)} = \mathrm{U}^{(\alpha)} \cup \{u_{(\alpha,0)}, u_{(\alpha,1)}\},$$

where the two snapshots $u_{(\alpha,0)}, u_{(\alpha,1)}$ are the ones that have been added by the restarted greedy algorithm. For $i \in \{0, 1\}$, we define

$$\mathrm{U}^{(\alpha,i)} := \mathrm{U}^{(\alpha)} \cup \{u_{(\alpha,i)}\}$$

and

$$V^{(\alpha,i)} := \begin{cases} \mathrm{span}\{\mathrm{U}^{(\alpha,i)}\} & \text{if } V = \text{Banach space} \\ \mathrm{Bar}(\Sigma_r, \mathrm{U}^{(\alpha,i)}) & \text{if } V = (\mathrm{Pr}, W_2) \end{cases}$$

with $r := |\mathrm{U}^{(\alpha,i)}|$. Next we decompose $\mathrm{Y}^{(\alpha)} = \mathrm{Y}^{(\alpha,0)} \cup \mathrm{Y}^{(\alpha,1)}$ according to

$$\mathrm{Y}^{(\alpha,0)} := \left\{y \in \mathrm{Y}^{(\alpha)} : \quad \mathrm{dist}(u(y), V^{(\alpha,0)}) \leqslant \mathrm{dist}(u(y), V^{(\alpha,1)})\right\}$$

and $\mathrm{Y}^{(\alpha,1)} := \mathrm{Y}^{(\alpha)} \setminus \mathrm{Y}^{(\alpha,0)}$. Then we define

$$e^{(\alpha,i)} := \mathrm{dist}(\mathcal{M}^{(\alpha,i)}, V^{(\alpha,i)}), \quad i = 0, 1,$$

and we update

$$A_k \leftarrow (A_k \setminus \{\alpha\}) \cup \{(\alpha, 0), (\alpha, 1)\},$$

where we have replaced the splitted index α by the refined sets $(\alpha, 0)$ and $(\alpha, 1)$. Once all the indices in A_{k-1} have been visited (and splitted if needed), we finally set

$$\mathrm{Y}_k = \cup_{\alpha \in A_k} \mathrm{Y}^{(\alpha)}$$
$$\mathrm{L}_k = \{V^{(\alpha)}\}_{\alpha \in A_k}.$$

If

$$e_k := \max_{\alpha \in A_k} e^{(\alpha)} \leqslant \varepsilon,$$

the algorithm stops and we set $K = k$, otherwise we increase the index k and repeat the inductive step.

This algorithm, referred to as $\mathcal{M}$-`based-split` in what follows, produces a library $\mathcal{L} = \mathrm{L}_K \in \mathcal{L}_{2^{K-1},n}$, where $n = K$ if V is a Banach space and $n = K + 1$ if $V = (\mathrm{Pr}, W_2)$. In this case, the total number of snapshots building the library corresponds to the number of nodes in the underlying tree, which may be much smaller than $\sum_{\alpha \in A_K} \dim(V^{(\alpha)})$. Indeed, many snapshots belong to several spaces, in particular the snapshot(s) associated to the root, i.e. $\mathrm{U}^{(1)}$, belong to all the spaces in $\mathcal{L}$.

We define the final approximation $\hat{u}_{\mathcal{L}}$ as

$$\hat{u}_{\mathcal{L}} : \mathrm{Y} \to \mathcal{L}, \quad \hat{u}_{\mathcal{L}}(y) \in V^{(\iota(y))},$$

where

$$\iota(y) = \{\alpha \in A_k \ : \ y \in \mathrm{Y}^{(\alpha)}\} \tag{30}$$

is defined similarly to the previous splitting strategy involving cartesian splittings of the parameter domain. In contrast to that approach, here the mapping ι does not have a simple characterization in general because the shape of the resulting subdomains in Y becomes very general (see Figures 3 and 6 for an illustration). Among the possible strategies to build this mapping, we may cite classification techniques from machine learning. We can also consider strategies based on metric embeddings where we learn distances of elements from $\mathcal{M}$ in V through distances in Y (see [15]).

Remark 3.1 Instead of making two steps in the restarted greedy algorithm, one can alternatively do more than two steps, and take the last two selected snapshots to perform the split into two subsets. In the numerical tests, this alternative is done for the KdV example considered in Section 4.2 (the restarted greedy algorithm makes three steps). Moreover, other strategies than `restarted_greedy` could be used for the selection of the additional snapshots. For instance, when V is a Hilbert space with inner product $(\cdot, \cdot)_V$, we could select all the elements for which the approximation error in $V^{(\alpha)}$ is a fraction (e.g., a half) of the largest error. Then among them, we pick the two that are the less correlated, namely v and w for which $(v, w)_V / (\|v\|_V \|w\|_V)$ is minimized. We have tested this approach for the diffusion problem of Section 4.1.1, but the results are not reported since they are similar to the ones obtained with two steps of greedy.

Remark 3.2 In the Hilbert setting, we can show that if the Kolmogorov n-width (5) of $\mathcal{M}$ decays algebraically (resp. exponentially), then so does the error dist$(\mathcal{M}, \mathcal{L})$ of the library $\mathcal{L}$ generated by $\mathcal{M}$`-based-split`. The proof follows [17, Corollary 3.3], but extended to local spaces and accounting for the fact that snapshots in $\mathrm{U}^{(\alpha)}$ may not belong to $\mathcal{M}^{(\alpha)}$. In other words, we can show that $\mathcal{M}$`-based-split` performs no worse than `plain-greedy`, which is expected due to (10). We leave further theoretical analysis of the performance of both $\mathcal{M}$`-based-split` and `Y-cart-split` for future work.

3.3 Advantages and disadvantages of each method

We discuss the advantages and limitations of each approach in the light of different aspects:

- **Stopping criteria:** A direct comparison of the algorithms $\mathcal{M}$`-based-split` and `Y-cart-split` is actually not trivial since their stopping criteria are not identical. Indeed, we only prescribe the target accuracy for $\mathcal{M}$`-based-split` (without control on the dimension of the spaces unless we add a restriction on the depth of the tree), while for `Y-cart-split` we prescribe ε and $n_{\max}$. As a result, we know in advance the maximum dimension of the local spaces and of course, the smaller $n_{\max}$ the more spaces in the library. One could consider modifications of `Y-cart-split` that would not involve $n_{\max}$. For example, in the spirit of hp-adaptivity, we could iterate with greedy algorithms over the splitted subdomains until the sum of their approximation errors becomes smaller than a fraction of the approximation error of the current parent. This option would be interesting to consider, but it has not been explored in our numerical examples.

- **Offline stage:** The construction of a library with `Y-cart-split` is much more time consuming than with $\mathcal{M}$`-based-split`. This is mainly due to the fact that we are testing all possible splits and selecting the best one in the sense of (28).
- **Online cost:** The numerical complexity to solve the parametric PDE for a given parameter $y \in \mathrm{Y}$ is related to the dimension $n(\alpha)$ of the local spaces $V^{(\alpha)}$. For both library approaches, this dimension is expected to be significantly smaller that the one for the `plain-greedy` as soon as we deal with problems that are not elliptic. So even if we deal with a collection of spaces, both library approaches are expected to provide significant gains in the online phase if one needs to compute many queries.
  In the Hilbert space setting, the online cost scales like $n(\alpha)^2$ (solution of a linear system with a dense matrix of size $n(\alpha) \times n(\alpha)$). For the Wasserstein space, evaluating Wasserstein barycenters with $n(\alpha)$ components scales proportionally to $n(\alpha)^2$ in dimension 1, and it becomes dependent on the number of discretization points $\mathcal{N}$ for larger dimensions (order $n(\alpha)\mathcal{N}^2 \log \mathcal{N}$ with state of the art algorithms based on entropic regularization [23]).
- **The mapping ι:** For a given $y \in \mathrm{Y}$, we need to decide which space of the library to use, namely construct a mapping $\iota : \mathrm{Y} \to A_k$. While this mapping is given by construction for `Y-cart-split`, we need to build it for $\mathcal{M}$`-based-split`. This can be done by using the training snapshots with an additional approximation procedure in the offline stage. Among the several options that one could consider, we may mention strategies based on learning metric embeddings as was introduced in [15]: we learn distances over elements in $\mathcal{M}$ by learning Euclidean distances in the parameter space. Then, for a new parameter $y \in \mathrm{Y}$, we select the space to use by searching for the element of the training set which is at closest distance (and for which we know that it is approximated by a certain space from $\mathcal{L}$). It would also be possible to train a classifier.

4 Numerical Examples

We illustrate the behavior of the algorithms for PDEs of different nature. In Section 4.1, we consider manifolds $\mathcal{M}$ connected to diffusion and convection-diffusion parametric PDEs defined on Hilbert spaces. In Section 4.2, we consider a parametric Korteweg-De Vries equation, and we view the generated manifold $\mathcal{M}$ as a subset of the Wasserstein space. We summarize the main results that emerge from these tests in Section 4.3, and we use the examples to discuss the advantages and disadvantages of each approach.

4.1 Diffusion and convection-diffusion problems in $V = H_0^1(\Omega)$

In this section, we set $\Omega = (0, 1)$ and we consider the Hilbert space $V = H_0^1(\Omega)$ endowed with the norm $\|v\|_V := \|v'\|_{L^2(\Omega)}$, $v \in H_0^1(\Omega)$. All the problems can be written in the form: given $y \in \mathrm{Y}$, find $u(y) \in V$ such that

$$a(u(y), v; y) = F(v), \quad \forall\, v \in V, \tag{31}$$

where $a(\cdot, \cdot; y)$ is a bilinear form parametrized by y which is uniformly coercive and continuous with constants $0 < r \leqslant R < \infty$. We let V_h be the space of piecewise linear ($\mathbb{P}_1$) finite elements based on a uniform partition of Ω with mesh size $h = 2^{-12}$, and consider the Galerkin projection $u_h(y) = P_{V_h}u(y) \in V_h$ of $u(y)$ to be the *truth solution*. Then, according to (16)-(17) and given the current approximation space $V_n \subset V_h$, we use the residual

$$\Delta_n(y) := \|\mathcal{R}_n(\cdot; y)\|_{V_h'}, \quad \mathcal{R}_n(v; y) := F(v) - a(P_{V_n}u(y), v; y)$$

instead of the error $\|u_h(y) - P_{V_n}u(y)\|_V$ to drive the greedy algorithms. We use a (deterministic) training set $\mathrm{Y}^{(N)} \subset \mathrm{Y}$ of finite dimension N for the greedy algorithms, and we *randomly* select one point in $\mathrm{Y}^{(N)}$ to generate the first snapshot instead of (2.1). Finally, we set the tolerance to $\varepsilon = 10^{-6}$.

4.1.1 Diffusion problems

We set $\mathrm{Y} = [-1, 1]^2$ (case $p = 2$) and consider the diffusion model problem

$$-\frac{\partial}{\partial x}\left(a(x, y)\frac{\partial u}{\partial x}(x, y)\right) = 1, \quad x \in (0, 1), \quad u(0, y) = u(1, y) = 0. \tag{32}$$

The problem has a weak formulation of the form (31) which reads: find $u \in H_0^1(\Omega)$ such that

$$\int_\Omega a(x, y)\frac{\partial u}{\partial x}(x, y)\frac{\partial v}{\partial x}(x)\,\mathrm{d}x = \int_\Omega v(x)\,\mathrm{d}x, \quad \forall v \in H_0^1(\Omega).$$

We next consider two examples corresponding to two different choices for the diffusion coefficients a.

Test 1 (labeled as Diff-1): We first choose the diffusion coefficient a defined for $y = (y_1, y_2) \in \mathrm{Y}$ and $x \in \Omega$ by

$$a(x, y) = 1 + \sum_{i=1}^{2} a_i(x)g_i(y_i) = 1 + \frac{\cos(2\pi x)}{\alpha\pi^2}y_1 + \frac{\cos(4\pi x)}{(2\pi)^2}y_2, \tag{33}$$

where $\alpha > \alpha_{\min} = \frac{4}{4\pi^2-1} \approx 0.104$ so that the problem is coercive. We examine the impact of coercivity by examining the different manifolds $\mathcal{M}$ generated by taking $\alpha = 1$ and $\alpha = 0.105$. The value $\alpha = 0.105$ corresponds to a degeneration of coercivity in the sense that r, R and the conditioning $\nu = R/r$ deviate from 1 ($r = 0.0097$, and $R = 1.9903$), see Figure 1a for some realizations of $a(\cdot, y)$. The value $\alpha = 1$ corresponds to a more benign coercive case with $r = 0.8733$ and $R = 1.1267$. The training set is $\mathrm{Y}^{(N)}$ with $N = 40000$ corresponding to the tensor product grid with 200 equally spaced points in each direction.

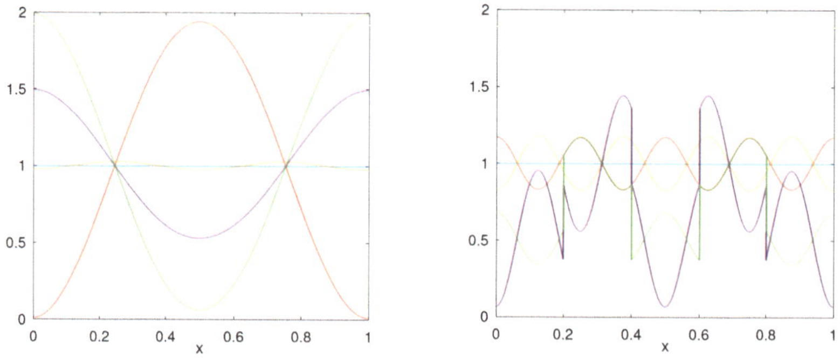

(a) Coefficient in (33) with $\alpha = 0.105$. (b) Coefficient characterized by (34)-(35).

Fig. 1: **Diffusion problems:** Diffusion coefficient $a(x, y)$ for the following values of the parameter y: $(0, 0)$, $(-1, -1)$, $(0, -1)$, $(0.5, 0.5)$, $(1, 1)$.

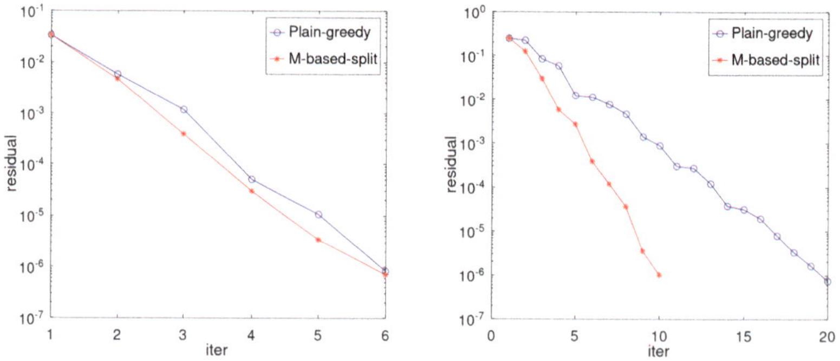

Fig. 2: **Diff-1 problem:** Comparison of the algorithms `plain-greedy` and $\mathcal{M}$-`based-split` with $\alpha = 1$ (left) and $\alpha = 0.105$ (right).

When $\alpha = 1$, the algorithm `plain-greedy` converges in 6 iterations while the algorithm $\mathcal{M}$-`based-split` generates a library with 11 spaces of dimension at most 6 (5 spaces of dimension 5 and 6 of dimension 6), see Figure 2-left for the evolution of the maximum residual as a function of the iteration k (at iteration k, all the RB spaces are of dimension at most k). Therefore, when coercivity is not degenerate there is no advantage of using the nonlinear reduced model. The situation is different when $\alpha \to \alpha_{\min}$ and coercivity becomes degenerate. We illustrate this by considering the manifold $\mathcal{M}$ generated when $\alpha = 0.105$. In this case, we need a space of dimension $n = 20$ to reach ε with the `plain-greedy` method while the maximum dimension of the space in the library is 10, see Figure 2-right. More precisely, the library contains 96 spaces. It is composed of $\{3, 6, 31, 37, 19\}$ spaces of dimension $\{6, 7, 8, 9, 10\}$. The spaces of the library are generated using a total of 261 snapshots. The partition of the parameter domain obtained at different steps is given in Figure 3.

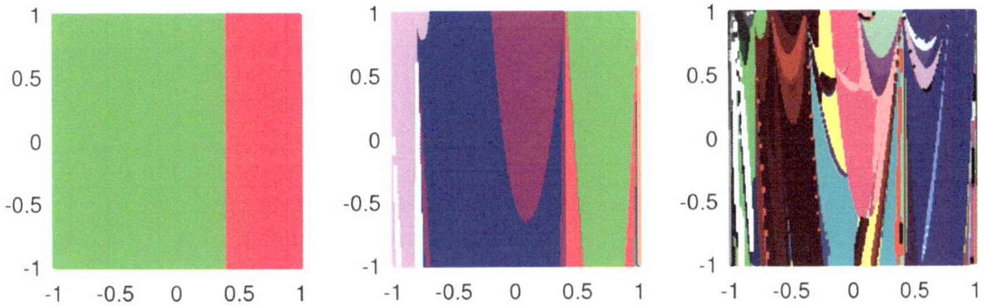

Fig. 3: **Diff-1 problem:** Partition of $Y^{(N)}$ obtained with algorithm $\mathcal{M}$-`based-split` after 2 iterations (left), 4 iterations (middle) and 10 iterations (right) in the case $\alpha = 0.105$.

For a comparison, we run the algorithm `Y-cart-split` with $n_{\max} = 10$. This algorithm produces a library with 12 spaces, see Figure 4 for the partition of the domain and the dimension of the local RB spaces. A summary of the main features of the reduced models for this problem is provided in Table 1 (see label Diff-1).

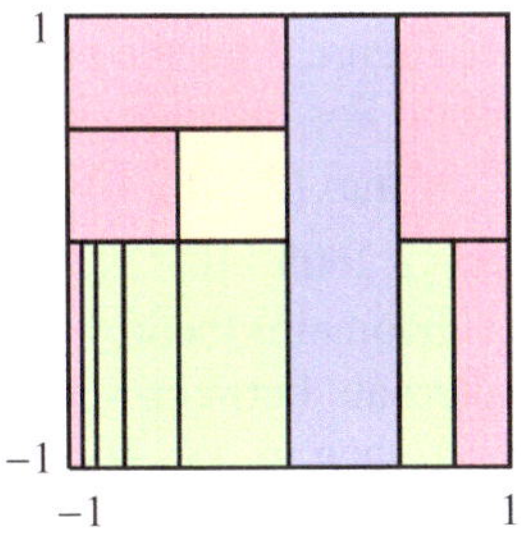

Fig. 4: **Diff-1 problem:** Final partition of $Y^{(N)}$ generated by the algorithm `Y-cart-split` with $n_{\max} = 10$ when $\alpha = 0.105$. The local spaces in the library are of dimension 6 (orange), 7 (green), 8 (blue) and 10 (red).

Test 2 (labeled as Diff-2): We next consider a different diffusion coefficient, namely a as in (33) but with

$$a_1(x) = 0.49\cos(8\pi x), \quad g_1(y_1) = ((1.4y_1)^2 - 0.8)^2 - 1 \tag{34}$$

and

$$a_2(x) = 0.49\chi_{A_1}(x), \quad g_2(y_2) = \chi_{B_1}(y_2) - \chi_{B_2}(y_2), \tag{35}$$

where χ denotes the indicator function and where $A_1 = [0, 0.2) \cup (0.4, 0.6) \cup (0.8, 1]$, $B_1 = (-0.2, 0.4)$ and $B_2 = (-0.7, -0.2) \cup (0.4, 1]$. We refer to Figure 5 for a plot of the functions $a_1(x)$, $a_2(x)$, $g_1(z)$ and $g_2(z)$ for $x \in [0, 1]$ and $z \in [-1, 1]$.

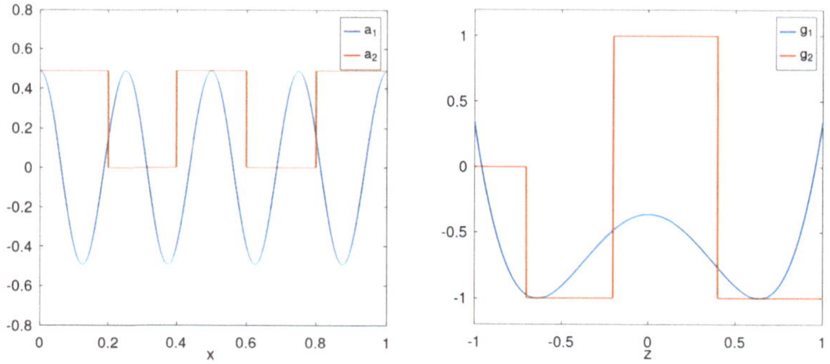

Fig. 5: **Diff-2 problem:** Plot of the functions defined (34) and (35).

In this case, the algorithm `plain-greedy` generates a space of dimension 18, while the library obtained with $\mathcal{M}$`-based-split` has 24 spaces of dimension at most 7 ($\{3, 11, 10\}$ spaces of dimension $\{5, 6, 7\}$). The spaces of the library are generated with 61 snapshots. The maximum residual at each step of the algorithm is given in Figure 6, which also contains the final partition associated to the library.

For this problem, the agorithm `Y-cart-split` with $n_{\max} = 7$ does not converge when we use (28) to determine which direction to split at each step. The direction y_2 is chosen most of the times, see Figure 7 for the partition of the parameter domain after a few steps of the algorithm. For instance, for $\mathrm{Y}^{(\alpha)} = [-1, 1] \times [0.75, 1]$, a split in y_1 gives the error $e_1^{(\alpha)} = \max\{2.2782 \cdot 10^{-4}, 9.6067 \cdot 10^{-5}\}$ while a split in y_2 gives the error $e_2^{(\alpha)} = \max\{2.2605 \cdot 10^{-4}, 2.2605 \cdot 10^{-4}\}$, and thus $p_{\text{split}} = 2$. However, the error on the two subdomains thereby generated is not decreased. If we set instead $n_{\max} = 10$ or we alternate between vertical ($p_{\text{split}} = 1$) and horizontal ($p_{\text{split}} = 2$) splits (starting with y_1, then y_2, y_1 and so on), then we obtain the results provided in Table 1 (see label Diff-2).

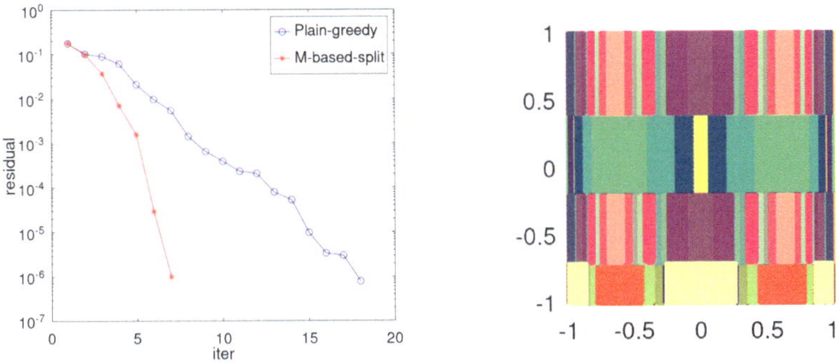

Fig. 6: **Diff-2 problem:** Comparison of the algorithms `plain-greedy` and $\mathcal{M}$`-based-split` (left) and final partition of $\mathrm{Y}^{(N)}$ generated by $\mathcal{M}$`-based-split` (right).

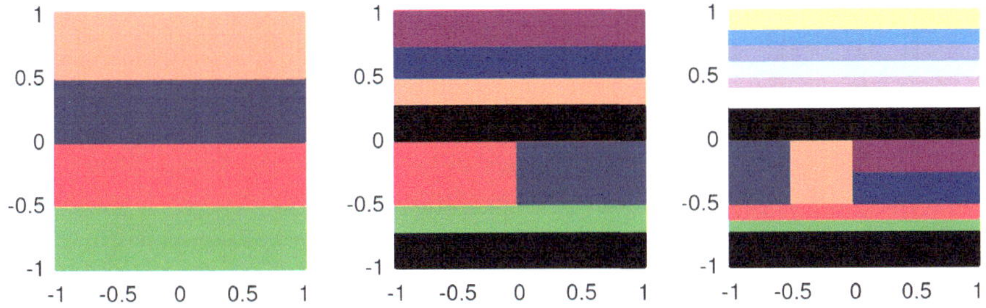

Fig. 7: **Diff-2 problem:** Partition of $\mathrm{Y}^{(N)}$ obtained with algorithm `Y-cart-split` after 3 iterations (left), 4 iterations (middle) and 5 iterations (right) when $n_{\max} = 7$. The target accuracy is reached on the subdomains in black and in white.

4.1.2 Convection-diffusion problem

Here $\mathrm{Y} = [y_{\min}, y_{\max}] = [0, 10000]$ and $u = u(x, y)$ is the (weak) solution to

$$-\frac{\partial^2 u}{\partial x^2} + y\frac{\partial u}{\partial x} = 1 \quad \text{in } \Omega, \quad u(0, y) = u(1, y) = 0. \tag{36}$$

For this problem we have $r = 1$ and $R = 1 + C_P y_{\max}$, where C_P denotes the Poincaré constant in $\|v\|_{L^2(\Omega)} \leqslant C_P \|v'\|_{L^2(\Omega)}$. Therefore, the condition number is such that $\nu = R/r \gg 1$ and we expect that the plain RB method will perform poorly (we refer to [24] for a strategy developed to keep the condition number close to 1). Here the training set $\mathrm{Y}^{(N)}$, $N = 10001$, consists of the integers from $y_{\min}$ to $y_{\max}$. We give in Figure 8 the evolution of the maximum residual at each step of the algorithm together with the location of all the selected parameters used to generate the final reduced model.

	Algorithm	$\|\mathcal{L}\|$	Dim. spaces	Nb. spaces	Nb. snap.	ε	$n_{\max}$
Diff-1 (α = 0.105)	plain-greedy	1	$\{20\}$	$\{1\}$	20	10^{-6}	∞
	$\mathcal{M}$-based-split	96	$\{6,7,8,9,10\}$	$\{3,6,31,37,19\}$	261	10^{-6}	∞
	Y-cart-split	12	$\{6,7,8,10\}$	$\{1,5,1,5\}$	99	10^{-6}	10
Diff-2	plain-greedy	1	$\{18\}$	$\{1\}$	18	10^{-6}	∞
	$\mathcal{M}$-based-split	24	$\{5,6,7\}$	$\{3,11,10\}$	61	10^{-6}	∞
	Y-cart-split	48	$\{8,9,10\}$	$\{6,36,6\}$	432	10^{-6}	10
	Y-cart-split*	314	$\{3,4,5,6,7\}$	$\{22,92,50,28,122\}$	1706	10^{-6}	7
Cv-Diff	plain-greedy	1	$\{27\}$	$\{1\}$	27	10^{-6}	∞
	$\mathcal{M}$-based-split	42	$\{5,6,7,8\}$	$\{3,11,10\}$	61	10^{-6}	∞
	Y-cart-split	11	$\{6,7\}$	$\{2,9\}$	75	10^{-6}	8
KdV	plain-greedy	1	$\{34\}$	$\{1\}$	34	10^{-3}	∞
	$\mathcal{M}$-based-split	59	$\{10,12,14,16\}$	$\{2,12,19,26\}$	208	10^{-3}	∞
	Y-cart-split	20	$\{10,13,14,15\}$	$\{1,2,10,7\}$	281	10^{-3}	15

Table 1: **Main features of the libraries obtained with the three different algorithms and for all the numerical examples**. Here, $|\mathcal{L}|$ is the cardinality of the library, 'Dim. spaces' gives the dimensions of the spaces contained in $\mathcal{L}$ and 'Nb. spaces' gives how many spaces of each dimension there are, 'Nb. snap.' is the amount of snapshots required to build all the spaces within $\mathcal{L}$, ε is the target accuracy of the algorithms, and $n_{\max}$ is the prescribed maximal dimension of the spaces. The results for Y-cart-split* correspond to the case where we alternate between vertical and horizontal splits.

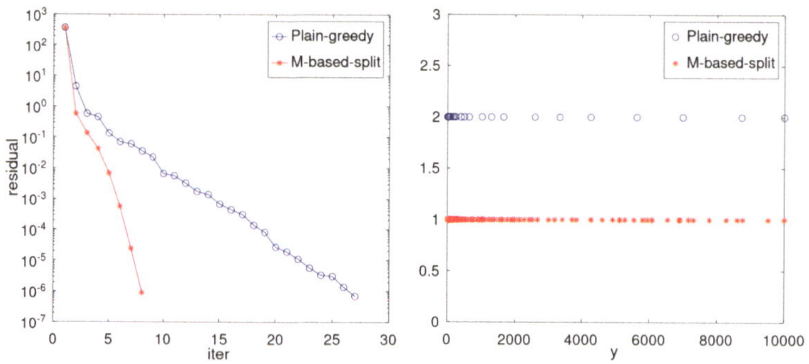

Fig. 8: **Cv-Diffproblem:** Comparison of the algorithms plain-greedy and $\mathcal{M}$-based-split.

The algorithm `plain-greedy` converges in 27 iterations, while the algorithm $\mathcal{M}$-`based-split` generates a library with 42 spaces of dimension at most 8 (4 spaces of dimension 5, 4 of dimension 6, and 5 of dimension 7) with a total of 111 snapshots. If we run the algorithm `Y-cart-split` with $n_{\max} = 8$, then the algorithm produces a library with 11 spaces, 2 of dimension 6 and 9 of dimension 7 and thus a total of 75 snapshots. Moreover, following the notation (26), the partition of the parameter domain is $\mathrm{Y} = \cup_{\alpha \in A} \mathrm{Y}^{(\alpha)}$ with

$$\mathrm{Y}^{(\alpha)} := y_{\max} \left[\frac{i}{2^k}, \frac{i+1}{2^k} \right], \quad \alpha = (k, i), \qquad A = \{(k, 1)\}_{k=1}^{10} \cup \{(10, 0)\}.$$

If we set instead $n_{\max} = 5$, then we obtain 22 spaces, all of dimension 5, with the partition corresponding to $A = \{(k, 2), (k, 3)\}_{k=2}^{10} \cup \{(11, i)\}_{i=0}^{3}$. The results obtained for this problem are summarized in Table 1 (see label Cv-Diff).

4.2 A KdV equation in $V = (\mathcal{P}_2(\Omega), W_2)$

We consider the same example from [11], which is a two-soliton solution of the Korteweg-de-Vries equation (KdV). Expressed in normalized units, the PDE reads for all $x \in \mathbb{R}$,

$$\partial_t u + 6u\partial_x u + \partial_x^3 u = 0.$$

The equation admits a general 2–soliton solution

$$u(x, t) = -2\partial_x^2 \log \det(I + A(x, t)), \tag{37}$$

where $A(x, t) \in \mathbb{R}^{2\times 2}$ is the interaction matrix whose components $a_{i,j}$ are

$$a_{i,j}(x, t) = \frac{c_i c_j}{k_i + k_j} \exp\left((k_i + k_j)x - (k_i^3 + k_j^3)t \right), \quad 1 \leqslant i, j \leqslant 2.$$

For any $t > 0$, the total mass is equal to

$$\int_{\mathbb{R}} u(t, x)\, \mathrm{d}x = 4(k_1 + k_2).$$

To illustrate the performance of our approaches, we set

$$T = 2.5 \cdot 10^{-3}, \ k_1 = 30 - k_2.$$

The parameter domain is[1]

[1] In [11], c_1 and c_2 are fixed and only two parameters (t, k_2) change. Here, we work with c_1 and c_2 as additional parameters.

$$\mathrm{Y} = \{(t, c_1, c_2, k_2) \in [0, 2.5.10^{-3}] \times [0.9, 1.1] \times [0.2, 0.4] \times [16, 22]\}$$

It follows that the total mass is equal to $m = 120$ for all the parametric solutions. We thus consider the solution manifold

$$\mathcal{M} = \{v(y)(x) = \frac{u(y)}{m}\mathrm{Leb}(\mathrm{d}x) \ : \ y \in \mathrm{Y}\} \subset \mathrm{Pr},$$

where we view the solutions as probability density functions dominated by the Lebesgue measure (denoted as $\mathrm{Leb}(\mathrm{d}x)$).

We set $\varepsilon = 10^{-3}$ as the final target accuracy, and run the 3 algorithms with the same discrete set $\mathcal{M}^{(N)}$ composed of $N = 2 \cdot 10^3$ training snapshots. The `plain-greedy` barycentric algorithm from Section (2.2.2) converges in $K = 34$ iterations, and Figure 9 records the convergence of $\mathrm{dist}(\mathcal{M}, V_n)$ as n increases. Here, $V_n = \mathrm{Bar}(\Sigma_n, \mathrm{U}_n)$ is the barycenter space obtained with that algorithm.

We next comment on the behavior of the $\mathcal{M}$-`based-split` algorithm. It converges in 6 iterations when we makes three steps in the restarted greedy algorithm. Figure 9 summarizes the approximation errors attained at each level of the tree. Figure 11 gives a similar type of information but in expanded form: it shows the intermediate accuracies that the algorithm reaches at each node. Figures 12a and 12b show the space dimension $n(\alpha) = \dim(V^{(\alpha)})$, and the number of training snapshots at each node. The output of this algorithm is a library with 59 spaces whose dimensions are $\{10, 12, 14, 16\}$ and the amount of spaces for each dimension is $\{2, 12, 19, 26\}$. The spaces are the ones that are located at the leaves of tree. The 59 spaces of the library are generated with 208 snapshots, and some of them are shared by several different spaces. The main features of the library are summarized in Table 1 (see label KdV). Figure 10 shows the approximation of a randomly selected snapshot (that was not used in the training set) with `plain-greedy` and $\mathcal{M}$-`based-split`. Visually, both approximations look identical and very close to the exact solution. However, $\mathcal{M}$-`based-split` works with a space of smaller dimension ($n = 10$ vs. $n = 34$).

To make a comparison with the Y-`cart-split` algorithm, we run it with $n_{\max} = 15$ which is the average dimension that the spaces from $\mathcal{M}$-`based-split` reach at convergence. In this case, Y-`cart-split` builds a library with 20 spaces whose dimensions are $\{10, 13, 14, 15\}$ and the amount of spaces with each dimension is $\{1, 2, 10, 7\}$. The spaces of the library are generated with 281 snapshots. Interestingly, even if this library is smaller in terms of cardinality with respect to the one built by $\mathcal{M}$-`based-split`, it requires more snapshots to be created. The detailed behavior of the Y-`cart-split` algorithm is summarized in Figures 13, 14a and 14b.

4.3 Main conclusion of the numerical tests

The summary of Table 1 illustrates that the superiority of one algorithm over another is relative, and depends on the criterion that is used. As predictited by the theory, the `plain-greedy` algorithm presents a fast decay for elliptic problems, and it is the

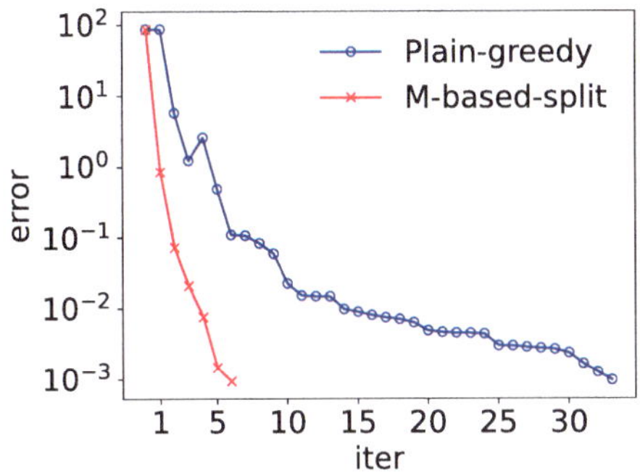

Fig. 9: **KdV problem:** Comparison of the algorithms `plain-greedy` and $\mathcal{M}$-`based-split`.

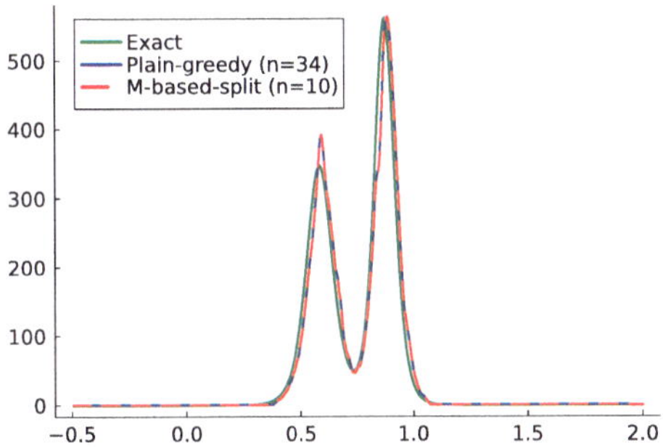

Fig. 10: **KdV problem:** Approximation of $u(y)$ for $y = (2.23.10^{-3}, 0.901, 0.386, 13.1)$ with `plain-greedy` and $\mathcal{M}$-`based-split` once the stopping criterion as been reached. Visually, both approximations look identical and very close to the exact solution. However, $\mathcal{M}$-`based-split` works with a space of smaller dimension ($n = 10$ vs. $n = 34$).

method of choice when coercivity is not degenerate (see results for Diff-1). As soon as one deviates from this setting, the other methods start becoming competitive. If one needs to work with local spaces of small dimension to reduce the online cost, `Y-cart-split` seems to be a good choice in most cases. This comes at the cost of having to store more snapshots to define the library (see the examples Diff-2, Cv-Diff and KdV). Thanks to its tree structure (namely one snaphots per node), the library generated by $\mathcal{M}$-`based-split` requires less snapshots to generate the spaces but their dimension is often slightly larger than the ones obtained with `Y-cart-split` in our examples. As already brought up, $\mathcal{M}$-`based-split` also comes with the advantage of being able to deal with very general parameter domains Y but the mapping ι of (30) is not trivial to deal with in practice.

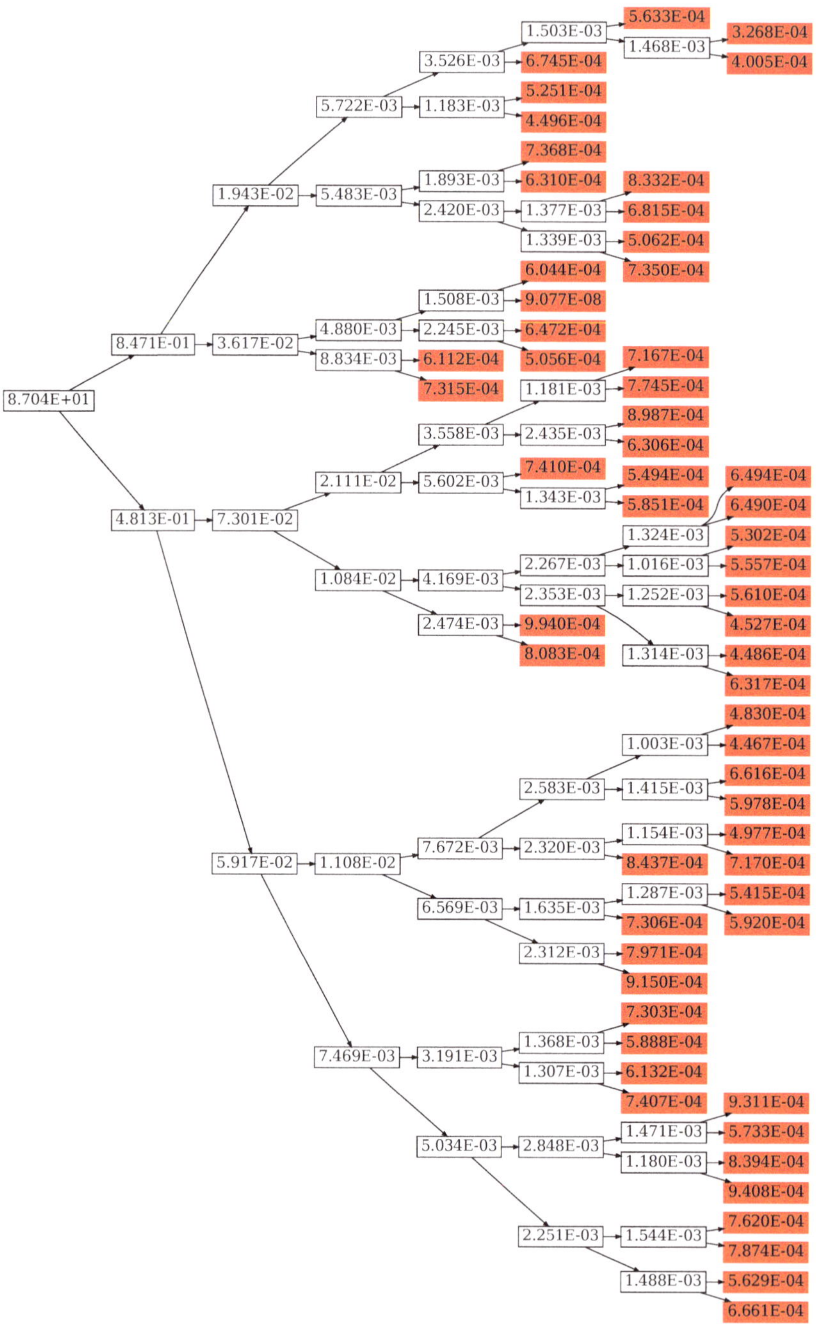

Fig. 11: **KdV problem:** Algorithm $\mathcal{M}$-`based-split`: accuracy $\mathrm{dist}(\mathcal{M}^{(\alpha)}, V^{(\alpha)})$ reached at each node. The spaces $V^{(\alpha)}$ that generate the library are located at the leaves of the tree (which are colored in orange).

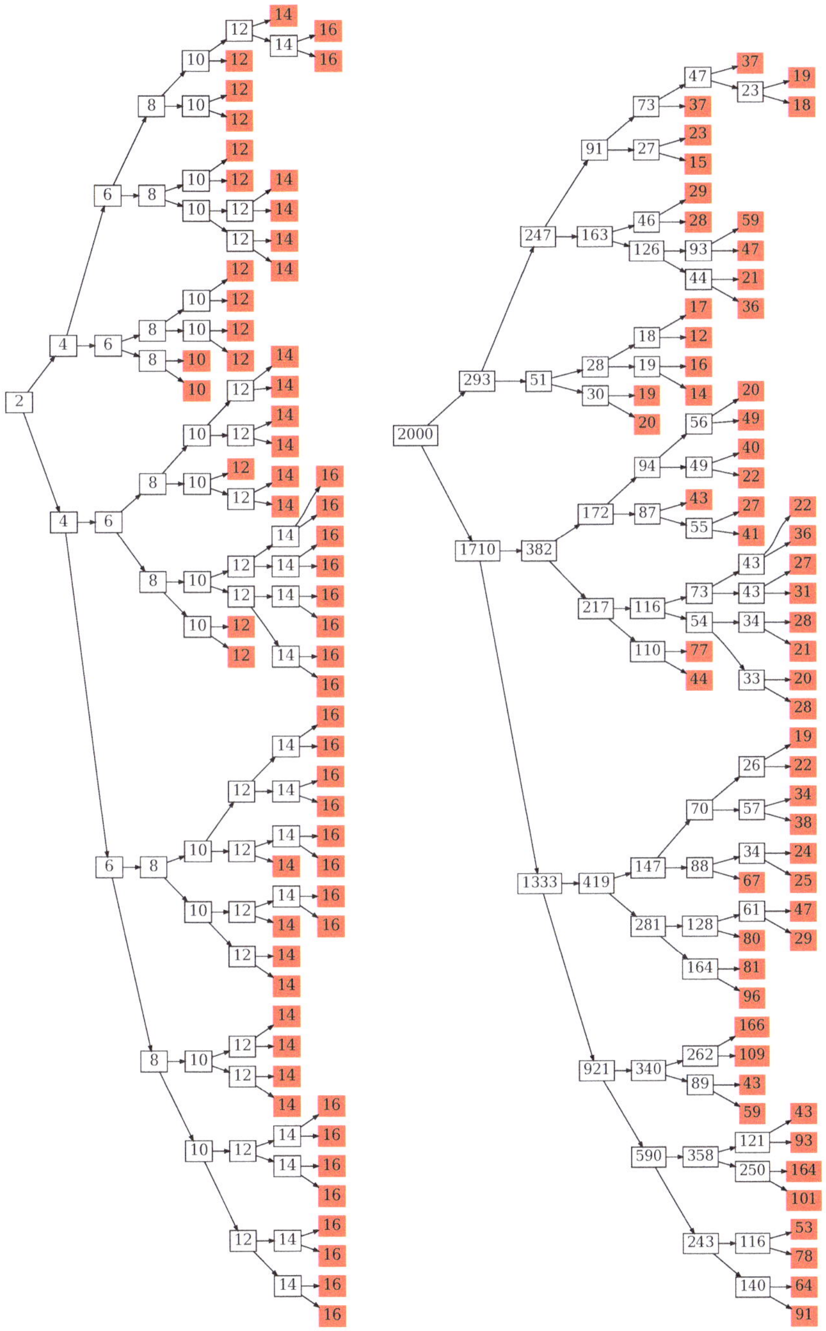

(a) Dimension $n(\alpha) = \dim(V^{(\alpha)})$ for each node.

(b) Number of training snapshots for each node.

Fig. 12: **KdV problem:** Algorithm $\mathcal{M}$-based-split: dimension and number of training snapshots in each tree node (leaves are colored in orange).

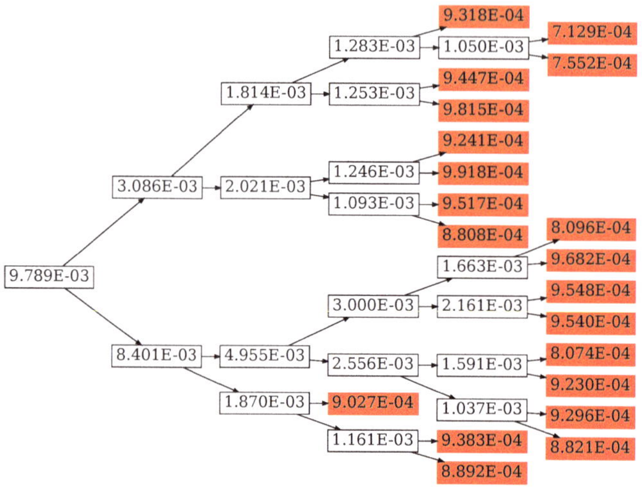

Fig. 13: **KdV problem:** Algorithm `Y-cart-split`: accuracy dist$(\mathcal{M}^{(\alpha)}, V^{(\alpha)})$ for each node (leaves are colored in orange).

5 Conclusion

The two different tree-based library approximations that we have explored in this work are very general, and can be applied to Hilbert, Banach and Wasserstein spaces. They can be formulated with minor modifications in general metric spaces. The most novel approach, called $\mathcal{M}$`-based-split` in the paper, appears to be a good trade-off between the amount of snapshots that are needed to define the spaces of the library, and their respective dimensions. More broadly, the numerical experiments show a mixed landscape where each algorithm has relative advantages with respect to the others. A theoretical study on the convergence rate of both `Y-cart-split` and $\mathcal{M}$`-based-split` is not entirely trivial and left for future work. Another element left for future work is the practical implementation of the mapping ι of (30) for $\mathcal{M}$`-based-split`.

Acknowledgments and disclosure of funding

The authors would like to thank Andrea Bonito, Albert Cohen, Ron DeVore, Peter Jantsch and Guergana Petrova for fruitful discussions.
This research is part of the programme DesCartes and is supported by the National Research Foundation, Prime Minister's Office, Singapore under its Campus for Re-

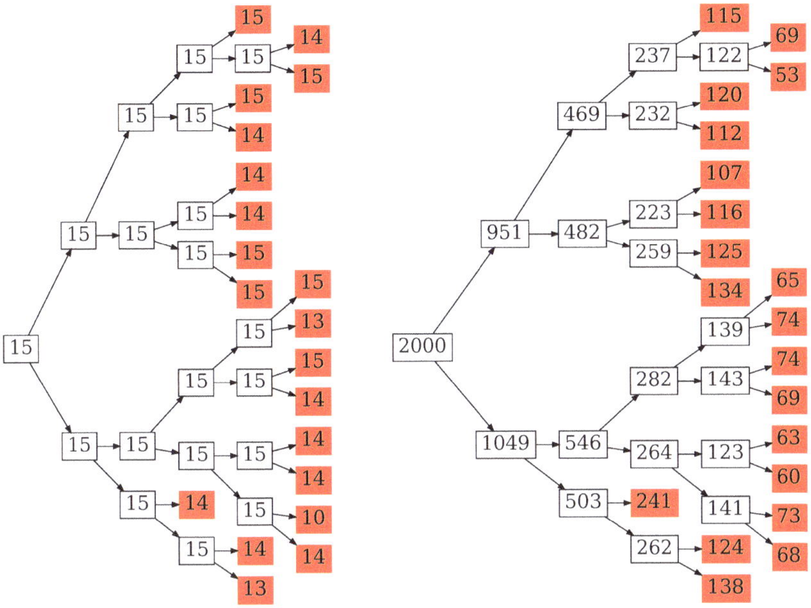

(a) Dimension $n(\alpha) = \dim(V^{(\alpha)})$ for each node.

(b) Number of training snapshots for each node.

Fig. 14: **KdV problem:** Algorithm `Y-cart-split`: dimension and number of training snapshots in each tree node (leaves are colored in orange).

search Excellence and Technological Enterprise (CREATE) programme. It is also supported by the NSERC Grant RGPIN-2021-04311.

References

1. A. Cohen and R. DeVore. Kolmogorov widths under holomorphic mappings. *IMA Journal of Numerical Analysis*, 36(1):1–12, 2016.
2. P. Benner, A. Cohen, M. Ohlberger, and K. Willcox. *Model Reduction and Approximation: Theory and Algorithms*, volume 15. SIAM, 2017.
3. J.L. Eftang, A.T. Patera, and E.M. Rønquist. An “hp” certified reduced basis method for parametrized elliptic partial differential equations. *SIAM Journal on Scientific Computing*, 32(6):3170–3200, 2010.
4. Y. Maday and B. Stamm. Locally adaptive greedy approximations for anisotropic parameter reduced basis spaces. *SIAM J. Scientific Computing*, 35(6):A2417–A2441, 2013.
5. A. Bonito, A. Cohen, R. DeVore, D. Guignard, P. Jantsch, and G. Petrova. Nonlinear methods for model reduction. *ESAIM: Mathematical Modelling and Numerical Analysis*, 55(2):507–531, 2021.

6. F. Galarce, J.F. Gerbeau, D. Lombardi, and O. Mula. Fast reconstruction of 3D blood flows from doppler ultrasound images and reduced models. *Computer Methods in Applied Mechanics and Engineering*, 375:113559, 2021.
7. A. Cohen, W. Dahmen, O. Mula, and J. Nichols. Nonlinear reduced models for state and parameter estimation. *SIAM/ASA Journal on Uncertainty Quantification*, 10(1):227–267, 2022.
8. T. Daniel, F. Casenave, N. Akkari, A. Ketata, and D. Ryckelynck. Physics-informed cluster analysis and a priori efficiency criterion for the construction of local reduced-order bases. *Journal of Computational Physics*, 458:111120, 2022.
9. A. Nouy and A. Pasco. Dictionary-based model reduction for state estimation. *Submitted*, 2023.
10. D. Guignard and P. Jantsch. Nonlinear approximation of high-dimensional anisotropic analytic functions. *Journal of Approximation Theory*, page 106040, 2024.
11. V. Ehrlacher, D. Lombardi, O. Mula, and F.-X. Vialard. Nonlinear model reduction on metric spaces. application to one-dimensional conservative PDEs in Wasserstein spaces. *ESAIM M2AN*, 54(6):2159–2197, 2020.
12. B. Battisti, T. Blickhan, G. Enchery, V. Ehrlacher, D. Lombardi, and O. Mula. Wasserstein model reduction approach for parametrized flow problems in porous media. *ESAIM: Proceedings and Surveys*, 73:28–47, 2023.
13. M. Dalery, G. Dusson, V. Ehrlacher, and A. Lozinski. Nonlinear reduced basis using mixture Wasserstein barycenters: application to an eigenvalue problem inspired from quantum chemistry. *Submitted*, 2023.
14. T. Blickhan. A registration method for reduced basis problems using linear optimal transport. *Submitted*, 2023.
15. M.-H. Do, J. Feydy, and O. Mula. Approximation and structured prediction with sparse Wasserstein barycenters. *arXiv preprint arXiv:2302.05356*, 2023.
16. P. Binev, A. Cohen, W. Dahmen, R. DeVore, G. Petrova, and P. Wojtaszczyk. Convergence rates for greedy algorithms in reduced basis methods. *SIAM Journal on Mathematical Analysis*, 43(3):1457–1472, 2011.
17. R. DeVore, G. Petrova, and P. Wojtaszczyk. Greedy algorithms for reduced bases in Banach spaces. *Constructive Approximation*, 37(3):455–466, 2013.
18. A. Cohen, W. Dahmen, R. DeVore, and J. Nichols. Reduced basis greedy selection using random training sets. *ESAIM: Mathematical Modelling and Numerical Analysis*, 54(5):1509–1524, 2020.
19. A. Quarteroni, A. Manzoni, and F. Negri. *Reduced Basis Methods for Partial Differential Equations*. Springer, 2016.
20. J.S. Hesthaven, G. Rozza, and B. Stamm. *Certified Reduced Basis Methods for Parametrized Partial Differential Equations*. SpringerBriefs in Mathematics. Springer, 2016.
21. C. Villani. *Topics in optimal transportation*. Number 58. American Mathematical Soc., 2003.
22. M. Agueh and G. Carlier. Barycenters in the Wasserstein space. *SIAM Journal on Mathematical Analysis*, 43(2):904–924, 2011.
23. G. Peyré and M. Cuturi. Computational optimal transport: With applications to data science. *Foundations and Trends® in Machine Learning*, 11(5-6):355–607, 2019.
24. W. Dahmen, , C. Plesken, and G. Welper. Double greedy algorithms: Reduced basis methods for transport dominated problems. *ESAIM: Mathematical Modelling and Numerical Analysis*, 48(3):623–663, 2014.

Samplets: Wavelet Concepts for Scattered Data

Helmut Harbrecht and Michael Multerer

Abstract This chapter is dedicated to recent developments in the field of wavelet analysis for scattered data. We introduce the concept of samplets, which are signed measures of wavelet type and may be defined on sets of arbitrarily distributed data sites in possibly high dimension. By employing samplets, we transfer well-known concepts known from wavelet analysis, namely the fast basis transform, data compression, operator compression and operator arithmetics to scattered data problems. Especially, samplet matrix compression facilitates the rapid solution of scattered data interpolation problems, even for kernel functions with nonlocal support. Finally, we demonstrate that sparsity constraints for scattered data approximation problems become meaningful and can efficiently be solved in samplet coordinates.

1 Introduction

Multiresolution methods and especially wavelet techniques have a long standing tradition and are a versatile tool in various fields. Applications comprise, among others, nonlinear approximation, image analysis, signal processing and machine learning, see for instance [16, 20, 25, 27, 55, 56] and the references therein. Starting from a signal, the pivotal idea of wavelet techniques is the splitting of this signal into its contributions relative to a hierarchy of scales. Such a *multiresolution analysis* starts from an approximation on a coarse scale and successively resolves details, that have not been captured so far, at finer scales. Wavelet techniques naturally accommodate data compression and adaptivity. The great success of wavelet techniques has specif-

Helmut Harbrecht
Departement Mathematik und Informatik, Universität Basel, Spiegelgasse 1, 4051 Basel, Schweiz, e-mail: `helmut.harbrecht@unibas.ch`

Michael Multerer
Istituto Eulero, Università della Svizzera italiana, Via la Santa 1, 6962 Lugano, Svizzera, e-mail: `michael.multerer@usi.ch`

R. DeVore, A. Kunoth (eds.), *Multiscale, Nonlinear and Adaptive Approximation II*,
https://doi.org/10.1007/978-3-031-75802-7_14

ically been triggered by the fast wavelet transform, which transforms a signal into its wavelet representation and back, that can be computed with linear cost in terms of the size of the wavelet basis, see [18, 54].

The original construction of wavelets is based on dilations and translations of a given mother wavelet. This way, a nested sequence of approximation spaces is obtained and the elements of this sequence are scaled copies of each other. This construction of wavelets is limited to structured data, such as uniform subdivisions of the real line. Adaptions to deal with bounded intervals have been suggested in e.g. [2, 17, 22], while wavelet constructions on manifolds are the topic of e.g. [24, 44, 75]. An extension to surface triangulations, has been suggested in [71], where (multi-)wavelets are constructed as linear combinations of functions at a fixed fine scale. The stability of the resulting *Tausch-White wavelet basis* follows from its orthonormality. Another approach to construct a multiresolution analysis on unstructured data, for example on graphs, are *diffusion wavelets*, see [19]. However, there is no linear cost bound for the computation of a diffusion wavelet basis.

This chapter is a survey on the generalization of the Tausch-White wavelet construction towards general scattered data. This is achieve by abstracting the construction [1, 71] towards discrete signed measures. The result is multiresolution analysis of discrete signed measures of localized support that we call *samplets*. Samplets are tailored to the underlying data sites and may be computed such that the exhibit *vanishing moments*, that is their associated measure integrals. Lowest order samplets, which resemble Haar wavelets on scattered data have been considered in the past for data compression in [62]. It is worth mentioning that the construction of samplets is not limited to the use of polynomial vanishing moments Indeed, it is easily be possible to adapt the presented concepts to primitives with different desired properties.

We present a general construction template for samplets with an arbitrary number of vanishing moments. This construction can always be performed with linear cost for a balanced cluster tree for the data sites, even for non-quasi-uniform data sites. The resulting basis is always orthonormal and hence stable. Due to the vanishing moments, the coefficients in the representation of scattered data with respect to samplet coordinates decay fast whenever the data values resemble a smooth function evaluated at the data sites. This straightforwardly enables data compression. In contrast, non-smooth regions in the data are indicated by large samplet coefficients. This, in turn, enables feature detection and extraction.

Similar to wavelet matrix compression [6, 21, 23, 67, 74], samplets are applicable for the compression of kernel matrices. Such matrices arise in the scattered data approximation context, compare [30, 46, 66, 76, 77, 78]. Kernel matrices are typically densely populated since the underlying reproducing kernels are nonlocal. Nonetheless, the kernels are typically *asymptotically smooth*, meaning that they behave like smooth functions apart from the diagonal. Hence, the discretization of such kernel matrices by samplets with vanishing moments results in quasi-sparse kernel matrices. They can be compressed such that only a sparse matrix remains. The resulting compression pattern has been derived in in [41]. Furthermore, as shown in [43], a respective matrix algebra can be defined.

Samplets provide a meaningful interpretation of *sparsity constraints* for scattered data, since the representation of data itself becomes sparse. Such constraints have a wide applicability in machine learning, statistics, as well as in signal processing. Examples for the latter are deblurring, feature selection and compressive sensing, see [11, 14, 29, 49, 70]. In practice, sparsity constraints are imposed by adding an ℓ^1-regularization term to the objective function. Dealing with this regularization in an efficient way is especially mandatory for *basis pursuit*, that is, for decomposing given data into an optimal superposition of dictionary elements, where optimal means having the smallest ℓ^1-norm of coefficients among all such decompositions, see, for example, [15, 57, 73]. We demonstrate that the basis pursuit problem can efficiently be solved within a the samplet basis and reconstruct scattered data using a dictionary of multiple kernels.

The rest of this chapter is organized as follows. In Section 2, multiresolution analyses for scattered data and the concept of samplets are introduced. The change of basis by means of the fast samplet transform and resulting applications are the topic of Section 3. Section 4 deals with scattered data approximation in reproducing kernel Hilbert spaces. Here, we especially introduce the samplet compression of kernel matrices and the efficient treatment of sparsity constraints by the semi-smooth Newton method.

Throughout this chapter, in order to avoid the repeated use of generic but unspecified constants, by $C \lesssim D$ we indicate that C can be bounded by a multiple of D, independently of parameters which C and D may depend on. Moreover, $C \gtrsim D$ is defined as $D \lesssim C$ and $C \sim D$ as $C \lesssim D$ and $D \lesssim C$.

2 Multiresolution analysis of scattered data

2.1 Samplet bases

Let $X := \{\boldsymbol{x}_1, \ldots, \boldsymbol{x}_N\} \subset \Omega$ denote a set of *data sites* within some bounded or unbounded region $\Omega \subset \mathbb{R}^d$. Associated to these data sites, we introduce the Dirac-δ-distributions

$$\delta_{\boldsymbol{x}_1}, \ldots, \delta_{\boldsymbol{x}_N} \in [C(\Omega)]', \tag{1}$$

where we endow $C(\Omega)$ with the sup-norm. The Dirac-δ-distributions satisfy

$$(f, \delta_{\boldsymbol{x}_i})_\Omega := \delta_{\boldsymbol{x}_i}(f) = f(\boldsymbol{x}_i) \text{ for all } f \in C(\Omega)$$

and serve as *information functionals* of some, maybe unknown, function $f \in C(\Omega)$. In this setting, the *data values* $f_i := (f, \delta_{\boldsymbol{x}_i})_\Omega$, $i = 1, \ldots, N$, amount to the available information of the function f, compare [59].

By the Riesz representation theorem, the dual space $[C(\Omega)]'$ is isometrically isomorphic to the space of regular, finitely additive signed measures of bounded variation on Ω, equipped with the total variation norm. We this space by $\mathrm{rba}(\Omega)$, as usual. For each $\ell \in [C(\Omega)]'$, there exists a signed measure $\mu_\ell \in \mathrm{rba}(\Omega)$ such that

$$\ell_\mu(f) = \int_\Omega f(\boldsymbol{x})\,\mathrm{d}\mu(\boldsymbol{x}) \text{ for all } f \in C(\Omega),$$

and vice versa, see [61]. In view of this identification, we interchangeably consider the Dirac δ-distributions from (1) as linear functionals or as measures.

The Dirac-δ-distributions span the linear subspace

$$\mathcal{X} := \mathrm{span}\{\delta_{\boldsymbol{x}_1}, \dots, \delta_{\boldsymbol{x}_N}\} \subset [C(\Omega)]'$$

containing all discrete and finite signed measures supported at the data sites in X. Endowed with the bilinear form $\langle\cdot,\cdot\rangle_{\mathcal{X}} : \mathcal{X} \times \mathcal{X} \to \mathbb{R}$ with $\langle \delta_{\boldsymbol{x}_i}, \delta_{\boldsymbol{x}_j}\rangle_{\mathcal{X}} = m_{i,j}$, where $\boldsymbol{M} = [m_{i,j}]_{i,j=1}^N \in \mathbb{R}^{N\times N}$ is a symmetric and positive definite matrix, the space $\mathcal{X}$ becomes a Hilbert space. For arbitrary $u', v' \in \mathcal{X}$, the inner product is given by

$$\langle u', v'\rangle_{\mathcal{X}} = \sum_{i,j=1}^N m_{i,j} u_i v_j, \text{ where } u' = \sum_{i=1}^N u_i \delta_{\boldsymbol{x}_i},\ v' = \sum_{i=1}^N v_i \delta_{\boldsymbol{x}_i}.$$

The canonical choice for $\boldsymbol{M}$ is the identity matrix, which renders the space $\mathcal{X}$ is isometrically isomorphic to $\mathbb{R}^N$ with the Euclidean inner product.

Given a *multiresolution analysis*

$$\mathcal{X}_0 \subset \mathcal{X}_1 \subset \cdots \subset \mathcal{X}_J = \mathcal{X},$$

we keep track of the increment of information between two consecutive levels j and $j+1$. Since there holds $\mathcal{X}_j \subset \mathcal{X}_{j+1}$, we may (orthogonally) decompose

$$\mathcal{X}_{j+1} = \mathcal{X}_j \overset{\perp}{\oplus} \mathcal{S}_j$$

for a certain *detail space* $\mathcal{S}_j$. In analogy to wavelet nomenclature, we call the elements of a basis of $\mathcal{X}_0$ *scaling distributions* and the elements of a basis of one of the spaces $\mathcal{S}_j$ *samplets*. This name is motivated by the idea that the basis distributions in $\mathcal{S}_j$ are supported at a small subsample or samplet of the data sites in X. The collection of the bases of $\mathcal{S}_j$ for $j = 0, \dots, J-1$ together with a basis of $\mathcal{X}_0$ is called a *samplet basis* for $\mathcal{X}$.

To enable data compression, we assume that the samplets have *vanishing moments* of order $q+1$, viz.,

$$(p, \sigma_{j,k})_\Omega = 0 \text{ for all } p \in \mathcal{P}_q, \tag{2}$$

where $\mathcal{P}_q := \mathrm{span}\{\boldsymbol{x}^{\boldsymbol{\alpha}} : \|\boldsymbol{\alpha}\|_1 \leqslant q\}$ denotes the space of all polynomials with total degree at most q. In view of larger spatial dimension d, we observe that the dimension of $\mathcal{P}_q$ is

$$m_q := \dim(\mathcal{P}_q) = \binom{q+d}{d} = \frac{(d+q)\cdots(d+1)}{q!} = \mathcal{O}(d^q). \tag{3}$$

Hence, the number of vanishing moment conditions only grows polynomially with respect to d.

Following the discrete signed measure interpretation of samplets, equation (2) implies that a samplet amounts to a quadrature formula, which annihilates polynomials up to order $q + 1$.

2.2 Cluster tree

To construct an orthonormal samplet basis with vanishing moments, see (2), we adapt the wavelet construction from [71] to the scattered data framework. The first step is to construct a hierarchy of subspaces of signed measures. To this end, we perform a hierarchical clustering of the set X with respect to the Euclidean distance. This approach amounts to a clustering of the Dirac-δ-distributions spanning $\mathcal{X}$ by means of the Wasserstein distance $W_p(\delta_{\boldsymbol{x}_i}, \delta_{\boldsymbol{x}_j}) = \|\boldsymbol{x}_i - \boldsymbol{x}_j\|_2$ for any $p \geqslant 1$.

Definition 2.1 Let $\mathcal{T} = (V, E)$ be a tree with vertices V and edges E. For $\tau \in V$, we denote the set of *children* of τ by $\operatorname{child}(\tau)$. The set of *leaves* is defined as $\mathcal{L}(\mathcal{T}) := \{\tau \in V : \operatorname{child}(\tau) = \emptyset\}$.

The tree $\mathcal{T}$ is a *cluster tree* for the set X, iff X is the root of $\mathcal{T}$ and all $\tau \in V \setminus \mathcal{L}(\mathcal{T})$ are disjoint unions of their children, viz., $\tau = \bigcup_{\tau' \in \operatorname{child}(\tau)} \tau'$. The *level* $\operatorname{level}(\tau)$ of $\tau \in \mathcal{T}$ is its distance from the root. Finally, the *depth* J of $\mathcal{T}$ is the maximum level of all clusters.

Remark 2.1 There exist different possibilities to construct a cluster tree for the set X. We focus on binary trees and remark that other options, such as 2^d-trees, may be considered with the obvious modifications.

Definition 2.1 provides a hierarchical structure for the set X. Even so, it does not provide guarantees for the cardinalities of the clusters. To account for this issue, we introduce the concept of a *balanced binary tree*.

Definition 2.2 Let $\mathcal{T}$ be a cluster tree for X with depth J. We call $\mathcal{T}$ a *balanced binary tree*, iff all clusters τ satisfy the following two conditions:

(*i*) The cluster τ has exactly two children if $\operatorname{level}(\tau) < J$. It has no children if $\operatorname{level}(\tau) = J$.

(*ii*) There holds $|\tau| \sim 2^{J - \operatorname{level}(\tau)}$, where $|\tau|$ denotes the number of points contained in τ.

A balanced binary tree can be constructed by *cardinality balanced clustering*. This means that the root cluster X is split into two child clusters of identical (or similar) cardinality. This process is repeated recursively for the resulting child clusters until their cardinality falls below a given threshold. For the subdivision, the cluster's bounding box is split along its longest edge such that the resulting two boxes both contain an equal number of points. Thus, as the cluster cardinality halves with each level, we obtain $O(\log N)$ levels in total. The cost per level is $O(N)$ if the subdivision is performed with respect to the median, compare [8]. The overall cost for constructing the cluster tree is therefore $O(N \log N)$.

We remark that a balanced tree is only required to guarantee the cost bounds for the presented algorithms. The error and compression estimates presented later on are robust in the sense that they are formulated directly in terms of the actual cluster sizes rather than the introduced cluster level.

2.3 Construction of the samplets

Given a cluster tree, we start by introducing a *two-scale transform* between basis elements on a cluster τ of level j. To this end, we represent scaling distributions $\boldsymbol{\Phi}_j^\tau = \{\varphi_{j,k}^\tau\}$ and samplets $\boldsymbol{\Sigma}_j^\tau = \{\sigma_{j,k}^\tau\}$ recursively as linear combinations of the scaling distributions $\boldsymbol{\Phi}_{j+1}^\tau$ of τ's child clusters. This amounts to the *refinement relations*

$$\varphi_{j,k}^\tau = \sum_{\ell=1}^{n_{j+1}^\tau} q_{j,\Phi,\ell,k}^\tau \varphi_{j+1,\ell}^\tau \text{ and } \sigma_{j,k}^\tau = \sum_{\ell=1}^{n_{j+1}^\tau} q_{j,\Sigma,\ell,k}^\tau \varphi_{j+1,\ell}^\tau \text{ with } n_{j+1}^\tau := |\boldsymbol{\Phi}_{j+1}^\tau|,$$

which may be written in matrix notation as

$$[\boldsymbol{\Phi}_j^\tau, \boldsymbol{\Sigma}_j^\tau] = \boldsymbol{\Phi}_{j+1}^\tau \boldsymbol{Q}_j^\tau = \boldsymbol{\Phi}_{j+1}^\tau \left[\boldsymbol{Q}_{j,\Phi}^\tau, \boldsymbol{Q}_{j,\Sigma}^\tau\right]. \tag{4}$$

To obtain vanishing moments and orthonormality, the transform $\boldsymbol{Q}_j^\tau$ is derived from an orthogonal decomposition of the *moment matrix* $\boldsymbol{M}_{j+1}^\tau \in \mathbb{R}^{m_q \times n_{j+1}^\tau}$, given by

$$\boldsymbol{M}_{j+1}^\tau := \begin{bmatrix} (\boldsymbol{x}^{\boldsymbol{0}}, \varphi_{j+1,1})_\Omega & \cdots & (\boldsymbol{x}^{\boldsymbol{0}}, \varphi_{j+1,n_{j+1}^\tau})_\Omega \\ \vdots & & \vdots \\ (\boldsymbol{x}^{\boldsymbol{\alpha}}, \varphi_{j+1,1})_\Omega & \cdots & (\boldsymbol{x}^{\boldsymbol{\alpha}}, \varphi_{j+1,n_{j+1}^\tau})_\Omega \end{bmatrix} = [(\boldsymbol{x}^{\boldsymbol{\alpha}}, \boldsymbol{\Phi}_{j+1}^\tau)_\Omega]_{|\boldsymbol{\alpha}|\leqslant q}.$$

Herein, m_q is given as in (3) and denotes the dimension of the space $\mathcal{P}_q(\Omega)$ of total degree polynomials.

In the original construction by Tausch and White, the matrix $\boldsymbol{Q}_j^\tau$ is obtained from the singular value decomposition of $\boldsymbol{M}_{j+1}^\tau$. For the construction of samplets, we follow the idea from [1] and rather employ the QR decomposition, which results in samplets with an increasing number of vanishing moments. We compute

$$(\boldsymbol{M}_{j+1}^\tau)^\top = \boldsymbol{Q}_j^\tau \boldsymbol{R} =: \left[\boldsymbol{Q}_{j,\Phi}^\tau, \boldsymbol{Q}_{j,\Sigma}^\tau\right] \boldsymbol{R} \tag{5}$$

The moment matrix for the cluster's scaling distributions and samplets is now given by

$$\begin{aligned}[\boldsymbol{M}_{j,\Phi}^{\tau}, \boldsymbol{M}_{j,\Sigma}^{\tau}] &= \Big[(\boldsymbol{x}^{\alpha}, [\boldsymbol{\Phi}_j^{\tau}, \boldsymbol{\Sigma}_j^{\tau}])_{\Omega}\Big]_{|\alpha|\leqslant q} \\ &= \Big[(\boldsymbol{x}^{\alpha}, \boldsymbol{\Phi}_{j+1}^{\tau}[\boldsymbol{Q}_{j,\Phi}^{\tau}, \boldsymbol{Q}_{j,\Sigma}^{\tau}])_{\Omega}\Big]_{|\alpha|\leqslant q} \\ &= \boldsymbol{M}_{j+1}^{\tau}[\boldsymbol{Q}_{j,\Phi}^{\tau}, \boldsymbol{Q}_{j,\Sigma}^{\tau}] \\ &= \boldsymbol{R}^{\top}.\end{aligned}$$

Since $\boldsymbol{R}^{\top}$ is a lower triangular matrix, the first $k-1$ entries in its k-th column are zero. This amounts to $k-1$ vanishing moments for the k-th distribution generated by the orthogonal transform $\boldsymbol{Q}_j^{\tau} = [\boldsymbol{Q}_{j,\Phi}^{\tau}, \boldsymbol{Q}_{j,\Sigma}^{\tau}]$. Defining the first m_q distributions as scaling distributions and the remaining ones as samplets, we obtain samplets with vanishing moments at least of order $q+1$.

The orthogonality of basis elements of two different clusters results from the non-overlapping supports, while orthogonality within a given branch is a consequence of the orthogonality of the transforms. A visualization of a scaling distribution and samplets on levels $j=1$ and $j=2$ with $q+1=3$ vanishing moments on $N=200$ uniformly distributed data sites is shown in Figure 1.

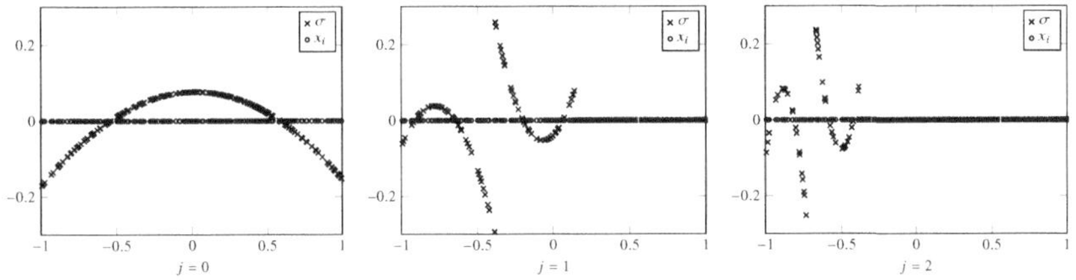

Fig. 1: Scaling distribution (left), samplet on level $j=1$ (middle) and samplet on level $j=2$ (right) for $N=200$ uniformly distributed data sites and $q+1=3$.

If we choose a minimum leaf size $|\tau| \geqslant m_{\hat{q}} \geqslant 2m_q$ for the cluster tree for some polynomial degree $\hat{q} > q$, we can even construct samplets whose number of vanishing moments increase successively from order $q+1$ up to order $\hat{q}+1$ without additional cost. This is advantageous since more vanishing moments typically improve the a-posteriori compression ratios of a given signal. Moreover, the preceding algebraic construction (5) of vanishing moments can easily be adapted to other primitives, for example, anisotropic total degree polynomial spaces. This particularly makes sense in the higher dimensional setting, when dimension weights are available.

Remark 2.2 Each cluster has at most a constant number of scaling distributions and samplets. For leaf clusters, this number is bounded by the leaf size. For non-leaf clusters, it is bounded by the number of scaling distributions from its child clusters. As there are two child clusters with a maximum of m_q scaling distributions each, we obtain the bound $2m_q$ for non-leaf clusters. If the cluster tree is balanced, resulting in a leaf size of $O(1)$, the above construction of the samplet basis has hence cost $O(N)$.

For leaf clusters, we define the scaling distributions by the Dirac measures supported at the points $\boldsymbol{x}_i$, viz., $\boldsymbol{\Phi}^\tau := \{\delta_{\boldsymbol{x}_i} : \boldsymbol{x}_i \in \tau, \tau \in \mathcal{L}(\mathcal{T})\}$. The scaling distributions of all clusters on a specific level j then generate the spaces

$$\mathcal{X}_j := \operatorname{span}\{\varphi_{j,k}^\tau : k \in I_j^{\Phi,\tau},\ \operatorname{level}(\tau) = j\}. \tag{6}$$

In contrast, the samplets span the detail spaces

$$\mathcal{S}_j := \operatorname{span}\{\sigma_{j,k}^\tau : k \in I_j^{\Sigma,\tau},\ \operatorname{level}(\tau) = j\}. \tag{7}$$

Combining the scaling distributions of the root cluster with all clusters' samplets gives rise to the samplet basis

$$\boldsymbol{\Sigma} := \boldsymbol{\Phi}^X \cup \bigcup_{\tau \in \mathcal{T}} \boldsymbol{\Sigma}^\tau. \tag{8}$$

2.4 Properties of the samplets

The properties of the samplet basis constructed in the previous paragraph are summarized in the next theorem, which can be inferred by adapting the corresponding results from [39, 71].

Theorem 2.1 *The spaces $\mathcal{X}_j$ defined in equation* (6) *form a multiresolution analysis*

$$\mathcal{X}_0 \subset \mathcal{X}_1 \subset \cdots \subset \mathcal{X}_J = \mathcal{X},$$

where the respective complement spaces $\mathcal{S}_j$ from (7) *satisfy*

$$\mathcal{X}_{j+1} = \mathcal{X}_j \overset{\perp}{\oplus} \mathcal{S}_j \textit{ for all } j = 0, 1, \ldots, J-1.$$

The associated samplet basis $\boldsymbol{\Sigma}$ defined in (8) *is an orthonormal basis in $\mathcal{X}$. In particular, there holds:*

(i) *The number of all samplets on level j behaves like 2^j.*
(ii) *The samplets have $q+1$ vanishing moments.*
(iii) *Each samplet is supported in a specific cluster τ.*

The key for data compression and feature detection is the following estimate which shows that the samplet coefficients decay with respect to the samplet's support size provided that the data result from the evaluation of a smooth function on the samplet's support, compare [41].

Lemma 2.1 *Given a samplet $\sigma_{j,k}$ with support τ and let $f \in C(\Omega)$ with $f \in C^{q+1}(O)$ for any open set $O \supset \tau$. Then, there holds*

$$|(f, \sigma_{j,k})_\Omega| \leqslant \sqrt{|\tau|}\left(\frac{d}{2}\right)^{q+1} \frac{\operatorname{diam}(\tau)^{q+1}}{(q+1)!} \|f\|_{C^{q+1}(O)}.$$

Hence, in case of smooth data, the samplet coefficients are small and can be set to zero without compromising the accuracy. Vice versa, a large samplet coefficient indicates that the data are singular in the region of the samplet's support. If the set X is *quasi-uniform* in the sense that the *separation radius* $q_X := \min_{i\neq j} \|\boldsymbol{x}_i - \boldsymbol{x}_j\|_2$ and the *fill distance* $h_{X,\Omega} := \sup_{\boldsymbol{x}\in\Omega} \min_{\boldsymbol{x}_i\in X} \|\boldsymbol{x} - \boldsymbol{x}_i\|_2$ satisfy $q_X \sim h_{X,\Omega}$, then there holds $\operatorname{diam}(\tau) \sim 2^{-\operatorname{level}(\tau)/d}$. In this case, Lemma 2.1 even guarantees the exponential decay of the samplet coefficients with the level provided that the underlying signal is smooth.

3 Scattered data analysis

3.1 Fast samplet transform

To transform between the samplet basis and the basis of Dirac measures, we introduce the *fast samplet transform* and its inverse. In accordance with [71], this transform and its inverse can be performed in linear cost. This result is well-known in case of wavelets and was crucial for their rapid development.

The mapping $f \mapsto \left[(f, \delta_{\boldsymbol{x}_1})_\Omega, \ldots, (f, \delta_{\boldsymbol{x}_N})_\Omega\right]^\top$ defines an operator $I\colon C(\Omega) \to \mathbb{R}^N$, called *information operator* in the context of optimal recovery, see [59]. By employing samplets, this information is represented in a multiresolution fashion $f \mapsto \left[(f, \sigma_{j,k})_\Omega\right]_{j,k}^\top \in \mathbb{R}^N$. More precisely, letting

$$f^\Delta := If \quad \text{and} \quad f^\Sigma := \left[(f, \sigma_{j,k})_\Omega\right]^\top, \tag{9}$$

the fast samplet transform computes

$$f^\Sigma = \boldsymbol{T} f^\Delta,$$

where $\boldsymbol{T} \in \mathbb{R}^{N\times N}$ is the orthogonal matrix containing the expansion coefficients of the samplet basis.

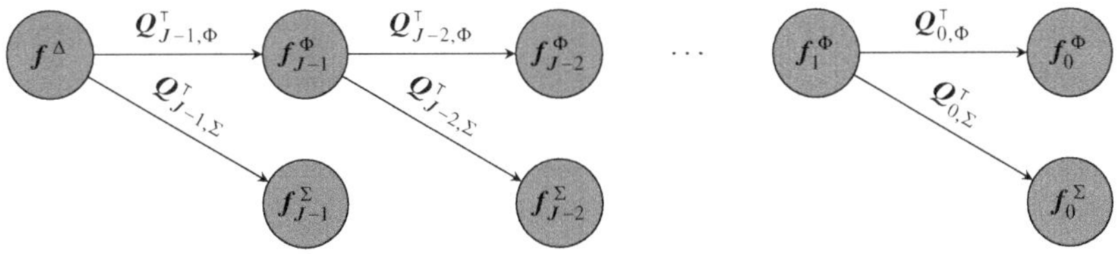

Fig. 2: Fishbone scheme of the fast samplet transform.

The actual implementation of the fast samplet transform is recursive and follows the fishbone scheme of the fast wavelet transform, see Figure 2 for a respective illustration of this scheme. The underlying recursion is based on the refinement relation (4), which translates to

$$(f, [\boldsymbol{\Phi}_j^\tau, \boldsymbol{\Sigma}_j^\tau])_\Omega = (f, \boldsymbol{\Phi}_{j+1}^\tau [\boldsymbol{Q}_{j,\Phi}^\tau, \boldsymbol{Q}_{j,\Sigma}^\tau])_\Omega = (f, \boldsymbol{\Phi}_{j+1}^\tau)_\Omega [\boldsymbol{Q}_{j,\Phi}^\tau, \boldsymbol{Q}_{j,\Sigma}^\tau]. \tag{10}$$

On the finest level, the entries of the vector $(f, \boldsymbol{\Phi}_J^\tau)_\Omega$ are exactly those of $\boldsymbol{f}^\Delta$. Recursive application of (10) therefore yields all the coefficients $(f, \boldsymbol{\Sigma}_j^\nu)_\Omega$, including $(f, \boldsymbol{\Phi}_0^X)_\Omega$, required for the representation of $\boldsymbol{f}^\Delta$ in the samplet basis.

The inverse transform is obtained in complete analogy by reversing the steps of the fast samplet transform: For each cluster, we compute

$$(f, \boldsymbol{\Phi}_{j+1}^\tau)_\Omega = (f, [\boldsymbol{\Phi}_j^\tau, \boldsymbol{\Sigma}_j^\tau])_\Omega [\boldsymbol{Q}_{j,\Phi}^\tau, \boldsymbol{Q}_{j,\Sigma}^\tau]^\intercal$$

to either obtain the coefficients of the child clusters' scaling distributions or, for leaf clusters, the coefficients $\boldsymbol{f}^\Delta$.

3.2 Data compression

In view of Lemma 2.1, samplets enable data compression by means of so-called *hard-thresholding*. Given the coefficients $\boldsymbol{f}^\Sigma$ of a transformed signal, we define $\mathrm{HT}_w(\boldsymbol{f}^\Sigma)$ as the operator which sets all entries f_i^Σ with $|f_i^\Sigma| < w$ to zero. To showcase this application of samplets, we consider the monthly ERA5 temperature data set, which provides the temperature two meters above the earth's surface. ERA5 is a reanalysis by the European Center for Medium-range Weather Forecasts (ECMWF) of global climate and weather for the past eight decades. Data is available from 1940 onwards, see [45]. We use the monthly data from 2022, which comprises, after removing `NaN` values, 1 038 240 data points per month. This results in 12 458 880 points in total. The original data format is World Geodesic System 1984 (WGS 84), which we transform by using the Robinson projection, see [64], to reduce the distortion and the projection angles. The bounding box for the data is $[-1.70 \cdot 10^7, 1.70 \cdot 10^7] \times [-8.63 \cdot 10^6, 8.63 \cdot 10^6] \times [0, 11]$. In particular, the data points are not located on a regular spatial grid and cluster towards the North and South pole. The fill distance with respect to the convex hull of the data points is $h_{X,\Omega} = 23\,700$ for each time slice, while the separation radius is $q_X = 7\,260$ for each time slice.

Figure 3 shows the reconstruction $\boldsymbol{T}^\intercal\big(\mathrm{HT}_w(\boldsymbol{f}^\Sigma)\big)$ of the hard-thresholded temperature for October 2022 with thresholds $w = 10^{-k}\|\boldsymbol{f}\|_2$ for $k = 2, 3, 4, 5$ and samplets with $q + 1 = 4$ vanishing moments. The threshold values result in 4, 87, 1 454, 15 688 non-zero coefficients, which amount to space savings of 99.9996%, 99.992%, 99.86%, and 98.49%, respectively. The associated relative errors in the Euclidean norm are 2.4%, 1.2%, 0.39%, and 0.11%.

The associated pointwise relative errors of the reconstruction are visualized in Figure 4. For the largest threshold, the maximum of the pointwise relative errors is 21%, while it is 2.2% for the smallest threshold.

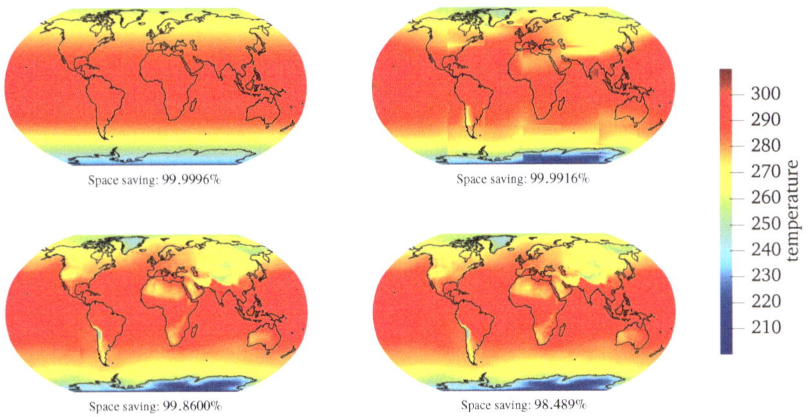

Fig. 3: Reconstruction of the temperature based on the hard-thresholded coefficient vector for the relative thresholds 10^{-k}, $k = 2, 3, 4, 5$ (from top left to bottom right). Notice that in case of $k = 2$ only the coarse level samplets remain after thresholding.

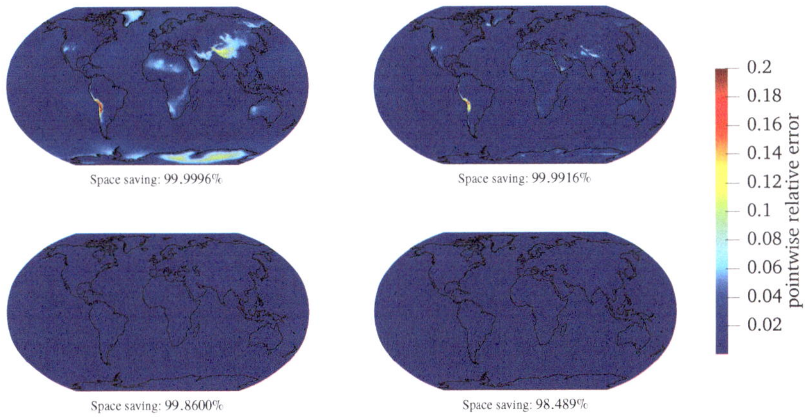

Fig. 4: Reconstruction error of the temperature with respect to the hard-thresholded coefficient vector for the relative thresholds 10^{-k}, $k = 2, 3, 4, 5$ (from top left to bottom right).

3.3 Adaptive subsampling

Consider data sites $X = \{\boldsymbol{x}_1, \ldots, \boldsymbol{x}_N\} \subset \Omega$ and corresponding data values $f^\Delta = If$. The cluster tree for X is denoted by $\mathcal{T}$ as before. We propose an adaptive subsampling strategy based on the idea of the *adaptive tree approximation* from [7]. Then, given a relevant subtree, we pursue an *entropy based approach* for sampling uniformly from the leaves of that subtree.

We start by generating a suitable subtree of $\mathcal{T}$ from the transformed data values $\boldsymbol{f}^{\Sigma} = \boldsymbol{T}\boldsymbol{f}^{\Delta}$. To this end, we apply tree-coarsening as developed in [7] and directly adapt it to our setting. The *energy* contained in a cluster $\tau \in \mathcal{T}$ is defined as the sum of energies of its children plus the squared Euclidean norm of the samplet coefficients belonging to τ, viz.,

$$e(\tau) := \sum_{\tau' \in \operatorname{child}(\tau)} e(\tau') + \sum_{\sigma \in \boldsymbol{\Sigma}^{\tau}} (f, \sigma)_{\Omega}^2. \tag{11}$$

Herein, we made the convention that $\boldsymbol{\Sigma}^X$ also contains the scaling distributions the coarsest level. The quantity $e(\tau)$ is the contribution of the subtree with root τ to the squared norm $\|\boldsymbol{f}^{\Sigma}\|_2^2$. In particular, we have $e(X) = \|\boldsymbol{f}^{\Sigma}\|_2^2$.

Based on the energies (11), we next define

$$\tilde{e}(\tau') := q(\tau) := \frac{\sum_{\mu \in \operatorname{child}(\tau)} e(\mu)}{e(\tau) + \tilde{e}(\tau)} \tilde{e}(\tau) \text{ for all } \tau' \in \operatorname{child}(\tau),$$

where we set $\tilde{e}(X) := e(X)$ for the root of the cluster tree. Given this modified energy, we perform the thresholding version of the second algorithm from [7] with threshold $w = \varepsilon^2 \|\boldsymbol{f}^{\Sigma}\|_2^2$. This results in a subtree$\mathcal{T}_w$ that approximates $\boldsymbol{f}^{\Sigma}$ up to a relative error of ε in the Euclidean norm. Since the algorithm always selects either none or all children of a given cluster, $\mathcal{T}_w$ is a cluster tree and its leaves $\mathcal{L}(\mathcal{T}_w)$ form a partition of X.

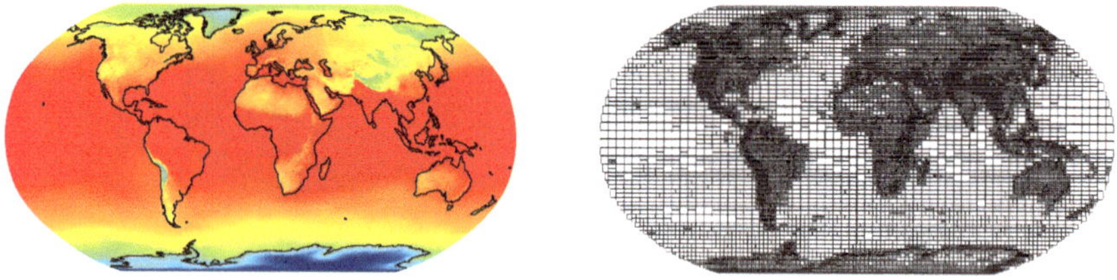

Fig. 5: Average temperature in October 2022 (left) and leafs of the corresponding adaptive tree with relative threshold $\varepsilon = 10^{-4}$ (right).

Given the partition generated by $\mathcal{L}(\mathcal{T}_w)$, we intend to generate a subsample $(\boldsymbol{x}_{i_j}, f_{i_j})$, $j = 1, \dots, n \leqslant N$ adapted to the features of the data. To this end, we pursue an *entropy based approach*: Let $\mathbb{P}$ be a probability distribution on a discrete finite set Z. Then, the *information content* of a sample $z \in Z$ is defined as $I_{\mathbb{P}}(z) := -\log \mathbb{P}(z)$. The *entropy* $H_{\mathbb{P}}$ is the expected information content

$$H_{\mathbb{P}} := \mathbb{E}_{\mathbb{P}}[I_{\mathbb{P}}] = -\sum_{z\in Z} \mathbb{P}(z) \log \mathbb{P}(z). \tag{12}$$

It is well-known that the entropy (12) is maximized if each element of Z is chosen with the same probability

$$\mathbb{P}(z) = \frac{1}{|Z|},$$

see the seminal work [69] for all the details. Therefore, to generate the adaptive subsample, we shall maximize the entropy for randomly selecting points from leafs by choosing a point $\boldsymbol{x} \in X$ from a certain leaf $\tau \in \mathcal{L}(\mathcal{T}_w)$ with probability

$$\mathbb{P}(\boldsymbol{x} \in \tau) = \frac{1}{|\mathcal{L}(\mathcal{T}_w)|}. \tag{13}$$

We like to emphasize that (13) can also be used to adaptively generate new samples $\boldsymbol{x} \in \Omega$ instead of considering only available points $\boldsymbol{x} \in X$.

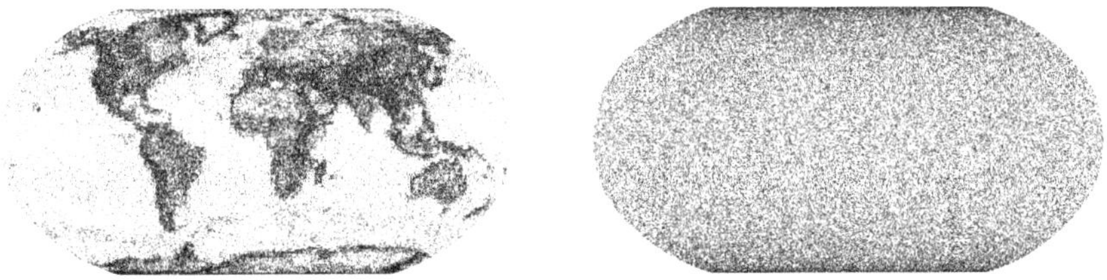

Fig. 6: Adaptive random subsample (left) and uniform random subsample (right).

To demonstrate the adaptive subsampling approach, we consider again the ERA5 temperature data set. Tree coarsening with relative threshold $\varepsilon = 10^{-4}$ for samplets with $q + 1 = 4$ vanishing moments yields the tree found in the right plot of Figure 5. The associated data are visualized in the accompanying left plot. The result of the subsampling using 100 000 points for the month October yields the point distribution seen in the left plot of Figure 6. In contrast, using the same amount of uniformly distributed samples yields the distribution in the right plot of Figure 6.

4 Scattered data approximation

4.1 Reproducing kernel Hilbert spaces

Let $\mathcal{H} \subset C(\Omega)$ be a Hilbert space with inner product $\langle\cdot,\cdot\rangle_{\mathcal{H}}$. Then, there holds $\delta_{\boldsymbol{x}} \in \mathcal{H}'$ for every $\boldsymbol{x} \in \Omega$. Consequently, by the Riesz representation theorem, there exists $(R\delta_{\boldsymbol{x}}) \in \mathcal{H}$ such that

$$\langle (R\delta_{\boldsymbol{x}}), f\rangle_{\mathcal{H}} = f(\boldsymbol{x}) \text{ for all } f \in \mathcal{H}. \tag{14}$$

The function $\kappa(\boldsymbol{x},\boldsymbol{y}) := (R\delta_{\boldsymbol{x}})(\boldsymbol{y})$ is the *reproducing kernel* for $\mathcal{H}$, which renders it a *reproducing kernel Hilbert space*, see [3, 76] for instance. Especially, the reproducing kernel is symmetric and positive definite in the following sense.

Definition 4.1 A symmetric kernel $\kappa\colon \Omega\times\Omega \to \mathbb{R}$ is *positive definite* on $\Omega \subset \mathbb{R}^d$, iff the *kernel matrix* $\boldsymbol{K} := [\kappa(\boldsymbol{x}_i, \boldsymbol{x}_j)]_{i,j=1}^N$ is symmetric and positive semi-definite for every choice of mutually distinct points $\boldsymbol{x}_1, \ldots, \boldsymbol{x}_N \in \Omega$ and any $N \in \mathbb{N}$. The kernel κ is *strictly positive definite*, iff $\boldsymbol{K}$ is even positive definite.

Associated to $X = \{x_1, \ldots, x_N\} \subset \Omega$, we introduce the subspace of kernel translates

$$\mathcal{H}_X := \operatorname{span}\{\phi_1, \ldots, \phi_N\} \subset \mathcal{H}, \tag{15}$$

where

$$\phi_i := \kappa(x_i, \cdot) \text{ for } i = 1, \ldots, N.$$

The subspace $\mathcal{H}_X$ is isometrically isomorphic to $\mathcal{X} = \operatorname{span}\{\delta_{x_1}, \ldots, \delta_{x_N}\} \subset \mathcal{H}'$ by the Riesz isometry. We identify

$$u' = \sum_{i=1}^N u_i \delta_{x_i} \in \mathcal{X}$$

with

$$Ru = \sum_{i=1}^N u_i \phi_i \in \mathcal{H}_X.$$

To approximate a general element $h \in \mathcal{H}$ by an element in $\mathcal{H}_X$, we compute its orthogonal projection. The latter is obtained by the solution of the Galerkin formulation

$$\langle s_h, v\rangle_{\mathcal{H}} = \langle h, v\rangle_{\mathcal{H}} \text{ for all } v \in \mathcal{H}_X. \tag{16}$$

Making the ansatz

$$s_h = \sum_{i=1}^N \alpha_i \phi_i \tag{17}$$

and choosing the basis of kernel translates as test functions in (16), we arrive at the equivalent interpolation problem $s_h(x_i) = h(x_i)$ for all $x_i \in X$ due to the reproducing property (14). The expansion coefficients $\alpha = [\alpha_i]_{i=1}^N$ from (17) can be retrieved by solving the linear system of equations

$$K\alpha = h, \text{ where } h := Ih = [(h, \delta_{x_i})_\Omega]_{i=1}^N, \tag{18}$$

Here, $K = [\kappa(x_i, x_j)]_{i,j=1}^N$ is the kernel matrix and I is the information operator from (9). It is known from both, Galerkin theory and optimal recovery, that s_h is a minimum norm solution in $\mathcal{H}$, see [10] and [58], respectively.

Depending on the choice of the kernel function, the linear system (18) of equations can be ill conditioned and a suitable regularization is required to obtain a solution. Traditionally, Tikhinov regularization is used and the system

$$(K + \mu \boldsymbol{I})\alpha = h \tag{19}$$

is solved for an appropriate regularization parameter $\mu > 0$. In Section 4.7, we address also ℓ^1-regularization with respect to the samplet basis, which is known to impose sparsity to the solution.

4.2 Samplets in reproducing kernel Hilbert spaces

Employing the Riesz isometry, we can embed a given samplet basis $\boldsymbol{\Sigma} = \{\sigma_{j,k}\}_{j,k}$ into a reproducing kernel Hilbert space. This idea follows the vast literature on the embedding of empirical distributions into reproducing kernel Hilbert spaces, see [60] for example.

Consider a samplet

$$\sigma_{j,k} = \sum_{i=1}^{N} \omega_{j,k,i} \delta_{\boldsymbol{x}_i} \subset \mathcal{H}'$$

where $\omega_{j,k,i}$, $i = 1, \ldots, N$, are the expansion coefficients of $\sigma_{j,k}$ with respect to the Dirac-δ-distributions in $\mathcal{X}$. The samplet can be identified with the function

$$\psi_{j,k} := \sum_{i=1}^{N} \omega_{j,k,i} \phi_i \in \mathcal{H}$$

by means of the Riesz isometry. The vanishing moment property (2) translates to

$$\langle \psi_{j,k}, h \rangle_{\mathcal{H}} = 0$$

for any $h \in \mathcal{H}$ which satisfies $h|_O \in \mathcal{P}_q$ for any open and convex set O with $\operatorname{supp}(\sigma_{j,k}) \subset O \cap \Omega$. Herein, we define the support of a samplet in the context of the support of measures according to

$$\operatorname{supp}(\sigma_{j,k}) := \{\boldsymbol{x}_i \in X : \omega_{j,k,i} \neq 0\}.$$

The functions $\{\psi_{j,k}\}_{j,k}$ span the subspace $\mathcal{H}_X$, see (15). In particular, defining

$$\mathcal{W}_j := \operatorname{span}\{\psi_{j,k}\}_k,$$

we obtain the primal multiresolution analysis

$$\mathcal{H}_X = \bigoplus_j \mathcal{W}_j.$$

Using the embedded samplets $\psi_{j,k}$ as ansatz- and test functions in the Galerkin formulation (16) yields the linear system of equations

$$\boldsymbol{K}^{\Sigma}\boldsymbol{\beta} = \boldsymbol{h}^{\Sigma}, \tag{20}$$

where

$$\begin{aligned} \boldsymbol{K}^{\Sigma} &= \left[(\kappa, \sigma_{j,k} \otimes \sigma_{j',k'})_{\Omega\times\Omega}\right]_{j,j',k,k'} \\ &= [\langle \psi_{j,k}, \psi_{j',k'}\rangle_{\mathcal{H}}]_{j,j',k,k'} \\ &= \boldsymbol{T}\boldsymbol{K}\boldsymbol{T}^{\intercal} \end{aligned} \tag{21}$$

and

$$\boldsymbol{h}^{\Sigma} = \left[(\sigma_{j,k}, h)_{\Omega}\right]_{j,k} = [\langle \psi_{j,k}, h\rangle_{\mathcal{H}}]_{j,k} = \boldsymbol{T}\boldsymbol{h}.$$

The solution $\boldsymbol{\beta}$ of the linear system (20) of equations is equivalent to the one of (18) by the transform

$$\boldsymbol{\beta} = \boldsymbol{T}\boldsymbol{\alpha} = \boldsymbol{T}\boldsymbol{K}^{-1}\boldsymbol{h}.$$

Noticing that the Gramian satisfies

$$[\langle \kappa(\boldsymbol{x}_i, \cdot), \kappa(\boldsymbol{x}_j, \cdot)\rangle_{\mathcal{H}}]_{i,j=1}^{N} = [\kappa(\boldsymbol{x}_i, \boldsymbol{x}_j)]_{i,j=1}^{N} = \boldsymbol{K},$$

we find

$$\langle \psi_{j,k}, \psi_{j',k'}\rangle_{\mathcal{H}} = \boldsymbol{\omega}_{j,k}^{\intercal}\boldsymbol{K}\boldsymbol{\omega}_{j',k'},$$

where we set $\boldsymbol{\omega}_{j,k} := [\omega_{j,k,i}]_{i=1}^{N}$ and $\boldsymbol{\omega}_{j',k'} := [\omega_{j',k',i}]_{i=1}^{N}$. Hence, the *dual basis* is given by

$$\widetilde{\psi}_{j,k} = \sum_{i=1}^{N} \widetilde{s}_{j,k,i}\phi_i, \text{ where } \widetilde{\boldsymbol{\omega}}_{j,k} := \boldsymbol{K}^{-1}\boldsymbol{\omega}_{j,k}.$$

Defining the spaces

$$\widetilde{\mathcal{W}}_j := \operatorname{span}\{\widetilde{\psi}_{j,k}\}_k$$

yields the dual multiresolution analysis

$$\mathcal{H}_X = \bigoplus_j \widetilde{\mathcal{W}}_j,$$

where $\mathcal{W}_j \perp \widetilde{\mathcal{W}}_{j'}$ for $j \neq j'$ since $\langle \psi_{j,k}, \widetilde{\psi}_{j',k'}\rangle_{\mathcal{H}} = \delta_{j,j'}\delta_{k,k'}$.

With respect to the dual basis, the interpolant (17) can be written as

$$s_h = \sum_{j,k} \beta_{j,k}\psi_{j,k} = \sum_{j,k} \langle \widetilde{\psi}_{j,k}, h\rangle_{\mathcal{H}}\psi_{j,k} = \sum_{j,k} \langle \psi_{j,k}, h\rangle_{\mathcal{H}}\widetilde{\psi}_{j,k}. \tag{22}$$

In view of the multiresolution representation of the information operator in $Ih = [(h, \sigma_{j,k})_{\Omega}]_{j,k} = [\langle h, \psi_{j,k}\rangle_{\mathcal{H}}]_{j,k}$, compare (9), the expansion with respect to the dual basis in (22) amounts to the multiresolution variant of the optimal recovery algorithm from [59].

In order to give a visual idea of such *embedded samplets* and the respective dual basis, we consider the Sobolev space $H^1(\mathbb{R})$, equipped with the usual norm

$$\|u\|_{H^1(\mathbb{R})}^2 = \|u\|_{L^2(\mathbb{R})}^2 + \|u'\|_{L^2(\mathbb{R})}^2.$$

Its reproducing kernel is given by $\kappa(s,t) = \frac{1}{2}\exp(|s-t|)$, see [5] for instance. Figure 7 shows an embedded scaling distribution (left plot) and two samplets samplets (middle and right plots) with $q+1 = 3$ vanishing moments, constructed for $N = 200$ uniformly distributed data sites.

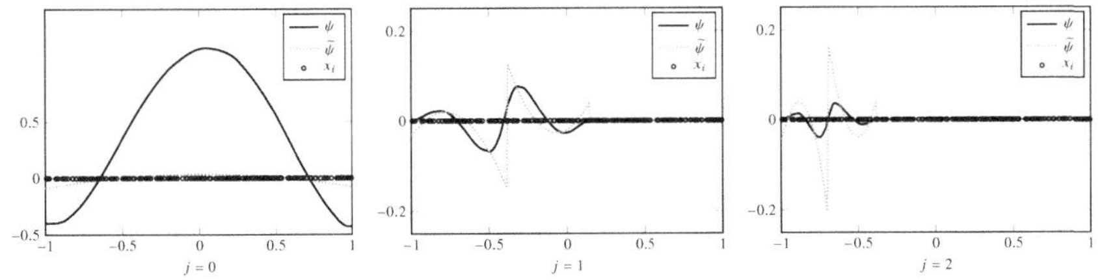

Fig. 7: $H^1(\mathbb{R})$-embedded primal and dual scaling distribution (left), samplet on level $j = 1$ (middle) and samplet on level $j = 2$ (right) for $N = 200$ uniformly distributed data sites and $q + 1 = 3$.

4.3 Asymptotically smooth kernels

It is well-known that nonlocal operators of Calderón-Zygmund type can be compressed in wavelet coordinates, see [6, 21, 67, 74, 75] for example. Analogous results are available for samplets in the context of kernel matrices of positive definite kernels.

The essential ingredient for the samplet compression of kernel matrices is the *asymptotical smoothness* property of the kernel κ. This means that there exists a $\rho > 0$ such that for all $(\boldsymbol{x}, \boldsymbol{y}) \in (\Omega \times \Omega) \setminus \Delta$

$$\left| \frac{\partial^{|\alpha|+|\beta|}}{\partial \boldsymbol{x}^{\alpha} \partial \boldsymbol{y}^{\beta}} \kappa(\boldsymbol{x}, \boldsymbol{y}) \right| \lesssim \frac{(|\alpha| + |\beta|)!}{\rho^{|\alpha|+|\beta|} \|\boldsymbol{x} - \boldsymbol{y}\|_2^{|\alpha|+|\beta|}} \tag{23}$$

uniformly in $\alpha, \beta \in \mathbb{N}^d$ apart from the diagonal $\Delta := \{(\boldsymbol{x}, \boldsymbol{y}) \in \Omega \times \Omega : \boldsymbol{x} = \boldsymbol{y}\}$.

A particular class of positive definite kernels which are asymptotically smooth are the *Matérn kernels* $\kappa(\boldsymbol{x}, \boldsymbol{y}) := k_\nu(\|\boldsymbol{x} - \boldsymbol{y}\|_2)$. They are known to be the reproducing kernels of the Sobolev spaces, see [76] for example, and are given by

$$k_\nu(r) := \frac{2^{1-\nu}}{\Gamma(\nu)}\left(\frac{\sqrt{2\nu}r}{\ell}\right)^\nu K_\nu\left(\frac{\sqrt{2\nu}r}{\ell}\right), \tag{24}$$

where $r = \|\boldsymbol{x} - \boldsymbol{y}\|_2 \geqslant 0$ is the Euclidean distance and $\ell > 0$ is the lengthscale parameter. Herein, K_ν is the modified Bessel function of the second kind of order ν and Γ is the gamma function. The parameter ν controls the smoothness of the kernel function, see, for example, [77]. In particular, we have the exponential kernel

$$k_{1/2}(r) = \exp\left(-\frac{r}{\ell}\right) \tag{25}$$

and the Gaussian kernel

$$k_\infty(r) = \exp\left(-\frac{r^2}{2\ell^2}\right).$$

For asymptotically smooth kernels, compare (23), we obtain the following result, which is the basis for the matrix compression introduced thereafter. The proof of this results is obtained in the traditional way by applying Taylor's expansion of the kernel function under consideration, using (23) to bound the remainder terms, compare [41].

Lemma 4.1 *Consider two samplets $\sigma_{j,k}$ and $\sigma_{j',k'}$ which exhibit $q+1$ vanishing moments and let the associated clusters τ and τ' be such that* $\operatorname{dist}(\tau,\tau') > 0$. *Then, for kernels satisfying* (23), *there holds that*

$$(\kappa, \sigma_{j,k} \otimes \sigma_{j',k'})_{\Omega\times\Omega} \lesssim \sqrt{|\tau||\tau'|}\frac{\operatorname{diam}(\tau)^{q+1}\operatorname{diam}(\tau')^{q+1}}{(\rho\operatorname{dist}(\tau,\tau')/d)^{2(q+1)}}. \tag{26}$$

4.4 Matrix compression

The decay estimate (26) immediately yields a compression strategy for kernel matrices in samplet representation, the proof of which is found in [41]. This compression differs from the wavelet matrix compression introduced in [21], since we do not exploit the decay of the samplet coefficients with respect to the level in case of smooth data. This enables us to also cover scattered data sets with arbitrarily distributed points. As a result, we use the same accuracy on all levels, which is similar to the setting in [6], but implies that the number $q+1$ of vanishing moments needs to be increased to arrive at a higher accuracy of the matrix compression. Fixing the accuracy seems however not to be an issue in view of the regularization of the kernel matrices in scattered data approximation.

Theorem 4.1 *Set all coefficients of the kernel matrix $\boldsymbol{K}^\Sigma$ from* (21) *to zero which satisfy the admissibility condition*

$$\operatorname{dist}(\tau,\tau') \geqslant \eta\max\{\operatorname{diam}(\tau),\operatorname{diam}(\tau')\}, \quad \eta > 0, \tag{27}$$

where τ is the cluster supporting $\sigma_{j,k}$ and τ' is the cluster supporting $\sigma_{j',k'}$, respectively. Then, there holds

$$\|\boldsymbol{K}^{\Sigma} - \boldsymbol{K}_{\varepsilon}^{\Sigma}\|_F \lesssim m_q(\rho\eta/d)^{-2(q+1)} N \log(N),$$

where m_q is given by (3).

Remark 4.1 In case of quasi-uniform points $\boldsymbol{x}_i \in X$, we have $\|\boldsymbol{K}^{\Sigma}\|_F \sim N$. Moreover, the term $\log(N)$ can be removed by a refined analysis in this case and we arrive at

$$\frac{\|\boldsymbol{K}^{\Sigma} - \boldsymbol{K}_{\varepsilon}^{\Sigma}\|_F}{\|\boldsymbol{K}^{\Sigma}\|_F} \lesssim m_q(\rho\eta/d)^{-2(q+1)}$$

while the compressed matrix has $O(m_q^2 N \log N)$ nonzero coefficients.

An illustration of the matrix pattern in case of the exponential kernel $k_{1/2}$ from (25) with lengthscale parameter $\ell = 0.1$ on the unit square using quasi-uniform samplets with $q + 1 = 4$ vanishing moments is found in the left plot of Figure 8. Especially, it has been observed in [40] that the compressed matrix can be reordered by means of nested dissection [32] such that a very sparse Cholesky decomposition is obtained. This is illustrated in the middle and right plots of Figure 8.

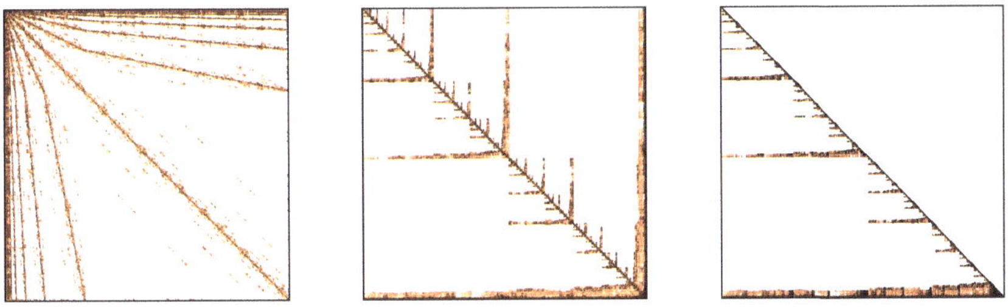

Fig. 8: Sparsity patterns of the samplet compressed exponential kernel on the unit square (left), the nested dissection reordering (middle), and the Cholesky factor (right). Each dot represents a matrix block of size 300 × 300. The entries per block are color coded, where lighter blocks have less entries.

4.5 Matrix assembly

Using the compression rule (27), we can now determine for a given pair of clusters whether the corresponding entries need to be calculated. As there are $O(N)$ clusters, naively checking the cut-off criterion for all pairs would still take $O(N^2)$ operations.

Hence, we require smarter means to determine the non-negligible cluster pairs. For this purpose, we first state the transitivity of the admissibility condition to child clusters, compare [21] for a proof.

Lemma 4.2 *Let* τ *and* τ' *be clusters satisfying the admissibility condition* (27). *Then, for the child clusters* τ_{child} *of* τ *and* τ'_{child} *of* τ', *we have*

$$\begin{aligned}\operatorname{dist}(\tau, \tau'_{\text{child}}) &\geqslant \eta \max\{\operatorname{diam}(\tau), \operatorname{diam}(\tau'_{\text{child}})\},\\ \operatorname{dist}(\tau_{\text{child}}, \tau') &\geqslant \eta \max\{\operatorname{diam}(\tau_{\text{child}}), \operatorname{diam}(\tau')\},\\ \operatorname{dist}(\tau_{\text{child}}, \tau'_{\text{child}}) &\geqslant \eta \max\{\operatorname{diam}(\tau_{\text{child}}), \operatorname{diam}(\tau'_{\text{child}})\}.\end{aligned}$$

The lemma tells us that we may omit cluster pairs whose parent clusters already satisfy the admissibility condition. This is essential for the efficient assembly of the compressed kernel matrix by means of $\mathcal{H}^2$-matrix techniques, see [37, 33]. This idea has already been proposed earlier in [1, 39, 50] in case of Tausch-White wavelets.

$\mathcal{H}^2$-matrices approximate the kernel interaction for sufficiently distant clusters τ and τ' in the sense of the admissibility condition (27) by means of a polynomial interpolant, see [9]. More precisely, given a suitable set of interpolation points $\{\boldsymbol{\xi}_t^\tau\}_t$ for each cluster τ with associated Lagrange polynomials $\{\mathcal{L}_t^\tau(\boldsymbol{x})\}_t$, we can approximate an admissible matrix block by interpolation:

$$[(\kappa, \delta_{\boldsymbol{x}} \otimes \delta_{\boldsymbol{y}})_{\Omega\times\Omega}]_{\boldsymbol{x}\in\tau,\boldsymbol{y}\in\tau'} \approx \sum_{s,t} \kappa(\boldsymbol{\xi}_s^\tau, \boldsymbol{\xi}_t^{\tau'})[(\mathcal{L}_s^\tau, \delta_{\boldsymbol{x}})_\Omega]_{\boldsymbol{x}\in\tau}[(\mathcal{L}_t^{\tau'}, \delta_{\boldsymbol{y}})_\Omega]_{\boldsymbol{y}\in\tau'}.$$

However, different from the $\mathcal{H}^2$-matrix setting, we shall consider this expansion also when the clusters τ and τ' are located on different levels of the cluster tree. By running recursively in a clever way through the samplet matrix, we arrive at an algorithm scheme that computes the compressed kernel matrix in loglinear overall cost. We skip further details here and refer the reader to [41] instead.

To demonstrate the capabilities of the samplet matrix compression, we consider a surface reconstruction problem, similarly to [13]. Given a planar triangulation resulting from a 3D scan of the Laokoon group (the scan is provided by the Statens Museum for Kunst), we generate uniform samples of the signed distance by using the about 500 000 vertices the surface mesh and another 250 000 random points within the bounding box of the Laokoon group. This results in $N = 750\,000$ data sites in total. The left image in Figure 9 shows a uniform subsample of size 100 000 of the data points. For the interpolation, we consider the exponential kernel $k_{1/2}$ from (25) with lengthscale parameter $\ell = 0.01$, where the data sites are rescaled to the hypercube $[0, 1]^3$. The kernel matrix is compressed by using samplets with $q + 1 = 4$ vanishing moments and the linear system (19) of equations is solved with the regularization parameter $\mu = 10^{-8}$ by the conjugate gradient method.

The interpolated signed distance function is then evaluated at a uniform grid of 125 000 000 points, that is 500 points per axis direction. The evaluation is performed by the fast multipole method developed in [42] using total degree polynomials of degree 6 for the interpolation of the kernel function in the farfield. The image in the middle of Figure 9 shows the levelsets of the signed distance function for the

Fig. 9: Signed distance function interpolation from a surface mesh of the Laokoon group. The left image shows a subsample of the signed distance function used for approximation, the image in the middle shows the levelsets of the signed distance function for the values $\{-5, -4, \ldots, 5\}$. The image on the right is the zero levelset with a zoom of the original surface mesh and the reconstruction of the right son's left hand.

values $\{-5, -4, \ldots, 5\}$. The right image shows the zero levelset, which corresponds to the desired surface. From the zoom of the right son's hand, it can be seen how the surface is smoothened in comparison to the original surface mesh.

4.6 Multiresolution kernel matrix algebra

Let $\Omega \subset \mathbb{R}^d$ be a smooth domain and assume that the set of sites $X \subset \Omega$ is *asymptotically distributed uniformly modulo one*, see [51]. This means, we have for every Riemann integrable function $f : \Omega \to \mathbb{R}$ that

$$\left| \int_\Omega f(\boldsymbol{x}) \,\mathrm{d}\boldsymbol{x} - \frac{|\Omega|}{N} \sum_{i=1}^{N} f(\boldsymbol{x}_i) \right| \to 0 \quad \text{as } N \to \infty.$$

Under this condition, the kernel matrix $\boldsymbol{K}$ corresponds to a Nyström discretization of an associated integral operator $\mathcal{K}$. Namely, the reproducing kernel $\kappa(\cdot,\cdot)$ gives rise to the compact integral operator

$$\mathcal{K} : L^2(\Omega) \to L^2(\Omega), \quad u \mapsto \int_\Omega \kappa(\cdot, \boldsymbol{y}) u(\boldsymbol{y}) \,\mathrm{d}\boldsymbol{y}. \tag{28}$$

For many practical applications, (28) constitutes a classical pseudo-differential operator of negative order $s < 0$, especially in case of Matérn kernels. We refer to, for example, [48, 65, 68, 72] for the details of this theory, including the subsequent developments. We are interested here in pseudo-differential operators $\mathcal{K}$ which belong to the subclass $OPS^s_{cl,1}$ of *analytic pseudo-differential operators*, see [65]. Their kernels are known to be asymptotically smooth, satisfying for all $(\boldsymbol{x},\boldsymbol{y}) \in (\Omega\times\Omega)\setminus\Delta$ the refined decay property

$$\left|\frac{\partial^{|\boldsymbol{\alpha}|+|\boldsymbol{\beta}|}}{\partial \boldsymbol{x}^{\boldsymbol{\alpha}}\partial \boldsymbol{y}^{\boldsymbol{\beta}}}\kappa(\boldsymbol{x},\boldsymbol{y})\right| \lesssim \frac{(|\boldsymbol{\alpha}|+|\boldsymbol{\beta}|)!}{\rho^{|\boldsymbol{\alpha}|+|\boldsymbol{\beta}|}\|\boldsymbol{x}-\boldsymbol{y}\|_2^{s+d+|\boldsymbol{\alpha}|+|\boldsymbol{\beta}|}} \tag{29}$$

uniformly in $\boldsymbol{\alpha},\boldsymbol{\beta}\in\mathbb{N}^d$. Note that, in case of reproducing kernels, we always have $s+d<0$ due to the continuity of the kernel such that (29) implies also (23), required for the samplet matrix compression in Section 4.4.

The addition $\mathbf{K}+\mathbf{K}'$ of two compressed kernel matrices in compressed format is obvious. For the computation of the matrix product $\mathbf{K}\cdot\mathbf{K}'$, we make use of the fact that it corresponds to the concatenation $K\circ K'$ of the underlying pseudo-differential operators since

$$\int_\Omega \kappa(\boldsymbol{x}_i,z)\,\kappa'(z,\boldsymbol{x}_j)\,\mathrm{d}z \approx \frac{|\Omega|}{N}\sum_{k=1}^N \kappa(\boldsymbol{x}_i,\boldsymbol{x}_k)\,\kappa'(\boldsymbol{x}_k,\boldsymbol{x}_j). \tag{30}$$

If $K\in OPS^s_{cl,1}$ and $K'\in OPS^{s'}_{cl,1}$, then there holds $K\circ K'\in OPS^{s+s'}_{cl,1}$ and, thus, $K\circ K'$ is compressible. In accordance with [43], we can therefore compute the matrix product $\mathbf{K}'\cdot\mathbf{K}$ in loglinear complexity on the given matrix pattern.

Consider next a symmetric and positive definite kernel function $\kappa(\cdot,\cdot)$ such that the associated pseudo-differential operator satisfies $K\in OPS^s_{cl,1}$ with $s+d<0$. As shown in [43], the inverse of $K+\mu\,\mathrm{Id}$ is of the form $\mu^{-1}\,\mathrm{Id}-K'$ with K' being likewise a pseudo-differential operator of class $OPS^s_{cl,1}$. Especially, the kernel function κ' which underlies the operator K' by the Schwartz kernel theorem, see [47] for instance, is symmetric, positive definite, and likewise asymptotically smooth. Thus, the inverse kernel matrix $(\boldsymbol{K}+\mu\boldsymbol{I})^{-1}$ is also compressible and can, in view of (30), be efficiently approximated by selected inversion, see [52], of the associated pattern, see [43] for the details.

More complicated matrix functions like powers of the kernel matrix or the matrix exponential become accessible, too, by using contour integrals, see [38]. Indeed, the kernel matrix algebra proposed in [43] has the property that the arithmetics is exact on the prescribed (fixed) pattern. We refer the reader to [43] for specific examples and results.

4.7 Samplet basis pursuit

Sparsity constraints are widely used in computational learning, statistics, and signal processing such as deblurring, feature selection and compressive sensing, see [11, 14, 29, 49, 70] for example. Sparsity constraints are imposed by adding an ℓ^1-penalty term to the functional to be minimized. However, such sparsity constraints make only sense if a basis is used for the discretization where the data become sparse. In the past, mostly wavelets bases, Fourier bases, or frames like curvelets, contourlets, and shearlets have been used as they are known to give raise to sparse representations, see [12, 26, 28, 31, 36] for example. However, such discretization concepts are based on regular grids and it is not obvious how to extend them to scattered data as they appear often in machine leaning.

In this section, we shall therefore discuss ℓ^1-regularization for scattered data approximation with respect to samplet coordinates. To this end, we consider the functional

$$\min_{\boldsymbol{\alpha}\in\mathbb{R}^N} \frac{1}{2}\|\boldsymbol{h} - \boldsymbol{K}\boldsymbol{\alpha}\|_2^2 + \sum_{i=1}^{N} w_i|\beta_i|, \text{ where } \boldsymbol{\beta} = \boldsymbol{T}\boldsymbol{\alpha}. \tag{31}$$

The weight vector $\boldsymbol{w} = [w_i]_i \in \mathbb{R}^N$ plays the role of the regularization parameter, where each coefficient is regularized individually. We refer to, e.g., [26, 53, 63] for the analysis of the regularizing properties and for appropriate parameter choice rules.

Numerical algorithms to solve the optimization problem (31) are based on soft-thresholding. Indeed, the sparsity constrained minimization problem (31) can be recast into the root finding problem

$$\boldsymbol{0} = \boldsymbol{\beta}^\star - \mathrm{SS}_{\gamma\boldsymbol{w}}\left(\boldsymbol{\beta}^\star + \gamma(\boldsymbol{K}^\Sigma)^\intercal(\boldsymbol{h}^\Sigma - \boldsymbol{K}^\Sigma\boldsymbol{\beta}^\star)\right), \tag{32}$$

where $\gamma > 0$ and

$$\mathrm{SS}_{\boldsymbol{w}}(\boldsymbol{v}) := \mathrm{sign}(\boldsymbol{v})\max\{\boldsymbol{0}, |\boldsymbol{v}| - \boldsymbol{w}\}$$

is the soft-shrinkage operator. Problem (32) can efficiently be solved by the semi-smooth Newton method, see [4, 34].

In the spirit of [15, 35], also a dictionary of multiple kernels can be employed in (31). Given the kernels $\kappa_1, \ldots, \kappa_L$, we are then looking for a sparse representation of the form

$$s_h = \sum_{\ell=1}^{L}\sum_{i=1}^{N} \alpha_i^{(\ell)}\kappa_\ell(\boldsymbol{x}_i, \cdot) = \sum_{\ell=1}^{L}\sum_{i=1}^{N} \beta_i^{(\ell)}\psi_i^{(\ell)}.$$

Setting

$$\boldsymbol{K} := [\boldsymbol{K}_1, \ldots, \boldsymbol{K}_L], \quad \boldsymbol{K}_\ell := [\kappa_\ell(\boldsymbol{x}_i, \boldsymbol{x}_j)]_{i,j=1}^N,$$

and

$$\boldsymbol{\alpha}^\intercal := [\boldsymbol{\alpha}_1^\intercal, \ldots, \boldsymbol{\alpha}_L^\intercal], \quad \boldsymbol{\alpha}_j \in \mathbb{R}^N,$$

this approach also amounts to (31) with obvious modifications. The most important difference to the original problem is, of course, that the matrix $\boldsymbol{K}$ is not quadratic any more.This means that the underlying linear system $\boldsymbol{K}\boldsymbol{\alpha} = \boldsymbol{h}$ of equations is underdetermined.

To illustrate the approach, we consider the sparse reconstruction of the temperature data from the ERA5 data set in space-time. We employ two space-time kernels, namely

$$\kappa_1(\boldsymbol{z}, \boldsymbol{z}') := k_{3/2}(\|\boldsymbol{x} - \boldsymbol{y}\|_2)k_{\mathrm{per}}(|t - t'|)$$

and

$$\kappa_2(\boldsymbol{z}, \boldsymbol{z}') := k_{1/2}(\|\boldsymbol{z} - \boldsymbol{z}'\|_2),$$

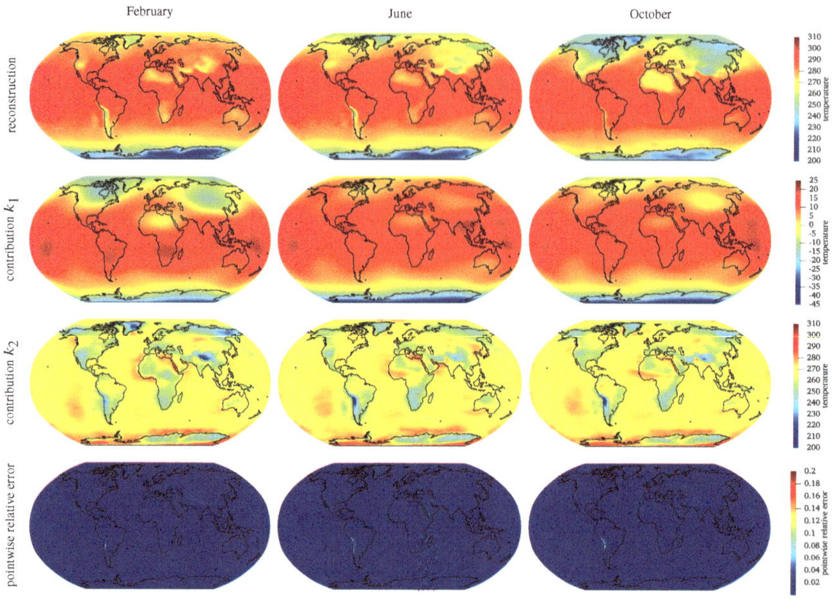

Fig. 10: Reconstruction of the average temperature for February, June, October 2022 (top row), contributions of k_1, k_2 (second and third row) and pointwise relative error (bottom row).

where $z := (x, t)$ and $z' := (y, t')$. Herein, the lengthscale parameter is set to $\ell = 0.2$ for $k_{3/2}$ and to $\ell = 0.01$ for $k_{1/2}$, compare (24). Moreover, we define the periodic Gaussian kernel $k_{\text{per}}(r) := e^{-50\sin^2(\pi r)}$. The kernel κ_1 is a tensor product kernel with relatively large lengthscale parameter. This kernel is intended to capture the smooth parts of the temperature distribution over time. The second kernel κ_2 is a quite rough exponential kernel in space time and chosen to capture sharp features.

For the numerical computation, the set of data sites is obtained by adaptively sampling 100 000 data sites per time step, compare Section 3.3. This results in $N = 1\,200\,000$ data sites in total. These data sites are rescaled to the unit hyper-cube $[0, 1]^3$. For the compression of the respective kernel matrices, we apply samplets with $q + 1 = 4$ vanishing moments. We compute the coefficients in (31) by using an iteratively regularized version of the semi-smooth Newton method, see [4] for details. The weight vector $\boldsymbol{w}$ is set to $w_i = 10^{-6}$ for all $i = 1, \ldots, N$. We obtain $\|\boldsymbol{\alpha}_1\|_0 = 671$ and $\|\boldsymbol{\alpha}_2\|_0 = 5\,883$ non-vanishing coefficients, resulting in a relative in-sample error of $8.18 \cdot 10^{-3}$ in the Euclidean norm, while the value of the functional is $4.76 \cdot 10^{-5}$. The pointwise relative error which is obtained on the full data set is smaller than 9.73% for all time steps.

The top row of Figure 10 shows the reconstruction evaluated for the full data set for the months February, June, and October. The second row shows the contribution of κ_1. As can be seen, κ_1 captures the coarse-scale structure of the temperature distribution with relatively few coefficients. The third row shows the contribution of κ_2. The kernel localizes at the coast lines and mountains and represents sharp features, which requires the major part of the non-zero coefficients. The pointwise relative approximation error is finally found in the last row.

References

1. D. Alm, H. Harbrecht, and U. Krämer. The $\mathcal{H}^2$-wavelet method. *J. Comput. Appl. Math.*, 267:131–159, 2014.
2. B.K. Alpert. A class of bases in L^2 for the sparse representation of integral operators. *SIAM J. Math. Anal.*, 24(1):246–262, 1993.
3. N. Aronszajn. Theory of reproducing kernels. *Trans. Amer. Math. Soc.*, 68(3):337–404, 1950.
4. D. Baroli, H. Harbrecht, and Multerer. Multiresolution kernel matrix algebra. *arXiv Preprints*, arXiv:2306.10180, 2023. to appear in IEEE Trans. Sign. Proc.
5. A. Berlinet and C. Thomas-Agnan. *Reproducing Kernel Hilbert Spaces in Probability and Statistics*. Springer Science & Business Media, New York, 2004.
6. G. Beylkin, R. Coifman, and V. Rokhlin. The fast wavelet transform and numerical algorithm. *Comm. Pure Appl. Math*, 44:141–183, 1991.
7. P. Binev and R.A. DeVore. Fast computation in adaptive tree approximation. *Numer. Math.*, 97:193–217, 2004.
8. M. Blum, R.W. Floyd, V.R. Pratt, R.L. Rivest, and R.E. Tarjan. Time bounds for selection. *J. Comput. Syst. Sci.*, 7(4):448–461, 1973.
9. S. Börm. *Efficient Numerical Methods for Mon-Local Operators: $\mathcal{H}^2$-Matrix Compression, algorithms and analysis*. European Mathematical Society, Zürich, 2010.
10. D. Braess. *Finite Elemente. Theorie, schnelle Löser und Anwendungen in der Elastizitätstheorie*. Springer, Berlin-Heidelberg, 2013.
11. E. Candès, J. Romberg, and T. Tao. Stable signal recovery from incomplete and inaccurate measurements. *Comm. Pure Appl. Math.*, 59(8):1207–1223, 2006.
12. E.J. Candes and D.L. Donoho. Curvelets: A surprisingly effective nonadaptive representation for objects with edges. In A. Cohen, C. Rabut, and L. Schumaker, editors, *Curves and Surface Fitting: Saint-Malo 1999*, page 105–120, Nashville, 2000. Vanderbilt University Press.
13. J.C. Carr, R.K. Beatson, J.B. Cherrie, T.J. Mitchell, W.R. Fright, B.C. McCallum, and T.R. Evans. Reconstruction and representation of 3D objects with radial basis functions. In *Proceedings of the 28th annual conference on Computer graphics and interactive techniques*, SIGGRAPH '01, pages 67–76, New York, 2001. Association for Computing Machinery.
14. S. Chen and D.L. Donoho. Basis pursuit. In *Proceedings of 1994 28th Asilomar Conference on Signals, Systems and Computers*, volume 1, pages 41–44, Pacific Grove, CA, USA, 1994. IEEE.
15. S.S. Chen, D.L. Donoho, and M.A. Saunders. Atomic decomposition by basis pursuit. *SIAM J. Sci. Comput.*, 20(1):33–61, 1998.
16. C.K. Chui. *An Introduction to Wavelets*. Academic Press, San Diego, 1992.
17. C.K. Chui and E. Quak. Wavelets on a bounded interval. *Numer. Meth. Approx. Theory*, 9:53–75, 1992.
18. A. Cohen. *Numerical Analysis of Wavelet Methods*. Elsevier, Amsterdam, 2003.
19. R.R. Coifman and M. Maggioni. Diffusion wavelets. *Appl. Comput. Harmon. Anal.*, 21(1):53–94, 2006.
20. W. Dahmen. Wavelet and multiscale methods for operator equations. *Acta Numer.*, 6:55–228, 1997.
21. W. Dahmen, H. Harbrecht, and R. Schneider. Compression techniques for boundary integral equations. Optimal complexity estimates. *SIAM J. Numer. Anal.*, 43(6):2251–2271, 2006.

22. W. Dahmen, A. Kunoth, and K. Urban. Biorthogonal spline wavelets on the interval – stability and moment conditions. *Appl. Comp. Harm. Anal.*, 6(2):132–196, 1999.
23. W. Dahmen, S. Prößdorf, and R. Schneider. Wavelet approximation methods for pseudodifferential equations II: Matrix compression and fast solution. *Adv. Comput. Math.*, 1(3):259–335, 1993.
24. W. Dahmen and R. Stevenson. Element-by-element construction of wavelets satisfying stability and moment conditions. *SIAM J. Numer. Anal.*, 37(1):319–352, 1999.
25. I. Daubechies. *Ten Lectures on Wavelets.* Society of Industrial and Applied Mathematics, Philadelphia, 1992.
26. I. Daubechies, M. Defrise, and C. De Mol. An iterative thresholding algorithm for linear inverse problems with a sparsity constraint. *Comm. Pure Appl. Math.*, 57:1413–1457, 2004.
27. R.A. DeVore. Nonlinear approximation. *Acta Numer.*, 7:51–150, 1998.
28. M.N. Do and M. Vetterli. The contourlet transform: An efficient directional multiresolution image representation. *IEEE Trans. Image Proc.*, 14:2091–2106, 2005.
29. D.L. Donoho. Compressed sensing. *IEEE Trans. Inf. Theory*, 52(4):1289–1306, 2006.
30. G.E. Fasshauer. *Meshfree Approximation Methods with MATLAB.* World Scientific, River Edge, 2007.
31. S. Foucart and H. Rauhut. *A Mathematical Introduction to Compressive Sensing.* Applied and Numerical Harmonic Analysis. Birkhäuser, New York, 2013.
32. A. George. Nested dissection of a regular finite element mesh. *SIAM J. Numer. Anal.*, 10(2):345–363, 1973.
33. K. Giebermann. Multilevel approximation of boundary integral operators. *Computing*, 67:183–207, 2001.
34. R. Griesse and D.A. Lorenz. A semismooth Newton method for Tikhonov functionals with sparsity constraints. *Inverse Problems*, 24(3):035007, 2008.
35. V. Guigue, A. Rakotomamonjy, and S. Canu. Kernel basis pursuit. In J. Gama, R. Camacho, P.B. Brazdil, A.M. Jorge, and L. Torgo, editors, *Machine Learning: ECML 2005*, pages 146–157, Berlin-Heidelberg, 2005. Springer.
36. K. Guo and D. Labate. Optimally sparse multidimensional representation using shearlets. *SIAM J. Math. Anal.*, 39(1):298–318, 2007.
37. W. Hackbusch and S. Börm. $\mathcal{H}^2$-matrix approximation of integral operators by interpolation. *Appl. Numer. Math.*, 43(1-2):129–143, 2002.
38. N. Hale, N.J. Higham, and L.N. Trefethen. Computing $\mathbf{A}^\alpha$, $\log(\mathbf{A})$, and related matrix functions by contour integrals. *SIAM J. Numer. Anal.*, 46(5):2505–2523, 2008.
39. H. Harbrecht, U. Kähler, and R. Schneider. Wavelet Galerkin BEM on unstructured meshes. *Comput. Vis. Sci.*, 8(3-4):189–199, 2005.
40. H. Harbrecht and M. Multerer. A fast direct solver for nonlocal operators in wavelet coordinates. *J. Comput. Phys.*, 428:110056, 2021.
41. H. Harbrecht and M. Multerer. Samplets: Construction and scattered data compression. *J. Comput. Phys.*, 471:111616, 2022.
42. H. Harbrecht, M. Multerer, and J. Quizi. The dimension weighted fast multipole method for scattered data approximation. *arXiv:2402.09531*, 2024.
43. H. Harbrecht, M. Multerer, O. Schenk, and C. Schwab. Multiresolution kernel matrix algebra. *arXiv Preprints*, arXiv:2211.11681, 2022.
44. H. Harbrecht and R. Schneider. Biorthogonal wavelet bases for the boundary element method. *Math. Nachr.*, 269(1):167–188, 2004.
45. H. Hersbach, B. Bell, P. Berrisford, S. Hirahara, A. Horányi, J. Muñoz-Sabater, J. Nicolas, C. Peubey, R. Radu, D. Schepers, A. Simmons, C. Soci, S. Abdalla, X. Abellan, G. Balsamo, P. Bechtold, G. Biavati, J. Bidlot, M. Bonavita, G. De Chiara, P. Dahlgren, D. Dee, M. Diamantakis, R. Dragani, J. Flemming, R. Forbes, M. Fuentes, A. Geer, L. Haimberger, S. Healy, R.J. Hogan, E. Hólm, M. Janisková, S. Keeley, P. Laloyaux, P. Lopez, C. Lupu, G. Radnoti, P. de Rosnay, I. Rozum, F. Vamborg, S. Villaume, and J.-N. Thépaut. The ERA5 global reanalysis. *Q. J. R. Meteorol.*, 146(730):1999–2049, 2020.
46. T. Hofmann, B. Schölkopf, and A.J. Smola. Kernel methods in machine learning. *Ann. Stat.*, 36(3):1171–1220, 2008.

47. L. Hörmander. *The Analysis of Linear Partial Differential Operators. I.* Classics in Mathematics. Springer, Berlin, 2003. Distribution theory and Fourier analysis, Reprint of the 1990 edition.
48. L. Hörmander. *The Analysis of Linear Partial Differential Operators. III.* Classics in Mathematics. Springer, Berlin, 2007. Pseudo-differential operators, Reprint of the 1994 edition.
49. G. James, D. Witten, T. Hastie, and R. Tibshirani. *An Introduction to Statistical Learning.* Springer Texts in Statistics. Springer, New York, 2013.
50. U. Kähler. *$\mathcal{H}^2$-Wavelet Galerkin BEM and Its Application to the Radiosity Equation.* Dissertation TU Chemnitz, Chemnitz, 2007.
51. G. Leobacher and F. Pillichshammer. *Introduction to Quasi-Monte Carlo Integration and Applications.* Springer International Publishing, Cham, 2010.
52. L. Lin, C. Yang, J. C. Meza, J. Lu, L. Ying, and W. E. Selinv—an algorithm for selected inversion of a sparse symmetric matrix. *ACM Trans. Math. Softw.*, 37(4):40, 2011.
53. D.A. Lorenz. Convergence rates and source conditions for tikhonov regularization with sparsity constraints. *J. Inverse Ill-Posed Probl.*, 16(5):463–478, 2008.
54. S.G. Mallat. A theory for multiresolution signal decomposition: The wavelet representation. *IEEE Trans. Pattern Anal. Mach. Intell.*, 2(7), 1989.
55. S.G. Mallat. *A Wavelet Tour of Signal Processing.* Academic Press, San Diego, 1999.
56. S.G. Mallat. Understanding deep convolutional networks. *Philos. Trans. R. Soc. A*, 374(2065):2015.0203, 2016.
57. S.G. Mallat and Z. Zhang. Matching pursuits with time-frequency dictionaries. *IEEE Trans. Sign. Proc.*, 41(12):3397–3415, 1993.
58. C.A. Micchelli. Orthogonal projections are optimal algorithms. *J. Approx. Theory*, 40(2):101–110, 1984.
59. C.A. Micchelli and T.J. Rivlin. A survey of optimal recovery. In C.A. Micchelli and T.J. Rivlin, editors, *Optimal Estimation in Approximation Theory*, pages 1–54, New York, 1977. Springer.
60. K. Muandet, K. Fukumizu, B. Sriperumbudur, and B. Schölkopf. Kernel mean embedding of distributions: A review and beyond. *Found. Trends Mach. Learn.*, 10(1-2):1–141, 2017.
61. J.T. Schwartz N.J. Dunford. *Linear Operators. Part I: General Theory.* Interscience Publishers, New York, 1958.
62. I. Ram, M. Elad, and I. Cohen. Generalized tree-based wavelet transform. *IEEE Trans. Signal Process.*, 59(9):4199–4209, 2011.
63. R. Ramlau and G. Teschke. A Tikhonov-based projection iteration for nonlinear ill-posed problems with sparsity constraints. *Numer. Math.*, 104(2):177–203, 2006.
64. A.H. Robinson. A new map projection: Its development and characteristics. *International yearbook of cartography*, 14(1974):145–155, 1974.
65. L. Rodino. *Linear Partial Differential Operators in Gevrey Spaces.* World Scientific Publishing Co. Inc., River Edge, NJ, 1993.
66. R. Schaback and H. Wendland. Kernel techniques: from machine learning to meshless methods. *Acta Numer.*, 15:543–639, 2006.
67. R. Schneider. *Multiskalen- und Wavelet-Matrixkompression: Analysisbasierte Methoden zur Lösung großer vollbesetzter Gleichungssysteme.* B.G. Teubner, Stuttgart, 1998.
68. R.T. Seeley. Topics in pseudo-differential operators. In *Pseudo-Differential Operators (C.I.M.E., Stresa, 1968)*, page 167–305. Edizioni Cremonese, Rome, 1969.
69. C.E. Shannon. A mathematical theory of communication. *Bell Syst. Tech. J.*, 27(3):379–423, 1948.
70. T. Tao and E.J. Candès. Near-optimal signal recovery from random projections: universal encoding strategies? *IEEE Trans. Inf. Theory*, 52(12):5406–5425, 2006.
71. J. Tausch and J. White. Multiscale bases for the sparse representation of boundary integral operators on complex geometry. *SIAM J. Sci. Comput.*, 24(5):1610–1629, 2003.
72. M.E. Taylor. *Pseudodifferential Operators*, volume 34 of *Princeton Mathematical Series*. Princeton University Press, Princeton, NJ, 1981.
73. J.A. Tropp. Greed is good: algorithmic results for sparse approximation. *IEEE Trans. Inf. Theory*, 50(10):2231–2242, 2004.

74. T. von Petersdorff and C. Schwab. Fully discrete multiscale Galerkin BEM. In W. Dahmen, A. Kurdila, and P. Oswald, editors, *Multiscale wavelet methods for PDEs*, pages 287–346. Academic Press, San Diego, 1997.
75. T. von Petersdorff, C. Schwab, and R. Schneider. Multiwavelets for second-kind integral equations. *SIAM J. Numer. Anal.*, 34(6):2212–2227, 1997.
76. H. Wendland. *Scattered Data Approximation.* Cambridge University Press, Cambridge, 2004.
77. C.K. Williams and C.E. Rasmussen. *Gaussian Processes for Machine Learning.* MIT Press, Cambridge, 2006.
78. C.K.I. Williams. Prediction with Gaussian processes. From linear regression to linear prediction and beyond. In M.I. Jordan, editor, *Learning in Graphical Models*, volume 89 of *NATO ASI Series (Series D: Behavioural and Social Sciences).* Springer, Dordrecht, 1998.

A novel multilevel approach for the efficient computation of random hyperbolic conservation laws

Michael Herty,
Adrian Kolb and
Siegfried Müller

Abstract A novel strategy for the efficient computation of moments to solutions of random hyperbolic conservation laws is presented. The random variables are considered as parameters to the equation but understood as additional (stochastic) directions. The key idea is to perform grid adaptation alternatingly in spatial and stochastic directions to compute the moments in stochastic direction and to select deterministically realizations, respectively. New realizations are computed on the adaptive spatial grid. For grid adaptation a multiresolution analysis is employed to perform data compression using different threshold values in spatial and stochastic directions.
For proof of concept, the efficiency of the new approach is verified by the 1D–Euler equations with uncertain initial data. The results are compared with the classical Monte–Carlo method showing the improved performance of the proposed method.

1 Introduction

While many physical problems are described by deterministic hyperbolic systems of conservation laws, see e.g., [15, 8, 24], some applications require the treatment of uncertainties. Those uncertainties may be modeled as random variables leading to random hyperbolic conservation laws. Typically, the interest is not in a single realization but on moments of the solution with respect to the random variable as e.g. the expectation or the variance. Several numerical approaches have been proposed for the efficient computation of such moments that are distinguished between non–intrusive (or sampling based) methods [23, 35, 41] and intrusive methods [6, 38]. In the case of the former, the stochastic moments are obtained by calculating a deterministic

Michael Herty, Adrian Kolb, Siegfried Müller,
Institut für Geometrie und Praktische Mathematik, RWTH Aachen University,
Templergraben 55, D-52056 Aachen, Germany
e-mail: herty,mueller@igpm.rwth-aachen.de, kolb@eddy.rwth-aachen.de

R. DeVore, A. Kunoth (eds.), *Multiscale, Nonlinear and Adaptive Approximation II*,
https://doi.org/10.1007/978-3-031-75802-7_15

solution for many randomly drawn samples and subsequent averaging. Since standard Monte–Carlo methods have a low convergence rate, more sophisticated variants are usually applied, e.g., [28, 3, 34]. Alternatively, collocation methods [42, 40, 14] choose sampling points as the roots of orthogonal function depending on the underlying probability density of the random variable. Applying quadrature rules yields the stochastic moments of the solution. Intrusive methods on the contrary represent the random solution in a series expansion using suitable orthogonal base functions [6, 43]. For the coefficients of the (truncated) series expansion a new deterministic system of partial differential equations is obtained using Galerkin projection. A major obstacle has been the loss of hyperbolicity [36, 9, 23, 32] requiring usually problem–dependent modifications of the method, see e.g., [17, 33, 13, 30, 31].

A different approach is considered in [37, 39, 26] treating the random variable as *additional* (spatial) variable and thus embedding the problem into a higher–dimensional, but *deterministic* partial differential equation. We denote these additional dimensions as stochastic dimensions to distinguish them from the original spatial dimensions. In [20, 22] it has been shown that for absolutely continuous random variables the solution to the higher–dimensional deterministic formulation coincides with the random entropy solution [27] of the original random problem. Given a deterministic formulation, classical numerical methods can be applied to compute numerical approximations to the solution. In particular, discontinuities in spatial and stochastic directions may lead to complex wave patterns that require proper resolution [4, 5, 1, 40, 36, 29, 27, 26, 30, 14, 25]. To this end, an efficient and reliable grid adaption strategy has been developed in [20, 22] where a classical multiresolution analysis (MRA) for performing data compression and, thus, grid adaptation, is tailored to the particular formulation of the random problem as higher–dimensional deterministic problem. Since the MRA is performed simultaneously in spatial and stochastic directions, it allows to investigate the interaction between the spatial scales and stochastic scales on the solution and the moments of the solution. For theoretical und numerical results we refer to [20, 22].

For the aforementioned higher–dimensional deterministic problem the approximation of the stochastic moments becomes infeasible even for a small number of random variables. To overcome this bottleneck, we develop a new multilevel strategy. The key idea is to apply grid adaptation separately in the spatial directions and the stochastic directions. Starting from an adaptive (spatial) grid for the moments, we compute in each element (cell) the local moments. Those are approximated by averages in the corresponding stochastic slice employing the probability density distribution of the random variables. For this purpose, we consider a finite number of realizations of the Cauchy problem. These realizations are evaluated in the local spatial positions corresponding to the stochastic slices. To select the samples for which we compute the realizations we perform local grid adaptation in the stochastic directions separately for *each* stochastic slice. Merging *all* locally adapted slices we determine *one* adaptive grid in the stochastic directions. The elements of this stochastic grid represent the samples for which we determine the realizations. This strategy is adaptive to the random variable since the sampling depends on the probability density function. Since the Cauchy problem is time–dependent, the moments depend on space and time. The evolution of the moments is accounted for by performing the time evolution of the realizations.

We emphasize that the above strategy can be realized using different discretizations for the approximation of the realizations, e.g., finite volume schemes and discontinuous Galerkin (DG) schemes, and different adaptation strategies for performing grid adaptation in the spatial and stochastic directions. In the present work, we present a unified concept where (i) the moments are discretized by DG spaces, (ii) the realizations are determined by applying a DG discretization [7] and (iii) the grid adaptation is performed using a multiresolution–based strategy [21, 12] and applying different threshold strategies for the spatial and stochastic directions. This allows to easily project data from the spatial grids to the stochastic slices and vice versa. Furthermore the influence of the threshold values on the moments and the realizations has been investigated in detail in [20, 22].

The outline of the current paper is as follows. In Sect. 2 we introduce the random Cauchy problem. The key ideas of the new multilevel strategy are illustrated by a simple example. This strategy is then realized in Sect. 3 by means of a unified concept based on DG spaces. In particular, we intertwine the MRA performed for the moments in spatial coordinates and for the slices in stochastic directions, respectively. The efficiency is validated in Sect. 4 for one configuration comparing the results with the classical Monte–Carlo method. Finally we summarize our findings in Sect. 5 and give an outlook on future work.

2 The random Cauchy problem

We are interested in conservation laws with uncertain initial data

$$\mathbf{u}_t(t,\boldsymbol{x};\omega_\xi) + \sum_{j=1}^{d} \frac{\partial}{\partial \boldsymbol{x}_j} \boldsymbol{f}_j(\mathbf{u}(t,\boldsymbol{x};\omega_\xi)) = 0,\ \boldsymbol{x} \in \Omega_x,\ \omega_\xi \in \Omega_\xi,\ t \in (0,T) \qquad (1a)$$

$$\mathbf{u}(0,\boldsymbol{x};\omega_\xi) = \mathbf{u}_0(\boldsymbol{x};\omega_\xi),\ \boldsymbol{x} \in \Omega_x \subset \mathbb{R}^d,\ \omega_\xi \in \Omega_\xi \subset \mathbb{R}^m \qquad (1b)$$

that have to be equipped with appropriate boundary conditions. Here, $\mathbf{u}(t,\boldsymbol{x};\omega_\xi) \in \mathbb{R}^q$ are the conserved variables, $\boldsymbol{f} \in C^1(\mathcal{D},\mathbb{R}^{q\times d})$ is the flux field defined on the set of admissible states $\mathcal{D} \subset \mathbb{R}^q$ and $T \in (0,\infty)$ is the final time. The initial condition (1b) is given by the $L^1(\Omega_x)$–valued random variable u_0. The variable ω_ξ denotes one realization of the random variable $\xi : \Omega \to \Omega_\xi$ defined on the probability space $(\Omega,\mathcal{F},\mathbb{P})$.

In general, we are not interested in the single realizations but in the moments of the random problem, i.e., averages over all realizations. Assuming that the probability distribution of ξ is an absolutely continuous random variable with respect to the Lebesgue measure there exists an essentially bounded probability density $p_\xi : \mathbb{R}^m \to [0,\infty)$ such that the expectation and the k–th centralized moments for $u \in L^1(\mathbb{R}^m,\mathbb{R})$ can be expressed as

$$\mathbb{E}[u] = \int_{\Omega_\xi} u(\xi)\, p_\xi(\xi)\, \mathrm{d}\xi, \quad \mathbb{M}^k[u] = \mathbb{E}\left[(u - \mathbb{E}[u])^k\right], \quad k \in \mathbb{N}. \tag{2}$$

In [27] it is proven for multi–dimensional scalar conservation laws that there exists a unique random entropy solution $u : \Omega_\xi \to C_b([0,T], L^1(\mathbb{R}^d))$ for a general probability space $(\Omega, \mathcal{F}, \mathbb{P})$, if the entropy solution exists for $\mathbb{P}$-a.s. $\omega \in \Omega$. In particular, the k–th moment of the random entropy solution exists provided the k-th stochastic moment of the initial condition (1b) exists.

For an example, we consider Burgers' equation with flux $f(u) = u^2/2$ and uncertain initial data $u_0(x,\xi) = e$ for $x < 0$ and $u_0(x,\xi) = e^\xi$ elsewhere. For each realization the solution is determined by a shock wave where the shock position $s(\xi) = (e + e^\xi)/2$ and the data on the right e^ξ depend on the realization. For an illustration see Fig. 1. Each horizontal line in the figure on the right corresponds to one realization ξ shown in the figure on the left determined by $u(t,x,\xi) = e$ for $x < s(\xi)$ and $u(t,x,\xi) = e^\xi$ elsewhere. Thus, for a uniform distribution $\xi \sim \mathcal{U}(0,1)$ the computation of the moments is straightforward. The expectation is shown at the bottom of Fig. 2.

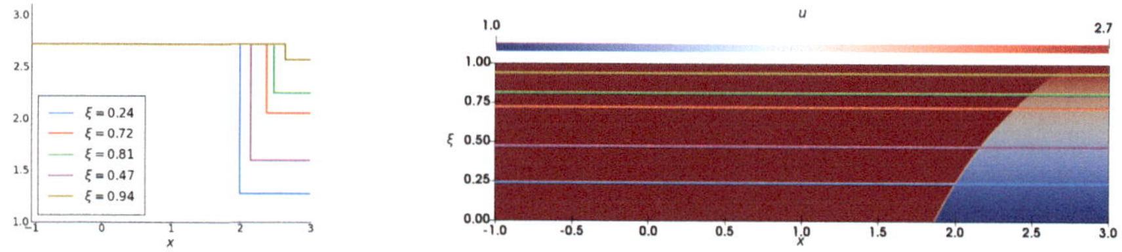

Fig. 1: 1D Burgers' equation with uncertain initial data corresponding to a shock wave at time $T = 1$: Single realizations (left), all realizations (right)

Typically, the solution of the realizations are not not known explicitly. To approximate the moments one might use sampling–type methods such as the Monte–Carlo method. However, these methods show a poor convergence rate. The goal of the present work is to design an efficient strategy for the computation of the moments of the random Cauchy problem (1). Here we briefly sketch the basic ideas by means of the above example. As can be seen in Fig. 2, the moments may suffer from kinks or even discontinuities in space that may move with time. Therefore it will be convenient to dynamically adapt the spatial resolution. For the problem at hand, this would lead to spatial grid refinement near the kink in the expectation, see Fig. 2 (bottom). For the new positions we then have to perform the averaging over all realizations, see the stochastic slices corresponding to the vertical green dashed lines in Fig. 2 (top). Due to the probability density p_ξ some realizations are more likely to happen than others. Therefore, realizations with a high probability density corresponding to the horizontal red dashed lines in Fig. 2 (top) should be used in the averaging process. The interplay between local grid adaptation in the moments and the selection of the realizations is realized by means of a multilevel strategy employing a multiresolution analysis performed in either spatial or stochastic direction.

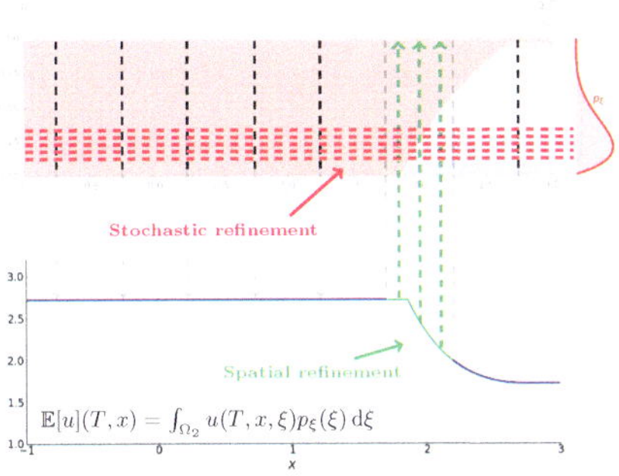

Fig. 2: 1D Burgers' equation with uncertain initial data corresponding to a shock wave at time $T = 1$: all realizations (top). expectation (bottom)

3 Multilevel Strategy for Random Conservation Laws

To realize the multilevel concept as outlined in the previous section, we first introduce a multiresolution analysis (MRA) for DG spaces in Sect. 3.1. A key ingredient in our approach is the exchange of spatial information corresponding to single realizations and the stochastic slices. This is described in Sect. 3.2. Finally, a time–step of the multilevel strategy for the computation of the temporal evolution of the moments is presented in Sect. 3.3.

3.1 Multiresolution analysis

DG spaces: For the approximation of the realizations to the random Cauchy problem (1) we will apply a discontinuous Galerkin (DG) scheme. Although not mandatory, it therefore will be convenient to represent the realizations, the moments (2) as well as the stochastic slices of the random Cauchy problem (1) as functions in some DG space. A DG function is a piecewise polynomial function defined on a discretization of the underlying space $\Omega \subset \mathbb{R}^n$. Here Ω represents either the spatial domain Ω_x or the stochastic space Ω_ξ, i.e., $n = d$ or $n = m$, respectively, and is assumed to be open and bounded with Lipschitz boundary. With regard to grid adaptation we introduce a sequence of nested DG spaces. Starting point is a hierarchy of nested grids $\mathcal{G}_l := \{V_\lambda\}_{\lambda \in \mathcal{I}_l}$, $l \in \mathbb{N}_0$, where for each level l the grid forms a discretization of the space Ω composed of pairwise disjoint open, bounded elements with Lipschitz boundary. To ensure nestedness of the grids each element V_λ, $\lambda \in \mathcal{I}_l$, on level l is assumed to be the union of elements V_μ, $\mu \in \mathcal{M}_\lambda$, in level $l+1$, where $\mathcal{M}_\lambda \subset \mathcal{I}_{l+1}$ is the refinement set of the cell V_λ. On this grid hierarchy we define the sequence $\mathcal{S} = \{S_l\}_{l\in\mathbb{N}_0}$ of DG spaces

$$S_l := \{u \in L^2(\Omega) : u|_{V_\lambda} \in \Pi_{p-1}(V_\lambda),\ \lambda \in \mathcal{I}_l\}$$

with $\Pi_{p-1}(V_\lambda) := \{q : V_\lambda \to \mathbb{R} : q(\boldsymbol{x}) = \sum_{\alpha\in\mathbb{N}_0^d,\ |\alpha|_\infty \leqslant p-1} u_\alpha \boldsymbol{x}^\alpha,\ \boldsymbol{x} \in V_\lambda,\ u_\alpha \in \mathbb{R}\}$ the local polynomial space with maximal degree $p-1$. Note that in contrast to standard literature, we define the maximal degree with respect to the l^∞–norm.

Multiresolution analysis: To perform dynamical local grid refinement and coarsening we employ a multiresolution analysis. For this purpose we note that the sequence of DG spaces is nested, i.e., $S_l \subset S_{l+1}$, due to the nestedness of the grid hierarchy. Thus, their exist complement spaces $W_l \subset S_{l+1}$ such that $S_{l+1} = S_l \oplus W_l$. By recursion, we obtain the *multiscale decomposition* $S_L = S_0 \oplus W_0 \oplus \cdots \oplus W_{L-1}$, i.e., each DG function $u_L \in S_L$ can be represented as the sum of orthogonal projections

$$u_L = u^0 + \sum_{l=0}^{L-1} d^l, \quad u^l = P_{S_l}(u^{l+1}) \in S_l, \quad d^l = P_{W_l}(u^{l+1}) \in W_l. \tag{3}$$

Since the spaces S_l and W_l are piecewise polynomials, the orthogonal projections are computed locally on each element by the local contributions

$$u_\lambda^l := u^l \cdot \chi_{V_\lambda} \in S_{l,\lambda} = \Pi_{p-1}(V_\lambda), \qquad d_\lambda^l := d^l \cdot \chi_{V_\lambda} \in W_{l,\lambda} \subset L^2(V_\lambda)$$

corresponding to the local DG space $S_{l,\lambda}$ and the local complement space $W_{l,\lambda}$, respectively. Here χ_{V_λ} is the indicator function on V_λ. Of particular interest are the local contributions of $W_{l,\lambda}$ referred to as details that may become small whenever the underlying data are locally smooth, cf. [10]. This motivates to discard small details from the multiscale decomposition of $u^L \in S_L$ performing *hard thresholding*. For this purpose, we introduce the notion of a $\|\cdot\|_\lambda$*–significant* local detail, i.e.,

$$\|d_\lambda^l\|_\lambda > \varepsilon_{\lambda,L} \tag{4}$$

with $\|\cdot\|_\lambda$ a local norm for the local complement space $W_{l,\lambda}$. The *local threshold values* are chosen such that

$$\sum_{l=0}^{L-1} \max_{\lambda\in\mathcal{I}_l} \left(\varepsilon_{\lambda,L}\|w\|_{L^\infty(V_\lambda^2)}\right) \leqslant \varepsilon_{\max}, \tag{5}$$

where $w \in L^\infty(\Omega)$ is an essentially bounded weight function. Later on, the weight w is chosen as $w = 1$ or $w = p_\xi$ when performing grid adaptation in spatial and stochastic direction referred to as *uniform* and *weighted* thresholding, respectively. This is motivated by the investigations in [20, 22].

The *set of significant details* $\mathcal{D}_{L,\varepsilon} \subset \bigcup_{l=0}^{L-1} \mathcal{I}_l$ is defined as the smallest set containing the indices of $\|\cdot\|_\lambda$–significant contributions. By this set we determine a sparse approximation for $u^L \in S_L$ defined by the *local multiscale representation*

$$u^{L,\varepsilon} := \sum_{\lambda\in\mathcal{I}_0} u_\lambda^0 + \sum_{l=0}^{L-1} \sum_{\lambda\in\mathcal{D}_{L,\varepsilon}\cap\mathcal{I}_l} d_\lambda^l. \tag{6}$$

Grid adaptation: The set of significant details $\mathcal{D}_{L,\varepsilon}$ induces a locally refined *adaptive grid* $\mathcal{G}_{L,\varepsilon} \subset \bigcup_{l=0}^{L} \mathcal{I}_l$. For this purpose, starting from the coarsest level we levelwise iterate through all active cells of the current level and refine a cell when there exists a significant detail to this cell, i.e., we replace the cell by its children. Note that this procedure only works if $\mathcal{D}_{L,\varepsilon}$ corresponds to a tree. The local multiscale representation (6) is equivalently rewritten in the *local single–scale representation*

$$u^{L,\varepsilon} = \sum_{l=0}^{L} \sum_{\lambda\in\mathcal{G}_{L,\varepsilon}\cap\mathcal{I}_l} u_\lambda^l \quad \text{with} \quad u_\lambda^l = \sum_{i\in\mathcal{P}} u_{l,\lambda,i}\phi_{l,\lambda,i} \in S_{l,\lambda}. \tag{7}$$

Prediction: Up to now, the MRA is performed on steady–state data only. If the data are dynamically changing by means of an evolution process, then the time evolution has to be performed on a grid that captures significant details on *both* the old and the new time step. To avoid recomputation of a time step it is convenient to predict an adequate grid from the data at the old time step. For this purpose prediction strategies have been designed inflating the set of significant details corresponding to the data of the old time step resulting in less sparse adaptive grids. Here we will always apply Harten's prediction strategy [19] to determine the prediction set $\tilde{\mathcal{D}}_{L,\varepsilon}$ and corresponding locally refined grid $\tilde{\mathcal{G}}_{L,\varepsilon}$. For details we refer to [10].

3.2 Exchange of realizations and stochastic slices

The aforementioned tools corresponding to the hierarchy of nested DG spaces, the multiresolution analysis and the grid adaptation are applied separately to the realizations in space and the stochastic slices, see the red and green dashed lines in Fig. 2, respectively. In the following the indices "1" and "2" refer to the spatial and stochastic directions, respectively. Although these directions are separated they are intertwined in an alternate procedure exchanging information between the realizations and the stochastic slices. For this purpose, we need further notation.

Scale–dependent quadrature points: On each cell $V_{\lambda_1}^1 \subset \Omega_1$, $\lambda_1 \in \mathcal{I}_\ell^1$, a polynomial $q_{\lambda_1} \in \Pi_{p_1-1}(V_{\lambda_1}^1)$ is uniquely determined by $\#p_1 := \#\Pi_{p_1-1}(V_{\lambda_1}^1)$ evaluations of some function $u(\cdot;\boldsymbol{x}_2)$ on $V_{\lambda_1}^1$ for fixed $\boldsymbol{x}_2 \in \Omega_2$. For each cell $V_{\lambda_1}^1$ in spatial direction we determine the set of nodes $Q_{\lambda_1}^1 := \{\boldsymbol{x}_{\lambda_1}^{(1)}, \ldots, \boldsymbol{x}_{\lambda_1}^{(\#p_1)}\}$ in $V_{\lambda_1}^1$. For instance, in 1D we choose the nodes of Gauss–Legendre quadrature [16] and transform them to $V_{\lambda_1}^1$ by an affine mapping. We denote

$$Q_\ell^1 := \bigcup_{\lambda_1 \in I_\ell^1} Q_{\lambda_1}^1$$

as the set of all quadrature nodes in spatial direction for $\ell \in \mathbb{N}_0$. Similarly, on each cell in stochastic direction $V_{\lambda_2}^2 \subset \Omega_2$ a polynomial $q_{\lambda_2} \in \Pi_{p_2-1}(V_{\lambda_2}^2)$ is uniquely determined by $\#p_2 := \#\Pi_{p_2-1}(V_{\lambda_2}^2)$ evaluations of some function $u(\boldsymbol{x}_1;\cdot)$. Again, we denoteas the set of quadrature nodes for a cell in stochastic direction $V_{\lambda_2}^2 \subset \Omega_2$, $\ell \in \mathbb{N}_0$ and we define the set of all quadrature nodes in stochastic direction as

$$Q_\ell^2 := \bigcup_{\lambda_2 \in I_\ell^2} Q_{\lambda_2}^2.$$

Expectation: To perform a MRA directly on the expected value, we approximate the expectation for fixed $\boldsymbol{x}_1 \in \Omega_1$ and given DG functions $u^\ell(\cdot;\boldsymbol{x}_2) \in S_\ell^1, \boldsymbol{x}_2 \in Q_\ell^2 \subset \Omega_2$, by the following quadrature formula

$$\mathbb{E}^\ell \equiv \mathbb{E}[u^\ell](\boldsymbol{x}_1) \approx \sum_{\lambda_2 \in I_\ell^2} \frac{|V_{\lambda_2}^2|}{2} \sum_{i=1}^{\#p_2} u^\ell(\boldsymbol{x}_1;\boldsymbol{x}_{\lambda_2}^{(i)}) p_\xi(\boldsymbol{x}_{\lambda_2}^{(i)}) w_2^{(i)}. \tag{8}$$

Here $w_2^{(i)} \in \mathbb{R}_{\geqslant 0}$ is the ith weight, $i = 1, \dots, \#p_2$, of the quadrature in stochastic direction. Note that the above strategy can also be applied to higher moments instead of the expectation.

Scale–dependent slices: To separate the spatial and stochastic scales on each level $l \in \mathbb{N}_0$, we introduce DG functions $u_{\boldsymbol{x}_2}^\ell \in S_\ell^1$ and $u_{\boldsymbol{x}_1}^\ell \in S_\ell^2$ on lower–dimensional DG spaces representing the problem in each direction separately, where the other direction only enters as a parameter. For fixed $\boldsymbol{x}_2 \in \Omega_2$, we define the functions in spatial direction $u_{\boldsymbol{x}_2}^\ell : \Omega_1 \to \mathbb{R}$, $\boldsymbol{x}_1 \mapsto u^\ell(\boldsymbol{x}_1;\boldsymbol{x}_2)$ and denote them as *realizations* of u. Similarly, for fixed $\boldsymbol{x}_1 \in \Omega_1$ we define the functions in stochastic direction to be $\overline{u}_{\boldsymbol{x}_1}^\ell : \Omega_2 \to \mathbb{R}$, $\boldsymbol{x}_2 \mapsto u^\ell(\boldsymbol{x}_1;\boldsymbol{x}_2)$ and denote them as *stochastic slices* of u. This allows to reconstruct a realization from the data of the stochastic slices when the stochastic grid changes, and we have to reconstruct a stochastic slice from the data of the realizations when the spatial grid changes.

Scale–dependent DG representations: For the local DG spaces $S_{l,\lambda}^k$, $k = 1, 2$, we introduce local orthogonal bases $\Phi_{l,\lambda}^k = \{\phi_{l,\lambda,i}^k : i \in \mathcal{P}_k\}$ $S_{l,\lambda}^k$, $k = 1, 2$ with $\mathcal{P}_k := \{1, \dots, \#p_k\}$. Then a time–dependent realization $u_{\boldsymbol{x}_2}^\ell(t,\cdot) \in S_\ell^1$ for fixed $\boldsymbol{x}_2 \in \Omega_2$ can be represented by

$$u_{\boldsymbol{x}_2}^\ell(t,\cdot) = \sum_{\lambda_1 \in I_1} \sum_{i_1 \in \mathcal{P}_1} u_{\lambda_1,i_1,\boldsymbol{x}_2}^\ell(t) \phi_{\lambda_1,i_1}^1(\cdot) \tag{9}$$

with coefficients

$$u_{\lambda_1,i_1,\boldsymbol{x}_2}^\ell(t) = \frac{|V_{\lambda_1}^1|}{2} \sum_{i \in \mathcal{P}_1} \overline{u}_{\boldsymbol{x}_1^{(i)}}^\ell(t,\boldsymbol{x}_2) \phi_{\lambda_1,i_1}^1(\boldsymbol{x}_1^{(i)}) w_1^{(i)}. \tag{10}$$

Analogously, a time–dependent stochastic slice $\overline{u}^{\ell}_{x_1}(t,\cdot) \in S^2_\ell$ for fixed $x_1 \in \Omega_1$ is determined by

$$\overline{u}^{\ell}_{x_1}(t,\cdot) = \sum_{\lambda_2 \in \mathcal{I}_2} \sum_{i_2 \in \mathcal{P}_2} \overline{u}^{\ell}_{\lambda_2,i_2,x_1}(t)\phi^2_{\lambda_2,i_2}(\cdot) \tag{11}$$

with coefficients

$$\overline{u}^{\ell}_{\lambda_2,i_2,x_1}(t) = \frac{|V^2_{\lambda_2}|}{2} \sum_{i \in \mathcal{P}_2} u^{\ell}_{x_2^{(i)}}(t,x_1)\phi^2_{\lambda_2,i_2}(x_2^{(i)})w_2^{(i)}. \tag{12}$$

Since we evaluate all realizations and stochastic slices in the quadrature points in x_1– and x_2–direction, respectively, we use the corresponding quadrature rules for the computation of the coefficients (10) and (12), respectively. We emphasize that in (10) the realizations are obtained by evaluating the stochastic slices $\overline{u}^{\ell}_{x_1^{(i)}}$ with respect to the quadrature points $x_2^{(i)}$ and in (12) the stochastic slices are obtained by evaluating the realizations $u^{\ell}_{x_2^{(i)}}$ with respect to the quadrature points $x_1^{(i)}$.

3.3 Time evolution

A single time step with the multilevel strategy is summarized in Alg. 9. In this procedure it is essential to keep track of all realizations and stochastic slices and synchronize the data accordingly when the number of realizations or the number of stochastic slices change. Since we only have fluxes in the spatial directions, we perform time evolution for all realizations using an SSP–RK scheme, cf. Sect. 4.1, whereby grid refinement is triggered only by local information of the expectation. Note that the algorithm can also be applied to a vector of (higher-order) moments instead of the expectation.

In the algorithm the maximum refinement level of the spatial and stochastic directions is fixed by $L \in \mathbb{N}$ for all time steps. For time step $n \in \mathbb{N}_0$ the set of all realizations is denoted by $\{u^{L,n}_{x_2}\}_{x_2 \in Q^2_{L,n}}$, where $Q^2_{L,n} \subset \Omega_2$ is the set of quadrature points in stochastic direction. Similarly, the set of all stochastic slices is defined by $\{\overline{u}^{L,n}_{x_1}\}_{x_1 \in Q^1_{L,n}}$, where $Q^1_{L,n} \subset \Omega_1$ is the set of quadrature points in spatial direction at time step n. The expectation is denoted by $\mathbb{E}^{L,n}$. Initially, for $n = 0$ these information can be provided by applying a bottom–up strategy to the random initial data according to [22].

Each time step of the multilevel algorithm consists of two major parts, namely, (i) time evolution and grid adaptation with respect to the expectation in spatial direction and (ii) update of the stochastic slices. We emphasize that the time evolution is performed separately for each realization, whereas grid adaptation is performed only on the expectation and not on the realizations. Thus, we may lose significant information for a single realization and we have to ensure that the numerical error in the stochastic moments caused by the suboptimal grid for each realization is as small as possible. Furthermore, the stochastic slices and realizations have to be synchronized correctly when changing the grid in the spatial and the stochastic directions.

Algorithm 9: Time step $n \longrightarrow n+1$:

Input: Expectation $\mathbb{E}^{L,n}$, realizations $\{u^{L,n}_{\boldsymbol{x}_2}\}_{\boldsymbol{x}_2 \in Q^2_{L,n}}$ and stochastic slices $\{\overline{u}^{L,n}_{\boldsymbol{x}_1}\}_{\boldsymbol{x}_1 \in Q^1_{L,n}}$ for time step n.

1. **Stochastic grid refinement:**
   Determine a predictive grid $\widetilde{\mathcal{G}}^2_{n+1}$ in stochastic direction by inflating the set of significant details applying Harten's prediction strategy.
2. **Update realizations:**
   Update all realizations using the data of the stochastic slices $\{\overline{u}^{L,n+1}_{\boldsymbol{x}_1}\}_{\boldsymbol{x}_1 \in Q^1_{L,n}}$ on grid $\widetilde{\mathcal{G}}^2_{n+1}$.
3. **Time evolution and spatial grid adaptation:**
   Perform time evolution for all realizations $\{u^{L,n}_{\boldsymbol{x}_2}\}_{\boldsymbol{x}_2 \in \widetilde{Q}^2_{L,n+1}}$ and determine new adaptive grid $\mathcal{G}^1_{n+1}$ by performing a MRA on the expectation $\widetilde{\mathbb{E}}^{L,n+1}$ using uniform thresholding, i.e., $w = 1$ in (5). Project all realizations $\{u^{L,n+1}_{\boldsymbol{x}_2}\}_{\boldsymbol{x}_2 \in \widetilde{Q}^2_{L,n+1}}$ on the new grid $\mathcal{G}^1_{n+1}$.
4. **Update stochastic slices:**
   Update all stochastic slices using the previous data from time step n and the new realizations $\{u^{L,n+1}_{\boldsymbol{x}_2}\}_{\boldsymbol{x}_2 \in Q^2_{L,n}}$.
5. **Stochastic grid coarsening:**
   Determine adaptive grid $\mathcal{G}^2_{n+1}$ in stochastic direction and stochastic slices $\{\overline{u}^{L,n+1}_{\boldsymbol{x}_1}\}_{\boldsymbol{x}_1 \in Q^1_{L,n}}$ using weighted thresholding, i.e., $w = p_\xi$ in (5).
6. **Update realizations:**
   Update all realizations using the previous data from time step n and the new stochastic slices $\{\overline{u}^{L,n+1}_{\boldsymbol{x}_1}\}_{\boldsymbol{x}_1 \in Q^1_{L,n}}$.
7. **Update expectation:**
   Calculate expectation $\mathbb{E}^{L,n+1}$ from the new realizations and stochastic slices on time step $n+1$.
8. **Increase time step** $n \longrightarrow n+1$.

Details on the single steps in Alg. 9 can be found in [22]. Nevertheless, the following remarks on the multilevel strategy might be useful: Since we perform a MRA on the expectation, the perturbation error of the expectation is controlled instead of the perturbation error of the realizations. Thus, realizations are not necessarily refined due to their local structure. This effect can be observed in the numerical results, cf. Sect. 4.

Furthermore, to stabilize a time step in Alg. 9, we need to limit the coefficients of all DG solutions to avoid oscillations at discontinuities. We emphasize that not only realizations have to be limited but also all stochastic slices due to the grid adaptation process, although we do not perform time evolution in the stochastic direction. We refer to Sect. 4.1 for more details. In contrast to previous work [12, 10], where the limiter is only applied in elements (cells) at the finest refinement level, we have to perform limiting in spatial direction for *all* realizations at all refinement levels. This is due to the weighted thresholding strategy applied to the realizations and, thus, discontinuities of a single realization are not necessarily detected and might not be resolved on the finest refinement level.

For an efficient implementation of the multilevel strategy it is essential to use appropriate data structures to track all current realizations and stochastic slices and their corresponding quadrature points to map them correctly. Since we separate both spatial and stochastic scales, we may choose different polynomial order in each direction. Additionally, the maximum refinement level in the spatial and stochastic direction do not have to coincide.

4 Numerical Results

We present numerical results for the multilevel strategy introduced in Sect. 3 to approximate the stochastic moments of hyperbolic random conservation laws with uncertain initial data. For this purpose, we first briefly summarize in Sect. 4.1 the solver used in the multilevel strategy (MLDET) to perform the time evolution of the realizations. For a proof of concept we then apply the multilevel strategy to the 1D-Euler equations with uncertain initial data depending on one random variable, cf. Sect. 4.2. The results are compared to computations with the classical Monte–Carlo method (MC) [41].

4.1 Numerical Method

To perform computations for the multilevel strategy (MLDET) we need to specify the following ingredients:

- The DG spaces S_l are composed of the local DG spaces $S_{l,\lambda} = \operatorname{span} \Phi_{l,\lambda}$, $\lambda \in \mathcal{G}_l$. These are spanned in 1D by Legendre polynomials. The corresponding local complement spaces $W_l = \operatorname{span} \Psi_{l,\lambda}$ are spanned by multiwavelets [2]. For a construction on multi–dimensional Cartesian grid hierarchies as used in our applications we refer to [12] providing us with the local orthogonal bases $\Phi_{l,\lambda} = \{\phi_{l,\lambda,i} : i \in \mathcal{P}\}$ and $\Psi_{l,\lambda} = \{\psi_{l,\lambda,i,e} : i \in \mathcal{P}, e \in \mathcal{E}^*_\lambda\}$ where $\mathcal{E}_\lambda := \{0, \ldots, \#\mathcal{M}_\lambda - 1\}$, $\mathcal{E}^*_\lambda := \mathcal{E}_\lambda \backslash \{0\}$, $\mathcal{P} := \{\alpha \in \mathbb{N}_0^n : \|\alpha\|_\infty \leqslant p - 1\}$.
- For data compression we apply hard thresholding according to (4). For a Cartesian grid hierarchy we choose the local threshold value

$$\varepsilon_{\lambda,L} = \frac{h_L}{h_l} \|w\|^{-1}_{L^\infty(V_\lambda)} \varepsilon_{\max}, \ \lambda \in \mathcal{I}_l \tag{13}$$

with h_l the uniform discretization on level l. Note that (5) holds true. In particular, we set $w = 1$ and $w = p_\xi$ to perform uniform thresholding and weighted thresholding, respectively. Then for a local norm in (4) we choose

$$\|d^l_\lambda\|_\lambda := \frac{1}{\sqrt{|V_\lambda|}} \max_{i\in\mathcal{P}, e\in\mathcal{E}^*_\lambda} \{|d_{l,\lambda,i,e}|\, \|\psi_{l,\lambda,i,e}\|_{L^2(V_\lambda)}\}. \tag{14}$$

- The expectation and the higher moments are represented by local DG spaces $S_{l,\lambda}$, $\lambda \in \mathcal{G}_{L,\varepsilon}$, spanned by quadratic polynomials, i.e., $p = 3$, corresponding to a locally refined grid $\mathcal{G}_{L,\varepsilon}$. These spaces change dynamically with time. Grid adaptation is performed by means of a MRA using uniform thresholding.
- To update the moments on the new time level, all realizations are evolved in time by means of a Runge–Kutta (RK) discontinuous Galerkin method [7] on the adaptive grid for the moments. The local DG spaces of the realizations are also spanned by quadratic polynomials, i.e., $p = 3$. For the time discretization an explicit third–order SSP–RK method with three stages for the time–discretization is used, cf. [18]. As numerical flux we choose the local Lax–Friedrichs flux with the Shu limiter [7].
- From the realizations in spatial direction we extract slices in stochastic directions. To determine which realizations have to be considered in the next time step we perform for each slice a MRA using weighted thresholding. We emphasize that the polynomial degree of the local DG spaces for the moments in the spatial direction and the slices in stochastic directions do not have to coincide.
- To perform the MRA for the moments as well as the slices we have to fix the global threshold value $\varepsilon_{\max}$ that is crucial for the efficiency and the accuracy of the overall scheme. Here we apply the heuristic strategy developed in [11] for Cartesian grid hierarchies
$$\varepsilon_{\max} = C_{\text{thresh}}\, h_L^\beta \tag{15}$$
with β the order of the discretization error. Since the data may exhibit discontinuities, the order of convergence is bounded by one, i.e, $\beta = 1$. The choice of C_{thresh} is problem–dependent and will be discussed below.

Alternatively, to approximate the moments (2) by he Monte–Carlo method (MC) we have to draw a large number of samples. For each sample we have to solve the Cauchy problem (1a) for fixed $\omega_\xi \in \Omega_\xi$. Following the work by [21] we combine the above DG solver with *multiresolution–based grid adaptation* to improve its efficiency. One time step consists of the following three steps, namely, (i) grid refinement by means of a MRA and applying Harten's prediction strategy, ii) time evolution with the aforementioned SSP–RK method and limiting performed only on the cells of the *highest* refinement level L and (iii) grid coarsening by means of a MRA and applying hard thresholding. Here the MRA in the grid refinement and grid coarsening is performed by means of uniform thresholding. The global threshold value is chosen as in the multilevel strategy. More details on the adaptive solver can be found in [12, 11].

For MC we fix $N = 16,000$ samples on each level $\ell = 0, \ldots, L$. We note that all samples in MC are performed on the highest refinement level $L = 6$. For MLDET we choose the maximum refinement level $L = 3, \ldots, 6$ for the spatial and stochastic directions, simultaneously.

4.2 Euler Equations with Uncertain Initial Data

We apply the multilevel strategy to the one–dimensional Euler equations

$$\frac{\partial \rho}{\partial t} + \frac{\partial \rho v}{\partial x} = 0, \quad \frac{\partial \rho v}{\partial t} + \frac{\partial(\rho v^2 + p)}{\partial x} = 0, \quad \frac{\partial \rho E}{\partial t} + \frac{\partial(v\,(\rho E + p))}{\partial x} = 0 \tag{16}$$

for $x \in \mathbb{R}$, $t > 0$, with uncertain Riemann initial data

$$\mathbf{u}(0, x; \omega_\xi) = \begin{cases} (1.0,\ 0.0,\ 0.5 + 2.5\omega_\xi)^T, & x < 0.5 \\ (0.125,\ 0.0,\ 0.25)^T, & x > 0.5 \end{cases}. \tag{17}$$

This corresponds to Sod's shock tube problem with uncertain initial pressure $p(0, x; \omega_\xi) = \omega_\xi + 0.2$ for $x < 0.5$ and $p(0, x; \omega_\xi) = 0.1$ elsewhere. Here $\rho \equiv \rho(t, x; \omega_\xi)$, $v \equiv v(t, x; \omega_\xi)$ and $E \equiv E(t, x; \omega_\xi)$ denote the density, momentum and total energy, respectively. The total energy can be written as $E := \frac{1}{2}v^2 + e$ where $e = e(t, x; \omega_\xi)$ is the internal energy of the system. We assume a *perfect gas*, i.e., $e = \frac{p}{(\gamma-1)\rho}$ with $\gamma = 1.4$.

We investigate the stochastic moments of a uniform random variable $\xi_1 \sim \mathcal{U}(0, 1)$ and a beta–distributed random variable $\xi_2 \sim \mathcal{B}(2, 5)$. We set the spatial domain $\Omega_1 := [0, 1]$ and the stochastic domain $\Omega_2 := [0, 1]$. We emphasize that the example is chosen such that the pressure on the left is higher than the pressure on the right for all drawn realizations. Thus, the gas flows from left to right for all realizations.

For the numerical simulations, we choose the maximum number of refinement levels $L = 6$ for the spatial and stochastic directions and we set the number of cells of the initial grid $M_{0,x} = M_{0,\xi} = 8$ for both directions. We use a third–order DG scheme for the stochastic and spatial direction, i.e., we set $p_1 = p_2 = 3$ and fix the final time $T = 0.2$. For the grid adaptation in spatial direction, we set $C_{thresh} = 0.1$ in (15). For the stochastic direction we perform weighted thresholding and set $C_{thresh,i} = \hat{C}/\|p_{\xi_i}\|_{L^\infty([0,1])}$ where p_{ξ_i} is the probability density of the random variable ξ_i for $i = 1, 2$. We fix $\hat{C} = 2.0$ in the stochastic direction. The high value in the spatial stochastic direction is sufficient whereas we have to choose a lower value in spatial direction to resolve discontinuities in spatial direction properly.

Computations in spatial direction using the multilevel strategy. In Fig. 3 we present the expectation, standard deviation and the corresponding refinement level in spatial direction for the Euler equations (16) with initial data (17) at time $T = 0.2$ for the uniform distribution ξ_1 and the beta distribution ξ_2, respectively. For both random variables, the resulting adaptive grid fits the local changes of the expectation. Nevertheless, we observe in the case with uniformly distributed initial pressure small oscillations at the contact discontinuity. This can especially be seen

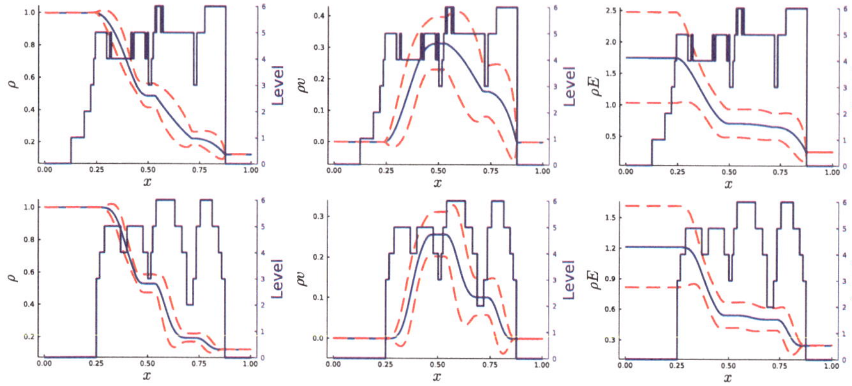

Fig. 3: Expectation (blue), standard deviation (red) and refinement level of the solution to the one–dimensional Euler equations (16) with uncertain initial data (17) at time $T = 0.2$ using up to $L = 6$ refinement levels. Left: Density ρ; Middle: Momentum ρv; Right: Density of energy ρE. Top: Uniform distribution $\mathcal{U}(0, 1)$; Bottom: Beta distribution $\mathcal{B}(2, 5)$.

in the standard deviation of the density ρ in Fig. 3 in case of the uniform distribution at $x = 0.5$. Thus, grid refinement is triggered more in this area. We will see that these oscillations do not have a significant impact on the L^1–error of the stochastic moments.

To study the effect of different random variables on the realizations, we show in Fig. 4 the state variables $(\rho, \rho v, \rho E)^T$ in spatial directions for fixed $x_2 \in \Omega_2$ of the uniform random variable and the beta–distributed random variable, respectively, at time $T = 0.2$. For illustration purposes, the points x_2 are chosen uniformly from the domain Ω_2. We note that for the density ρ lower lines for $x < 0.5$ correspond to a higher stochastic value $x_2 \in \Omega_2$, whereas higher lines correspond to larger stochastic values $x_2 \in \Omega_2$ for the momentum ρv and density of energy ρE. For the uniform random variable, the majority of the realizations are resolved properly, thus kinks and discontinuities are on the highest refinement level, whereas areas with no significant local changes are resolved on a coarser grid. Nevertheless, not all discontinuities are resolved properly, as can be seen for example in the density ρ for the uniform random variable at $x \approx 0.75$. This justifies the choice of the small value for C_{thresh} in the spatial direction since otherwise, the error would become too large. Although the discontinuity is not resolved properly for a small number of realizations, it does not have a significant impact on the expectation. Thus, the value for C_{thresh} in spatial direction has to be chosen carefully, to control the error of not properly refined realizations. For the beta–distributed random variable ξ_2, we observe that for higher $x_2 \in \Omega_2$ the spatial projections are looking poor compared to realizations with lower $x_2 \in \Omega_2$. This can be expected due to the weighted thresholding in stochastic direction. Since the probability density p_{ξ_2} is low for $x_2 > 0.75$, the resolution for the corresponding realizations is poor, too. Nevertheless, since these realizations have almost no effect on the stochastic moments, the error in the stochastic moments is small. Although weighted thresholding only affects grid refinement in the stochastic direction, it also has an effect on the spatial solution due to the interaction of the different scales.

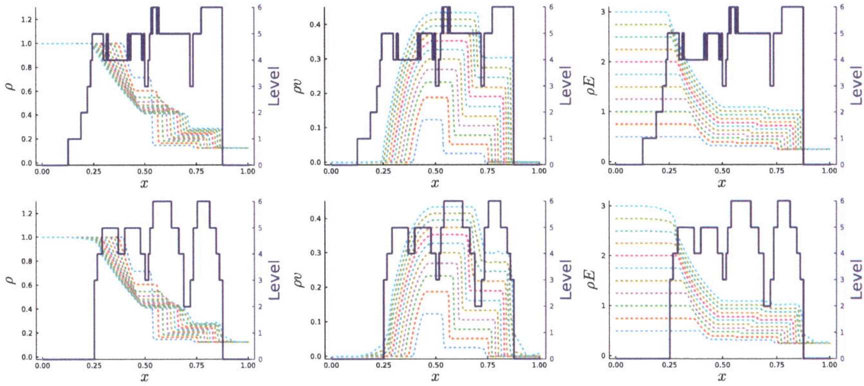

Fig. 4: Projection in spatial direction for fixed $\boldsymbol{x}_2 \in \Omega_2$ of the solution to the one–dimensional Euler equations (16) with uncertain initial data (17) at time $T = 0.2$ and up to $L = 6$ refinement levels. Left: Density ρ; Middle: Momentum ρv; Right: Density of energy ρE. Top: Uniform distribution $\mathcal{U}(0, 1)$; Bottom: Beta distribution $\mathcal{B}(2, 5)$.

Computations in stochastic direction using the multilevel strategy. In Fig. 5 we present projections in the stochastic direction for some fixed spatial variables $x_1 \in \Omega_1$. As before, x_1 is chosen uniformly from Ω_1 to illustrate all effects in the stochastic slices. We additionally show the corresponding refinement level in stochastic direction and the corresponding probability density. As can be seen in Fig. 4, a discontinuity occurs in stochastic direction for all $x_2 \in \Omega_2$. Since a discontinuity in one stochastic slice is sufficient to trigger grid refinement in all stochastic slices, we have for the uniformly distributed initial pressure a fully refined grid on refinement level $L = 5$. Since we have chosen $\hat{C} = 2.0$ in $C_{thresh,i}$ in stochastic direction, we do not fully refine the stochastic slices on the highest refinement level $L = 6$. Nevertheless, discontinuities are well resolved.

Due to weighted thresholding in stochastic direction, we have for the beta–distributed random variable less grid refinement for $x_2 > 0.75$, although we have discontinuities for all realizations. Since the probability density p_{ξ_2} has its maximum at $x_2 \approx 0.2$, the grid is fully refined on the highest refinement level $L = 6$. Thus, in areas with high local probability density grid refinement is triggered faster than in areas with low local probability density.

Comparison of multilevel strategy and MC method. To study the efficiency of the multilevel strategy, we measure the L^1–error of the expectation and variance with respect to the needed degrees of freedom (DoF). The results are shown in

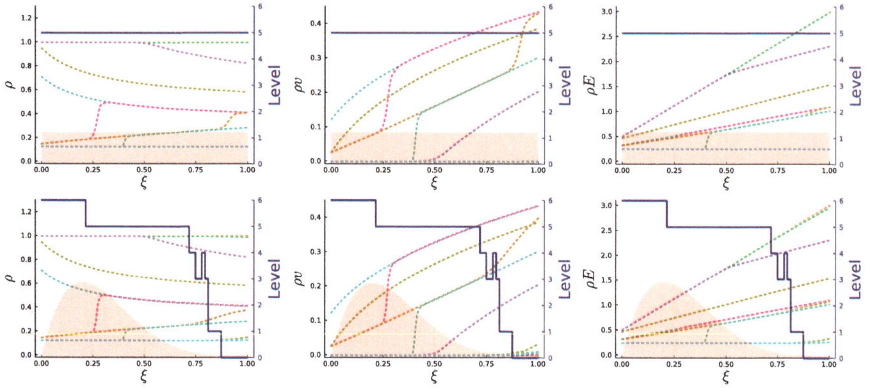

Fig. 5: Projection in stochastic direction for fixed $\boldsymbol{x}_1 \in \Omega_1$ of the solution to the one–dimensional Euler equations (16) with uncertain initial data (17) at time $T = 0.2$ and up to $L = 6$ refinement levels. Left: Density ρ; Middle: Momentum ρv; Right: Density of energy ρE. Top: Uniform distribution $\mathcal{U}(0, 1)$; Bottom: Beta distribution $\mathcal{B}(2, 5)$.

Fig. 6 for the uniform distribution and the beta–distributed initial pressure, respectively. The stochastic moments of the multilevel strategy (MLDET) are compared with the stochastic moments obtained by the Monte–Carlo (MC) method. For a reference solution we perform a two–dimensional computation for the corresponding higher–dimensional deterministic problem on a two-dimensional grid with $L = 10$ refinement levels and uniform thresholding that is significantly more efficient than the Monte–Carlo method. For details we refer to [22]. All Monte–Carlo simulations are performed on a grid with $L = 6$ refinement levels and $M_0 = 8$ cells on the initial grid. Additionally, we perform multiresolution–based grid adaptation with $C_{\text{thresh}} = 0.1$ for each sample in the Monte–Carlo simulation. To study the efficiency of the Monte–Carlo method we vary the number of needed samples. In this case, we use up to $N = 16{,}000$ samples to approximate the stochastic moments. We emphasize that no refinement strategy is used in stochastic direction for the Monte–Carlo simulations. The simulations of the multilevel strategy are performed on different maximum refinement levels. Here, we choose $L = 3, \ldots, 6$ refinement levels. The final time is set to $T = 0.2$. In the following, we study the L^1–error of the stochastic moments of the state variables $(\rho, \rho v, \rho E)$.

We call a method more efficient, if the number of DoFs is smaller for a fixed L^1–error and, thus, needs less computational effort to realize a certain error in the stochastic moments. We observe that we need significantly more DoFs using MC to approximate the expectation than MLDET. This is observed in all investigated quantities for both random variables ξ_1 and ξ_2. Approximating the expectation, the Monte–Carlo approach has poor efficiency compared to the multilevel strategy. On the other hand, the Monte–Carlo simulation is comparable to MLDET in approximating the variance of the density ρ and the momentum ρv. For the density of energy

ρE the approximation of the variance using MC is also poor compared to MLDET. We emphasize that these results also hold on the coarsest refinement level $L = 3$. Thus, MLDET performs significantly better than MC on a coarse refinement level.

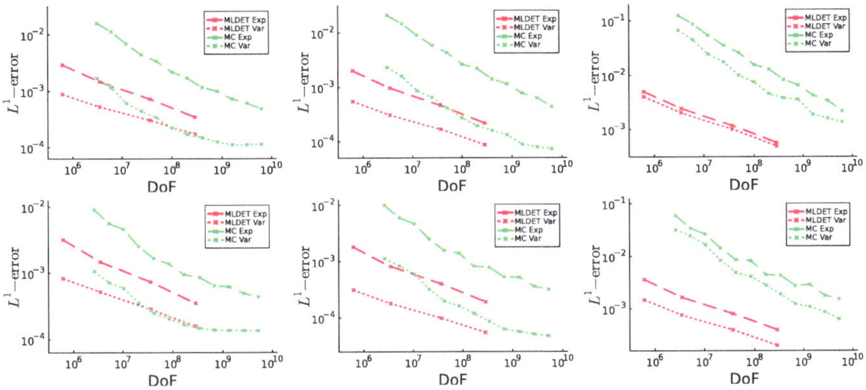

Fig. 6: L^1–error of the stochastic moments with respect to the total number of degrees of freedom of the one–dimensional Euler equations (16) with uniformly distributed and beta–distributed initial pressure (17) at time $T = 0.2$ using MC and MLDET. Left: Density ρ; Middle: Momentum ρv; Right: Density of energy ρE. Top: Uniform distribution $\mathcal{U}(0, 1)$; Bottom: Beta distribution $\mathcal{B}(2, 5)$.

5 Conclusion

A new multilevel strategy to compute moments of random hyperbolic conservation laws with uncertain initial data has been proposed. Grid adaptation is performed separately in spatial direction and stochastic direction. The data on the stochastic slices are determined for fixed spatial coordinates by projecting the solution of realizations onto the slices in the stochastic direction. For grid adaptation we have applied a multiresolution analysis performing uniform and weighted thresholding for the moments and the stochastic slices, respectively. We emphasize that other grid adaptation concepts, for instance, based on error estimators, can be used as an alternative in the same framework.

For a proof of concept the multilevel strategy has been applied to 1D Euler equations with one random variable in the initial data. The results have been compared with the classical Monte–Carlo method. The multilevel strategy has been verified to be significantly more efficient than the Monte–Carlo method, i.e., less DoFs are needed to realize the same accuracy. In [22] investigations for other test configurations and comparisons with the multilevel Monte–Carlo method can be found confirming the above findings.

In future work, we will extend the multilevel strategy to *multidimensional* problems with *multiple* uncertainties both in the initial data and the fluxes. For this purpose, the multilevel strategy has to be parallelized. Investigations with multiple CPU cores have to be considered and a more detailed performance analysis will be performed. Furthermore, to deal with the curse of dimensionality we may apply *sparse tensor strategies*, cf. [27], on the stochastic space to reduce the number of realizations in the multilevel strategy.

Acknowledgements The authors thank the Deutsche Forschungsgemeinschaft (DFG, German Research Foundation) for the financial support through Deutsche Forschungsgemeinschaft (DFG, German Research Foundation) by 320021702/GRK2326 Energy, Entropy, and Dissipative Dynamics (EDDY).

References

1. Abgrall, R., Mishra, S.: Uncertainty quantification for hyperbolic systems of conservation laws. In: Handbook of Numerical Analysis, pp. 507–544. Elsevier (2017). DOI 10.1016/bs.hna.2016.11.003
2. Alpert, B.: A class of bases in L^2 for the sparse representation of integral operators. SIAM J. Math. Anal. **24**, 246–262 (1993)
3. Badwaik, J., Klingenberg, C., Risebro, N.H., Ruf, A.M.: Multilevel Monte Carlo finite volume methods for random conservation laws with discontinuous flux. ESAIM: Mathematical Modelling and Numerical Analysis **55**(3), 1039–1065 (2021). DOI 10.1051/m2an/2021011
4. Barth, T.: On the propagation of statistical model parameter uncertainty in CFD calculations. Theoretical and Computational Fluid Dynamics **26**(5), 435–457 (2011). DOI 10.1007/s00162-011-0221-2
5. Barth, T.: Non-intrusive uncertainty propagation with error bounds for conservation laws containing discontinuities. In: Uncertainty Quantification in Computational Fluid Dynamics, pp. 1–57. Springer International Publishing (2013). DOI 10.1007/978-3-319-00885-1_1
6. Cameron, R.H., Martin, W.T.: The orthogonal development of non-linear functionals in series of Fourier-Hermite functionals. The Annals of Mathematics **48**(2), 385 (1947). DOI 10.2307/1969178
7. Cockburn, B., Shu, C.W.: The Runge-Kutta discontinuous Galerkin method for conservation laws V. Journal of Computational Physics **141**(2), 199–224 (1998). DOI 10.1006/jcph.1998.5892
8. Dafermos, C.M.: Hyperbolic Conservation Laws in Continuum Physics. Springer (2016)
9. Després, B., Poëtte, G., Lucor, D.: Robust uncertainty propagation in systems of conservation laws with the entropy closure method. In: Uncertainty Quantification in Computational Fluid Dynamics, pp. 105–149. Springer International Publishing (2013). DOI 10.1007/978-3-319-00885-1_3
10. Gerhard, N.: An adaptive multiresolution discontinuous Galerkin scheme for conservation laws. PhD dissertation, RWTH Aachen (2017). DOI 10.18154/RWTH-2017-06869
11. Gerhard, N., Iacono, F., May, G., Müller, S., Schäfer, R.: A high-order discontinuous Galerkin discretization with multiwavelet-based grid adaptation for compressible flows. J. Sci. Comput. **62**(1), 25–52 (2015). DOI 10.1007/s10915-014-9846-9
12. Gerhard, N., Müller, S.: Adaptive multiresolution discontinuous Galerkin schemes for conservation laws: multi-dimensional case. Comp. Appl. Math. **35**(2), 321–349 (2016). DOI 10.1007/s40314-014-0134-y
13. Gerster, S., Herty, M., Sikstel, A.: Hyperbolic stochastic Galerkin formulation for the p-system. Journal of Computational Physics **395**, 186–204 (2019). DOI 10.1016/j.jcp.2019.05.049

14. Giesselmann, J., Meyer, F., Rohde, C.: A posteriori error analysis and adaptive non-intrusive numerical schemes for systems of random conservation laws. BIT Numerical Mathematics **60**(3), 619–649 (2020). DOI 10.1007/s10543-019-00794-z
15. Godlewski, E.: Numerical Approximation of Hyperbolic Systems of Conservation Laws. Springer, New York (1996)
16. Golub, G.H., Welsch, J.H.: Calculation of Gauss quadrature rules. Mathematics of Computation **23**(106), 221–230 (1969). DOI 10.1090/s0025-5718-69-99647-1
17. Gottlieb, D., Xiu, D.: Galerkin method for wave equations with uncertain coefficients. Communications in Computational Physics **3**, 505–518 (2008)
18. Gottlieb, S., Shu, C.W., Tadmor, E.: Strong stability-preserving high-order time discretization methods. SIAM Review **43**(1), 89–112 (2001). DOI 10.1137/s003614450036757x
19. Harten, A.: Multiresolution algorithms for the numerical solution of hyperbolic conservation laws. Communications on Pure and Applied Mathematics **48**, 1305–1342 (1995)
20. Herty, M., Kolb, A., Müller, S.: Multiresolution analysis for stochastic hyperbolic conservation laws. IMA Journal of Numerical Analysis (2023). DOI 10.1093/imanum/drad010. URL `https://doi.org/10.1093/imanum/drad010`
21. Hovhannisyan, N., Müller, S., Schäfer, R.: Adaptive multiresolution discontinuous Galerkin schemes for conservation laws. Mathematics of Computation **83**(285), 113–151 (2014)
22. Kolb, A.: Multiresolution-based grid adaptation for hyperbolic conservation laws with uncertain initial data. PhD dissertation, RWTH Aachen (2024). DOI 10.18154/RWTH-2024-00677
23. Le Maître, O.P., Knio, O.M.: Spectral Methods for Uncertainty Quantification. Springer-Verlag (2010)
24. Leveque, R.J.: Numerical Methods for Conservation Laws. Birkhäuser Basel (2008)
25. Meyer, F., Rohde, C., Giesselmann, J.: A posteriori error analysis for random scalar conservation laws using the stochastic Galerkin method. IMA Journal of Numerical Analysis **40**(2), 1094–1121 (2019). DOI 10.1093/imanum/drz004
26. Mishra, S., Risebro, N.H., Schwab, C., Tokareva, S.: Numerical solution of scalar conservation laws with random flux functions. SIAM/ASA Journal on Uncertainty Quantification **4**(1), 552–591 (2016). DOI 10.1137/120896967
27. Mishra, S., Schwab, C.: Sparse tensor multi-level Monte Carlo finite volume methods for hyperbolic conservation laws with random initial data. Mathematics of Computation **81**(280), 1979–2018 (2012). DOI 10.1090/s0025-5718-2012-02574-9
28. Mishra, S., Schwab, C., Šukys, J.: Multilevel Monte Carlo finite volume methods for shallow water equations with uncertain topography in multi-dimensions. SIAM Journal on Scientific Computing **34**(6), B761–B784 (2012). DOI 10.1137/110857295
29. Nordström, J.: Conservative finite difference formulations, variable coefficients, energy estimates and artificial dissipation. Journal of Scientific Computing **29**(3), 375–404 (2005). DOI 10.1007/s10915-005-9013-4
30. Öffner, P., Glaubitz, J., Ranocha, H.: Stability of correction procedure via reconstruction with summation-by-parts operators for Burgers equation using a polynomial chaos approach. ESAIM: Mathematical Modelling and Numerical Analysis **52**(6), 2215–2245 (2018). DOI 10.1051/m2an/2018072
31. Pettersson, P., Iaccarino, G., Nordström, J.: A stochastic Galerkin method for the Euler equations with Roe variable transformation. Journal of Computational Physics **257**, 481–500 (2014). DOI 10.1016/j.jcp.2013.10.011
32. Poëtte, G., Després, B., Lucor, D.: Uncertainty quantification for systems of conservation laws. Journal of Computational Physics **228**(7), 2443–2467 (2009). DOI 10.1016/j.jcp.2008.12.018
33. Pulch, R., Xiu, D.: Generalised polynomial chaos for a class of linear conservation laws. Journal of Scientific Computing **51**(2), 293–312 (2011). DOI 10.1007/s10915-011-9511-5
34. Risebro, N.H., Schwab, C., Weber, F.: Multilevel Monte Carlo front-tracking for random scalar conservation laws. BIT Numerical Mathematics **56**(1), 263–292 (2015). DOI 10.1007/s10543-015-0550-4
35. Robert, C.P., Casella, G.: Monte Carlo Statistical Methods. Springer (2004)

36. Schlachter, L., Schneider, F., Kolb, O.: Weighted essentially non-oscillatory stochastic Galerkin approximation for hyperbolic conservation laws. Journal of Computational Physics **419**, 109663 (2020). DOI 10.1016/j.jcp.2020.109663
37. Schwab, C., Tokareva, S.: High order approximation of probabilistic shock profiles in hyperbolic conservation laws with uncertain initial data. ESAIM: Mathematical Modelling and Numerical Analysis **47**(3), 807–835 (2013). DOI 10.1051/m2an/2012060
38. Sullivan, T.J.: Introduction to Uncertainty Quantification. Springer-Verlag (2016)
39. Tokareva, S.: Stochastic finite volume methods for computational uncertainty quantification in hyperbolic conservation laws. Ph.D. thesis, ETH Zurich (2013). DOI 10.3929/ETHZ-A-009965237
40. Wan, X., Karniadakis, G.E.: Multi-element generalized polynomial chaos for arbitrary probability measures. SIAM Journal on Scientific Computing **28**(3), 901–928 (2006). DOI 10.1137/050627630
41. Xiu, D.: Numerical Methods for Stochastic Computations: A Spectral Method Approach. Princeton University Press, Princeton, N.J (2010)
42. Xiu, D., Hesthaven, J.S.: High-order collocation methods for differential equations with random inputs. SIAM Journal on Scientific Computing **27**(3), 1118–1139 (2005). DOI 10.1137/040615201
43. Xiu, D., Karniadakis, G.E.: The Wiener–Askey polynomial chaos for stochastic differential equations. SIAM Journal on Scientific Computing **24**(2), 619–644 (2002). DOI 10.1137/s1064827501387826

On the Construction of Bases and Frames with Applications

Kamen G. Ivanov, Gerard Kerkyacharian, George Kyriazis, and Pencho Petrushev

Abstract This article is a survey of results on the construction of bases and frames for Besov and Triebel-Lizorkin spaces (B- and F-spaces) in various classical and nonclassical settings. We first review our small-perturbation method for construction of bases for B- and F-spaces on $\mathbb{R}^d$ based on wavelets and show some of its applications to nonlinear approximation. Then we describe the construction of frames for B- and F-spaces in the classical setting on $\mathbb{R}^d$, on the sphere in $\mathbb{R}^d$, and in a general setting that covers a number of classical and nonclassical setups such as on $\mathbb{R}^d$, the interval, sphere, ball, simplex as well as Riemannian manifolds and Lee groups. Finally, we present a small perturbation method for construction of frames for B- and F-spaces in a general setting. Several applications of this scheme are presented.

Kamen G. Ivanov
Institute of Mathematics and Informatics, Bulgarian Academy of Sciences, Sofia, Bulgaria, e-mail: kamen@math.bas.bg; Supported by Grant KP-06-N62/4 of the Fund for Scientific Research of the Bulgarian Ministry of Education and Science.

Gerard Kerkyacharian
University Paris Diderot-Paris 7, LPMA, Paris, France, e-mail: kerk@math.univ-paris-diderot.fr

George Kyriazis
Department of Mathematics and Statistics, University of Cyprus, 1678 Nicosia, Cyprus, e-mail: kyriazis@ucy.ac.cy

Pencho Petrushev
Department of Mathematics, University of South Carolina, Columbia, SC 29208, USA, e-mail: pencho@math.sc.edu; Supported by Grant KP-06-N62/4 of the Fund for Scientific Research of the Bulgarian Ministry of Education and Science.

R. DeVore, A. Kunoth (eds.), *Multiscale, Nonlinear and Adaptive Approximation II*,
https://doi.org/10.1007/978-3-031-75802-7_16

1 Introduction

Bases and frames play a fundamental role in Harmonic analysis and, in particular, in the theory of Besov and Triebel-Lizorkin spaces as well as in Approximation theory. The point is that most distribution spaces such as Besov and Triebel-Lizorkin spaces can be characterized by the magnitude of the respective basis or frame coefficients. Therefore, using bases or frames one can simplify, extend, and unify a variety of results in Harmonic analysis and Approximation theory. For instance, one obtains conditions for boundedness of linear operators on Besov and Triebel-Lizorkin spaces by considering conditions for matrices on the respective sequence spaces with applications to e.g. multipliers, interpolation, dual spaces.

In this survey we focus on the construction of bases and frames and their use for characterization of Besov and Triebel-Lizorkin spaces (B- and F-spaces) in various classical and nonclassical settings. More specifically we are interested in bases and frames for B- and F-spaces with smoothness parameter s in the range $|s| \leqslant A$ for large A. We will term this sort of bases and frames *good bases and frames*. There are three general features that characterize a *good basis or frame* in the context of Besov and Triebel-Lizorkin spaces on $\mathbb{R}^d$ or a subset of $\mathbb{R}^d$, namely,

(i) an excellent space localization,
(ii) high smoothness, and
(iii) a number of vanishing moments

of the base or frame elements. Thus we are in principle interested in bases and frames with these properties or natural substitutes of some of them depending on the particular setting.

Bases. Recall that a sequence $\{\psi_n\}_{n\geqslant 1} \subset \mathfrak{B}$ is a basis for the Banach or quasi-Banach space $\mathfrak{B}$ over $\mathbb{C}$ or $\mathbb{R}$ if each $f \in \mathfrak{B}$ has a unique representation in the form

$$f = \sum_{n\geqslant 1} c_n\psi_n, \quad c_n \in \mathbb{C} \quad \text{or} \quad c_n \in \mathbb{R}, \tag{1}$$

where the convergence is in the norm of $\mathfrak{B}$, and $\{\psi_n\}$ is a unconditional basis if it is a basis and the convergence in (1) is unconditional, i.e. every rearrangement also converges. We are only interested in **good bases** which means unconditional bases for Besov spaces B_p^{sq} or Triebel-Lizorkin spaces F_p^{sq} with sufficiently wide range for the smoothness parameter s. Note that all classical orthogonal systems are not good bases in this sense. This is the case for the trigonometric system $\{e^{2\pi kx}\}$, Jacobi polynomials on $[-1, 1]$, spherical harmonics, orthogonal polynomials on the ball and simplex as well as the Hermite and Laguerre functions. It is quite easy to manufacture orthonormal bases by applying the Gram-Schmidt process. Such an example for instance is the Franklin system on $[0, 1]$ created by orthogonalization of the constant 1 and the head functions obtained by integration of the Haar functions. However, it is very hard to construct good bases. Yet the problem becomes harder if one wants the basis elements to have some additional properties. The basis elements of a good basis need to obey conditions (i)–(iii) above. The Franklin basis elements lacks smoothness and enough vanishing moments. Therefore, it is not good. Observe that the Haar basis is also not a good one because the Haar functions lack smoothness and vanishing moments. The requirement of uniqueness of representation (1) plus conditions (i)–(iii) let along other features are very stringent making it difficult as a rule to construct good bases. A huge success in this respect was the construction of wavelet bases by Y. Meyer, I. Daubechies, and many others, see [34], [9], [51], and the references therein.

Our approach to construction of good bases utilizes a small perturbation method. It works in all settings where there is a good basis. This method originated in [43] and was further refined in [30] in the classical case on $\mathbb{R}^d$, where a perturbation of the tensor-product Meyer's wavelets was used. The gist of this scheme is to approximate the wavelets by new basis functions obeying conditions (i)–(iii) above in a specific economical manner which uses (small) ε approximation budget. We describe this method in Section 3 below. The method is quite flexible in allowing to prescribe the nature and properties of the new basis elements. For example, good rational bases on $\mathbb{R}^d$ were constructed in [43, 30]. More precisely, the new small perturbation technique has been applied in [43, 30] for the construction of bases consisting of functions that are linear combinations of a fixed number of shifts and dilates of a single function Φ. This function has to be smooth enough and with sufficiently rapid decay. For example, the rational function $\Phi(x) = (1 + |x|^2)^{-N}$ generates the above mentioned rational basis. Another interesting example is the Gaussian $\Phi(x) = e^{-|x|^2}$. A good rational basis has been constructed in [31] for holomorphic Besov and Triebel-Lizorkin spaces on the unit disc in $\mathbb{C}$ by small perturbation of Meyer's two hump holomorphic wavelet basis [34, Section 6.11]. In [43] a good basis was constructed on the interval $[0, 1]$, where compactly supported wavelets (see e.g. [5, 1]) were used. Note that Ciesielski's spline bases [3, 4] can also be used in this case. Also, good bases for the space VMO have been constructed in [20] and utilized for rational approximation in BMO on $\mathbb{R}^d$. The circle $\mathbb{T}$ is also amenable to our small perturbation method for construction of good bases, where 1-periodic Meyer's wavelets (see [34, Section 3.11] or [51, Section 2.5]) are to be used.

Frames. Good frames for Besov and Triebel-Lizorkin spaces are an alternative to good bases. Several important points are to be made regarding frames. First, just as good bases good frames can be used for complete characterization of Besov and Triebel-Lizorkin spaces with the full range of indexes. Second, good frames are much easier to construct than bases. Third, good frames can be constructed in a variety of setting such as on $[-1, 1]$ with Jacobi weights, on the sphere, ball, simplex as well as Riemannian manifolds and Lee groups, where there are no good bases as for now. Fourth, it is easy to prescribe the properties of the constructed frames. The main disadvantage of frames over bases is that the representations of functions or distributions using frames are not unique. We will describe two methods for construction of frames.

The method of Frazier and Jawerth for construction of frames. This method has been used for construction of frames on $\mathbb{R}^d$ in [12, 13, 14]. The same method was first utilized for construction of frames in settings other then $\mathbb{R}^d$ on the sphere in [36, 37]. After that it was used for construction of frames on $[-1, 1]$ with Jacobi weights in [44, 32], on the ball in [45, 33], on product spaces in [19], on conic and double hyperbolic domains in [52, 53], in the context of Hermite and Laguerre expansions in [46] and [27], respectively. Below we present this method on $\mathbb{R}^d$, on the unit sphere $\mathbb{S}^{d-1}$ in $\mathbb{R}^d$, and in a general setting of a doubling metric measure space in the presence of a non-negative self-adjoint operator whose heat kernel has Gaussian space localization and the Markov property, developed in [6, 24]. This general setup covers the settings mentioned above and many others, see [6]. The method relies on the following

Localization principle. Let (X, μ, ρ) be a metric measure space and suppose that there is an orthogonal decomposition of $L^2(X, \mu)$:

$$L^2(X,\mu) = \bigoplus_{n=0}^{\infty} \mathcal{V}_n,$$

where $\mathcal{V}_n$ is a subspace of L^2 of dimension $\dim \mathcal{V}_n \sim n^d$, $d \geqslant 1$. Let $P_n(x, y)$ be the kernel of the orthogonal projector $\mathrm{Proj}_n : L^2(X, \mu) \to \mathcal{V}_n$, i.e.

$$(\mathrm{Proj}_n f)(x) = \int_X P_n(x,y) f(y) d\mu, \quad f \in L^2(X,\mu).$$

Clearly, $P_n(x, y)$ can be written in the form $P_n(x, y) = \sum_{j=1}^{\dim \mathcal{V}_n} p_j(x)\overline{p_j(y)}$, where $\{p_j\}$ is an orthonormal basis for $\mathcal{V}_n$.

Suppose $\varphi \in C^\infty(\mathbb{R})$ is real-valued and even, and φ is compactly supported or $\varphi \in \mathcal{S}$ (the Schwartz class of rapidly decreasing C^∞ functions on $\mathbb{R}$). Define

$$L_N(x,y) := \sum_{j=0}^{\infty} \varphi\Big(\frac{j}{N}\Big) P_j(x,y), \quad N \in \mathbb{N}.$$

Denote $B(x,r) := \{y \in X : \rho(x,y) < r\}$ and set $V(x,r) := \mu(B(x,r))$.

Our general localization principle asserts that for all "natural orthogonal systems", the kernel $L_N(x, y)$ from above has nearly exponential space localization; more specifically, for any $M > 0$ there exists a constant $c = c(M)$ such that

$$|L_N(x,y)| \leqslant \frac{cN^d\big(1 + N\rho(x,y)\big)^{-M}}{[V(x,N^{-1})V(y,N^{-1})]^{1/2}}, \quad x, y \in X. \tag{2}$$

Observe that the localization principle is valid for the trigonometric system $\{e^{2\pi nx}\}$ and all of the above mentioned settings. For the case of spherical harmonics, see Theorem 5.1 in Section 5 below. In the general setting of a doubling metric measure space with a non-negative self-adjoint operator whose heat kernel has Gaussian space localization and the Markov property the localization principle is expressed in terms of localization of kernels of spectral multipliers, see Theorem 6.1 in Section 6 below. The localization principle on $\mathbb{R}^d$ simply says that the Fourier transform $\widehat{\phi}$ of each function ϕ in the Schwartz class $\mathcal{S}(\mathbb{R}^d)$ belongs to $\mathcal{S}(\mathbb{R}^d)$.

The localization principle has several important implications:

(a) It allows to naturally introduce spaces of distributions in various nonclassical settings and, in particular, to develop Besov and Triebel-Lizorkin spaces with full range of indexes.

(b) It makes it possible to construct good frames, which can be used for characterization of Besov and Triebel-Lizorkin spaces in term of respective sequence spaces. In turn, frames and the associated sequence spaces allow to handle more easily hosts of problems such as for interpolation, Hörmander multipliers, and dual spaces.

(c) It allows to develop Approximation theory in nonclassical settings.

For details see Sections 5, 6 below.

Small perturbation method for construction of good frames. This method was a natural progression of the idea for construction of bases from [43]. It has been used for construction of frames with small shrinking support on the sphere in [30], and in the context of harmonic Besov and Triebel-Lizorkin spaces on the ball with an application to nonlinear n-term approximation from shifts of the Newtonian kernel in [18]. We will describe this method in a general setting following the development in [10], where compactly supported frames were constructed.

Applications of the newly constructed good bases and frames mainly to nonlinear approximation theory will be given.

Outline. This paper is organised as follows: In Section 2, tensor-product Meyer's wavelets are recalled and it is shown how they are used for characterization of Besov and Triebel-Lizorkin spaces and BMO on $\mathbb{R}^d$. Our small perturbation method for construction of bases on $\mathbb{R}^d$ is given in Section 3, where concrete new bases are described as well as bases in terms of shifts and dilates of a single function are given. In Section 4, the construction of good frames and the frame characterization of Besov and Triebel-Lizorlin spaces on $\mathbb{R}^d$ are given. Similar construction and results on the unit sphere in $\mathbb{R}^d$ are described in Section 5, and in a general setting in Section 6. Section 7 is devoted to our small perturbation method for construction of frames in a general setting.

2 Wavelets are good bases

In this section we show that the tensor-product Meyer's wavelets on $\mathbb{R}^d$ form a good basis.

2.1 Tensor-product Meyer's wavelets on $\mathbb{R}^d$

We will use the dyadic cubes on $\mathbb{R}^d$ to index the elements of the tensor-product Meyer's wavelet system Ψ that we describe next. Denote by $\mathscr{D}_j$ the set of all dyadic cubes of side-length 2^{-j}, i.e. $I \in \mathscr{D}_j$ if it is of the form

$$I = \Big[\frac{k_1}{2^j}, \frac{k_1+1}{2^j}\Big) \times \cdots \times \Big[\frac{k_d}{2^j}, \frac{k_d+1}{2^j}\Big), \quad k_\nu \in \mathbb{Z}. \tag{3}$$

Set $\mathcal{D} := \cup_{j\in\mathbb{Z}}\mathcal{D}_j$.

For any $I \in \mathcal{D}$ we denote by $\ell(I)$ its side length and by x_I its lower left-corner, e.g. $x_I = (k_1 2^{-j}, \dots, k_d 2^{-j})$ in the case of I from (3). Also, for any function g on $\mathbb{R}^d$ we denote

$$g_I(x) := |I|^{-1/2} g\Big(\frac{x - x_I}{\ell(I)}\Big), \quad I \in \mathcal{D}. \tag{4}$$

We denote by φ and ψ the scaling function and mother wavelet of Meyer's wavelet system on $\mathbb{R}$, see [34]. Set $\psi^0 := \varphi$, $\psi^1 := \psi$. Let E be the set of all nonzero vertices of $[0,1]^d$ and set $E_0 := E \cup \{(0,\dots,0)\}$. For $e = (e_1,\dots,e_d) \in E_0$ we define

$$\psi^e(x) := \prod_{j=1}^d \psi^{e_j}(x_j). \tag{5}$$

The tensor-product Meyer's wavelet basis Ψ is defined by

$$\Psi := \{\psi_I^e : I \in \mathcal{D}, e \in E\}.$$

We next specify some basic properties of the wavelet system Ψ.

P1. Ψ is an orthonormal wavelet system, i.e.

$$\langle \psi_I^e, \psi_{I'}^{e'}\rangle = \delta_{I,I'}\delta_{e,e'}, \quad I, I' \in \mathcal{D}, \quad e, e' \in E.$$

Also, $\langle \psi_I^0, \psi_{I'}^0\rangle = \delta_{I,I'}$, $I, I' \in \mathcal{D}_j$, $j \in \mathbb{Z}$.

P2. Ψ is an orthonormal basis for $L^2(\mathbb{R}^d)$, that is, for any $f \in L^2(\mathbb{R}^d)$

$$f = \sum_{I\in\mathcal{D}, e\in E} \langle f, \psi_I^e\rangle \psi_I^e \quad \text{and} \quad \|f\|_{L^2} = \Big(\sum_{I\in\mathcal{D}, e\in E} |\langle f, \psi_I^e\rangle|^2\Big)^{1/2}.$$

P3. For any $M > 0$ and $\alpha \in \mathbb{N}^d$ there exists a constant $c = c(\alpha, M)$ such that

$$|\partial^\alpha \psi_I^e(x)| \leqslant \frac{c 2^{j(d/2+|\alpha|)}}{(1 + 2^j|x - x_I|)^M}, \quad x \in \mathbb{R}^d, \ I \in \mathcal{D}_j, \ e \in E_0,$$

and

$$\int_{\mathbb{R}^d} \psi_I^e(x) x^\alpha dx = 0, \quad I \in \mathcal{D}, \ e \in E.$$

The following inhomogeneous version of Meyer's wavelets is sometimes used:

$$\Psi_0 := \{\psi_I^0 : I \in \mathcal{D}_0\} \cup \{\psi_I^e : I \in \cup_{j>0}\mathcal{D}_j, e \in E\}. \tag{6}$$

We refer the reader to [51, Section 5] for more details and proofs, see also [34].

2.2 Besov and Triebel-Lizorkin spaces on $\mathbb{R}^d$

We next recall the definition of the homogeneous and inhomogeneous Besov and Triebel-Lizorkin spaces on $\mathbb{R}^d$ and their wavelet characterization. We refer the reader to [39, 50] and also [12, 13, 14] for the theory of Besov and Triebel-Lizorkin spaces.

Let the function $\Psi \in \mathcal{S}(\mathbb{R}^d)$ be such that

$$\operatorname{supp} \widehat{\Psi} \subset \{\xi \in \mathbb{R}^d : 2^{-1} \leqslant |\xi| \leqslant 2\} \text{ and } |\widehat{\Psi}(\xi)| \geqslant c > 0 \text{ if } 3/5 \leqslant |\xi| \leqslant 5/3. \quad (7)$$

We will use the notation $\Psi_\nu(\cdot) := 2^{\nu d}\Psi(2^\nu \cdot)$. Further denote by $\mathcal{P}$ the set of all algebraic polynomials on $\mathbb{R}^d$ and by $\mathcal{S}'/\mathcal{P}$ the set of tempered distributions modulo polynomials.

Given $s \in \mathbb{R}$ and $0 < p, q \leqslant \infty$, *the homogeneous Besov space* $\dot{B}_p^{sq}$ is defined as the set of all $f \in \mathcal{S}'/\mathcal{P}$ such that

$$\|f\|_{\dot{B}_p^{sq}} := \Big(\sum_{\nu \in \mathbb{Z}} \big[2^{\nu s}\|\Psi_\nu * f\|_{L^p}\big]^q\Big)^{1/q} < \infty,$$

where the ℓ^q-norm is replaced by the sup-norm if $q = \infty$.

Similarly, given $s \in \mathbb{R}$, $0 < p < \infty$, and $0 < q \leqslant \infty$, *the homogeneous Triebel-Lizorkin space* $\dot{F}_p^{sq}$ is defined as the set of all $f \in \mathcal{S}'/\mathcal{P}$ such that

$$\|f\|_{\dot{F}_p^{sq}} := \Big\|\Big(\sum_{\nu \in \mathbb{Z}} \big[2^{\nu s}|\Psi_\nu * f|\big]^q\Big)^{1/q}\Big\|_{L^p} < \infty.$$

Furthermore, for $s \in \mathbb{R}$, $0 < q \leqslant \infty$, *the homogeneous Triebel-Lizorkin space* $\dot{F}_\infty^{sq}$ is defined as the set of all $f \in \mathcal{S}'/\mathcal{P}$ such that

$$\|f\|_{\dot{F}_\infty^{sq}} := \sup_{I \in \mathcal{D}} \Big(\frac{1}{|I|}\int_I \sum_{\nu \geqslant -\log_2 \ell(I)} \big[2^{\nu s}|\Psi_\nu * f(x)|\big]^q dx\Big)^{1/q} < \infty.$$

Note that the above definitions are independent of the specific selection of the function Ψ with the required properties.

Sometimes for local problems the **inhomogeneous Besov and Triebel-Lizorkin spaces** are more appropriate. Let $\Phi \in \mathcal{S}(\mathbb{R}^d)$ be such that

$$\operatorname{supp} \widehat{\Phi} \subset \{\xi \in \mathbb{R}^d : |\xi| \leqslant 2\} \quad \text{and} \quad |\widehat{\Phi}(\xi)| \geqslant c > 0 \text{ if } |\xi| \leqslant 5/3.$$

Let $s \in \mathbb{R}$ and $0 < p, q \leqslant \infty$. Using also the notation from above (see (7)) *the inhomogeneous Besov space* B_p^{sq} is defined as the set of all $f \in \mathcal{S}'$ such that

$$\|f\|_{B_p^{sq}} := \|\Phi * f\|_{L^p} + \Big(\sum_{\nu \in \mathbb{Z}} \big[2^{\nu s}\|\Psi_\nu * f\|_{L^p}\big]^q\Big)^{1/q} < \infty.$$

Let $s \in \mathbb{R}$, $0 < p < \infty$, and $0 < q \leqslant \infty$. *The inhomogeneous Triebel-Lizorkin space* F_p^{sq} is defined as the set of all $f \in \mathcal{S}'$ such that

$$\|f\|_{F_p^{sq}} := \|\Phi * f\|_{L^p} + \Big\|\Big(\sum_{\nu\in\mathbb{Z}} \big[2^{\nu s}|\Psi_\nu * f|\big]^q\Big)^{1/q}\Big\|_{L^p} < \infty.$$

In addition, given $s \in \mathbb{R}$ and $0 < q < \infty$, *the inhomogeneous Triebel-Lizorkin space* F_∞^{sq} is defined as the set of all $f \in \mathcal{S}'$ such that

$$\|f\|_{F_\infty^{sq}} := \|\Phi * f\|_{L^\infty} + \sup_{I\in\mathscr{D}} \Big(\frac{1}{|I|}\int_I \sum_{\nu\geqslant -\log_2 \ell(I)} [2^{\nu s}|\Psi_\nu * f(x)|]^q dx\Big)^{1/q} < \infty.$$

Recall the following identifications:

$$L^p \sim F_p^{02} \sim \dot{F}_p^{02} \quad \text{if } 1 < p < \infty \quad \text{and} \quad H^p \sim \dot{F}_p^{02} \quad \text{if } 0 < p \leqslant 1,$$

where H^p stands for the Hardy space on $\mathbb{R}^d$. For a similar identification of BMO, see Section 2.3 below.

Wavelet decomposition of homogeneous Besov and Triebel-Lizorkin spaces. We next present the characterization of the homogeneous Besov and Triebel-Lizorkin spaces on $\mathbb{R}^d$ in terms of tensor-product Meyer's wavelet coefficients.

To state this result we now introduce the associated sequence spaces.

Definition 2.1 (a) For $s \in \mathbb{R}$ and $0 < p, q \leqslant \infty$ the space $\dot{b}_p^{sq}$ is defined as the collection of all sequences $h = \{h_{Ie}\}_{Ie\in\mathscr{D}\times E}$ of complex numbers such that

$$\|h\|_{\dot{b}_p^{sq}} := \Big(\sum_{\nu\in\mathbb{Z}} \Big[\sum_{I\in\mathscr{D}_\nu, e\in E} (|I|^{-s/d-1/2+1/p}|h_{Ie}|)^p\Big]^{q/p}\Big)^{1/q} < \infty.$$

(b) For $s \in \mathbb{R}$, $0 < p < \infty$, and $0 < q \leqslant \infty$ the space $\dot{f}_p^{sq}$ is defined as the collection of all sequences $h = \{h_{Ie}\}_{Ie\in\mathscr{D}\times E}$ of complex numbers such that

$$\|h\|_{\dot{f}_p^{sq}} := \Big\|\Big(\sum_{I\in\mathscr{D}, e\in E} (|I|^{-s/d}|h_{Ie}|\tilde{\mathbb{1}}_I)^q\Big)^{1/q}\Big\|_{L^p} < \infty,$$

where $\tilde{\mathbb{1}}_I(x) := |I|^{-1/2}\mathbb{1}_I(x)$ is the L^2-normalized characteristic function of I.

(c) For $s \in \mathbb{R}$ and $0 < q \leqslant \infty$ the space $\dot{f}_\infty^{sq}$ is defined as the collection of all sequences $h = \{h_{Ie}\}_{Ie\in\mathscr{D}\times E}$ of complex numbers such that

$$\|h\|_{\dot{f}_\infty^{sq}} := \sup_{J\in\mathscr{D}} \Big(\frac{1}{|J|}\int_J \sum_{I\subset J, e\in E} \big(|I|^{-s/d}|h_{Ie}|\tilde{\mathbb{1}}_I(x)\big)^q dx\Big)^{1/q} < \infty.$$

Above as usual the L^p, L^q, ℓ^p, ℓ^q norms are replaced by the respective sup-norms if $p = \infty$ or $q = \infty$.

We also introduce the following analysis and synthesis operators:

$$S_\psi : f \to \{\langle f, \psi_I^e\rangle\}_{Ie\in\mathscr{D}\times E}, \quad T_\psi : \{h_{Ie}\}_{Ie\in\mathscr{D}\times E} \to \sum_{Ie\in\mathscr{D}\times E} h_{Ie}\psi_I^e. \tag{8}$$

Theorem 2.1 *Let $s \in \mathbb{R}$ and $0 < p, q \leqslant \infty$. Then $S_\psi : \dot{B}_p^{sq} \to \dot{b}_p^{sq}$ is a bounded isomorphism with bounded inverse $T_\psi : \dot{b}_p^{sq} \to \dot{B}_p^{sq}$. Therefore, each $f \in \dot{B}_p^{sq}$ has a unique representation in the form*

$$f = \sum_{I \in \mathcal{D}} \sum_{e \in E} \langle f, \psi_I^e \rangle \psi_I^e,$$

where the convergence is in the sense of $\mathcal{S}'/\mathcal{P}$ (and unconditional in $\dot{B}_p^{sq}$ if $p, q \neq \infty$) and

$$\|f\|_{\dot{B}_p^{sq}} \sim \|S_\psi f\|_{\dot{b}_p^{sq}}.$$

The same is true for $\dot{F}_p^{sq}$ and $\dot{f}_p^{sq}$.

Observe that the inhomogeneous Besov and Triebel-Lizorkin spaces B_p^{sq} and F_p^{sq} have characterizations similar to the one from Theorem 2.1 using the inhomogeneous version of tensor-product Meyer's wavelets from (6).

For more details on Besov and Triebel-Lizorkin spaces and proofs of the above results, see [28, 12, 13, 14].

2.3 Wavelet characterization of BMO and VMO

The space of functions of bounded mean oscillation BMO plays an important role in Harmonic analysis. The space BMO on $\mathbb{R}^d$ is defined as the set of all functions $f \in L^1_{\text{loc}}(\mathbb{R}^d)$ such that

$$\|f\|_{\text{BMO}} := \sup_B \frac{1}{|B|} \int_B |f(x) - \text{Avg}_B f| dx < \infty, \quad \text{Avg}_B f := \frac{1}{|B|} \int_B f(x) dx, \tag{9}$$

where the sup is over all balls B in $\mathbb{R}^d$ and $|B|$ stands for the volume of B. The reader is referred to [47, Chapter 4], [15, Chapter 6], [17, Chapter 7], and [51, Section 6.2] for the basics of BMO.

Note that BMO is a Banach space, but it is not separable. For this reason in Approximation theory one works with functions in the space VMO of vanishing mean oscillation, defined as the closure of $C_0(\mathbb{R}^d)$ in the BMO norm (see Section 4 in [7]). VMO is a closed subspace of BMO.

As is well known (see [13, Section 5] and also [19, Theorem 3.3]) BMO $= F_\infty^{02}$ with equivalent norms

$$\|f\|_{\text{BMO}} \sim \|f\|_{F_\infty^{02}} \sim \|\{\langle f, \psi_I^e \rangle\}\|_{\mathfrak{f}_\infty^{02}}, \quad \forall f \in \text{BMO}.$$

The following wavelet characterization of VMO is obtained in [20, Theorem 3.8].

Theorem 2.2 *Let $f \in$ BMO and denote*

$$\tilde{M}(f, J) := \Big(\frac{1}{|J|} \sum_{I \in \mathcal{D}, I \subset J, e \in E} |\langle f, \psi_I^e \rangle|^2 \Big)^{1/2}, \quad J \in \mathcal{D}.$$

Then $f \in \mathrm{VMO}$ *if and only if*

$$\lim_{a\to 0} \sup_{J\in\mathcal{D}, |J|\leqslant a} \tilde{M}(f,J) = 0,$$

$$\lim_{a\to \infty} \sup_{J\in\mathcal{D}, |J|\geqslant a} \tilde{M}(f,J) = 0, \quad \textit{and}$$

$$\lim_{a\to \infty} \sup_{J\in\mathcal{D}_j, |x_J|\geqslant a} \tilde{M}(f,J) = 0, \quad \forall j \in \mathbb{Z}.$$

3 Construction of bases by small perturbation

In this section we present our small perturbation method for construction of bases. The idea for this construction was first conceived in [43] and further refined in [29] and then implemented in [20] and [31]. We will present the version of the small perturbation scheme on $\mathbb{R}^d$, where tensor-product Meyer's wavelets are used as a backbone in the construction.

Setting and conditions on the old basis. As elsewhere in this article here $\mathcal{S} = \mathcal{S}(\mathbb{R}^d)$ stands for the Schwartz class of test function on $\mathbb{R}^d$ and $\mathcal{S}' = \mathcal{S}'(\mathbb{R}^d)$ is the space of tempered distributions. We assume that $\mathcal{Y}$ is a collection of quasi-Banach spaces $\mathfrak{B} = \mathfrak{B}(\mathbb{R}^d)$ of distributions with quasi-norms $\|\cdot\|_{\mathfrak{B}}$, which are continuously embedded in $\mathcal{S}'$. Also, we assume that $\mathcal{S}$ is a dense subset of each $\mathfrak{B} \in \mathcal{Y}$.

Furthermore, we assume that there exists a collection $\mathcal{Y}_d$ of quasi-Banach complex-valued sequence spaces $\mathfrak{b} = \mathfrak{b}(\mathcal{D} \times E)$, with quasi-norms $\|\cdot\|_{\mathfrak{b}}$, such that every $\mathfrak{B} \in \mathcal{Y}$ is associated with a space $\mathfrak{b} \in \mathcal{Y}_d$.

Targeted application of this construction is to the homogeneous Besov and Triebel-Lizorkin spaces $\dot{B}^{sq}_p$, $\dot{F}^{sq}_p$ on $\mathbb{R}^d$ with indexes (s,p,q) from the range

$$Q(A) := \{(s,p,q) : |s| \leqslant A,\ A^{-1} \leqslant p \leqslant A,\ A^{-1} \leqslant q < \infty\}, \quad A > 1, \tag{10}$$

and VMO. For the B- and F-spaces $\dot{B}^{sq}_p$, $\dot{F}^{sq}_p$ the sets $\mathcal{Y}$ and $\mathcal{Y}_d$ are defined by

$$\mathcal{Y} = \{\dot{B}^{sq}_p : (s,p,q) \in Q(A)\} \quad \textit{and} \quad \mathcal{Y}_d = \{\dot{b}^{sq}_p : (s,p,q) \in Q(A)\},$$

and similarly for $\dot{F}^{sq}_p$, where $A > 1$ is fixed.

The old basis. We stipulate that the tensor-product Meyer's wavelets $\Psi = \{\psi^e_I\}$ (see §2.1) form a unconditional basis for each space $\mathfrak{B} \in \mathcal{Y}$, which we specify in the following. For this we use the *analysis* and *synthesis* operators S_ψ and T_ψ from (8).

A1. The analysis operator $S_\psi : \mathfrak{B} \to \mathfrak{b}$ is a bounded isomorphism with bounded inverse $T_\psi : \mathfrak{b} \to \mathfrak{B}$. Therefore, each $f \in \mathfrak{B}$ has a unique representation

$$f = \sum_{Ie\in\mathcal{D}\times E} \langle f, \psi^e_I\rangle \psi^e_I, \tag{11}$$

where the series converge unconditionally in $\mathfrak{B}$.

It is also assumed that the norms of the operators S_ψ and T_ψ are uniformly bounded relative to $\mathfrak{B} \in \mathcal{Y}$ and $\mathfrak{b} \in \mathcal{Y}_d$ by a constant $C = C(\mathcal{Y}, \mathcal{Y}_d, \Psi) > 1$. Thus, for all $\mathfrak{B} \in \mathcal{Y}$ and $f \in \mathfrak{B}$ we have

$$C^{-1}\|f\|_{\mathfrak{B}} \leqslant \|S_\psi f\|_{\mathfrak{b}} = \|\{\langle f, \psi_I^e\rangle\}\|_{\mathfrak{b}} \leqslant C\|f\|_{\mathfrak{B}}.$$

A2. In addition, we assume that each $\mathfrak{b} \in \mathcal{Y}_d$ obeys the conditions:

(a) Compactly supported sequences belong to $\mathfrak{b}$ and are dense in $\mathfrak{b}$.
(b) Let $h = \{h_{Ie}\}_{Ie\in\mathcal{D}\times E} \in \mathfrak{b}$ and $\{h_j\}_{j\geqslant 1}$ be any ordering of the sequence h. Denote by $h^{(m)}$ the truncated sequence defined by

$$h_j^{(m)} := h_j \quad \text{if } j > m \quad \text{and} \quad h_j^{(m)} := 0 \quad \text{if } j \leqslant m.$$

We stipulate that $\|h^{(m)}\|_{\mathfrak{b}} \to 0$ as $m \to \infty$.

Construction of new bases. We next construct a new basis $\Theta = \{\theta_I^e\}_{Ie\in\mathcal{D}\times E}$ for all spaces $\mathfrak{B} \in \mathcal{Y}$ by small perturbation of Meyer's wavelet basis Ψ.

For a collection $\Theta = \{\theta_I^e\}_{Ie\in\mathcal{D}\times E} \subset \cap\{\mathfrak{B} : \mathfrak{B} \in \mathcal{Y}\}$ of real-valued functions $\theta_I^e \in \mathfrak{B}$, $Ie \in \mathcal{D} \times E$, we introduce the matrix

$$\mathbf{A} := \{a(Ie, I'e')\}_{Ie,I'e'\in\mathcal{D}\times E}, \quad a(Ie, I'e') := \langle \theta_{I'}^{e'}, \psi_I^e\rangle.$$

From (11) it follows that

$$\theta_I^e = \sum_{I'e'\in\mathcal{D}\times E} a(I'e', Ie)\psi_{I'}^{e'}, \quad Ie \in \mathcal{D} \times E,$$

where the convergence is unconditional in $\mathfrak{B}$ for each $\mathfrak{B} \in \mathcal{Y}$.

Our **main assumption** here is that the system Θ can be constructed so that the operator $\mathbf{A} : \mathfrak{b} \to \mathfrak{b}$ is invertible and

$$\|\mathbf{A}\|_{\mathfrak{b}\to\mathfrak{b}} \leqslant c^\star, \quad \|\mathbf{A}^{-1}\|_{\mathfrak{b}\to\mathfrak{b}} \leqslant c^\star, \quad \forall \mathfrak{b} \in \mathcal{Y}_d,$$

where $c^\star > 0$ is a uniform constant. We denote by $b(Ie, I'e')$ the entries of $\mathbf{A}^{-1}$, i.e.

$$\mathbf{A}^{-1} =: \{b(Ie, I'e')\}_{Ie,I'e'\in\mathcal{D}\times E}.$$

Theorem 3.1 *Under the assumptions from above for any $f \in \mathfrak{B}$ there exists a unique sequence $d(f) = \{d_{Ie}(f)\}$, namely,*

$$d_{Ie}(f) := \sum_{I'e'\in\mathcal{D}\times E} b(Ie, I'e')\langle f, \psi_{I'}^{e'}\rangle$$

such that

$$f = \sum_{Ie\in\mathcal{D}\times E} d_{Ie}(f)\theta_I^e,$$

where the convergence is unconditional in $\mathfrak{B}$, and uniformly

$$\|f\|_{\mathfrak{B}} \sim \|\{d(f)\}\|_{\mathfrak{b}}.$$

For the proof of this theorem, see the proofs of Theorems 3.1, 3.2 in [30] and Theorem 4.1 in [20].

3.1 Construction of concrete new bases on $\mathbb{R}^d$

In this section we apply the method from above to the construction of new bases consisting of diadic dilations and translations of smooth well localized functions, see [43, 29, 20].

Main construction: Let $0 < \varepsilon < 1$ and assume $\Theta^e := \{\theta^e : e \in E\}$ is a collection of real-valued functions in $C^r(\mathbb{R}^d)$, $r \geqslant 1$, satisfying the following condition:

$$|\partial^\alpha \psi^e(x) - \partial^\alpha \theta^e(x)| \leqslant \varepsilon(1+|x|)^{-M}, \quad |\alpha| \leqslant r,\ e \in E,\ M > 0, \tag{12}$$

$$\int_{\mathbb{R}^d} x^\alpha \theta^e(x)dx = 0, \quad |\alpha| < r, \quad e \in E, \tag{13}$$

where ψ^e, $e \in E$, are the tensor-product Meyer's wavelets from (5).

The next theorem shows that under appropriate conditions on r, M, and ε the collection of functions

$$\Theta := \{\theta_I^e : I \in \mathscr{D},\ e \in E\} \tag{14}$$

is an unconditional basis for homogeneous Besov and Triebel-Lizorkin spaces and VMO. Above θ_I^e is the dilated version of θ^e, see (4).

Theorem 3.2 *Fix $A > 1$. For sufficiently large r and M and sufficiently small $\varepsilon > 0$, depending only on A, the new system Θ, defined in* (14), *is a unconditional basis for the homogeneous Besov and Triebel-Lizorkin spaces $\dot{B}_p^{sq}$, $\dot{F}_p^{sq}$ for $(s,p,q) \in Q(A)$ (see* (10)*), and* VMO *in the following sense: For any $f \in \dot{B}_p^{sq}$ there exists a unique sequence $d(f) = \{d_{Ie}(f)\}$, namely,*

$$d(f) := \mathbf{A}^{-1}c(f), \quad c_{Ie}(f) := \langle f, \psi_I^e \rangle,$$

where $\mathbf{A} := \{a(Ie, I'e')\}$ with $a(Ie, I'e') := \langle \theta_{I'}^{e'}, \psi_I^e \rangle$, such that

$$f = \sum_{Ie \in \mathscr{D}\times E} d_{Ie}(f)\theta_I^e,$$

where the convergence is unconditional in $\dot{B}_p^{sq}$, and uniformly

$$\|f\|_{\dot{B}_p^{sq}} \sim \|\{d(f)\}\|_{\dot{b}_p^{sq}}.$$

The same is true for $\dot{F}_p^{sq}$ and VMO.

We next describe the main steps in the proof of this theorem. Observe first that by rescaling (12) and (13) yield that for $I \in \mathscr{D}$ and $e \in E$

$$|\partial^\alpha \psi_I^e(x) - \partial^\alpha \theta_I^e(x)| \leqslant \varepsilon |I|^{-1/2-|\alpha|/d}\Big(1 + \frac{|x - x_I|}{\ell(I)}\Big)^{-M}, \quad |\alpha| \leqslant r,$$

$$\int_{\mathbb{R}^d} x^\alpha \theta_I^e(x)dx = 0, \quad |\alpha| < r.$$

One uses the above to derive the following (see [28, Lemma 2.1])

Lemma 3.1 *Assume that conditions* (12) *and* (13) *are satisfied, where* $r \geqslant 1$ *and* $M \geqslant d + r + \delta$ *for some* $\delta > 0$. *Then for* $(I, e) \neq (I', e')$ *we have*

$$|a(Ie, I'e')| \leqslant c\varepsilon \min\Big\{\frac{\ell(I)}{\ell(I')}, \frac{\ell(I')}{\ell(I)}\Big\}^{r+d/2}\Big(1 + \frac{|x_I - x_{I'}|}{\max\{\ell(I), \ell(I')\}}\Big)^{-M}$$

and

$$|1 - a(Ie, Ie)| \leqslant c\varepsilon,$$

where the constant $c > 0$ *depends only on* d *and* δ.

Using the above lemma and the machinery of the almost diagonal operators (see [12, 13]) one can show that for sufficiently large r and M and sufficiently small $\varepsilon > 0$, depending only on A, the operator $\mathbf{A}$ with associated matrix

$$\mathbf{A} := \{a(Ie, I'e')\}_{Ie, I'e' \in \mathcal{D}\times E}, \quad a(Ie, I'e') := \langle \theta_{I'}^{e'}, \psi_I^e \rangle,$$

is invertible on $\dot{b}_p^{sq}$ and

$$\|\mathbf{A}\|_{\dot{b}_p^{sq}\to\dot{b}_p^{sq}} \leqslant c^\star, \quad \|\mathbf{A}^{-1}\|_{\dot{b}_p^{sq}\to\dot{b}_p^{sq}} \leqslant c^\star, \quad (s, p, q) \in Q(A),$$

where $c^\star$ depends only on A. The same holds for $\dot{f}_p^{sq}$ and VMO.

Finally, one applies Theorem 3.1 to complete the proof of Theorem 3.2.

3.2 New bases on $\mathbb{R}^d$ in terms of shifts and dilates of a single function

Here we show how one can construct new bases whose elements are linear combinations of small numbers of dilates and shifts of a single smooth and well localized function Φ. The most prominent examples of such functions are

$$\Phi(x) := (1 + |x|^2)^{-N} \text{ (rational function)} \quad \text{and} \quad \Phi(x) := \exp(-|x|^2) \text{ (Gaussian)}.$$

For details and proofs, see [43, 29, 20].

Assumptions. Let $\Phi \in C^{r+1}(\mathbb{R}^d)$, $r \geqslant 1$, satisfy the following conditions:

$$|\partial^\alpha \Phi(x)| \leqslant c(1 + |x|)^{-M^\diamond}, \quad |\alpha| \leqslant r + 1, \quad M^\diamond > r + d, \quad \text{and} \tag{15}$$

$$\int_{\mathbb{R}^d} \Phi(x)dx = 1. \tag{16}$$

Denote by Θ_K the set of all functions of the form

$$\theta(x) = \sum_{k=1}^{K} c_k \Phi(ax + b_k), \quad x \in \mathbb{R}^d,\ a, c_k \in \mathbb{R},\ b_k \in \mathbb{R}^d.$$

Theorem 3.3 *Fix $A > 1$. Under assumptions (15)–(16) for sufficiently large r, K, and $M^\diamond$ ($M^\diamond > r + d$) there exist functions $\{\theta^e : e \in E\} \subset \Theta_K$ such that*

$$\Theta := \{\theta_I^e : I \in \mathcal{D}, e \in E\},$$

is a unconditional basis for the Besov spaces $\dot{B}_p^{sq}$, $(s, p, q) \in Q(A)$, in the following sense: For any $f \in \dot{B}_p^{sq}$ there exists a unique sequence $d(f) = \{d_{Ie}(f)\}$, namely,

$$d(f) := \mathbf{A}^{-1} c(f), \quad c_{Ie}(f) := \langle f, \psi_I^e \rangle,$$

where $\mathbf{A} := \{a(Ie, I'e')\}$ with $a(Ie, I'e') := \langle \theta_{I'}^{e'}, \psi_I^e \rangle$, such that

$$f = \sum_{Ie \in \mathcal{D} \times E} d_{Ie}(f) \theta_I^e,$$

where the convergence is unconditional in $\dot{B}_p^{sq}$, and uniformly

$$\|f\|_{\dot{B}_p^{sq}} \sim \|\{d(f)\}\|_{\dot{b}_p^{sq}}.$$

In particular, there exists a unconditional basis for the Besov space $\dot{B}_p^{sq}$ of the form

$$R := \{R_I^e : I \in \mathcal{D}, e \in E\}, \quad \text{where} \quad R^e(x) = \sum_{k=1}^{K} \frac{c_k}{(1 + |ax + b_k|^2)^N}, \quad N > M^\diamond/2.$$

The above holds also for $\dot{F}_p^{sq}$ and VMO.

The above theorem follows readily using Theorem 3.2 and the following

Proposition 3.1 *Let $M^\diamond > M > d$ and assume that the function $\Phi \in C^{r+1}(\mathbb{R}^d)$, $r \geqslant 1$, satisfies conditions (15)–(16). Then for any $\varepsilon > 0$ there exist $K \geqslant 1$ and functions $\theta^e \in \Theta_K$, $e \in E$, such that*

$$|\partial^\alpha \psi^e(x) - \partial^\alpha \theta^e(x)| \leqslant \varepsilon(1 + |x|)^{-M}, \quad |\alpha| \leqslant r, \quad e \in E, \quad \text{and}$$
$$\int_{\mathbb{R}^d} x^\alpha \theta^e(x) dx = 0, \quad |\alpha| < r, \quad e \in E.$$

For the proof of this proposition see the proof of Theorem 4.1 in [29].

Application to rational approximation in H^p and BMO. Denote by $\mathcal{R}_n$ the set of all rational functions of the form

$$R(x) = \sum_{j=1}^{n} \frac{c_j}{(1 + |a_j x + b_j|^2)^N}, \quad x \in \mathbb{R}^d,\ a_j, c_j \in \mathbb{R},\ b_j \in \mathbb{R}^d,$$

where N is from Theorem 3.3. We define

$$R_n(f)_p := \inf_{R \in \mathcal{R}_n} \|f - R\|_{H^p} \quad \text{and} \quad R_n(f)_{\text{BMO}} := \inf_{R \in \mathcal{R}_n} \|f - R\|_{\text{BMO}},$$

where H^p is the Hardy space on $\mathbb{R}^d$ if $0 < p \leqslant 1$ and $H^p = L^p$ if $1 < p < \infty$.

Theorem 3.4 (Jackson estimate) *(a) If $f \in B_\tau^{s\tau}$, $1/\tau = s/d + 1/p$, $0 < p < \infty$, then $f \in H^p$ and*

$$R_n(f)_p \leqslant cn^{-s/d} \|f\|_{B_\tau^{s\tau}}, \quad n \geqslant 1, \tag{17}$$

(b) If $f \in B_\tau^{s\tau}$, $1/\tau = s/d$, then $f \in$ VMO and

$$R_n(f)_{\text{BMO}} \leqslant cn^{-s/d} \|f\|_{B_\tau^{s\tau}}, \quad n \geqslant 1. \tag{18}$$

This theorem follows readily by Theorem 3.3 and well known techniques for nonlinear n-term approximation from bases, see [29] and [20].

Note that in dimension $d = 1$ Theorem 3.4 implies Jackson estimates for rational approximation on $\mathbb{R}$ in H^p, $0 < p < \infty$, and BMO, companion to the respective Jackson estimates of A. Pekarskii [41] for rational approximation on the disc and $[0, 1]$.

Remark 3.1 Theorem 3.3 also yields Jackson estimates like (17) and (18) for nonlinear n-term approximation from function of the form

$$g(x) = \sum_{k=1}^{n} c_j e^{-|a_j x + b_j|^2}, \quad x \in \mathbb{R}^d,\ a_j, c_j \in \mathbb{R},\ b_j \in \mathbb{R}^d.$$

It should be pointed out that above the Gaussian $e^{-|x|^2}$ can be replaced by any sufficiently smooth and well localized function Φ. Of course, the range for the rate of approximation would depend on the smoothness and localization of Φ.

4 Construction of good frames in the classical case on $\mathbb{R}^d$

In this section we recall the φ-transform of Frazier and Jawerth which is a powerful tool in the theory of Besov and Triebel-Lizorkin spaces on $\mathbb{R}^d$. We refer the reader to [12, 13, 14] for details and proofs.

Let the functions $\varphi, \psi \in \mathcal{S}(\mathbb{R}^d)$ be such that

$$\operatorname{supp} \widehat{\varphi}, \widehat{\psi} \subset \{\xi \in \mathbb{R}^d : 2^{-1} \leqslant |\xi| \leqslant 2\},$$
$$|\widehat{\varphi}(\xi)|, |\widehat{\varphi}(\xi)| \geqslant c > 0 \ \text{ if } \ 3/5 \leqslant |\xi| \leqslant 5/3,$$

and

$$\sum_{\nu \in \mathbb{Z}} \overline{\widehat{\varphi}(2^\nu \xi)} \widehat{\psi}(2^\nu \xi) = 1 \quad \text{if} \quad \xi \neq 0.$$

Here $\widehat{\varphi}$, $\widehat{\psi}$ stand for the Fourier transforms of φ, ψ. For any function g on $\mathbb{R}^d$ we introduce the notation $g_\nu(x) := 2^{\nu d} g(2^\nu x)$, $\nu \in \mathbb{Z}$. The following Calderón reproducing formula follows from above

$$f = \sum_{\nu \in \mathbb{Z}} \tilde{\varphi}_\nu * \psi_\nu * f, \quad f \in \mathcal{S}'/\mathcal{P}, \tag{19}$$

where $\tilde{\varphi}_\nu(x) := \overline{\varphi_\nu(-x)}$.

As in Section 2.1 we denote by $\mathcal{D}$ the set of all diadic cubes in $\mathbb{R}^d$. With the notation from (4) the pair of dual frames of Frazier and Jawerth is defined by

$$\{\psi_I : I \in \mathcal{D}\} \quad \text{and} \quad \{\varphi_I : I \in \mathcal{D}\}.$$

The following identity follows from (19) applying the Shannon sampling theorem

$$f = \sum_{I \in \mathcal{D}} \langle f, \varphi_I \rangle \psi_I, \quad f \in \mathcal{S}'/\mathcal{P}. \tag{20}$$

Note that $\operatorname{supp} \widehat{\varphi}_I, \widehat{\psi}_I \subset \{\xi \in \mathbb{R}^d : 2^{\nu-1} \leqslant |\xi| \leqslant 2^{\nu+1}\}$, $I \in \mathcal{D}_\nu$, and for any $M > 0$ and $\alpha \in \mathbb{N}_0^d$

$$|\partial^\alpha \psi_I(x)|, |\partial^\alpha \varphi_I(x)| \leqslant \frac{c|I|^{-1/2-|\alpha|/d}}{(1 + \ell(I)^{-1}|x - x_I|)^M}, \quad x \in \mathbb{R}^d, \ I \in \mathcal{D},$$

where the constant $c > 0$ depends on d, α, and M.

One can easily construct a frame like the above where $\varphi = \psi$. Then (20) takes the form $f = \sum_{I \in \mathcal{D}} \langle f, \psi_I \rangle \psi_I$.

For the frame characterization of the homogeneous Besov and Triebel-Lizorkin spaces we now introduce the associated sequence spaces.

Definition 4.1 (a) For $s \in \mathbb{R}$ and $0 < p, q \leqslant \infty$ the space $\dot{b}_p^{sq}$ is defined as the collection of all sequences $h = \{h_I\}_{I \in \mathcal{D}}$ of complex numbers such that

$$\|h\|_{\dot{b}_p^{sq}} := \Big(\sum_{\nu \in \mathbb{Z}} \Big[\sum_{I \in \mathcal{D}_\nu} (|I|^{-s/d-1/2+1/p} |h_I|)^p \Big]^{q/p} \Big)^{1/q} < \infty.$$

(b) For $s \in \mathbb{R}$, $0 < p < \infty$, and $0 < q \leqslant \infty$ the space $\dot{f}_p^{sq}$ is defined as the collection of all sequences $h = \{h_I\}_{I \in \mathcal{D}}$ of complex numbers such that

$$\|h\|_{\dot{f}_p^{sq}} := \Big\| \Big(\sum_{I \in \mathcal{D}} (|I|^{-s/d} |h_I| \tilde{\mathbb{1}}_I)^q \Big)^{1/q} \Big\|_{L^p} < \infty,$$

where $\tilde{\mathbb{1}}_I(x) := |I|^{-1/2} \mathbb{1}_I(x)$ is the L^2-normalized characteristic function of I.

(c) For $s \in \mathbb{R}$ and $0 < q \leqslant \infty$ the space $\dot{f}_\infty^{sq}$ is defined as the collection of all sequences $h = \{h_I\}_{I \in \mathcal{D}}$ of complex numbers such that

$$\|h\|_{\dot{f}^{sq}_{\infty}} := \sup_{J\in\mathscr{D}} \left(\frac{1}{|J|}\int_J \sum_{I\subset J} \left(|I|^{-s/d}|h_I|\tilde{\mathbb{1}}_I(x)\right)^q dx\right)^{1/q} < \infty.$$

Above as usual the L^p, L^q, ℓ^p, ℓ^q norms are replaced by the respective sup-norms if $p=\infty$ or $q=\infty$.

The next theorem establishes the frame characterization of the homogeneous Besov and Triebel-Lizorkin spaces. To state this results we introduce the following analysis and synthesis operators:

$$S_\varphi : f \to \{\langle f, \varphi_I\rangle\}_{I\in\mathscr{D}}, \quad T_\psi : \{h_I\}_{I\in\mathscr{D}} \to \sum_{I\in\mathscr{D}} h_I\psi_I.$$

Theorem 4.1 *Let $s\in\mathbb{R}$ and $0<p,q\leqslant\infty$. The operators $S_\varphi : \dot{B}^{sq}_p \to \dot{b}^{sq}_p$ and $T_\psi : \dot{b}^{sq}_p \to \dot{B}^{sq}_p$ are bounded and $T_\psi \circ S_\varphi = I$ on $\dot{B}^{sq}_p$. Hence,*

$$f = \sum_{I\in\mathscr{D}} \langle f, \varphi_I\rangle \psi_I, \tag{21}$$

where the convergence is in $\mathcal{S}'/\mathcal{P}$, and unconditional in the norm of $\dot{B}^{sq}_p$ if $p,q\neq\infty$. Consequently, for $f\in\mathcal{S}'/\mathcal{P}$ we have $f\in\dot{B}^{sq}_p$ if and only if $S_\varphi f\in\dot{b}^{sq}_p$. Furthermore,

$$\|f\|_{\dot{B}^{sq}_p} \sim \|S_\varphi f\|_{\dot{b}^{sq}_p}, \quad \forall f\in\dot{B}^{sq}_p.$$

The same is true for $\dot{F}^{sq}_p$.

Remark 4.1 One can easily modify the construction of Frazier and Jawerth to produce an inhomogeneous version of the frames from above. In this case the frame elements will be indexed by the dyadic cubes from $\mathscr{D}_+ := \cup_{\nu\geqslant 0}\mathscr{D}_j$. Just as above the resulting dual frames $\{\psi_I : I\in\mathscr{D}_+\}$, $\{\varphi_I : I\in\mathscr{D}_+\}$ can be used for characterisation of the inhomogeneous Besov and Triebel-Lizorkin spaces B^{sq}_p and F^{sq}_p.

Remark 4.2 Comparing Theorems 4.1 with Theorem 2.1 one may decide that the two families $\{\psi_I : I\in\mathscr{D}\}$, $\{\varphi_I : I\in\mathscr{D}\}$ form a pair of dual bases for Besov and Triebel-Lizorkin spaces. But this is not the case because there is no uniqueness in representation (21). Each of the sequences $\{\psi_I\}$ and $\{\varphi_I\}$ is redundant. However, $\{\psi_I\}$ and $\{\varphi_I\}$ are frames in $L^2(\mathbb{R}^d)$, that is, for each $f\in L^2$ identity (21) is valid in L^2 and

$$c_1\|f\|_{L^2} \leqslant \left(\sum_{I\in\mathscr{D}} |\langle f, \varphi_I\rangle|^2\right)^{1/2} \leqslant c_2\|f\|_{L^2}, \quad \forall f\in L^2, \tag{22}$$

where $c_1, c_2 > 0$ are constants independent of f. The same is true with the roles of $\{\varphi_I\}$ and $\{\psi_I\}$ interchanged. The above follows from the fact that L^p can be identified with $\dot{F}^{02}_p$ when $1<p<\infty$, see [14, p. 42] or [13]. Furthermore, it is easy to show that if by construction $\varphi=\psi$ and hence $\{\varphi_I\}=\{\psi_I\}$, then $\{\psi_I\}$ is a tight frame, i.e. $\|f\|_{L^2} = \left(\sum_{I\in\mathscr{D}} |\langle f, \psi_I\rangle|^2\right)^{1/2}$, see [13].

5 Construction of good frames (needlets) on the sphere

In this section we present the construction of good frames on the unit sphere $\mathbb{S}^{d-1}$ in $\mathbb{R}^d$ from [36, 37]. This construction uses the idea for construction of frames of Frasier and Jawerth [12, 13] and was the first such construction in a setting different from $\mathbb{R}^d$.

Spherical harmonics. The spherical harmonics are the main vehicle in constructing good frames on $\mathbb{S}^{d-1}$. See [8, 35, 48] for the basics of spherical harmonics.

Denote by $\mathcal{H}_k$ the space of all spherical harmonics of degree k on $\mathbb{S}^{d-1}$. As is well known the spaces $\mathcal{H}_k$, $k = 0, 1, \dots$, are orthogonal and $L^2(\mathbb{S}^{d-1}) = \bigoplus_{k\geqslant 0} \mathcal{H}_k$. The dimension of $\mathcal{H}_k$ is $N(k, d) \sim k^{d-1}$.

Let $\{Y_{k\nu} : \nu = 1, \dots, N(k, d)\}$ be a real-valued orthonormal basis for $\mathcal{H}_k$. Then the kernel of the orthogonal projector onto $\mathcal{H}_k$ is given by

$$Z_k(x \cdot y) = \sum_{\nu=1}^{N(k,d)} Y_{k\nu}(x) Y_{k\nu}(y), \quad x, y \in \mathbb{S}^{d-1}.$$

As is well known (see e.g. [8, Theorem 1.2.6])

$$Z_k(x \cdot y) = \frac{k+\mu}{\mu \omega_d} C_k^\mu(x \cdot y), \quad x, y \in \mathbb{S}^{d-1}, \quad \mu := \frac{d-2}{2}, \ d > 2,$$

where C_k^μ is the Gegenbauer (ultraspherical) polynomial of degree k normalized by $C_k^\mu(1) = \binom{k+2\mu-1}{k}$. The polynomials C_k^μ, $k = 0, 1, \dots$, are orthogonal in the space $L^2([-1, 1], w)$ with weight $w(u) := (1 - u^2)^{\mu-1/2}$, see [49, p. 80, (4.7.1)].

Kernels on the sphere $\mathbb{S}^{d-1}$ of the form

$$\Lambda_N(x \cdot y) := \sum_{k=0}^{\infty} \varphi(k/N) Z_k(x \cdot y), \quad x, y \in \mathbb{S}^{d-1}, \quad N \geqslant 1, \tag{23}$$

play a key role in what follows.

Theorem 5.1 *Let $\varphi \in C^\infty(\mathbb{R})$ be even and compactly supported or in the Schwartz class $\mathcal{S}(\mathbb{R})$. Then for any $\nu \in \mathbb{N}$ and $M > 0$ there exists a constant $c > 0$ such that for each $N \geqslant 1$ the kernel Λ_N from* (23) *obeys*

$$|\Lambda_N^{(\nu)}(x \cdot y)| \leqslant \frac{cN^{d-1+2\nu}}{(1 + N\rho(x, y))^M}, \quad x, y \in \mathbb{S}^{d-1},$$

where $\rho(x, y) :=$ arccos$(x \cdot y)$ is the geodesic distance between $x, y \in \mathbb{S}^{d-1}$.

This theorem confirms the localization principle from the introduction (see (2)). For its proof, see [36, Theorem 3.5] and [37, Lemmas 2.4, 2.6], also [21].

Maximal δ-nets and cubature formulas on the sphere. The construction of frames on $\mathbb{S}^{d-1}$ relies on cubature formulas, constructed using maximal δ-nets on $\mathbb{S}^{d-1}$.

Definition. Given $\delta > 0$ we say that a finite set $\mathcal{Z} \subset \mathbb{S}^{d-1}$ is a maximal δ-net on $\mathbb{S}^{d-1}$ if (i) $\rho(\zeta_1, \zeta_2) \geqslant \delta$ for all $\zeta_1, \zeta_2 \in \mathcal{Z}$, $\zeta_1 \neq \zeta_2$, and (ii) $\cup_{\zeta \in \mathcal{Z}} B(\zeta, \delta) = \mathbb{S}^{d-1}$. Here $B(\zeta, \delta)$ is the spherical cap (ball) centered at ζ of radius δ.

For every maximal δ-net $\mathcal{Z}$ it is easy to construct (see [6, Proposition 2.5]) a disjoint partition $\{A_\zeta\}_{\zeta \in \mathcal{Z}}$ of $\mathbb{S}^{d-1}$ consisting of measurable sets such that

$$B(\zeta, \delta/2) \subset A_\zeta \subset B(\zeta, \delta), \quad \zeta \in \mathcal{Z}.$$

Cubature formulas on $\mathbb{S}^{d-1}$. Let $\Xi_j \subset \mathbb{S}^{d-1}$ be a maximal δ_j-net on $\mathbb{S}^{d-1}$ with $\delta_j := \gamma 2^{-j+1}$, $0 < \gamma < 1$, $j \geqslant 1$. In [36, Theorem 4.3] it is shown that there exist γ $(0 < \gamma < 1)$, depending only on d, and weights $\{w_\xi\}_{\xi \in \Xi_j}$, satisfying

$$c^{-1} 2^{-j(d-1)} \leqslant w_\xi \leqslant c 2^{-j(d-1)}, \quad \xi \in \Xi_j, \tag{24}$$

with constant $c > 0$ depending only on d, such that the cubature formula

$$\int_{\mathbb{S}^{d-1}} f(x) d\sigma(x) \approx \sum_{\xi \in \Xi_j} w_\xi f(\xi) \tag{25}$$

is exact for all spherical harmonics f of degree $\leqslant 2^{j+1}$, $j \geqslant 1$. Above $d\sigma$ stands for the Lebesgue measure on $\mathbb{S}^{d-1}$.

Construction of good frames on $\mathbb{S}^{d-1}$. We next present the construction of the frame (needlets) on $\mathbb{S}^{d-1}$ from [37]. Note that in dimension $d = 2$ the Meyer's periodic wavelets (see [34] and also [51]) form a basis with the desired properties.

The first step in the construction of needlets on $\mathbb{S}^{d-1}$, $d > 2$, is the selection of a real-valued function $\varphi \in C^\infty(\mathbb{R}_+)$ with the properties: $\operatorname{supp} \varphi \subset [1/2, 2]$, $0 \leqslant \varphi \leqslant 1$, $\varphi(u) \geqslant c > 0$ for $u \in [3/5, 5/3]$, $\varphi^2(u) + \varphi^2(u/2) = 1$ for $u \in [1, 2]$, and hence $\sum_{\nu=0}^\infty \varphi^2(2^{-\nu} u) = 1$ for $u \in [1, \infty)$. Set

$$L_0 := Z_0, \quad \text{and} \quad L_j := \sum_{k=0}^\infty \varphi\Big(\frac{k}{2^{j-1}}\Big) Z_k, \quad j \geqslant 1.$$

It is easy to see that $f = \sum_{j=0}^\infty L_j * L_j * f$ for every $f \in \mathcal{S}'$ (convergence in $\mathcal{S}'$). Here $\mathcal{S}'$ is the space of distributions on $\mathbb{S}^{d-1}$, defined below.

Since the cubature formula (25) is exact for spherical harmonics of degree $\leqslant 2^{j+1}$ we have

$$L_j * L_j(x \cdot y) = \int_{\mathbb{S}^{d-1}} L_j(x \cdot \eta) L_j(\eta \cdot y) d\sigma(\eta) = \sum_{\xi \in \Xi_j} w_\xi L_j(x \cdot \xi) L_j(\xi \cdot y),$$

which allows to discretize $f = \sum_{j=0}^\infty L_j * L_j * f$ and obtain

$$f = \sum_{j=0}^\infty \sum_{\xi \in \Xi_j} \langle f, \psi_\xi \rangle \psi_\xi, \quad \forall f \in \mathcal{S}' \quad \text{(convergence in } \mathcal{S}'),$$

$$\psi_\xi(x) := w_\xi^{1/2} L_j(\xi \cdot x), \quad \xi \in \Xi_j,\ j \geqslant 0.$$

We set $\Xi := \cup_{j\geqslant 0}\Xi_j$ assuming that equal points from different sets Ξ_j are distinct points in Ξ so that Ξ can be used as an index set. This completes the construction of the system $\Psi = \{\psi_\xi\}_{\xi\in\Xi}$.

Observe that the frame elements $\{\psi_\xi\}$ are not only band limited, but also have excellent localization on $\mathbb{S}^{d-1}$. From the properties of φ and Theorem 5.1 and (24) it follows that (see also [36, 37]) for any $M > 0$

$$|\psi_\xi(x)| \leqslant c2^{(j-1)(d-1)/2}(1 + 2^{j-1}\rho(x,\xi))^{-M}, \quad x \in \mathbb{S}^{d-1},\ \xi \in \Xi_j,\ \ j \geqslant 0.$$

Moreover, the localization of ψ_ξ can be improved to sub-exponential as shown in [21, Theorem 5.1].

Besov and Triebel-Lizorkin spaces on $\mathbb{S}^{d-1}$. The Besov and Triebel-Lizorkin spaces on $\mathbb{S}^{d-1}$ in general are spaces of distributions. As test functions we use the class $\mathcal{S} := C^\infty(\mathbb{S}^{d-1})$ of all functions ϕ on $\mathbb{S}^{d-1}$ such that

$$\|Z_k * \phi\|_2 \leqslant c(\phi, m)(1 + k)^{-m}, \quad \forall k, m \geqslant 0.$$

Here $Z_k * \phi(x) := \int_{\mathbb{S}^{d-1}} Z_k(x \cdot y)\phi(y)d\sigma(y)$. The topology on $\mathcal{S}$ is defined by the sequence of norms

$$P_m(\phi) := \sum_{k=0}^{\infty}(k+1)^m\|Z_k * \phi\|_2.$$

The space $\mathcal{S}' := \mathcal{S}'(\mathbb{S}^{d-1})$ of distributions on $\mathbb{S}^{d-1}$ is defined as the space of all continuous linear functionals on $\mathcal{S}$. For any $f \in \mathcal{S}'$ we define $Z_k * f$ by

$$Z_k * f(x) := \langle f, \overline{Z_k(x \cdot \bullet)}\rangle = \langle f, Z_k(x \cdot \bullet)\rangle, \tag{26}$$

where on the right f is acting on $\overline{Z_k(x \cdot y)} = Z_k(x \cdot y)$ as a function of y (Z_k is real-valued). Observe that the representation

$$f = \sum_{k=0}^{\infty} Z_k * f, \quad \forall f \in \mathcal{S}'$$

holds with convergence in distributional sense.

Definition 5.1 Let $s \in \mathbb{R}$, $0 < q \leqslant \infty$, and φ satisfy the conditions: $\varphi \in C^\infty(\mathbb{R}_+)$, $\operatorname{supp}\varphi \subset [1/2, 2]$, and $|\varphi(u)| \geqslant c > 0$ for $u \in [3/5, 5/3]$. For any $f \in \mathcal{S}'$ set

$$\Phi_0 * f = Z_0 * f, \quad \Phi_j * f = \sum_{k=0}^{\infty}\varphi\Big(\frac{k}{2^{j-1}}\Big)Z_k * f,\ j \geqslant 1.$$

(a) The Besov space $\mathcal{B}_p^{sq} := \mathcal{B}_p^{sq}(\mathbb{S}^{d-1})$, $0 < p \leqslant \infty$, is defined as the set of all distributions $f \in \mathcal{S}'$ such that

$$\|f\|_{\mathcal{B}_p^{sq}} := \Big(\sum_{j=0}^{\infty}\Big(2^{sj}\|\Phi_j * f\|_{L^p(\mathbb{S}^{d-1})}\Big)^q\Big)^{1/q} < \infty,$$

(*b*) The Triebel-Lizorkin space $\mathcal{F}_p^{sq} := \mathcal{F}_p^{sq}(\mathbb{S}^{d-1})$, $0 < p < \infty$, is defined as the set of all distributions $f \in \mathcal{S}'$ such that

$$\|f\|_{\mathcal{F}_p^{sq}} := \Big\|\Big(\sum_{j=0}^{\infty}(2^{sj}|\Phi_j * f(\cdot)|)^q\Big)^{1/q}\Big\|_{L^p(\mathbb{S}^{d-1})} < \infty.$$

In the above norms we make the usual modification when $p = \infty$ or $q = \infty$.

We next define the Besov and Triebel-Lizorkin sequence spaces $\mathfrak{b}_p^{sq}$ and $\mathfrak{f}_p^{sq}$ associated to Ξ.

Definition 5.2 Let $s \in \mathbb{R}$, $0 < p, q \leqslant \infty$. Then $\mathfrak{b}_p^{sq} := \mathfrak{b}_p^{sq}(\Xi)$ is defined as the space of all complex-valued sequences $h := \{h_\xi\}_{\xi\in\Xi}$ such that

$$\|h\|_{\mathfrak{b}_p^{sq}} := \Big(\sum_{j=0}^{\infty}\Big[2^{j[s+(d-1)(1/2-1/p)]}\Big(\sum_{\xi\in\Xi_j}|h_\xi|^p\Big)^{1/p}\Big]^q\Big)^{1/q} < \infty.$$

Definition 5.3 Let $s \in \mathbb{R}$, $0 < p < \infty$, and $0 < q \leqslant \infty$. Then $\mathfrak{f}_p^{sq} := \mathfrak{f}_p^{sq}(\Xi)$ is defined as the space of all complex-valued sequences $h := \{h_\xi\}_{\xi\in\Xi}$ such that

$$\|h\|_{\mathfrak{f}_p^{sq}} := \Big\|\Big(\sum_{\xi\in\Xi}[|B_\xi|^{-s/(d-1)-1/2}|h_\xi|\mathbb{I}_{B_\xi}(\cdot)]^q\Big)^{1/q}\Big\|_{L^p} < \infty.$$

Here $B_\xi := B(\xi, \gamma 2^{-j+1})$, $\xi \in \Xi_j$, where γ is used in the selection of Ξ_j, $|B_\xi|$ is the measure of B_ξ and $\mathbb{I}_{B_\xi}$ is the characteristic function of B_ξ.

The main result here asserts that $\{\psi_\xi\}_{\xi\in\Xi}$ is a good frame for Besov and Triebel-Lizorkin spaces on the sphere. To state this result we introduce the following *analysis* and *synthesis* operators: $S_\psi : f \mapsto \{\langle f, \psi_\xi\rangle\}_{\xi\in\Xi}$, $T_\psi : \{h_\xi\}_{\xi\in\Xi} \mapsto \sum_{\xi\in\Xi} h_\xi\psi_\xi$.

Theorem 5.2 *Let $s \in \mathbb{R}$ and $0 < p, q < \infty$. The operators $S_\psi : \mathcal{B}_p^{sq} \to \mathfrak{b}_p^{sq}$ and $T_\psi : \mathfrak{b}_p^{sq} \to \mathcal{B}_p^{sq}$ are bounded, and $T_\psi \circ S_\psi = I$ on $\mathcal{B}_p^{sq}$. Hence, if $f \in \mathcal{S}'$, then $f \in \mathcal{B}_p^{sq}$ if and only if $\{\langle f, \psi_\xi\rangle\}_{\xi\in\Xi} \in \mathfrak{b}_p^{sq}$, and*

$$f = \sum_{\xi\in\Xi}\langle f, \psi_\xi\rangle\psi_\xi \quad \textit{and} \quad \|f\|_{\mathcal{B}_p^{sq}} \sim \|\{\langle f, \psi_\xi\rangle\}\|_{\mathfrak{b}_p^{sq}}.$$

The same is true for $\mathcal{F}_p^{sq}$ and $\mathfrak{f}_p^{sq}$.

For details and proofs, see [37, Theorems 4.5, 5.5].

Remark 5.1 A careful examination of the proofs in [37] shows that the operators S_ψ and T_ψ are uniformly bounded on the respective spaces with parameters

$$(s,p,q) \in Q(A) := \{(s,p,q) : |s| \leqslant A,\ A^{-1} \leqslant p \leqslant A,\ A^{-1} \leqslant q < \infty\}, \quad A > 1,$$

that is, all constants that appear in the equivalences in Theorem 5.2 depend only on A, d, and φ, if $(s,p,q) \in Q(A)$.

6 Construction of good frames in a general setting

The construction of good frames in a number of settings such as on $[-1,1]$ with Jacobi weights, on the sphere, ball, simplex as well as Riemannian manifolds and Lee groups can be put under the umbrella of a construction in a general setting that we describe in this section.

The general setting. We will operate in the following general setting [6, 24]:

I. It is assumed that (X,ρ,μ) is a metric measure space, where (X,ρ) is a locally compact complete metric space with distance $\rho(\cdot,\cdot)$ and μ is a positive regular Borel measure. We will use the notation $B(x,r) := \{y \in X : \rho(x,y) < r\}$ for the balls in X and introduce the abbreviated notation

$$V(x,r) := \mu(B(x,r)).$$

The following *doubling volume condition* is assumed: There exists a constant $\hat{c}_0 > 1$ such that

$$0 < V(x,2r) \leqslant \hat{c}_0 V(x,r) < \infty \quad \text{for } x \in X \text{ and } r > 0. \tag{27}$$

II. The space (X,ρ,μ) is accompanied by a non-negative self-adjoint *operator* L, mapping real-valued to real-valued functions, whose *domain* $D(L)$ is dense in $L^2(X,d\mu)$. It is assumed that the associated to L semigroup $P_t = e^{-tL}$ consists of integral operators with (heat) kernel $p_t(x,y)$ obeying the conditions:

(a) *Gaussian upper bound:* There exist constants $\hat{c}_1,\ \hat{c}_2 > 0$ such that

$$|p_t(x,y)| \leqslant \frac{\hat{c}_1 \exp\big(-\frac{\rho^2(x,y)}{\hat{c}_2 t}\big)}{\big[V(x,\sqrt{t})V(y,\sqrt{t})\big]^{1/2}} \quad \text{for } x,y \in X,\ t > 0. \tag{28}$$

(b) *Hölder continuity:* There exists a constant $\alpha > 0$ such that

$$\big|p_t(x,y) - p_t(x,y')\big| \leqslant \hat{c}_1 \Big(\frac{\rho(y,y')}{\sqrt{t}}\Big)^{\alpha} \frac{\exp\big(-\frac{\rho^2(x,y)}{\hat{c}_2 t}\big)}{\big[V(x,\sqrt{t})V(y,\sqrt{t})\big]^{1/2}} \tag{29}$$

for $x,y,y' \in X$ and $t > 0$, whenever $\rho(y,y') \leqslant \sqrt{t}$.

(c) *Markov property:*

$$\int_X p_t(x,y)\,d\mu(y) = 1 \quad \text{for } x \in X, t > 0.$$

(*d*) *Reverse doubling condition:* There exists a constant $\hat{c}_3 > 1$ such that

$$V(x, 2r) \geqslant \hat{c}_3 V(x, r) \quad \text{for } x \in X \text{ and } r > 0.$$

(*e*) *Non-collapsing condition:* There exists a constant $\hat{c}_4 > 0$ such that

$$\inf_{x \in X} V(x, 1) \geqslant \hat{c}_4.$$

A natural effective realization of the above setting appears in the framework of strictly local regular Dirichlet spaces with a complete intrinsic metric, where one has to only verify the local Poincaré inequality and the global doubling condition on the measure. For more details, see [6]. Two-sided Gaussian bounds for the heat kernels on the unit ball and simplex in $\mathbb{R}^d$ generated by classical differential operators whose eigenfunctions are algebraic polynomials are obtained in [25, 26]. The case of $[-1, 1]$ with Jacobi weights is covered in [6] with a sharper version obtained in [38]. A key point is that situations where the above setting applies are quite common, which becomes evident from the examples given in [6].

Prolate spheroidal wave functions (PSWF). In [2] Gaussian upper and lower bounds and Hölder continuity are established for the heat kernel associated to the prolate spheroidal wave functions of order zero. These results are obtained by application of a general perturbation principle using the fact that the PSWF operator is a perturbation of the Legendre operator. Furthermore, Gaussian bounds are obtained for the heat kernels associated to generalized univariate PSWFs and PSWFs on the unit ball in $\mathbb{R}^d$.

In what follows we review all results that will be needed from [6, 24].

Functional calculus. Since L is a non-negative self-adjoint operator by Spectral theory there exists a unique spectral resolution E_λ associated to L, consisting of orthoprojections in $L^2(X, d\mu)$ such that

$$L = \int_0^\infty \lambda dE_\lambda.$$

Moreover, for any measurable and bounded function F on $[0, \infty)$ the operator (spectral multiplier) $F(L)$, defined by

$$F(L) := \int_0^\infty F(\lambda) dE_\lambda,$$

is bounded on $L^2(X, d\mu)$ and self-adjoint.

We now come to the principle localization estimates for the kernels of operators of the form $F(\delta\sqrt{L})$ in the setting of this section, see [24, Theorem 3.1].

Theorem 6.1 *Assume $F \in C^k(\mathbb{R})$, $k \geqslant 2d+1$, $d := \log_2 \hat{c}_0$, is real-valued and even, and* supp $F \subset [-R, R]$ *for some $R \geqslant 1$. Then $F(\delta\sqrt{L})$, $\delta > 0$, is an integral operator bounded from $L^p(X, d\mu)$ to $L^p(X, d\mu)$, $1 \leqslant p \leqslant \infty$, with kernel $\mathcal{K}_{F(\delta\sqrt{L})}(x, y)$ satisfying*

$$|\mathscr{K}_{F(\delta\sqrt{L})}(x,y)| \leqslant \frac{c_k\left(1+\frac{\rho(x,y)}{\delta}\right)^{-k}}{\left[V(x,\delta)V(y,\delta)\right]^{1/2}} \quad \textit{and} \tag{30}$$

$$|\mathscr{K}_{F(\delta\sqrt{L})}(x,y) - \mathscr{K}_{F(\delta\sqrt{L})}(x,y')| \leqslant \frac{c_k'\left(\frac{\rho(y,y')}{\delta}\right)^{\alpha}\left(1+\frac{\rho(x,y)}{\delta}\right)^{-k}}{\left[V(x,\delta)V(y,\delta)\right]^{1/2}} \tag{31}$$

if $\rho(y,y') \leqslant \delta$. *Above* $c_k = R^d\left((ck)^k\|F\|_\infty + c^k R^k \|F^{(k)}\|_\infty\right)$ *and* $c_k' = R^\alpha c_k$, *where* c *depends on the constants* $\hat{c}_0, \hat{c}_1, \hat{c}_2$ *from* (27)–(29). *In addition,*

$$\int_X \mathscr{K}_{F(\delta\sqrt{L})}(x,y)d\mu(y) = F(0), \quad \forall x \in X.$$

Remark 6.1 Observe that if in Theorem 6.1 we assume that $F \in \mathcal{S}(\mathbb{R})$, then estimates (30) and (31) are valid for an arbitrary $k > 0$, see [24, Corollary 3.5]. Furthermore, the localization in estimates (30) and (31) can be improved to sub-exponential if F is selected with "small derivatives", see Theorem 3.6 in [24].

Theorem 6.1 should be regarded as another manifestation of the *localization principle* put forward in the introduction, see (2).

Distributions. Distributions are naturally defined in the general setting of this section.

Definition 6.1 The class of **test functions** $\mathcal{S} = \mathcal{S}(L)$ associated to L is defined as the set of all complex-valued functions $\phi \in \cap_{m\geqslant 1} D(L^m)$ such that

$$\mathcal{P}_{k,m}(\phi) := \sup_{x\in X}(1+\rho(x,x_0))^k|L^m\phi(x)| < \infty, \quad \forall k,m \in \mathbb{N}_0. \tag{32}$$

Here $x_0 \in X$ is selected arbitrarily and fixed once and for all.

Clearly, the definition of the class $\mathcal{S}$ is independent of x_0.

It is sometimes convenient to work with a directed family of norms. For this reason we introduce the following norms on $\mathcal{S}(L)$:

$$\mathcal{P}^\star_{k,m}(\phi) := \max_{0\leqslant \nu\leqslant m} \mathcal{P}_{k,\nu}(\phi).$$

Observe that the vector space $\mathcal{S} := \mathcal{S}(L)$ with the natural topology induced by the semi-norms $\mathcal{P}_{k,m}(\phi)$, $k,m \in \mathbb{N}_0$, is a Fréchet space.

Definition 6.2 The space $\mathcal{S}' = \mathcal{S}'(L)$ of **distributions induced by** $\boldsymbol{L}$ is defined as the set of all continuous linear functionals on $\mathcal{S}$ and the action of $f \in \mathcal{S}'$ on $\overline{\phi} \in \mathcal{S}$ will be denoted by $\langle f, \phi\rangle := f(\overline{\phi})$, which is consistent with the inner product on $L^2(X)$.

Clearly, for any $f \in \mathcal{S}'$ there exist constants $k,m \in \mathbb{N}_0$ and $c > 0$ such that

$$|\langle f,\phi\rangle| \leqslant c \max_{0\leqslant\nu\leqslant m} \mathcal{P}_{k,\nu}(\phi) = c\mathcal{P}^\star_{k,m}(\phi), \quad \forall \phi \in \mathcal{S}.$$

Spectral spaces. The spectral spaces are an analogue of band-limited functions.

Definition 6.3 The spectral space Σ_t, $t > 0$, is defined by

$$\Sigma_t := \left\{g \in \mathcal{S}' : \theta(\sqrt{L})g = g, \quad \forall \theta \in \mathcal{A}_t\right\},$$

where $\mathcal{A}_t := \{\theta \in C_0^\infty(\mathbb{R}) : \theta(\pm\lambda) = \theta(\lambda) \text{ and } \theta(\lambda) = 1 \text{ for } \lambda \in [-t, t]\}$.

Although the spectral space Σ_t is defined as a class of distributions in fact it can be identified with a family of slowly growing continuous functions.

Proposition 6.1 *Let $g \in \Sigma_t$, $t > 0$. Then there exist constants $N, c > 0$ such that*

$$|g(x)| \leqslant cV(x_0, 1)^{-1}(1 + \rho(x, x_0))^N, \quad x \in X, \quad \text{and}$$

$$|g(x) - g(x')| \leqslant cV(x_0, 1)^{-1}(1 + \rho(x, x_0))^N \rho(x, x')^\alpha, \quad \text{if } \rho(x, x') \leqslant 1.$$

Here $\alpha > 0$ is the constant from (29) *and $x_0 \in X$ is from the definition of test functions, see* (32).

The next proposition provides examples of spectral functions.

Proposition 6.2 *Let $\varphi \in C^\infty(\mathbb{R})$ be real-valued and even, and* $\operatorname{supp} \varphi \subset [-t, t]$, *$t > 0$. Then $\mathcal{K}_{\varphi(\sqrt{L})}(\cdot, y) \in \Sigma_t$ and for any $f \in \mathcal{S}'$ we have $\varphi(\sqrt{L})f \in \Sigma_t$.*

6.1 Frames

We now describe the frames developed in [24]. There are two versions of the frame constriction.

Frames in the case when the spectral spaces $\{\Sigma_\lambda\}$ have the polynomial property. Consider the case when the spectral spaces have the following "polynomial property": There exists a constant $k > 1$ such that

$$\Sigma_t \cdot \Sigma_t \subset \Sigma_{kt}, \quad \text{i.e.} \quad f, g \in \Sigma_t \Longrightarrow fg \in \Sigma_{kt}. \tag{33}$$

The construction begins with the introduction of a pair of real-valued even functions $\Psi_0, \Psi \in C^\infty(\mathbb{R})$ with the following properties:

$$\begin{aligned}
&\operatorname{supp} \Psi_0 \subset [-b, b], \quad \operatorname{supp} \Psi \cap \mathbb{R}_+ \subset [b^{-1}, b], \quad 0 \leqslant \Psi_0, \Psi \leqslant 1,\\
&\Psi_0(u) \geqslant c > 0, \;\; u \in [0, b^{3/4}], \quad \Psi(u) \geqslant c > 0, \;\; u \in [b^{-3/4}, b^{3/4}],\\
&\Psi_0(u) = 1 \quad u \in [0, 1], \quad \Psi_0^2(u) + \sum_{j \geqslant 1} \Psi^2(b^{-j}u) = 1, \quad u \in \mathbb{R},
\end{aligned}$$

where $b > 1$ is a constant. Set $\Psi_j(u) := \Psi(b^{-j}u)$. Then $\sum_{j \geqslant 0} \Psi_j^2(u) = 1$, $u \in \mathbb{R}$, which leads to the following Calderón type decomposition

$$f = \sum_{j \geqslant 0} \Psi_j^2(\sqrt{L}) f, \quad f \in \mathcal{S}'. \tag{34}$$

The key observation is that the polynomial property (33) of the spectral spaces allows to discretize the above expansion and as a result to obtain the desired frame. More specifically, by construction $\Psi_j(\sqrt{L})$ is a kernel operator whose kernels has almost exponential space localization and $\mathcal{K}_{\Psi_j(\sqrt{L})}(x, \cdot) \in \Sigma_{b^{j+1}}$. Now, choose Ξ_j ($j \geqslant 0$) to be a maximal δ-net on X with $\delta := \gamma k^{-1} b^{-j-1} \sim b^{-j}$, where $\gamma > 0$ is sufficiently small. As is shown in [24], Section 4.1, there exists a cubature formula of the form

$$\int_X f(x) d\mu(x) \sim \sum_{\xi \in \Xi_j} w_{j\xi} f(\xi),$$

where $\frac{2}{3} V(\xi, \delta/2) \leqslant w_{j\xi} \leqslant 2V(\xi, \delta)$, which is exact for all $f \in \Sigma_{kb^{j+1}}$. From Proposition 6.2 and (33) it follows that $\mathcal{K}_{\Psi_j(\sqrt{L})}(x, \cdot)\mathcal{K}_{\Psi_j(\sqrt{L})}(\cdot, y) \in \Sigma_{kb^{j+1}}$ and using the cubature formula from above we obtain

$$\begin{aligned} \Psi_j(\sqrt{L}) \mathcal{K}_{\Psi_j(\sqrt{L})}(x, y) &= \int_X \mathcal{K}_{\Psi_j(\sqrt{L})}(x, u) \mathcal{K}_{\Psi_j(\sqrt{L})}(u, y) d\mu(u) \\ &= \sum_{\xi \in \Xi_j} w_{j\xi} \mathcal{K}_{\Psi_j(\sqrt{L})}(x, \xi) \mathcal{K}_{\Psi_j(\sqrt{L})}(\xi, y). \end{aligned} \tag{35}$$

Now, the frame elements are defined by

$$\psi_\xi(x) := \sqrt{w_{j\xi}} \mathcal{K}_{\Psi_j(\sqrt{L})}(x, \xi), \quad \xi \in \Xi_j, \ j \geqslant 0. \tag{36}$$

Set $\Xi := \cup_{j \geqslant 0} \Xi_j$. It will be convenient to use Ξ as an index set and for this equal points from different Ξ_j's will be regarded as distinct element of Ξ. Thus the frame is define by $\{\psi_\xi\}_{\xi \in \Xi}$. From (34), (35), and (36) it readily follows that

$$f = \sum_{\xi \in \Xi} \langle f, \psi_\xi \rangle \psi_\xi, \quad f \in \mathcal{S}'. \tag{37}$$

Furthermore, $\{\psi_\xi\}_{\xi \in \Xi}$ is a tight frame for L^2, i.e. (37) holds for each $f \in L^2$ and $\|f\|_2^2 = \sum_{\xi \in \Xi} |\langle f, \psi_\xi \rangle|^2$.

Remark 6.2 Observe that in many of the specific cases covered by the general setting from this section the polynomial property (33) is valid. In particular, this is the case of the interval $[-1, 1]$ with Jacobi weights, the sphere, ball, simplex, on conic and double hyperbolic domains and in the context of Hermite and Laguerre expansions.

Frames in the general case. The construction now is carried out in two steps.
Construction of Frame # 1. The construction begins with a cut-off function Φ with the following properties: $\Phi \in C^\infty(\mathbb{R}_+)$, $0 \leqslant \Phi \leqslant 1$, $\Phi(u) = 1$ for $u \in [0, 1]$, and $\operatorname{supp} \Phi \subset [0, b]$, where $b \geqslant 2$ is a constant, see [24]. The constant b will remain fixed throughout. Set $\Psi(u) := \Phi(u) - \Phi(bu)$.

An important point is that the function Φ can be selected so that the operators $\Phi(\delta\sqrt{L})$ and $\Psi(\delta\sqrt{L})$ are integral operators whose kernels $\mathscr{K}_{\Phi(\delta\sqrt{L})}(x,y)$ and $\mathscr{K}_{\Psi(\delta\sqrt{L})}(x,y)$ have sub-exponential space localization, namely, for any $x,y \in X$

$$|\mathscr{K}_{\Phi(\delta\sqrt{L})}(x,y)|, |\mathscr{K}_{\Psi(\delta\sqrt{L})}(x,y)| \leqslant c\frac{\exp\big\{-\kappa\big(\frac{\rho(x,y)}{\delta}\big)^{\beta}\big\}}{(|B(x,\delta)||B(y,\delta)|)^{1/2}},$$

where $0<\beta<1$, $\kappa, c>0$, and β can be selected as close to 1 as we wish. Furthermore, $\mathscr{K}_{\Phi(\delta\sqrt{L})}(x,y)$ and $\mathscr{K}_{\Psi(\delta\sqrt{L})}(x,y)$ are Hölder continuous (see [24]).

Setting

$$\Psi_0(u) := \Phi(u) \quad \text{and} \quad \Psi_j(u) := \Psi(b^{-j}u), \quad j \geqslant 1,$$

we have $\Psi_j \in C^{\infty}(\mathbb{R}_+)$, $0 \leqslant \Psi_j \leqslant 1$, $\operatorname{supp}\Psi_0 \subset [0,b]$, $\operatorname{supp}\Psi_j \subset [b^{j-1}, b^{j+1}]$, $j \geqslant 1$, and $\sum_{j\geqslant 0}\Psi_j(u) = 1$ for $u \in \mathbb{R}_+$. Hence we have the following decomposition

$$f = \sum_{j\geqslant 0}\Psi_j(\sqrt{L})f \quad \text{for } f \in \mathcal{S}'.$$

For $j \geqslant 0$ we let $\Xi_j \subset X$ be a maximal δ_j-net on X with $\delta_j := \gamma b^{-j-2}$ and suppose $\{A_\xi\}_{\xi\in\Xi_j}$ is a companion disjoint partition of X consisting of measurable sets such that $B(\xi,\delta_j/2) \subset A_\xi \subset B(\xi,\delta_j)$, $\xi \in \Xi_j$. Here $\gamma>0$ is a sufficiently small constant. The jth level frame elements ψ_ξ are defined by

$$\psi_\xi(x) := |A_\xi|^{1/2}\Psi_j(\sqrt{L})(x,\xi), \quad \xi \in \Xi_j.$$

Let $\Xi := \cup_{j\geqslant 0}\Xi_j$. Then $\{\psi_\xi\}_{\xi\in\Xi}$ is Frame #1.

The construction of a dual frame $\{\tilde{\psi}_\xi\}_{\xi\in\Xi}$ is much more involved; we refer the reader to §4.3 in [24] for the details.

We next describe the main properties of $\{\psi_\xi\}_{\xi\in\Xi}$ and $\{\tilde{\psi}_\xi\}_{\xi\in\Xi}$.

Proposition 6.3 (a) Localization: *For any $0<\hat{\kappa}<\kappa/2$ there exists a constant $\hat{c}>0$ such that for any $\xi \in \Xi_j$, $j \geqslant 0$,*

$$|\psi_\xi(x)|, |\tilde{\psi}_\xi(x)| \leqslant \hat{c}|B(\xi,b^{-j})|^{-1/2}\exp\big\{-\hat{\kappa}(b^j\rho(x,\xi))^{\beta}\big\}.$$

Also, if $\rho(x,y) \leqslant b^{-j}$, then

$$|\psi_\xi(x)-\psi_\xi(y)| \leqslant \hat{c}|B(\xi,b^{-j})|^{-1/2}(b^j\rho(x,y))^{\alpha}\exp\big\{-\hat{\kappa}(b^j\rho(x,\xi))^{\beta}\big\}$$

and the same inequality holds for $\tilde{\psi}_\xi$.

(b) Representation: *For any $f \in \mathcal{S}'$*

$$f = \sum_{\xi\in\Xi}\langle f,\tilde{\psi}_\xi\rangle\psi_\xi = \sum_{\xi\in\Xi}\langle f,\psi_\xi\rangle\tilde{\psi}_\xi \quad \text{in } \mathcal{S}'.$$

For more details, see [24].

6.2 Besov and Triebel-Lizorkin spaces

The Besov and Triebel-Lizorkin spaces associated with the operator L are in general spaces of distributions. To handle possible anisotropic geometries there are two types of Besov (B) and Triebel-Lizorkin (F) spaces introduced in [24]: (i) classical B-spaces $B_p^{sq} = B_p^{sq}(L)$ and F-spaces $F_p^{sq} = F_p^{sq}(L)$, and (ii) nonclassical B-spaces $\tilde{B}_p^{sq} = \tilde{B}_p^{sq}(L)$ and F-spaces $\tilde{F}_p^{sq} = \tilde{F}_p^{sq}(L)$. We next recall them. Let the functions $\varphi_0, \varphi \in C^\infty(\mathbb{R})$ be even real-valued and with the properties

$$\operatorname{supp}\varphi_0 \subset [-2,2], \quad |\varphi_0(\lambda)| \geqslant c > 0 \ \text{ for } \lambda \in [-2^{3/4}, 2^{3/4}],$$

$$\operatorname{supp}\varphi \cap \mathbb{R}_+ \subset [1/2, 2], \quad \text{and} \quad |\varphi(\lambda)| \geqslant c > 0 \ \text{ for } \lambda \in [2^{-3/4}, 2^{3/4}].$$

Then $|\varphi_0(\lambda)| + \sum_{j\geqslant 1} |\varphi(2^{-j}\lambda)| \geqslant c > 0$, $\lambda \in \mathbb{R}_+$. Set $\varphi_j(\lambda) := \varphi(2^{-j}\lambda)$ for $j \geqslant 1$.

Definition 6.4 Let $s \in \mathbb{R}$ and $0 < p, q \leqslant \infty$.

(a) The Besov space $B_p^{sq} = B_p^{sq}(L)$ is defined as the set of all $f \in \mathcal{S}'$ such that

$$\|f\|_{B_p^{sq}} := \Big(\sum_{j\geqslant 0} \Big(2^{sj}\|\varphi_j(\sqrt{L})f(\cdot)\|_{L^p}\Big)^q\Big)^{1/q} < \infty.$$

(b) The Besov space $\tilde{B}_p^{sq} = \tilde{B}_p^{sq}(L)$ is defined as the set of all $f \in \mathcal{S}'$ such that

$$\|f\|_{\tilde{B}_p^{sq}} := \Big(\sum_{j\geqslant 0} \Big(\|V(\cdot, 2^{-j})^{-s/d}\varphi_j(\sqrt{L})f(\cdot)\|_{L^p}\Big)^q\Big)^{1/q} < \infty.$$

Definition 6.5 Let $s \in \mathbb{R}$, $0 < p < \infty$, and $0 < q \leqslant \infty$.

(a) The Triebel-Lizorkin space $F_p^{sq} = F_p^{sq}(L)$ is defined as the set of all $f \in \mathcal{S}'$ such that

$$\|f\|_{F_p^{sq}} := \Big\|\Big(\sum_{j\geqslant 0} \Big(2^{js}|\varphi_j(\sqrt{L})f(\cdot)|\Big)^q\Big)^{1/q}\Big\|_{L^p} < \infty.$$

(b) The Triebel-Lizorkin space $\tilde{F}_p^{sq} = \tilde{F}_p^{sq}(L)$ is defined as the set of all $f \in \mathcal{S}'$ such that

$$\|f\|_{\tilde{F}_p^{sq}} := \Big\|\Big(\sum_{j\geqslant 0} \Big(V(\cdot, 2^{-j})^{-s/d}|\varphi_j(\sqrt{L})f(\cdot)|\Big)^q\Big)^{1/q}\Big\|_{L^p} < \infty.$$

Above as usual the ℓ^q-norm is replaced by the sup-norm if $q = \infty$.

Frame characterization of Besov and Triebel-Lizorkin spaces. One of the main results in [24] asserts that the Besov and Triebel-Lizorkin spaces from above can be characterized in terms of respective sequence norms of the frame coefficients of distributions, using the frames $\{\psi_\xi\}_{\xi\in\Xi}$, $\{\tilde{\psi}_\xi\}_{\xi\in\Xi}$ from §6.1.

To state this result we next introduce the sequence spaces b_p^{sq}, $\tilde{b}_p^{sq}$, and f_p^{sq}, $\tilde{f}_p^{sq}$, associated with the B- and F-spaces. As before $\Xi := \cup_{j\geqslant 0}\Xi_j$ will denote the sets of the centers of the frame elements and $\{A_\xi\}_{\xi\in\Xi_j}$ will be the associated partitions of X; $b \geqslant 2$ will be the constant from §6.1.

Definition 6.6 Let $s \in \mathbb{R}$ and $0 < p, q \leqslant \infty$.

(*a*) b_p^{sq} is defined as the space of all complex-valued sequences $h = \{h_\xi\}_{\xi\in\Xi}$ such that

$$\|h\|_{b_p^{sq}} := \Big(\sum_{j\geqslant 0} b^{jsq}\Big[\sum_{\xi\in\Xi_j}\big(V(\xi, b^{-j})^{1/p-1/2}|h_\xi|\big)^p\Big]^{q/p}\Big)^{1/q} < \infty.$$

(*b*) $\tilde{b}_p^{sq}$ is defined as the space of all complex-valued sequences $h = \{h_\xi\}_{\xi\in\Xi}$ such that

$$\|h\|_{\tilde{b}_p^{sq}} := \Big(\sum_{j\geqslant 0}\Big[\sum_{\xi\in\Xi_j}\big(V(\xi, b^{-j})^{-s/d+1/p-1/2}|h_\xi|\big)^p\Big]^{q/p}\Big)^{1/q} < \infty.$$

Definition 6.7 Suppose $s \in \mathbb{R}$, $0 < p < \infty$, and $0 < q \leqslant \infty$.

(*a*) f_p^{sq} is defined as the space of all complex-valued sequences $h = \{h_\xi\}_{\xi\in\Xi}$ such that

$$\|h\|_{f_p^{sq}} := \Big\|\Big(\sum_{j\geqslant 0} b^{jsq}\sum_{\xi\in\Xi_j}\big[|h_\xi|\tilde{\mathbb{I}}_{A_\xi}(\cdot)\big]^q\Big)^{1/q}\Big\|_{L^p} < \infty.$$

(*b*) $\tilde{f}_p^{sq}$ is defined as the space of all complex-valued sequences $h = \{h_\xi\}_{\xi\in\Xi}$ such that

$$\|h\|_{\tilde{f}_p^{sq}} := \Big\|\Big(\sum_{\xi\in\Xi}\big[|A_\xi|^{-s/d}|h_\xi|\tilde{\mathbb{I}}_{A_\xi}(\cdot)\big]^q\Big)^{1/q}\Big\|_{L^p} < \infty.$$

To state the result we will use the analysis and synthesis operators defined by $S_{\tilde{\psi}} : f \to \{\langle f, \tilde{\psi}_\xi\rangle\}_{\xi\in\Xi}$ and $T_\psi : \{h_\xi\}_{\xi\in\Xi} \to \sum_{\xi\in\Xi} h_\xi\psi_\xi$. Here the roles of $\{\psi_\xi\}$ and $\{\tilde{\psi}_\xi\}$ can be interchanged.

Theorem 6.2 [24] *Let $s \in \mathbb{R}$ and $0 < p, q \leqslant \infty$. (a) The operators $S_{\tilde{\psi}} : B_p^{sq} \to b_p^{sq}$ and $T_\psi : b_p^{sq} \to B_p^{sq}$ are bounded and $T_\psi \circ S_{\tilde{\psi}} = I$ on B_p^{sq}. Consequently, for $f \in \mathcal{S}'$ we have $f \in B_p^{sq}$ if and only if $\{\langle f, \tilde{\psi}_\xi\rangle\}_{\xi\in\Xi} \in b_p^{sq}$. Moreover, if $f \in B_p^{sq}$, then $\|f\|_{B_p^{sq}} \sim \|\{\langle f, \tilde{\psi}_\xi\rangle\}\|_{b_p^{sq}}$.*
(b) The operators $S_{\tilde{\psi}} : \tilde{B}_p^{sq} \to \tilde{b}_p^{sq}$ and $T_\psi : \tilde{b}_p^{sq} \to \tilde{B}_p^{sq}$ are bounded and $T_\psi \circ S_{\tilde{\psi}} = I$ on $\tilde{B}_p^{sq}$. Hence, $f \in \tilde{B}_p^{sq}$ if and only if $\{\langle f, \tilde{\psi}_\xi\rangle\}_{\xi\in\Xi} \in \tilde{b}_p^{sq}$. Furthermore, if $f \in \tilde{B}_p^{sq}$, then $\|f\|_{\tilde{B}_p^{sq}} \sim \|\{\langle f, \tilde{\psi}_\xi\rangle\}\|_{\tilde{b}_p^{sq}}$.

The same is true for the spaces F_p^{sq} and $\tilde{F}_p^{sq}$.

Remark 6.3 In [16] homogeneous Besov and Triebel-Lizorkin have been developed in the general setting from this section. We focus only on the inhomogeneous B- and F-spaces in this section because most of our examples are inhomogeneous.

7 Small perturbation method for construction of frames

The purpose of this section is to develop in general a small perturbation method for construction of frames in the case when there exists a pair of dual frames $\{\psi_\xi\}$, $\{\tilde{\psi}_\xi\}$ for a quasi-Banach space $\mathfrak{B}$ of distributions (or a collection $\mathscr{Y}$ of spaces $\mathfrak{B}$).

Assumptions in the case of a single space $\mathfrak{B}$. Assume (X, ρ, μ) is a metric measure space and $\mathcal{S} \subset L^2(X, \mu)$ is a linear space of test functions on X equipped with a locally convex topology induced by a sequence of norms or semi-norms. Let $\mathcal{S}'$ be the dual of $\mathcal{S}$ consisting of all continuous linear functionals on $\mathcal{S}$. Further, we assume that $\mathfrak{B} \subset \mathcal{S}'$ with norm $\|\cdot\|_{\mathfrak{B}}$ is a quasi-Banach space of distributions, which is continuously embedded in $\mathcal{S}'$ and $\mathcal{S} \subset \mathfrak{B}$.

The old pair of frames. We stipulate the existence of a pair of dual frames $\{\psi_\xi\}_{\xi\in\Xi}$, $\{\tilde{\psi}_\xi\}_{\xi\in\Xi}$ in $\mathfrak{B}$, where $\psi_\xi, \tilde{\psi}_\xi \in \mathcal{S}$ and Ξ is a countable index set, with the following properties:

A1. For any $f \in \mathfrak{B}$

$$f = \sum_{\xi\in\Xi} \langle f, \tilde{\psi}_\xi \rangle \psi_\xi = \sum_{\xi\in\Xi} \langle f, \psi_\xi \rangle \tilde{\psi}_\xi, \tag{38}$$

where the two series converge unconditionally in $\mathfrak{B}$ and hence in $\mathcal{S}'$.

A2. Consider the following *analysis* and *synthesis* operators:
$S_{\tilde{\psi}} : f \mapsto (\langle f, \tilde{\psi}_\xi \rangle)_{\xi\in\Xi}$ and $T_\psi : (h_\xi)_{\xi\in\Xi} \mapsto \sum_{\xi\in\Xi} h_\xi \psi_\xi$. The condition is that there exists a quasi-Banach (complex) sequence space $\mathfrak{b}$ with quasi-norm $\|\cdot\|_{\mathfrak{b}}$ such that:

(i) the operator $S_{\tilde{\psi}} : \mathfrak{B} \mapsto \mathfrak{b}$ is bounded, and

(ii) for any sequence $h = (h_\xi)_{\xi\in\Xi} \in \mathfrak{b}$ the series $\sum_{\xi\in\Xi} h_\xi \psi_\xi$ converges unconditionally in $\mathfrak{B}$ and $T_\psi : \mathfrak{b} \to \mathfrak{B}$ is bounded. Furthermore, the roles of ψ and $\tilde{\psi}$ can be interchanged.

Therefore, for any $f \in \mathfrak{B}$ we have $(\langle f, \tilde{\psi}_\xi \rangle)_{\xi\in\Xi} \in \mathfrak{b}$, $(\langle f, \psi_\xi \rangle)_{\xi\in\Xi} \in \mathfrak{b}$, and $\|f\|_{\mathfrak{B}} \sim \|(\langle f, \tilde{\psi}_\xi \rangle)\|_{\mathfrak{b}} \sim \|(\langle f, \psi_\xi \rangle)\|_{\mathfrak{b}}$.

In addition, we assume that $\mathfrak{b}$ obeys the conditions:

A3. (i) For any sequence $(h_\xi)_{\xi\in\Xi} \in \mathfrak{b}$, $\|(h_\xi)\|_{\mathfrak{b}} = \|(|h_\xi|)\|_{\mathfrak{b}}$.
(ii) Let $h = (h_\xi)_{\xi\in\Xi} \in \mathfrak{b}$ and assume $(h_{\xi_j})_{j\geqslant 1}$ is any ordering of the terms of the sequence h. Set $\Xi_m := \{\xi_j : j \geqslant m\}$ and define the *truncated* sequence $h^{(m)} \in \mathfrak{b}$ by

$$h_\xi^{(m)} := h_\xi \ \text{ if } \ \xi \in \Xi_m \quad \text{and} \quad h_\xi^{(m)} := 0 \ \text{ if } \ \xi \in \Xi \setminus \Xi_m.$$

The condition is that $\|h^{(m)}\|_{\mathfrak{b}} \to 0$ as $m \to \infty$.

From this it follows that compactly supported sequences are dense in $\mathfrak{b}$.

A4. The operator with matrix

$$A := (a_{\xi,\eta})_{\xi,\eta\in\Xi}, \quad a_{\xi,\eta} := \langle \psi_\eta, \psi_\xi \rangle$$

is bounded on $\mathfrak{b}$, i.e. $\|A\|_{\mathfrak{b}\mapsto\mathfrak{b}} \leqslant c < \infty$.

Construction of new frames. Next we construct a new pair of dual frames $\{\theta_\xi\}_{\xi\in\Xi}$, $\{\tilde{\theta}_\xi\}_{\xi\in\Xi}$ for $\mathfrak{B}$, where Ξ is the index set from above, in two steps: We first construct the new system $\{\theta_\xi\}_{\xi\in\Xi}$ to approximate well $\{\psi_\xi\}_{\xi\in\Xi}$ in terms of the size of the inner products $\langle\psi_\eta - \theta_\eta, \psi_\xi\rangle$ and be well localized in terms of $\langle\theta_\eta, \psi_\xi\rangle$. More precisely, we **assume** that $\theta_\xi \in \mathfrak{B}$, $\xi \in \Xi$, can be constructed so that the operators with matrices

$$B := (b_{\xi,\eta})_{\xi,\eta\in\Xi}, \quad b_{\xi,\eta} := \langle\theta_\eta, \psi_\xi\rangle,$$
$$D := (d_{\xi,\eta})_{\xi,\eta\in\Xi}, \quad d_{\xi,\eta} := \langle\psi_\eta - \theta_\eta, \psi_\xi\rangle$$

are bounded on $\mathfrak{b}$, i.e. $\|B\|_{\mathfrak{b}\mapsto\mathfrak{b}} \leqslant c$, and more importantly for a sufficiently small $\varepsilon > 0$ (to be determined later on)

$$\|D\|_{\mathfrak{b}\mapsto\mathfrak{b}} \leqslant \varepsilon. \tag{39}$$

We introduce two operators:

$$T_d h := \sum_{\xi\in\Xi} h_\xi \theta_\xi, \quad h = (h_\xi)_{\xi\in\Xi} \in \mathfrak{b}, \quad \text{and} \tag{40}$$

$$T f := \sum_{\xi\in\Xi} \langle f, \tilde{\psi}_\xi\rangle \theta_\xi, \quad f \in \mathfrak{B}. \tag{41}$$

The following properties of the operators T_d and T are instrumental for the construction of the new frames.

(a) The operators T_d and T are well defined and bounded, that is,

$$\|T_d h\|_{\mathfrak{B}} \leqslant c\|h\|_{\mathfrak{b}}, \quad \forall h \in \mathfrak{b} \quad \text{and} \quad \|Tf\|_{\mathfrak{B}} \leqslant c\|f\|_{\mathfrak{B}}, \quad \forall f \in \mathfrak{B}.$$

Furthermore, the series in (40)-(41) converge unconditionally in $\mathfrak{B}$ and hence in S'.

(b) If ε in (39) is sufficiently small then $\|I - T\|_{\mathfrak{B}\mapsto\mathfrak{B}} < 1$ and hence T^{-1} exists and

$$\|T^{-1}\|_{\mathfrak{B}\mapsto\mathfrak{B}} \leqslant c < \infty.$$

(c) The operator H with matrix $\mathbf{H} := (\langle T^{-1}\psi_\eta, \tilde{\psi}_\xi\rangle)_{\xi,\eta\in\Xi}$ is bounded on $\mathfrak{b}$.

Construction of the dual frame $\{\tilde{\theta}_\xi\}$. For any $\xi \in \Xi$ we define the linear functional $\tilde{\theta}_\xi$ by

$$\tilde{\theta}_\xi(f) = \langle f, \tilde{\theta}_\xi\rangle := \sum_{\eta\in\Xi} \langle T^{-1}\psi_\eta, \tilde{\psi}_\xi\rangle\langle f, \tilde{\psi}_\eta\rangle \quad \text{for } f \in \mathfrak{B}. \tag{42}$$

From property (c) above and **A2** it follows that for $f \in \mathfrak{B}$

$$|\tilde{\theta}_\xi(f)| = |\langle f, \tilde{\theta}_\xi\rangle| \leqslant \|H\|_{\mathfrak{b}\mapsto\mathfrak{b}}\|(\langle f, \tilde{\psi}_\eta\rangle)\|_{\mathfrak{b}} \leqslant c\|f\|_{\mathfrak{B}}.$$

Thus, $\tilde{\theta}_\xi$ is a bounded linear functional on $\mathfrak{B}$, i.e. $\tilde{\theta}_\xi \in \mathfrak{B}'$.

Further, by property (b) for $f \in \mathfrak{B}$ we have $T^{-1}f \in \mathfrak{B}$ and using property (a)

$$f = T(T^{-1}f) = \sum_{\xi\in\Xi} \langle T^{-1}f, \tilde{\psi}_\xi\rangle\theta_\xi. \tag{43}$$

On the other hand, from the fact that T^{-1} is a bounded operator on $\mathfrak{B}$ and (38) it follows that for any $f \in \mathfrak{B}$ we have $T^{-1}f = \sum_{\eta\in\Xi}\langle f,\tilde{\psi}_\eta\rangle T^{-1}\psi_\eta$, where the series converges unconditionally in $\mathfrak{B}$ and hence in $\mathcal{S}'$. This and the fact that $\tilde{\psi}_\xi \in \mathcal{S}$ imply

$$\langle T^{-1}f,\tilde{\psi}_\xi\rangle = \sum_{\eta\in\Xi}\langle T^{-1}\psi_\eta,\tilde{\psi}_\xi\rangle\langle f,\tilde{\psi}_\eta\rangle = \langle f,\tilde{\theta}_\xi\rangle. \tag{44}$$

Here the series converges unconditionally and hence absolutely because of the unconditional convergence of the former series. From (43)–(44) we infer

$$f = \sum_{\xi\in\Xi}\langle f,\tilde{\theta}_\xi\rangle\theta_\xi, \quad f \in \mathfrak{B}, \tag{45}$$

where $\langle f,\tilde{\theta}_\xi\rangle$ is defined in (42); the convergence is unconditional in $\mathfrak{B}$.

Theorem 7.1 *If ε in* (39) *is sufficiently small, the above defined system $\{\theta_\xi\}$ is a frame for $\mathfrak{B}$ in the following sense: For any $f \in \mathfrak{B}$*

$$f = \sum_{\xi\in\Xi}\langle f,\tilde{\theta}_\xi\rangle\theta_\xi,$$

where the convergence is unconditional in $\mathfrak{B}$, and

$$\|f\|_{\mathfrak{B}} \sim \|(\langle f,\tilde{\theta}_\xi\rangle)\|_{\mathfrak{b}} \tag{46}$$

with constants of equivalence independent of f.

Moreover, the operator T_d defined by $T_dh := \sum_{\xi\in\Xi}h_\xi\theta_\xi$ for sequences of complex numbers $h = (h_\xi)_{\xi\in\Xi}$ is bounded as a map $T_d : \mathfrak{b} \mapsto \mathfrak{B}$.

For the proof of this theorem, see the proof of Theorem 3.5 in [10].

Construction of frames for classes of quasi-Banach spaces. Let $\mathscr{Y}$ be a collection of quasi-Banach spaces $\mathfrak{B}$ of distributions and assume that $\{\psi_\xi\}_{\xi\in\Xi}$, $\{\tilde{\psi}_\xi\}_{\xi\in\Xi}$ is a pair of dual frames, just as in the assumptions of this section, for each $\mathfrak{B} \in \mathscr{Y}$. We denote by $\mathscr{Y}_d$ the class consisting of the respective sequence spaces $\mathfrak{b}$.

Now, our *main assumption* is that all constants in the conditions on the old frames are uniform with respect to $\mathfrak{B} \in \mathscr{Y}$ and $\mathfrak{b} \in \mathscr{Y}_d$, i.e. they are the same for all $\mathfrak{B} \in \mathscr{Y}$ and $\mathfrak{b} \in \mathscr{Y}_d$.

In the construction of the new frames $\{\theta_\xi\}_{\xi\in\Xi}$, $\{\tilde{\theta}_\xi\}_{\xi\in\Xi}$ for $\mathfrak{B} \in \mathscr{Y}$ our main assumption now is that the constants are also uniform. Thus we assume that $\theta_\xi \in \mathcal{S}$, $\xi \in \Xi$, can be constructed so that $\|B\|_{\mathfrak{b}\mapsto\mathfrak{b}} \leqslant c$ and $\|D\|_{\mathfrak{b}\mapsto\mathfrak{b}} \leqslant \varepsilon$ for all $\mathfrak{b} \in \mathscr{Y}_d$, where $\varepsilon > 0$ is sufficiently small.

A careful examination of the arguments shows that under the above assumptions Theorem 7.1 holds for all $\mathfrak{B} \in \mathscr{Y}$, where the constants in (46) are independent of $\mathfrak{B}$ as well; they may depend on $\mathscr{Y}$, $\mathscr{Y}_d$, and the constants from the assumptions.

Applications. The small perturbation method from this section has been used in [30] for construction of good frames on the sphere $\mathbb{S}^{d-1}$ with small shrinking support, and in [18] for construction of highly localized frames for Besov and Triebel-Lizorkin spaces on the sphere $\mathbb{S}^{d-1}$ whose elements are linear combinations of a fixed number of shifts of the Newtonian kernel. The latter result was utilized for nonlinear n-term approximation from shifts of the Newtonian kernel. In [10] the results of this section were used for construction of good frames with compactly supported elements with small shrinking supports in the general setting from Section 6 and as application on $[-1, 1]$ in the context of Jacobi polynomials.

References

1. Andersson, L., Hall, N., Jawerth, B., Peters, P.: Wavelets on closed subsets of the real line, in "Recent Advances in Wavelet Analysis," Wavelet Analysis and Its Applications, Vol. 3, pp. 1–61, Academic Press, Boston, MA, 1994.
2. Bonami, A., Kerkyacharian, G., Petrushev, P.: Gaussian Bounds for the Heat Kernel Associated to Prolate Spheroidal Wave Functions with Applications. Constr. Approx. **57**, 351–403 (2023)
3. Ciesielski, Z.: Constructive function theory and spline systems. Studia Math. **53**, 277–302 (1975)
4. Ciesielski, Z., Domsta, J.: Construction of an orthogonal basis in $C^m(I^d)$ and $W_p^m(I^d)$. Studia Math. **41**, 211–224 (1972)
5. Cohen, A., Daubechies, I., Vial, P.: Wavelets on the interval and fast wavelet transforms. Appl. Comput. Harmon. Anal. textbf1, 54–81 (1993)
6. Coulhon, T., Kerkyacharian, G., Petrushev, P.: Heat kernel generated frames in the setting of Dirichlet spaces. J. Fourier Anal. Appl. **18**, 995–1066 (2012)
7. Coifman, R., Weiss, G.: Extensions of Hardy spaces and their use in analysis. Bull. Amer. Math. Soc. **83**, 569–645 (1977)
8. Dai, F., Xu, X.: Approximation theory and harmonic analysis on spheres and balls. Springer (2015)
9. Daubechies, I.: Ten Lectures on wavelets. SIAM, Philadelphia (1992)
10. Dekel, S., Kerkyacharian, G., Kyriazis, G., Petrushev, P.: Compactly supported frames for spaces of distributions associated with nonnegative self-adjoint operators. Studia Math. **225**, 115–163 (2014)
11. DeVore, R.: Nonlinear approximation. Acta numerica, 51–150, Acta Numer. 7, Cambridge Univ. Press, Cambridge (1998)
12. Frazier, M., Jawerth, B.: Decomposition of Besov Spaces. Indiana Univ. Math. J. **34**, 777–799 (1985)
13. Frazier, M., Jawerth, B.: A discrete transform and decompositions of distribution spaces. J. of Funct. Anal. **93**, 34–170 (1990)
14. Frazier, M., Jawerth, B., Weiss, G.: Littlewood-Paley theory and the study of function spaces. CBMS 79 (1991), AMS.
15. Garnett, J.: Bounded analytic functions. Academic Press, New York (1981)
16. Georgiadis, A.G., Kyriazis, G., Kerkyacharian, G., Petrushev, P.: Atomic and moleculiar decomposition of homogeneous spaces of distributions associated to non-negative self-adjoint operators. J. Fourier Anal. Appl. **25**, 3259–3309 (2019)
17. Grafakos, L.: Classical and modern Fourier analysis. Pearson Education, Inc., Upper Saddle River, NJ (2004)
18. Ivanov, K. G., Petrushev, P.: Nonlinear n-term approximation of harmonic functions from shifts of the Newtonian kernel. Trans. Amer. Math. Soc. **373**, 3117–3176 (2020)
19. Ivanov, K. G., Petrushev, P.: Nonlinear wavelet approximation in BMO. Constr. Approx. **57**, 631–661 (2023)
20. Ivanov, K. G., Petrushev, P.: On the rational approximation in BMO. Submitted to Proc. Intern. Conf. "Constructive Theory of Functions, Lozenets 2023".

21. Ivanov, K. G., Petrushev, P., Xu, Y.: Sub-exponentially localized kernels and frames induced by orthogonal expansions. Math. Z. **264**, 361–397 (2010)
22. Ivanov, K. G., Petrushev, P., Xu, Y.: Decomposition of spaces of distributions induced by tensor product bases. J. Funct. Anal. **263**, 1147–1197 (2012)
23. John, F., Nirenberg, L.: On functions of bounded mean oscilation. Comm. Pure and Appl. Math. **14**, 415–426 (1961)
24. Kerkyacharian, G., Petrushev, P.: Heat kernel based decomposition of spaces of distributions in the framework of Dirichlet spaces. Trans. Amer. Math. Soc. **367**, 121–189 (2015)
25. Kerkyacharian, G., Petrushev, P., Xu, Y.: Gaussian bounds for the weighted heat kernels on the interval, ball, and simplex. Constr. Approx. **51**, 73–122 (2020)
26. Kerkyacharian, G., Petrushev, P., Xu, Y.: Gaussian bounds for the heat kernels on the ball and simplex: Classical approach. Studia Math. **250**, 235–252 (2020)
27. Kerkyacharian, G., Petrushev, P., Picard, D., Xu, Y.: Decomposition of Triebel-Lizorkin and Besov spaces in the context of Laguerre expansions. J. Funct. Anal. **256**, 1137–1188 (2009)
28. Kyriazis, G.: Decomposition systems for function spaces. Studia Math. **157**, 133–169 (2003)
29. Kyriazis, G., Petrushev, P.: New bases for Triebel-Lizorkin and Besov spaces. Trans. Amer. Math. Soc. **354**, 749–776 (2002)
30. Kyriazis, G., Petrushev, P.: On the construction of frames for spaces of distributions. J. Funct. Anal. **257**, 2159–2187 (2009)
31. Kyriazis, G., Petrushev, P.: Rational bases for spaces of holomorphic functions in the disc. J. Lond. Math. Soc. **89**, 434–460 (2014)
32. Kyriazis, G., Petrushev, P., Xu, Y.: Jacobi decomposition of weighted Triebel-Lizorkin and Besov spaces. Studia Math. **186**, 161–202 (2008)
33. Kyriazis, G., Petrushev, P., Xu, Y.: Decomposition of weighted Triebel-Lizorkin and Besov spaces on the ball. Proc. London Math. Soc. **97**, 477–513 (2008)
34. Meyer, Y.: Wavelets and operators. Cambridge University Press, Cambridge (1992)
35. Müller, C.: Spherical harmonics, Lecture Notes in Mathematics. Vol. 17, Springer Verlag, Berlin (1966).
36. Narcowich, F., Petrushev, P., Ward, J.: Localized tight frames on spheres. SIAM J. Math. Anal. **38**, 574–594 (2006)
37. Narcowich, F., Petrushev, P., Ward, J.: Decomposition of Besov and Triebel-Lizorkin spaces on the sphere. J. Funct. Anal. **238**, 530–564 (2006)
38. Nowak, A., Sjögren, P.: Sharp estimates of the Jacobi heat kernel. Studia Math. **218**, 219–244 (2013)
39. Peetre, J.: New thoughts on Besov spaces. Duke Univ. Math. Series, Duke University, Durham, NC (1976)
40. Pekarskii, A.: Inequalities of Bernstein type for the derivatives of rational functions and inverse theorems of rational approximation. Mat. Sbornik **124(166)**, 571–588 (1984)
41. Pekarskii, A.: Classes of analytic functions determined by the best rational approximation in H_p. Mat. Sbornik **127 (169)**, 3–20 (1985)
42. Pekarskii, A.: Approximation by rational functions with free poles. East Journal on Approximations, **13**, 1–93 (2007)
43. Petrushev, P.: Bases consisting of rational functions of uniformly bounded degrees or more general functions. J. Funct. Anal. **174**, 18–75 (2000)
44. Petrushev, P., Xu, Y.: Localized polynomial frames on the interval with Jacobi weights. J. Fourier Anal. Appl. **11**, 557–575 (2005)
45. Petrushev, P., Xu, Y.: Localized polynomial frames on the ball. Constr. Approx. **27**, 121–148 (2008)
46. Petrushev, P., Xu, Y.: Decomposition of spaces of distributions induced by Hermite expansions. J. Fourier Anal. and Appl. **14**, 372–414 (2008)
47. Stein, E.: Harmonic analysis: real-variable methods, orthogonality, and oscillatory integrals. Princeton University Press, Princeton, NJ (1993)
48. Stein, E., Weiss, G.: Fourier analysis on Euclidean spaces. Princeton University Press, Princeton, New Jessey (1971)

49. Szegö, G.: Orthogonal polynomials. Amer. Math. Soc. Colloq. Publ. Vol. 23, Amer. Math. Soc. Providence (1975)
50. Triebel, H.: Theory of function spaces. Monographs in Math. Vol. 78, Birkhäuser, Verlag, Basel (1983)
51. Wojtaszczyk, P.: A mathematical introduction to wavelets. Cambredge University Press, Cambredge (1997)
52. Xu, Y.: Approximation and localized polynomial frame on conic domains. J. Funct. Anal. **281**, no. 12, Paper No. 109257, 94 pp. (2021)
53. Xu, Y.: Approximation and localized polynomial frame on double hyperbolic and conic domains. Constr. Approx. **57**, 921–976 (2003)

Towards Continuous Mathematical Models for the Analysis of Classes of Deep Neural Networks

Angela Kunoth,
Mathias Oster and
Reinhold Schneider

Abstract Our goal is to develop and formulate continuous mathematical models for the analysis of deep neural networks, in order to

a) provide a convergence analysis,
b) conduct an optimization analysis with respect to the optimal choice and computation of parameters,
c) study the role of overparametrization.

The tools used here are control theory, Hamilton-Jacobi-Bellman equations, Barron functions as well as different optimization algorithms.

1 Introduction

In recent years, an incredible development in machine learning displayed the power of neural networks based on efficient back-propagation and optimization algorithms as well as powerful hardware. All these allowed for very complicated neural network architectures, see [9, 18, 21, 27] for an introduction to neural networks from a mathematical perspective.

A neural network has a *compositional structure*, i.e., one can write it as a composition of so-called *layers* Ψ^i, i.e.,

Angela Kunoth
Department of Mathematics and Computer Science, Division of Mathematics, University of Cologne, Weyertal 86–90, 50931 Cologne, Germany, e-mail: akunoth@uni-koeln.de

Mathias Oster
IGPM, RWTH Aachen, Templergraben 55, 52062 Aachen, Germany, e-mail: oster@igpm.rwth-aachen.de

Reinhold Schneider
Department of Mathematics, TU Berlin, Str. des 17. Junis 135, 10623 Berlin, Germany, e-mail: schneidr@math.tu-berlin.de

R. DeVore, A. Kunoth (eds.), *Multiscale, Nonlinear and Adaptive Approximation II*,
https://doi.org/10.1007/978-3-031-75802-7_17

$$f^N(x) := (\Psi^{L+1} \circ \Psi^L \circ \cdots \circ \Psi^1)(x) \tag{1}$$

with input $x \in \Omega \subset \mathbb{R}^d$. For so-called fully connected neural networks each layer, indexed by i, is of the form

$$\Psi^i(z) := \sigma_i(A_i z + b_i), \quad i \in \{1, \ldots, L+1\}. \tag{2}$$

Here, σ_i is an *activation function*, A_i are called *weights*, b_i are *biases* (their precise form and dimension are detailed later), and the input z for Ψ^i is the output from the previous layer $i-1$. The amount of *(interior) layers* L is called the *depth* of the network. A network is called *deep* if $L > 1$, i.e., it has at least 2 interior layers, else it is called *shallow*. For typical neural networks, in addition to its length L, another important parameter is the *width* W^i of each layer, i.e., the number of neurons per layer.

This very versatile structure and the non-linear parametrization, allowing for high expressitivity, stimulated extensive research, for example, in image recognition [36], or the solution of high-dimensional PDEs (partial differential equations) [7].

In order to find the best weights and biases for a given task like regression, one usually employs gradient descent based schemes, named *learning* in the machine learning community. For that, given some input-target pairs (x_m, y_m), $m = 1, \ldots, M$, learning a network can be understood as an optimization problem with the so-called mean squared error

$$\min_{(A_i, b_i),\, i=1,\ldots,L} \sum_{m=1}^{M} |y_m - f^N(x_m)|^2, \tag{3}$$

as objective. The functional in (3) is one example of a *loss function*; many others are used in practice.

Neural networks are highly non-linear, non-convex functions. Thus, they are difficult to analyze and optimize as such, entailing a great variety of heuristic optimization strategies often lacking global convergence results.

As the function classes of neural networks have a non-linear structure, from an approximation theoretical point of view, they might allow one to approximate high dimensional functions without the curse of dimensionality. This could be, for example, in cases for which linear models suffer from bad scaling, e.g., if the Kolmogorov n-widths are slowly decaying, see, e.g., [5, 15]. Hence, discovering the interplay of neural network approximations and solutions to scientific computing tasks is central in contemporary high dimensional applied mathematics.

One way to improve our understanding of certain classes of neural networks is to pass to some kind of continuous limit with respect to depth and width.

Let us first consider the interesting case of the so-called *Barron functions* which provide in some sense 'infinitely wide' but shallow neural networks, meaning $W = \infty$. With an activation function σ, Barron functions admit an integral representation

$$f(x) = \int a\sigma(Ax + b)\, \mathrm{d}\mu(a, A, b)$$

where the probability measure μ on the space of weights and biases is the new unknown. In [24], explicit representation formulas were used to show that solutions to certain PDEs lie in Barron spaces or in multilayer spaces if the data for the PDE lie in these spaces. One can, therefore, deduce that these solutions can be efficiently represented using neural networks even for PDEs in high spatial dimensions.

In a different fashion, one can model the limit $L \to \infty$ for some classes of neural networks, called ResNets, by making use of the notion of neural ordinary differential equations (ODEs) via exploiting a dynamical system-like structure to model compositions of functions. This can be seen as follows. If the layer function $\Psi^i(z)$ takes the special form

$$\Psi^i(z) = z + \sigma_i(A_i z + b_i), \tag{4}$$

the neural network is called residual neural network. This particular architecture is reminiscent to an explicit Euler discretization and inspired the so-called Neural ODEs, i.e., ODEs of the form

$$\frac{d}{dt} z(t) = \sigma(A(t)z + b(t)).$$

In this way, Neural ODEs can be considered 'infinitely deep' (but finitely wide) neural networks.

In summary, Barron functions represent 'infinitely wide' nets, and Neural ODEs give 'infinitely deep' nets. Both concepts will be detailed in the following sections.

The most straightforward next step is to combine the notion of Barron functions and Neural ODEs, leading to the so-called flow induced spaces (notion explained below) introduced in [22].

For the first steps of an analysis in this paper we follow this path and reformulate (3) subject to (1) with (4) as a *continuous model* with respect to depth and width. Firstly, we interpret (4) for each i as discretization of an ordinary differential equation (ODE)

$$\begin{aligned} \dot{x}(t) &= v(t, x(t)), \quad t \in (0, T], \\ x(0) &= x. \end{aligned} \tag{5}$$

over a time interval with initial time 0 and final time T. On the right hand side, we use a *vector field* $v : [0, T] \times \mathbb{R}^d \to \mathbb{R}^d$ given by

$$v(t, x) := \int_{\mathcal{V}} a\, \sigma(Ax + b)\, d\mu^2(t; a, A, b) \tag{6}$$

where $\mathcal{V} \subset \mathbb{R}^{d\times 1} \times \mathbb{R}^{1\times d} \times \mathbb{R}$, i.e., we replace a shallow neural net by a Barron function with activation function $\sigma : \mathbb{R} \to \mathbb{R}$. Here, $a \in \mathbb{R}^{d\times 1}$ a parameter function, $A \in \mathbb{R}^{1\times d}$ is a weight and $b \in \mathbb{R}$ a bias. The choice of $\mathcal{V}$ is based on the finite-dimensional network input $x \in \Omega \subset \mathbb{R}^d$ for A, b to be well-defined. Moreover, we assume $\mu^2(t) \in \mathcal{P}_2(\mathcal{V})$, the set of Borel probability measures on $\mathcal{V}$ with bounded second moment. We explain this choice later. Then (3) generalizes to an optimal control task as described below.

In order to exploit optimal control theoretic tools, an important reinterpretation of (5) for all initial values simultaneously is to consider the induced *flow* z at time $t \in [0, T]$ as a map

$$\begin{aligned} x \mapsto z(t, x) &= x(t) \\ z(0, x) &= x, \end{aligned} \tag{7}$$

with $z \in C^0([0, T], C^0(\Omega, \mathbb{R}^d))$, that is, as an abstract ODE in the Banach space $C^0(\Omega, \mathbb{R}^d)$.

Now we can interpret the regression of a function g_{target} by a generalized deep neural network as a Banach space-valued optimal control problem with measure-valued controls. To this end, set $X = C^0(\Omega, \mathbb{R}^d)$ with compact $\Omega \subset \mathbb{R}^d$. Consider the *optimal control problem*

$$\min_{\substack{z \in C^0([0,T],X) \\ (\mu_1,\mu_2) \in \mathcal{P}_2(\mathcal{U}) \times \mathcal{Y}([0,T],\mathcal{V})}} \mathcal{J}(z, \mu^1, \mu^2) \tag{8}$$

with

$$\mathcal{J}(z, \mu^1, \mu^2) := \int_\Omega \left| g_{\text{target}}(x) - \int_{\mathcal{U}} a\sigma(Az(T, x) + b) \mathrm{d}\mu^1(a, A, b) \right|^2 \mathrm{d}\rho(x) \tag{9}$$

subject to

$$\begin{aligned} \frac{d}{dt} z(t, x) &= \int_{\mathcal{V}} f(z(t, x), v) \mathrm{d}\mu^2(t; v), \quad t \in (0, T], \\ z(0, x) &= x. \end{aligned} \tag{10}$$

Here $\mathcal{U} \subset \mathbb{R}^{m_1}, \mathcal{V} \subset \mathbb{R}^{m_2}$ are compact. ρ is some probability measure absolutely continuous w.r.t the Lebesgue measure on Ω and $g_{\text{target}} \in C^2(\Omega)$ is a given target function to be matched in a least-squares sense. We assume $f \in C^2(\mathbb{R}^d \times \mathbb{R}^{m_2}, \mathbb{R}^d)$ to be bounded with bounded derivatives, and z is the flow corresponding to (6). The space $\mathcal{Y}([0, T], \mathcal{V})$ is the space of all Young measures with bounded second moment on $[0, T] \times \mathcal{V}$ which can be identified with the set of all measurable mappings $\mu^2 : [0, T] \to \mathcal{P}_2(\mathcal{V})$.

Notice that if we choose the Ansatz space for our function approximation as

$$\mathcal{M} = \left\{ x \mapsto \int_{\mathcal{U}} a\sigma(Az(T, x) + b) \, \mathrm{d}\mu^1(a, A, b) \right. \\ \left. \text{s.t.} \quad \dot{z} = \int_{\mathcal{V}} a\sigma(Az + b) \, \mathrm{d}\mu^2(t; a, A, b), \quad z(0, x) = x \right\},$$

then the cost functional is the ρ-weighted L^2 discrepancy, i.e.,

$$\mathcal{J}(z, \mu^1, \mu^2) = \|g_{\text{target}} - g_{\mathcal{M}}\|^2_{L^2(\Omega;\rho)}$$

for some appropriate $g_{\mathcal{M}} \in \mathcal{M}$. If the measures happen to be sums of Dirac measures, the optimal control problem is the standard function recovery of machine learning with neural networks. We wish to point out that, introducing two different measures and activation functions, allows us to understand this optimal control problem as a generalization of deep neural networks with distinguished last layers.

In summary, this provides a completely continuous formulation of function recovery in machine learning. In [23], gradient flow equations have been already employed in the case of a shallow network.

In [19], this optimal control problem, its gradient flow, and the connection to neural networks have been analyzed. The authors established the existence of a unique (adjusted) gradient flow by a local fixed point argument and continued the solution globally. They also showed that, if the PDE corresponding to the gradient flow has a stationary solution with support in all of $\mathbb{R}^m$, the stationary point is a global minimum. This way, they found an explanation for why over-parameterized residual networks generalize well. They use $\mu \in C^0([0,T], \mathcal{P}_2(\mathbb{R}^m))$ as controls. They stress that their parametrization deviates from the gradient flow notion of [3] as they have a continuous family of probability measures to which the results of [3] cannot be applied directly. Hence, they call their gradient flow *adjusted*.

In this article we allow for more general controls from $\mathcal{Y}([0,T], \mathcal{V})$. We will partly reestablish the findings proving the existence of a gradient flow from [19] with the tools of [3] and [38] by interpreting the controls as Young measures instead of a continuous family of probability measures. This will allow us to use the framework of gradient flow in Wasserstein space as in [3] tailored to Young measures. Thus, we can establish the existence of gradient flows that have the same governing equations as in [19]. Uniqueness is not shown here but could be established if one can show some kind of relaxed convexity of the cost functional w.r.t. the Young measures. We would also like to point out that the use of Young measures instead of time-continuous families allows us to establish the well-posedness of the optimal control problem, i.e., the existence of minimizers with standard tools from optimization. This is due to the fact that the Young measures are closed in the fibered Wasserstein space. Thus, compactness and continuity arguments guarantee the existence of minimizers. We also think that allowing for measurable controls is closer to the implementations in practice where no dependence between the weights of consecutive layers is enforced.

Formulating the function recovery by neural networks as an abstract optimal control problem with measure-valued control not only allows one to analyze the role of overparametrization in deep learning but also establishes a bridge to the rich literature of control theory and invites to use tools like controllablity to understand expressivity. Furthermore, it provides clues on how to develop learning algorithms that adapt the architecture of neural networks. To obtain a fully adaptive scheme, very advanced a posteriori estimaters would be necessary that not yet have been developed. However, in order to have a first glimpse at the potential of the continuous generalization of neural networks we provide a naïve algorithm with a simple numerical example as outlook at the end of these notes.

2 Barron Spaces and Neural ODEs

In the following two sections, we will introduce the notions of Barron functions and Neural ODEs in more detail and report some numerical analysis results that have been established in these function classes. Thereafter, we will combine these two approaches and reiterate some finding in flow-induced spaces from a different perspective.

2.1 Barron Spaces

From a theoretical point of view, establishing the universal approximation theorem, e.g. [28], for neural networks was a major breakthrough in the study of the expressivity of neural networks: one could show that any continuous function on a compact domain can be approximated by neural networks. However, the number of parameters might be large and subject to the 'curse of dimensionality', i.e., the number of weights and biases might grow exponentially with the input dimensions. There are numerous works analyzing situations when deep neural networks can break the curse of dimensionality, see for example [4, 6, 8, 16, 29, 31, 41].

For shallow networks, i.e., networks with few layers, one can identify a function space that has nice approximation properties with respect to neural networks. This means that the number of weights gives a Monte Carlo-like error decrease independent of the input dimension.

We consider a generalized shallow neural network as follows: for $x \in \Omega \subset \mathbb{R}^d$ define

$$f(x) := \int_{\mathcal{U}} a\,\sigma(Ax+b)\, d\mu(a, A, b) \tag{11}$$

where $\sigma : \mathbb{R} \to \mathbb{R}$ is the activation function which is typically the ReLU, i.e., $\sigma(x) = \max(x, 0)$. Here $a \in \mathbb{R}$, $A \in \mathbb{R}^{1\times d}$, $b \in \mathbb{R}$ and the parameter set $\mathcal{U} \subset \mathbb{R} \times \mathbb{R}^{1\times d} \times \mathbb{R}$.

For the measure, we assume $\mu \in \mathcal{P}_2(\mathcal{U})$ where $\mathcal{P}_2(\mathcal{U})$ is the set of *probability (Borel) measures* on $\mathcal{U}$ with bounded second moment, i.e.

$$\mathcal{P}_2(\mathcal{U}) = \{\mu \in \mathcal{P}(\mathcal{U}) : \int_{\mathcal{U}} |u|^2 \mathrm{d}\mu(u) < \infty\}.$$

We define a *Barron 'norm'*

$$\|f\|_{\mathcal{B}} := \inf_{\mu \in \mathcal{P}_2(\mathcal{U})} \left\{ \int_{\mathcal{U}} |a|(\|A\| + |b|)\, \mathrm{d}\mu(a, A, b) : \; f \text{ defined as in (11)} \right\} \tag{12}$$

where $\|A\|$ denotes the operator norm corresponding to the Euclidean norm $|x|$ of a vector $x \in \mathbb{R}^d$.

This representation is used to introduce a class of functions called Barron functions [13, 22, 23]

$$\mathcal{B} = \{f \in L^2(\Omega) : \|f\|_{\mathcal{B}} < \infty\},$$

which are well-approximated by shallow neural networks. The idea is also known as the mean-field limit of over-parameterized system, [14, 33].

Remark 2.1 Using different integrands, as for example

$$\int_{\mathcal{V}} a \exp((A^{-1}(x-b))^2)\, \mathrm{d}\mu(a, A, b),$$

we can also analyze Gaussian mixture models from this point of view. Thus, the idea of integral representations provides a fairly general tool, see also the variation spaces of [17, 42, 43, 44]. One can furthermore observe that certain kernel methods also have this form.

The key idea is that, instead of learning the weights and biases of the neural network, we learn a family of measures according to which the weights and biases are distributed. In the case of shallow networks, this leads to convex optimization tasks and opens up new perspectives on the training of neural networks [25]. Note that by choosing the probability measures to be sums of Dirac measures, the neural networks are contained in the Barron spaces.

Relation to Shallow Neural Networks

A shallow neural network with n *neurons* is obtained from the Barron representation by approximating the measure μ with a sum of n Dirac delta distributions (particle approximation)

$$\mu_n := \frac{1}{n} \sum_{i=1}^{n} \delta_{a_i, A_i, b_i}, \tag{13}$$

with the abbreviation

$$\delta_{a_i, A_i, b_i} := \delta(a-a_i)\, \delta(A-A_i)\, \delta(b-b_i) \tag{14}$$

for $(a_i, A_i, b_i) \in \mathcal{U} \subset \mathbb{R} \times \mathbb{R}^{1\times d} \times \mathbb{R}$.
With this, we can define an approximation of f in (11) as

$$f_n(x) = \frac{1}{n} \sum_{i=1}^{n} a_i\, \sigma(A_i x + b_i). \tag{15}$$

Here, f_n can be viewed as a Monte Carlo approximation of the continuous version of f if the weights and biases (a_i, A_i, b_i) are drawn randomly according to μ_n. This allows us to derive convergence estimates similar to the ones of Monte Carlo integration.

If the measure μ is supported on a compact set $\mathcal{U} \subset \mathbb{R}^{m_1}$ with diameter R, there exists f_n such that

$$\|f - f_n\|^2_{L_\infty(\Omega)} \leqslant c\,\frac{d^2R^2}{n}\,\|f\|^2_{\mathcal{B}},$$
$$\|f - f_n\|^2_{L_2(\Omega)} \leqslant \frac{3}{n}\,\|f\|^2_{\mathcal{B}}.$$

Similar results hold for other spaces, e.g., $H^1(\Omega)$, see, e.g., [24].

Note that there is no curse of dimensionality directly appearing in the error estimates. However, the Barron norm of certain functions classes might depend on the state dimension in an exponential fashion.

2.2 Neural ODEs: Continuous Model

Barron functions allow one to have a notion of 'infinitely wide' neural network. On the other hand, one still needs to understand the role of depths in neural network architectures. One particular design, the residual network, inspired the idea of so-called neural ODEs [11, 12, 45]. In the residual network architecture, the individual layers have the form

$$\Psi^i(x) = x + \sum_{k=1}^{n} a^i_k \sigma(A^i_k x + b^i_k),$$

where $a^i_k \in \mathbb{R}^{d\times 1}$, $A^i_k \in \mathbb{R}^{1\times d}$, $b^i_k \in \mathbb{R}$. By re-scaling as

$$x^{i+1} := \Psi^i(x^i) = x^i + \tau \sum_{k=1}^{n} a^i_k \sigma(A^i_k x^i + b^i_k),$$

this reminds of the explicit Euler discretization of the ODE

$$\frac{d}{dt}x = \sum_{k=1}^{n} a_k \sigma(A_k(t)x + b_k(t)).$$

This point of view allows one to employ well-known methods from dynamical system theory, in particular methods with adaptive time step sizes, see, e.g., [1]. As a time discretization with step length $O(\tau)$ of the ODE will produce a neural net with depth $O(\frac{1}{\tau})$, the Neural ODE can be seen as the infinite depth limit of a residual neural network.

This time-continuous formulation of neural networks leads to the interpretation of the optimization of neural networks as an optimal control problem [10, 20, 30]. In [20], the input-target pairs (x, y) are jointly distributed according to some (unknown) density ρ. Given some cost functions c_T, ℓ the learning problem is phrased as

$$\inf_{\theta\in L^\infty([0,T],\Theta)} \mathbb{E}_\rho \left[c_T(z(T), y) + \int_0^T \ell(z(t), \theta(t))\, dt \right] \tag{16}$$

subject to $\frac{d}{dt}z = v(t, z(t), \theta(t))$ (v is a shallow neural network here) with initial condition x distributed jointly with y according to ρ. Here Θ is some subset of $\mathbb{R}^m$ from which the weights and biases are taken. The terminal cost function c_T will be specified later, for a prototypical example. The running cost ℓ represent a cost for the transport of the states x and is often used as a regularization parameter.

Then the dynamics $t \mapsto x(t)$ is used to introduce the law of the transported distributions, and the optimal control problem is analyzed as an abstract optimal control problem with states in Wasserstein space. They prove an infinite dimensional HJB equation and a Pontryagin Maximum principle.

It is now a very natural next step to combine the concepts of Barron functions and Neural ODEs into the notion of a flow-induced space.

2.3 Neural ODEs with Barron vector fields

In [22] flow-induced functions, i.e., ODEs with a Barron vector field, have been introduced. There it has been shown that they can be approximated by neural networks with certain rates depending on the depth of the neural network. This can be seen as follows. Define

$$\begin{aligned} \frac{d}{dt}N(t) &= (\mathbb{E}_{\mu_t}[(|a|\,|A|)^2])^{\frac{1}{2}}N(t) \\ N(0) &= [1, \ldots, 1]^T, \end{aligned}$$

where $|B|^2$ is understood element-wise for a matrix B. Furthermore, define

$$\|\rho\|_{\text{Lip}} := \inf\{L : |\mathbb{E}_{\rho_t} a\sigma(Az)] - \mathbb{E}_{\rho_s} a\sigma(Az)| \leqslant L|t-s||z|\}.$$

Defining an appropriate measure of complexity $\|\cdot\|_{\tilde{D},2}$ as in [22], as

$$\|f\|_{\tilde{D},2} = \inf\left\{|\alpha|^T N(T) + \|N(T)\|_1 - D + \|\rho_t\|_{\text{Lip}} : f(x) = \alpha^T z(T,x)\right\},$$

it can be shown that for every f with $\|f\|_{\tilde{D},2} < \infty$ where $D > d+1$ and $\delta \in (0,1)$, there is a L-layer neural network f_L such that

$$\|f - f_L\|_{L^2}^2 \leqslant \frac{C\|f\|_{\tilde{D},2}}{L^{1-\delta}}.$$

Let us give a slightly changed definition of flow-induced spaces by using a different norm-like measure of complexity. Consider $\mathcal{W}$ as the space

$$\mathcal{W} = \Big\{z(T,\cdot) \in C^0(\Omega, \mathbb{R}^d) : \text{ there is } \mu \in \mathcal{Y}([0,T], \mathcal{V}) \text{ such that}$$
$$\frac{d}{dt}z(t,x) = \int_\Omega a\sigma(Az(t,x)+b)\,\mathrm{d}\mu(t; a, A, b),\; z(0,x) = x\Big\}.$$

We equip $\mathcal{W}$ with the 'norm'

$$\|v\|_{\mathcal{W}} := \inf_{\mu\in\mathcal{Y}([0,T],\mathcal{V})}\left\{\int_0^T\int_\Omega |a|(\|A\|+|b|)\,\mathrm{d}\mu(t;a,A,b)\,\mathrm{d}t : \right. \\ \left. v(t,x) = \int_\Omega a\sigma(Ax-b)\,\mathrm{d}\mu(t;a,A,b)\right\} \\ = \int_0^T \|v(t,\cdot)\|_{\mathcal{B}}\,\mathrm{d}t.$$

Remark 2.2 Notice that $\mathcal{W}$ is not necessarily a linear space.

Remark 2.3 Observe that this representation is highly non-unique. For example, for all $t_1 < t_2 \in [0,T]$ with $t_2 - t_1 = \tau$, we get that the family of measures defined on $\mathbb{R}\times\mathbb{R}\times\mathbb{R}$ of the form

$$\mu_{t_1,t_2}(t;a,A,b) = \begin{cases}\delta_{0,0,0} & \text{for } t < t_1 \text{ and } t > t_2\\ \mu_2(a,A,b) & \text{for } t_1 \leqslant t \leqslant t_2\end{cases}$$

for some fixed measure μ_2 give rise to the same function $z(T,\cdot)$. Even more, if σ is a polynomial up to degree p, all families of measures with the same (time-dependent) moments up to order p will induce the same coordinate transform z.

We say that $f \in \mathcal{D} := \mathcal{B}\times\mathcal{W}$ if there exist $\tilde{f}_{\mathcal{B}}$ and $z(T,\cdot) \in \mathcal{W}$ such that $f_{\mathcal{D}} := \tilde{f}_{\mathcal{B}} \circ z(T,\cdot)$. Let us define a 'norm' on the product space via

$$\|f_{\mathcal{D}}\|_{\mathcal{D}} := \|\tilde{f}_{\mathcal{B}}\|_{\mathcal{B}} + \|v\|_{\mathcal{W}}$$

where v is the vector field inducing z.

If we assume that $\mu \in PC^1([0,T],\mathcal{V})$, then the error of the numerical solution of the ODE when using explicit Euler schemes scales as $O(\tau)$, and the errors of discretization with discrete measures provide an error like $\leqslant cn^{-1/2}e^{T\mathcal{L}}$ due to Gronwall's inequality. This can be seen as follows.

First we observe that we have for activation functions σ' with bounded derivatives

$$|v(t,y) - v(t,x)| \leqslant \|\sigma'\|_\infty \|v(t,\cdot)\|_{\mathcal{B}}|x-y| \;=: \mathcal{L}(t)|x-y|.$$

Now let us consider discrete fields $v^{\varepsilon}(t,\cdot)$ induced by measures $\frac{1}{n}\sum_{i=1}^n \delta_{a_i,A_i,b_i}$ with n- samples. According to [32] there is an L_∞-error bound

$$\|v^{\varepsilon}(t,\cdot) - v(t,\cdot)\|_{L_\infty} \leqslant \frac{d^3R^2}{\sqrt{n}}\|v(t,\cdot)\|_{\mathcal{B}}$$

for measures supported on a domain with radius R.

Let $0 = t_0 < t_1 < \cdots < t_L = T$ be an equidistant subdivision of $[0,T]$ respecting the piecewise in time structure of v and $\tau = t_1 - t_0$. Define

$$x_k = x_{k-1} + \tau v(t_k, x_{k-1}), \quad x_0 = x,$$

keeping in mind that we will choose $\tau = T/L$ later. Then we obtain an error bound at the final layer L as follows.

Theorem 2.1 *Let $L \in \mathbb{N}$ and define $e_k = x_k - x_k^\epsilon$ where x^ϵ is induced by the approximated vector field v^ϵ and set the time step-length $\tau := \frac{T}{L}$. Then we have*

$$\begin{aligned}|e_L| &\leqslant \frac{c}{\sqrt{n}}\left(\prod_{k=1}^{L-1}(1+\tau\mathscr{L}(t_k)) + \tau\sum_{k=1}^{L-1}\|v(t_k,\cdot)\|_{\mathcal{B}}\right)\\ &\leqslant \frac{c}{\sqrt{n}}\left(\left(1+e^{\int_0^T \mathscr{L}(s)\mathrm{d}s}\right)\int_0^T \|v(s)\|_{\mathcal{B}}\,\mathrm{d}s\right)\\ &\leqslant \frac{c}{\sqrt{n}}\left(1+e^{\|\sigma'\|_\infty\|f\|_{\mathcal{D}}}\right)\|f\|_{\mathcal{D}}\end{aligned}$$

where $c = c(d, R)$.

Proof. Let us consider the two (time) sequences x_k and y_k with

$$x_{k+1} := x_k + \tau v(t_k, x_k)\ ,\quad y_{k+1} := y_k + \tau v^{\varepsilon}(t_k, y_k)$$

with the same initial value $x_0 = y_0 = x$, and set $e_k = x_k - y_k$. Then we conclude

$$\begin{aligned}e_{k=1} = x_{k+1} - y_{k+1} &= x_k - y_k + \tau(v(t_k, x_k) - v^{\varepsilon}(t_k, y_k))\\ &= e_k + \tau\left[v(t_k, y_k+e_k) - v(t_k, y_k) + v(t_k, y_k) - v^{\varepsilon}(t_k, y_k)\right]\end{aligned}$$

Since $|v(t_k, y_k + e_k) - v(t_k, y_k)| \leqslant \mathscr{L}(t_k)|e_k|$ we obtain

$$|e_{k+1}| \leqslant |e_k| + \tau\mathscr{L}(t_k)|e_k| + \tau\varepsilon_k = (1+\tau\mathscr{L}(t_k))|e_k| + \tau\varepsilon_k$$

where $\varepsilon_k := \|v^{\varepsilon}(t_k,\cdot) - v(t_k,\cdot)\|_{L_\infty}$. Iterating this inequality L times yields the estimate

$$\begin{aligned}|e_L| &\leqslant (1+\tau\mathscr{L}(t_{L-1})|e_{L-1}| + \tau\varepsilon_{L-1}\\ &\leqslant (1+\tau\mathscr{L}(t_{L-1}))(1+\tau\mathscr{L}(t_{L-2}))|e_{L-2}| + (1+\tau\mathscr{L}(t_{L-2}))\tau\varepsilon_{L-2} + \tau\varepsilon_{L-1}\\ &\leqslant \prod_{k=1}^{L-1}(1+\tau\mathscr{L}(t_k))|e_1| + \sum_{k=1}^{L-1}\tau\varepsilon_k + \sum_{k=0}^{L-1}\tau\varepsilon_k\prod_{i=0}^{k}\tau\mathscr{L}(t_{L-1-i}).\end{aligned}$$

Recalling $|e_0| = 0$ and taking now $\tau = \frac{T}{L}$ and $L \to \infty$, we obtain a uniform upper bound

$$|e_L| \leqslant \lim_{L\to\infty}\sum_{k=0}^{L-1}(1+\Pi_{i=0}^k\tau\mathscr{L}(t_{L-1-i}))\tau\varepsilon_k \leqslant \frac{c}{\sqrt{n}}((1+e^{\int_0^T \mathscr{L}(s)\mathrm{d}s})\int_0^T \|v(s)\|_{\mathcal{B}}\mathrm{d}s)\,.$$

□

So far we have discussed the error accumulation when approximating the flux.

Proposition 2.1 *Since the forward Euler method has consistency order $p = 1$, there exists $L_0 \leqslant L$, $\tau := \frac{T}{L}$ such that the discrete error ε_L between the approximated flow and the Euler discretization converges with order $p = 1$, (see e.g. [32]), i.e.,*

$$\varepsilon_L \leqslant C(T, \mathcal{L})\frac{T}{L}.$$

Therefore the number of required level is

$$L \geqslant \max\left\{L_0, \frac{\varepsilon_L}{TC(T, \mathcal{L})}\right\}.$$

Notice that we need sufficiently small time steps $\tau > 0$ to provide strong stability, even for dissipative systems. In case of dissipative systems, we may use a semi-implicit Euler scheme

$$x^{i+1} = Ax^i + N(x^i)$$

to reduce the number of layers. Alternatively, one can also use higher order schemes.

However, present situation is somewhat different from the classical discretization of ODEs. Commonly, a given initial value problem is by definition the ground truth and the discretization is the approximation. In our case the neural network is in some sense the ground truth and the continuous model an approximation. This becomes more pronounced when considering the associated optimization and optimal control problems respectively, where the adjoint equation defines the optimal control and thereby a gradient-descent like scheme in the control space whereas the back propagation provides the correct gradient for optimizing the (true and discrete) network. However, the back-propagation only is in a certain sense the discretization of the adjoint problem if one uses appropriate discretization schemes for the forward equation. Thus, the solution of the adjoint equation provides only an approximation of the "correct" gradient.

Another point is our assumption of piecewise-smoothness in PC^1 in the time-interval. We assume the driving field is smooth (at least C^1) inside the time intervals (t_k, t_{k+1}), but this is practically hard to satisfy, especially as we will not know the time subdivision ahead of the optimization. We do not elaborate more on these details, and conjecture that we may obtain the convergence rate of the Euler-Mayurana scheme, which is $\varepsilon_L \leqslant C(T, \mathcal{L})T\frac{1}{\sqrt{L}}$ when relaxing the regularity assumptions and higher order (DE) schemes would not provide a better order of convergence. We believe that this might be a realistic conjecture.

3 Abstract Regression Problem

Now we can interpret the regression of a function by a generalized deep neural network as an Banach space-valued optimal control problem with measure-valued controls. In particular, we will focus on networks of the type $f_1 \circ f_2$ where f_1 is a Barron function, i.e. the generalisation of a shallow neural network, and f_2 is a Neural ODE with Barron vector fields. This corresponds to the idea that we search for a coordinate transform, parametrized bz f_2 such that we can represent a function in the new coordinates easily by a shallow neural network f_1.

Consider the optimal control problem

$$\min_{\substack{y\in C^0([0,T],C^0(\Omega,\mathbb{R}^d))\\(\mu_1,\mu_2)\in\mathcal{P}_2(\mathcal{U})\times\mathcal{Y}([0,T],\mathcal{V})}} \mathcal{J}(y,\mu^1,\mu^2) = \int_\Omega h(x, \int_\mathcal{U} g_T(x, y(T,x), u)\mathrm{d}\mu^1(u))\mathrm{d}\rho(x)$$

$$\text{s.t.} \quad \frac{d}{dt}y(t,x) = \int_\mathcal{V} f(y(t,x), v)\,\mathrm{d}\mu^2(t;v) \quad y(0,x) = x \tag{17}$$

where $\mathcal{U} \subset \mathbb{R}^{m_1}, \mathcal{V} \subset \mathbb{R}^{m_2}$ are compact and $\Omega \subset \mathbb{R}^d$ is compact. Assume $g_T \in C^2(\Omega\times\mathbb{R}^d\times R^{m_1})$, $f \in C^2(\mathbb{R}^d\times R^{m_2}, \mathbb{R}^d)$ and $h \in C^2(\Omega\times\mathbb{R})$ bounded with bounded derivatives. Furthermore, assume $h \geqslant 0$. If we choose

$$f(y,\mu_t^2) = \int_\mathcal{V} a\sigma_2(Ay+b)\,\mathrm{d}\mu^2(t;a,A,b)$$

and if we want to recover some function g_{target}, choosing

$$h(x, y(T,x), \mu^1) = |g_{\text{target}}(x) - \int_\mathcal{U} a\sigma(Ay(T,x)+b)\mathrm{d}\mu^1(a,A,b)|^2$$

gives an optimal control problem in terms of Barron functions. If the measures happen to be sums of Dirac delta distributions, the optimal control problem is the standard function recovery of machine learning with neural networks.

We say that $(y,\mu^2) \in C([0,T], C^0(\Omega,\mathbb{R}^d)) \times \mathcal{Y}([0,T],\mathcal{V})$ is admissible and a solution to $\frac{d}{dt}y = \int_\mathcal{V} f(y,v)\mathrm{d}\mu^2(t;v)$ if they fulfill

$$y(T,x) = x + \int_0^T \int_\mathcal{V} f(y(t,x),v)\mathrm{d}\mu^2(t;v)\mathrm{d}t$$

for all initial values $x \in \Omega$. We will be primarily interested in the case

$$y(T,x) = x + \int_0^T \int_\mathcal{V} a\sigma(Ay(t,x)+b)\mathrm{d}\mu^2(t;a,A,b)\mathrm{d}t.$$

Remark 3.1 Very importantly, in particular with view on the gradient flow equation in the subsequent section, the compactness assumptions on $\mathcal{U}$ and $\mathcal{V}$ can be replaced by a boundedness assumption of the measures. In particular, let us consider the following alternative.

Let $C > 0$ be fixed. Note that, instead of $\mathcal{V}$ being compact, we can also look at $\mu \in \mathbb{V} \subset \mathcal{Y}([0,T],\mathbb{R}^m)$, with W the Wasserstein-2 metric,

$$\mathbb{V} = \{\mu \in \mathcal{Y}([0,T],\mathbb{R}^m) : \sup_{t\in[0,T]} W(u(t),\delta_0) < C\}.$$

Then any sequence in $\mathbb{V}$ is tight, c.f. [34], which will give us enough compactness to prove existence of minimizers.

Note that $\mathcal{Y}([0,T],K) \subset \mathbb{V}$ for $K \subset \mathbb{R}^m$ compact as $W_2^2(\mu,\delta_0) = \int_K x^2 \mathrm{d}\mu \leqslant \sup_{x\in K} x^2 \int_K \mathrm{d}\mu = \sup_{x\in K} x^2$.

3.1 Existence and First order Necessary Conditions

The formulation of the regression problem as an optimal control problem raises the question on the well-posedness, i.e. the existence of minimizing control. The advantage of using Young measures instead of continuous in time families of measures as in [19] is that one can employ classical techniques from optimization to verify the existence of minimizers.

A very first observation is, that the narrow convergence of the Young measures implies the pointwise convergence of the vector fields.

Lemma 3.1 ([34]) *Let $\{\mu_n\}$ be a narrowly convergent family of measures in $\mathcal{Y}([0,T],\mathcal{V})$ and $y \in C^0([0,T],C^0(\Omega,\mathbb{R}^d))$. Then it holds for all $t \in [0,T]$*

$$\lim_{n\to\infty} \sup_{x\in\Omega} \left| \int_{[0,t]\times\mathcal{V}} f(y(s,x),v)\mathrm{d}\mu_n(s,v) - \int_{[0,t]\times\mathcal{V}} f(y(s,x),v)\mathrm{d}\mu(s,v)\right| = 0.$$

Therefore, convergence of controls implies convergence of the induced trajectories. Now let us turn to the existence of optimal controls and first-order necessary conditions of optimality. In order to show the existence of minimizers, we will resort to some compactness arguments. By assumption, c.f. also Remark 3.1, this will be fulfilled as by Prokhorov's theorem tightness implies sequential compactness, see [34] for more details.

Furthermore, assuming h and g_T to be continuous implies lower semi-continuity of the cost functional, again we refer to [34] for more details. Compactness together with continuity allows us to prove existence and first order necessary conditions.

Theorem 3.1 ([34]) *Let f, g fulfill the above assumptions. Assume there is at least one admissible triple*

$$(\tilde{y},\tilde{\mu}^1,\tilde{\mu}^2) \in C^0([0,T],C^0(\Omega,\mathbb{R}^d)) \times \mathcal{P}_2(\mathcal{U}) \times \mathcal{Y}([0,T],\mathcal{V})$$

such that $\mathcal{J}(\tilde{y},\tilde{\mu}^1,\tilde{\mu}^2) < \infty$.

a) Then there exists at least one optimal triple (y,μ^1,μ^2).
b) Furthermore, for any optimal triple, there exists an adjoint state

$$p \in C_{w^*}([0,T],(C^0(\Omega,\mathbb{R}^d)^*)$$

such that Pontryagin's maximum principle is fulfilled, i.e.

1. $y(t,x) = x + \int_0^T \int_{\mathcal{V}} f(y(t,x),v)\mathrm{d}\mu^2(t;v)\mathrm{d}t,$

2. $p(t) = \partial_y h(\cdot, \int_{\mathcal{U}} g(\cdot, y(T,\cdot), u) \mathrm{d}\mu^1(u))$
$\quad + \int_t^T \int_{\mathcal{V}} \partial_y f(y(s,\cdot), v) \mathrm{d}\mu^2(s; v))^* p(s) \mathrm{d}s$
3. $\langle p(t), \int_{\mathcal{V}} f(y(t,\cdot), v) \mathrm{d}\mu^2(t; v) \rangle_{X^*,X} \geqslant \langle p(t), \int_{\mathcal{V}} f(y(t,\cdot), v) \mathrm{d}\nu) \rangle_{X^*,X}$ *for all* $\nu \in \mathcal{P}_2(\mathcal{V})$ *for a.e.* $t \in [0,T]$.
4. $[\int_\Omega \nabla_y h(x, \int_{\mathcal{U}} g(x, y(T,x), u) \mathrm{d}\mu^1(u)) \nabla_u g(x, y(T,x), u) \mathrm{d}\rho(x)](u) = 0$ *for* μ^1 *a.e.* $u \in \mathcal{U}$.

Remark 3.2 Formally, the necessary optimality conditions are equal to the standard Pontryagin's Maximums Principle (PMP) in finite dimensional optimal control. A closer inspection reveals that if ρ is discrete, then this is actually equal to solving the foward and backward equations of standard PMP for all initial data x_0.

Remark 3.3 Notice in particular that here the use of Young measures facilitates the use of standard optimization tools to obtain the existence results which is not readily available for the continuous families used in [19].

3.2 Related Gradient Flow equations

One of the remaining questions is how to find critical points of the optimization problem (17). Training a neural network usually employs gradient descent-like methods. The time-continuous formulation of gradient descent is the so-called gradient flow. For the continuous model of neural networks, we then need to establish the existence to a solution to the gradient flow equation in fibered Wasserstein space.

In recent years, the Wasserstein metric on probability measures with bounded moments has gained a great deal of attention [37]. Based on the observations of [35], it became apparent that the Wasserstein-2 space has almost a Riemannian manifold-like structure, i.e. one can introduce reasonable notions of tangent spaces, as well as that some evolution equations are gradient flows of corresponding potentials with respect to the Wasserstein metric.

It turns out that absolutely continuous curves can be identified with solutions to the transport equation. On the other hand, defining a concise notion of derivative in Wasserstein space is rather intricate, see [2] for Wasserstein and [38, 34] for fibered Wasserstein space; however, as a rule of thumb on can say

$$\nabla \mathcal{J} = \nabla_u \frac{\delta J(\mu)}{\delta \mu}$$

where $\frac{\delta J(\mu)}{\delta \mu}$ is the functional derivative, see [14]. With the tools of [2] one can then establish the existence and form of solutions to the gradient flow equations to our regression problem.

Theorem 3.2 ([34]) *Let* $\mu_0^1 \in \mathcal{P}_2(\mathcal{U})$ *and* $\mu_0^2 \in \mathcal{Y}([0,T], \mathcal{V})$. *Then there is an absolutely continuous curve* $\boldsymbol{\mu} : s \mapsto (\mu_s^1, \mu_s^2) \in \mathcal{P}_2(\mathcal{U}) \times \mathcal{Y}([0,T], \mathcal{V})$ *with corresponding velocities* v_s^1, v_s^2 *such that*

1. *we have*
$$\partial_s \mu_s^1 + \operatorname{div}(v_s^1 \mu_s^1) = 0$$
and
$$\partial_s \mu_s^2 + \operatorname{div}_v(v_s^2 \mu_s^2) = 0.$$
2. $\lim_{s\to 0} \mu_s^1 = \mu_0^1$ *and* $\lim_{s\to 0} \mu_s^2 = \mu_0^2$ *in the narrow sense of* $\mathcal{P}_2(\mathcal{U})$ *and* $\mathcal{Y}([0,T],\mathcal{V})$ *respectively.*
3. *The function* $s \mapsto \mathcal{J}(\mu_s^1, \mu_s^2)$ *is non-increasing.*
4. $(v_s^1, v_s^2) = \nabla \mathcal{J}(\mu_s^1, \mu_s^2)$.

Remark 3.4 In the case of the continuous formulation of neural networks, it holds

$$v_s^1(a, A, b) = \int_\Omega 2(h(x) - a\sigma(Ay_s(T, x) + b))\nabla_{a,A,b} a\sigma(Ay_s(T, x) + b)\mathrm{d}\rho(x)$$

and

$$v_2^s(t; a, A, b) = \int_\Omega p(s, t, x) \cdot \nabla_{a,A,b} a\sigma(Ay_s(t, x) + b)\mathrm{d}\rho(x)$$

where

$$\frac{d}{dt} y_s(t, x) = \int_\mathcal{V} a\sigma(Ay_s(t, x) + b)\mathrm{d}\mu_s^2(t; a, A, b)$$

and

$$\frac{d}{dt} p(s, t, x) = \int_\mathcal{V} \nabla_y a\sigma(Ay_s(t, x) + b)\mathrm{d}\mu_s^2(t; a, A, b)p(s, t, x)$$

and

$$p(s, T, x) = \int_\mathcal{U} 2(h(x) - a\sigma(Ay_s(T, x) + b))\nabla_y a\sigma(Ay_s(T, x) + b)\mathrm{d}\mu_s^1.$$

Note that the divergence is w.r.t a, A, b. Recalling Corollary 4.10 from [34], this especially implies that if e.g. $\mu_s^i = \sum_{n=1}^m \alpha_n^i \delta_{u_n}$ and if we solve the underlying ODE by explicit Euler we get $\mu_{s+h}^1 \approx \sum_{n=1}^m \alpha_n^i \delta_{u_n + h\, v_s^i(u_i)}$, i.e. an ordinary gradient update step. Note, that this is also true for the Young measure by Corollary 4.11 in [34].

Remark 3.5 Note that the formulas given here agree with the formulas in [19] and the formulas in [23] when we set $\sigma = 0$.

Remark 3.6 For smooth activation functions there is one crucial observation that can be drawn from the gradient flow equation: If our initial guess for the measures μ^1 and μ^2 is a (family of) sum of Diracs, then the gradient flow will give no insights on how to increase the number of Diracs. This is due to the fact that the vector field of the transport equation will be at least Lipschitz and thus Diracs cannot be split, i.e. the number of Diracs can only non-increase over time.

Remark 3.7 Instead of learning the weights and biases, or in other words Dirac measures, one can try to approximate the measures with classical methods. Very likely, one would assume here some regularity of the Young measure w.r.t time, as for example absolute continuity. This would be again a high-dimensional approximation problem, but one could hope that the measures will exhibit some structures that can be exploited. Possible obstructions are the constraints of positivity and L^1 normalization. However, they can be circumvented by using the so-called Koopman-von Neuman dynamics, which describes the time evolution of one of the square roots of the density. The associated operator is orthogonal in L^2 and the L^1 normalization is replaced by a L^2 normalization that is much more accessible for example for tensor train approximation methods.

This approach of learning the measure instead of the weights and biases also would open up new perspectives on how to design adaptive algorithms for neural networks, i.e. how to change the architecture of the net, e.g. width, and depth, in an online fashion.

Another important question is if there is a stationary point $\lim_{s\to\infty} \mu(s)$. It is a priori not clear if the gradient flow curve has a limit point. For a Hilbertian setting in [39], the convergence of the Gradient flow in the standard optimal control setting was only shown for initial controls with weak (time) derivative. Very likely, one will need to assume some regularity of the Young measure w.r.t time also in our case. Another very popular approach to investigate the convergence of gradient flows is with the help of the Kurdyka-Łojasiewicz-Simon (KLS) inequality. Notably, the connection between curves of maximal slope with finite length and a version of the KLS inequality on general metric spaces has been investigated in [26, 40].

4 Very First Numerical Examples and Outlook

One of the advantages of the generalization of deep neural networks to abstract optimal control problems with measure-valued controls is that it offers new perspectives on how to adapt the architecture of the neural network during the training. One way to achieve this is to actually learn an approximation of the underlying measures directly.

An important observation to be drawn from Theorem 3.2, see Remark 3.6, is that if one starts with a classical Neural ODE, i.e., sum of Diracs for the measures, one will stay in that framework and the numbers of neurons cannot grow (with the help of information gained by the gradient flow). Therefore, using singular measures as initial guess does not allow for adaptive algorithms based on the gradient flow ansatz and, thus, one should consider absolutely continuous measures.

One ansatz would be to use Gaussian mixture models as it is easy to sample from them and as there are well-analyzed algorithms on how to recover a Gaussian mixture model from samples. The key idea of this very first naïve implementation is the following assumption.

If one approximates a Gaussian mixture by a sum of Diracs, e.g. by sampling, the transport of the Gaussian mixture and the transport of the sum of Diracs along the underlying transport equation might stay close to each other for short times. Therefore, one could assume that the Diracs after one time step will approximate the transported Gaussians after one time step well. Thus, if one recovers a Gaussian mixture from the transported sum of Diracs, this is a good approximation of the transported original Gaussian mixture. In order to transport the sample one can use the standard tools for neural networks implemented in efficient libraries like TensorFlow2 or PyTorch as the transport can be approximated by a gradient step. Then the remaining task is to recover a new Gaussian mixture from data points. This can be done for example by Expectation Maximation algorithms. The whole procedure is condensed in Algorithm 10. Notice that we can freely choose how many Diracs we sample to approximate the Gaussian mixture, i.e., if we had an a posteriori decision rule, we could start an adaptive scheme at this point. Deriving such rules is a difficult task and under investigation.

Algorithm 10: Basic algorithm

Require $\mu_0 = \sum \alpha_i \mu_i^G$ with Gaussians μ_i^G and $\sum \alpha_1 = 1$, $\alpha_i \geqslant 0$
While until some user-specified criteria

1. sample weights and biases from μ_i
2. calculate loss and gradient with respect to the neural network given by sampled weights and biases
3. update weights and biases by gradient step
4. use new weights and biases as 'sample data' of transported density
5. set μ_{i+1} as Gaussian mixture fitting the 'sample data' by Expectation Maximization

Output: μ_{final}

In Figure 1 one can see the reconstruction of the function

$$f : \mathbb{R} \to \mathbb{R}, f(x) = x^2$$

with shallow net and one ResNet block with final linear layer respectively with 10000 neurons per layer after 2000 iterations of the basic Algorithm reaching a training and test loss of $O(10^{-4})$ and $O(0.5 \cdot 10^{-4})$ respectively. In each layer the measures were parameterized by a sum of two Gaussians with fixed variance. As one can see, the neural network is able to capture the function, which is not always possible for naïve implementations of fully connected neural networks. Furthermore, the complexity of the approximation space is much lower, i.e. 12-18 dimensional instead of ten thousands, when going to the Barron picture.

However, for really high-dimensional problems one is now facing the task of parameterizing measures on high-dimensional state spaces. This is also a very difficult problem. So the key ingredient will be now to identify underlying structures that will allow for a low-complexity parameterization of the measures while a direct approximation will suffer from the curse of dimensionality. This falls in line with the philosophy, that there will be no efficient and accurate black box solvers in high-dimensions.

After introducing this continuous model for neural networks the most direct open questions are about the convergence of the gradient flow and the relation of the

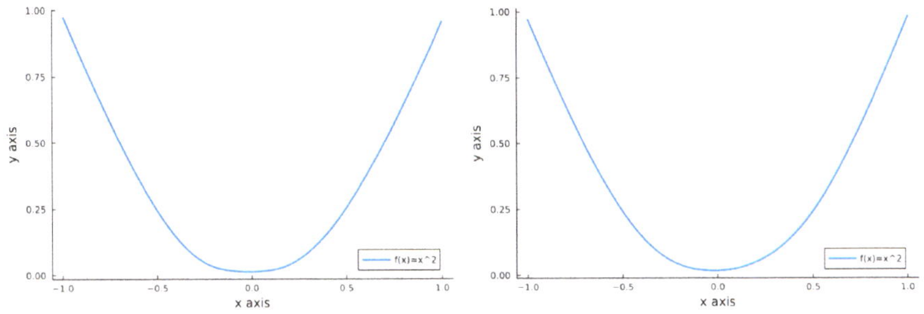

Fig. 1: Reconstruction of the univariate square function with a neural network with 2 respectively 3 layers and 10000 neurons per layer after training with basic Algorithm 10

basic Algorithm 10 to the solution of the gradient flow from a numerical analysis perspective.

Acknowledgements

Mathias Oster acknowledges funding by the Deutsche Forschungsgemeinschaft (DFG, German Research Foundation) - Project number 442047500 through the Collaborative Research Center "Sparsity and Singular Structures" (SFB 1481 - B01).

References

1. Joubine Aghili and Olga Mula. An optimal control framework for adaptive neural odes. *Advances in Computational Mathematics*, 50(3):52, 2024.
2. Luigi Ambrosio, Maria Colombo, and Alessio Figalli. Existence and uniqueness of maximal regular flows for non-smooth vector fields. *Archive for Rational Mechanics and Analysis*, 218(2):1043–1081, 11 2015.
3. Luigi Ambrosio, Nicola Gigli, and Guiseppe Savaré. *Gradient Flows in Metric Spaces and in the Space of Probability Measures*. Birkhäuser, second edition, 2005.
4. Francis Bach. Breaking the curse of dimensionality with convex neural networks. *Journal of Machine Learning Research*, 18(19):1–53, 2017.
5. Andrew R. Barron. Universal approximation bounds for superpositions of a sigmoidal function. *IEEE Transactions on Information Theory*, 39(3):930–945, 1993.
6. Benedikt Bauer and Michael Kohler. On deep learning as a remedy for the curse of dimensionality in nonparametric regression. *The Annals of Statistics*, 47(4):2261 – 2285, 2019.
7. Christian Beck, Martin Hutzenthaler, Arnulf Jentzen, and Benno Kuckuck. An overview on deep learning-based approximation methods for partial differential equations. *Discrete and Continuous Dynamical Systems - B*, 28(6):3697–3746, 2023.
8. Julius Berner, Philipp Grohs, and Arnulf Jentzen. Analysis of the generalization error: Empirical risk minimization over deep artificial neural networks overcomes the curse of dimensionality

in the numerical approximation of black–scholes partial differential equations. *SIAM Journal on Mathematics of Data Science*, 2(3):631–657, 2020.
9. Julius Berner, Philipp Grohs, Gitta Kutyniok, and Philipp Petersen. The modern mathematics of deep learning. In *Mathematical Aspects of Deep Learning*, pages 1–111. Cambridge University Press, 12 2022.
10. Benoît Bonnet, Cristina Cipriani, Massimo Fornasier, and Hui Huang. A measure theoretical approach to the mean-field maximum principle for training neurodes. *Nonlinear Analysis*, 227:113161, 2023.
11. Bo Chang, Lili Meng, Eldad Haber, Lars Ruthotto, David Begert, and Elliot Holtham. Reversible architectures for arbitrarily deep residual neural networks. *Proceedings of the AAAI Conference on Artificial Intelligence*, 32(1), 04 2018.
12. Ricky T. Q. Chen, Yulia Rubanova, Jesse Bettencourt, and David Duvenaud. Neural ordinary differential equations. In *Advances in Neural Information Processing Systems 31 (NeurIPS 2018)*, NIPS'18, page 6572–6583, Red Hook, NY, USA, 2018. Curran Associates Inc.
13. Ziang Chen, Jianfeng Lu, and Yulong Lu. On the representation of solutions to elliptic pdes in barron spaces. In *Neural Information Processing Systems*, 2021.
14. Lénaïc Chizat and Francis Bach. On the global convergence of gradient descent for over-parameterized models using optimal transport. In *Advances in Neural Information Processing Systems 31 (NeurIPS 2018)*, NIPS'18, page 3040–3050, Red Hook, NY, USA, 2018. Curran Associates Inc.
15. Wolfgang Dahmen. Compositional sparsity, approximation classes, and parametric transport equations, 2022.
16. Jérôme Darbon, Gabriel P. Langlois, and Tingwei Meng. Overcoming the curse of dimensionality for some hamilton–jacobi partial differential equations via neural network architectures. *Research in the Mathematical Sciences*, 7(3):20, Jul 2020.
17. Ronald A. DeVore. Nonlinear approximation. *Acta Numerica*, 7:51–150, 1998.
18. Ronald A. DeVore, Boris Hanin, and Guergana Petrova. Neural network approximation. *Acta Numerica*, 30:327–444, 2021.
19. Zhiyan Ding, Shi Chen, Qin Li, and Stephen J. Wright. Overparameterization of deep resnet: Zero loss and mean-field analysis. *Journal of Machine Learning Research*, 23(48):1–65, 2022.
20. Weinan E, Jiequn Han, and Qianxiao Li. A mean-field optimal control formulation of deep learning. *Research in the Mathematical Sciences*, 6(1):10, 12 2018.
21. Weinan E, Chao Ma, Stephan Wojtowytsch, and Lei Wu. Towards a mathematical understanding of neural network-based machine learning: what we know and what we don't, 2020.
22. Weinan E, Chao Ma, and Lei Wu. The barron space and the flow-induced function spaces for neural network models. *Constructive Approximation*, 55(1):369–406, 02 2022.
23. Weinan E. and Stephan Wojtowytsch. Representation formulas and pointwise properties for Barron functions. *Calculus of Variations and Partial Differential Equations*, 61(2):46, 02 2022.
24. Weinan E and Stephan Wojtowytsch. Some observations on high-dimensional partial differential equations with Barron data. In Joan Bruna, Jan Hesthaven, and Lenka Zdeborova, editors, *Proceedings of the 2nd Mathematical and Scientific Machine Learning Conference*, volume 145 of *Proceedings of Machine Learning Research*, pages 253–269. PMLR, 16–19 Aug 2022.
25. Patrick Gelß, Aizhan Issagali, and Ralf Kornhuber. Fredholm integral equations for function approximation and the training of neural networks, 2023.
26. Daniel Hauer and Jose Mazon. Kurdyka-Łojasiewicz-Simon inequality for gradient flows in metric spaces. *Transactions of the American Mathematical Society*, 372, 07 2017.
27. Catherine F. Higham and Desmond J. Higham. Deep learning: An introduction for applied mathematicians. *SIAM Review*, 61(4):860–891, 2019.
28. Kurt Hornik, Maxwell Stinchcombe, and Halbert White. Multilayer feedforward networks are universal approximators. *Neural Networks*, 2(5):359–366, 1989.
29. Martin Hutzenthaler, Arnulf Jentzen, Thomas Kruse, and Tuan Anh Nguyen. A proof that rectified deep neural networks overcome the curse of dimensionality in the numerical approximation of semilinear heat equations. *SN Partial Differential Equations and Applications*, 1(2):10, Apr 2020.

30. Noboru Isobe and Mizuho Okumura. Variational formulation of ode-net as a mean-field optimal control problem and existence results. *arXiv preprint arXiv:2303.05924*, 2023.
31. Arnulf Jentzen, Diyora Salimova, and Timo Welti. A proof that deep artificial neural networks overcome the curse of dimensionality in the numerical approximation of kolmogorov partial differential equations with constant diffusion and nonlinear drift coefficients. *Communications in Mathematical Sciences*, 19(5):1167–1205, 2021.
32. Stig Larsson and Vidar Thomée. *Partial differential equations with numerical methods*, volume 45 of *Texts in Applied Mathematics*. Springer-Verlag, Berlin, 2009. Paperback reprint of the 2003 edition.
33. Song Mei, Andrea Montanari, and Phan-Minh Nguyen. A mean field view of the landscape of two-layer neural networks. *Proceedings of the National Academy of Sciences*, 115(33):E7665–E7671, 2018.
34. Mathias Oster. *On semi-global optimal control problems and their applications in machine learning*. PhD thesis, Fakulty II - Mathematics and Natural Sciences, 2023.
35. Felix Otto. The geometry of dissipative evolution equations: The porous medium equation. *Communications in Partial Differential Equations*, 26(1-2):101–174, 2001.
36. Myeongsuk Pak and Sanghoon Kim. A review of deep learning in image recognition. In *2017 4th International Conference on Computer Applications and Information Processing Technology (CAIPT)*, pages 1–3, 2017.
37. Victor M Panaretos and Yoav Zemel. *An Invitation to Statistics in Wasserstein Space*. SpringerBriefs in Probability and Mathematical Statistics. Springer International Publishing : Imprint: Springer, Cham, 1st ed. 2020. edition, 2020.
38. Jan Peszek and David Poyato. Heterogeneous gradient flows in the topology of fibered optimal transport, 2023.
39. Allessandro Scagliotti. A gradient flow equation for optimal control problems with end-point cost. *Journal of Dynamical and Control Systems*, 29(1):521–568, 2023.
40. Lorenzo Dello Schiavo, Jan Maas, and Francesco Pedrotti. Local conditions for global convergence of gradient flows and proximal point sequences in metric spaces, 2023.
41. Jonathan W. Siegel. Optimal approximation rates for deep relu neural networks on sobolev and besov spaces. *Journal of Machine Learning Research*, 24(357):1–52, 2023.
42. Jonathan W. Siegel and Jinchao Xu. Approximation rates for neural networks with general activation functions. *Neural Networks*, 128:313–321, 2020.
43. Jonathan W. Siegel and Jinchao Xu. Characterization of the variation spaces corresponding to shallow neural networks, 2021.
44. Jonathan W. Siegel and Jinchao Xu. Sharp bounds on the approximation rates, metric entropy, and n-widths of shallow neural networks. *Foundations of Computational Mathematics*, 11 2022.
45. Takeshi Teshima, Koichi Tojo, Masahiro Ikeda, Isao Ishikawa, and Kenta Oono. Universal approximation property of neural ordinary differential equations, 2020.

Unstoppable Mathematicians

Dominique Picard

Version originale

On était quatre, dans ma maison, à boire du thé, peut-être aussi des bières, et à parler de nos jeunes années.

Alim, Paul et Alice étaient arrivés là, indépendamment, mais tous bloqués par le traffic aérien. Quelques heures, un certain temps, on ne savait pas. Ils avaient préféré venir chez moi plutôt que d'aller à l'hotel. Et je dois dire, ça me faisait plaisir.

Pour être tout à fait honnête, ils aimaient tous venir me voir, mais ils aimaient peut-être encore plus ma maison, très cosy -la maison de ma grand-mère en fait- et le chien Charly qui savait mettre tout le monde à l'aise.

L'atmosphère était très décontractée, même s'ils ne se connaissaient pas vraiment les uns les autres.

À un moment donné, la discussion s'était orientée vers une question particulière : lorsqu'on est mathématicien, peut-on arrêter de faire des mathématiques ou restera-t-on mathématicien toute sa vie ?

Je les regardais. J'avais envie de savoir ce que chacun aurait à dire sur ce sujet là. Pour l'instant, ils s'observaient. Alice semblait un peu fermée. Manifestement, elle était très attentive mais sans vouloir prendre part à la discussion. Je pensais que, face à ce sujet, Paul serait probablement celui qui aurait le plus envie d'expliquer sa vie, ses choix. En fait, c'est Alim qui a lancé la discussion comme si c'était un enjeu majeur pour lui.

- Vous savez, à l'âge que j'ai, on commence à tirer des conclusions et moi, je pense que j'ai raté beaucoup - vraiment beaucoup - d'opportunités importantes. Non pas pour devenir célèbre ou quoi que ce soit de ce genre, mais pour aller beaucoup plus loin dans ce que je faisais.

Université Paris-Cité, Laboratoire de Probabilités, Statistique et Modélisation (LPSM, UMR 8001) Bâtiment Sophie Germain, 8 place Aurélie Nemours 75013 Paris, e-mail: dominiquepicard@gmail.com

R. DeVore, A. Kunoth (eds.), *Multiscale, Nonlinear and Adaptive Approximation II*,
https://doi.org/10.1007/978-3-031-75802-7_18

En fait, en Turquie, avec l'éducation que j'ai reçue à Istanbul puis à Paris - ma famille était riche et cultivée - j'étais parfaitement préparé pour écrire un grand nombre de choses. Je veux parler de mathématiques, bien sur, mais aussi de la littérature, de la poésie, des romans... En fait, je suis un romancier qui n'a pas écrit une seule ligne. Je n'ai jamais eu d'ambition, j'aimais beaucoup comprendre les mathématiques et communiquer ma façon de les appréhender. Et j'étais très doué. Je peux le dire sans immodestie. C'est du passé. Totalement.

A l'âge de 26 ans, j'avais une énorme culture en mathématiques, j'étais vraiment brillant, avec de grandes capacités. Mais j'ai eu une sorte d'accident.

Je venais de terminer mes études, que j'ai faites très vite, trop vite - c'était une erreur, mais des erreurs, j'en ai faites beaucoup...- je commençais à faire de la recherche. Mon directeur de thèse - un homme très respecté - m'avait donné un article à lire pendant les vacances. Il s'agissait d'un article très riche, profond et intéressant sur l'approximation trigonométrique avec des taux d'erreur exponentiels, rédigé par un très jeune chercheur, Wolgang Dahmen. Il devait avoir le même age que moi ou pas beaucoup plus. J'ai appris en deux mois ce que très peu de gens peuvent assimiler en si peu de temps.

J'ai ensuite rédigé un rapport que j'ai remis à mon directeur de thèse. En fait, il a été très impressionné. Il ne s'attendait manifestement pas à quelque chose d'aussi bien structuré et d'aussi bien écrit.

Il était tellement enthousiaste que j'ai obtenu un poste presque immédiatement. A Paderborn. J'ai même eu le choix. Paderborn ou Paris, j'ai choisi Paderborn, parce que mon Allemand était meilleur que mon Français. Même si j'avais fait mes études au lycée français d'Istanbul. Tradition familiale sans doute.

Et trois mois plus tard, j'ai littéralement explosé. Cela fait bien longtemps maintenant, mais je ne sais toujours pas ce qu'il s'est réellement passé. L'explication la plus probable c'est un burn-out. C'est la chose la plus logique : j'avais travaillé si dur pour mes études et ces derniers mois que mon cerveau a fait comme un court-circuit pour refuser de continuer à fonctionner à ce régime là.

J'ai pensé qu'il pouvait s'agir d'un empoisonnement, j'ai envisagé beaucoup de choses. Dans ma famille, il y a beaucoup de maladies mentales : mon oncle est mort très jeune, on pense qu'il s'est suicidé, mon plus jeune frère est devenu schizophrène...

Il m'a fallu du temps pour réaliser que j'avais une sorte de maladie. Lorsque j'ai perdu la capacité de faire des maths, je ne savais pas si c'était à cause de mon cerveau ou de ma façon de vivre. Je pensais que les choses s'amélioreraient avec une vie plus équilibrée, une femme et des enfants. Malheureusement, ce ne fut le cas.

J'ai mis longtemps à comprendre ne serait-ce que le fait que mon cerveau ne stockait plus les choses comme avant. Quand j'étais jeune, je pouvais lire des quantités et cela restait dans ma tête pour toujours. Aujourd'hui, si je lis dix pages de mathématiques, je les oublie en une semaine.

Et ça n'est jamais revenu, je n'ai jamais refait des maths avec cette totale disponibilité que j'avais à 26 ans. À un moment donné, il a bien fallu se rendre à l'évidence. Je me suis dit : ne fais que les maths que tu aimes, ne fais pas souffrir ton cerveau.

Et c'est ce que j'ai fait. J'ai continué à exercer mon métier, à enseigner les mathématiques à des étudiants. J'ai bien aimé les étudiants, surtout ceux des premières années. C'est tellement important.

J'ai également écrit des livres. J'ai toujours beaucoup écrit : même lorsque j'étais un très jeune étudiant, j'écrivais déjà des polys pour expliquer les choses à mes amis.

Ma famille avait une énorme bibliothèque à Istanbul. C'est mon oncle qui achetait la plupart des livres. Probablement pour nous, car il est mort sans enfant. Il fumait beaucoup. Beaucoup de gens dans ma famille fumaient. Du mauvais tabac.

Aujourd'hui, j'ai une grande bibliothèque dans ma propre maison, bien plus grande que celle que nous avons à Istanbul. Je suis un passionné de livres. Je passe tout mon temps à lire, à écrire, à essayer de trouver des ouvrages mathématiques rares et à m'occuper de ma maison d'édition. En fait j'ai créé deux maisons d'édition, une en Allemagne et une autre à Istanbul. Malheureusement celle d'Istanbul n'a pas fonctionné parce que je n'étais pas sur place. Je suis un amoureux des livres.

Je pense que ce que j'ai fait n'est pas déshonorant. J'ai "fait le job" comme on dirait, j'ai même été utile, pour certaines choses... Mais certainement pas autant que j'aurais pu l'être sans cet accident.

Alice avait écouté Alim avec une empathie évidente. Elle a hésité avant de lui demander :

- As-tu déjà essayé de voir un psychothérapeute ? Peut-être que cette maladie de ton cerveau pourrait être guérie ou traitée.

- De toutes façons, c'est trop tard maintenant, mais bien sûr, j'ai vu de nombreux thérapeutes. Quand mon jeune frère a été diagnostiqué schizophrène, toute la famille a été traitée. Je ne sais pas trop pourquoi, comme si les médecins pratiquaient une sorte de dératisation... Je me souviens de la psychiatre qui m'a dit : "Je ne peux rien pour vous parce que vous êtes complètement immature. Vous êtes un enfant." J'ai pris ça pour un compliment.

Alim riait.

- Beaucoup de gens me disent que pas mal de choses ont mal tourné dans ma vie à cause de ce frère schizophrène dont je me suis occupé : ma carrière, mon mariage qui s'est terminé par un divorce... Mais je pense que ce n'est pas vrai. Je considère qu'il est dans ma vie comme une bénédiction. Je l'aime énormément.

Mais nous parlions de mathématiciens qui ont complètement quitté les mathématiques pour faire autre chose. Je ne suis pas exactement dans cette catégorie. J'ai juste fonctionné en sous-régime. Mais toi, Alice, tu l'es, il me semble. Tu veux bien nous expliquer ?

Alice s'était renfoncée dans son fauteuil. J'ai eu l'impression qu'elle n'était pas à l'aise pour parler. Néanmoins, elle commença, bravement.

- Eh bien, Alim, mon histoire effectivement est un peu différente. Tu le sais peut-être, j'étais mathématicienne, il y a longtemps maintenant. C'est-à-dire, je n'étais probablement pas aussi brillante que ce que tu décris, mais j'ai fait des mathématiques "raisonnables" je dirais. Ce n'était pas une passion pour moi et il n'y avait pas de tradition familiale même pas un environnement intellectuel comme tu le décris. J'ai été la premiere personne à faire des mathématiques dans ma famille. Pire, je pense qu'ils étaient complètement agnostiques à ce que je faisais. Ce sont tous des gens très physiques, attirés par toutes sortes de sports. Mais les activités intellectuelles, ils ne les intègrent tout simplement pas dans leur vie. Ils ne sont pas stupides non plus, mais ils ne comprennent pas ce que la recherche en mathématiques peut signifier, sans parler d'en vivre.

En fait, j'étais très heureuse de faire des mathématiques. Fière. J'étais plutôt respectée dans de département de mathématiques en Caroline du Sud, où j'étais. J'aimais travailler, lire des papiers, aller à des réunions, visiter des pays, avoir de nombreuses discussions mathématiques. Tout cela me fascinait. D'ailleurs, c'est drôle, tu as mentionné Wolfgang Dahmen. Je me souviens que j'avais commencé à travailler sur ses travaux. Il me faisait fantasmer, il était super brillant et en plus champion de Taekwondo... Je ne l'ai jamais rencontré. Pourtant il venait de temps en temps en Caroline du Sud. Dommage, ça aurait peut être changé ma vie ! En tout cas, j'ai épousé un collègue universitaire, un physicien. Peut-être pas la meilleure de mes idées, mais bon, c'est la vie... Nous étions dans la même université et, au bout de quelques années, il s'est passé une chose qui n'a rien d'exceptionnel. Il m'a quittée pour une autre femme. Une autre collègue, plus jeune que moi et plus brillante, ou du moins plus ambitieuse. Je ne suis peut-être pas tout à fait juste envers elle. C'est loin maintenant...

Elle souriait mais s'était arrêtée de parler pendant un moment, avait bu une gorgée de thé comme pour se donner du courage :

- À ce moment-là, je me suis sentie plutôt mal. Je ne savais pas quoi faire, et ce que j'ai fait n'était peut-être pas la chose la plus intelligente. C'est difficile à dire...

Quoi qu'il en soit, je suis partie, j'ai démissionné du département et j'ai déménagé dans la ville où se trouvait ma sœur. Plus au nord, pas loin de DC. J'ai emmené mon fils avec moi. Il avait 4 ans.

Paul a réagi presque immédiatement :

- Mais personne ne t'a dit que tu avais des droits, que tu ne devais simplement t'en aller, t'enfuir, comme si c'était toi qui étais en faute ?

- Oui, oui, beaucoup de gens me l'ont dit, surtout le chef du département de mathématiques. Il m'a vraiment encouragée à rester. Non. C'était ma décision à moi. Je ne me sentais plus chez moi là-bas. J'ai senti que si je restais, je ne me dégagerai pas de cette sorte de sentiment d'humiliation que j'avais. Je devais partir. Comme je l'ai dit, ce n'était peut-être pas la décision la plus sage, mais en fait, je ne l'ai jamais regrettée.

Pour parer au plus pressé, j'ai dû trouver un moyen de gagner de l'argent. Mon mari m'en donnait un peu pour l'éducation de son fils, mais il n'a jamais été très généreux. Sa contribution était loin d'être suffisante. Je n'avais pas d'argent de coté, il fallait donc que je trouve quelque chose au plus vite. Ma soeur a été super. Elle nous a dépanné : on a eu un foyer où se poser et beaucoup de soutien moral. C'est une femme très active, très énergique. Elle m'a tout de suite encouragée à oublier tout ça, à tourner la page.

Un peu comme Alim, je me suis demandée ce que je pouvais faire pour me réconforter et me faire plaisir. La réponse est venue assez vite avec le yoga. J'avais une bonne souplesse physique et je faisais du yoga depuis l'adolescence. Je sentais que je pourrais être un bon professeur. Il y avait un studio de yoga pas loin de la

maison de ma sœur. On a discuté, ils ont été d'accord pour me donner un mois d'essai. Je travaille toujours au même studio. Je suis maintenant l'une des propriétaires. Financièrement, tout va bien.
Paul s'est mis à rire :
- Du point de vue financier c'est sans doute mieux que d'être restée au département de mathématiques.
- Oui, bien mieux, c'est certain.

Alim était impatient. Il voulait parler. Il avait presque du mal à laisser Alice finir sa phrase.
- Mais vraiment, tu as arrêté les maths et tu t'es tournée vers une activité physique ? comme ça, en un instant ? sans remords ?
- Je ne dirais pas que c'était en un instant. Il m'a fallu un certain temps pour réaliser que la situation entre mon mari et moi était irréversible. Il m'a fallu encore quelques semaines pour décider que la meilleure solution était de déménager. Puis encore quelques semaines pour trouver un emploi...
- Mais est ce que tu lis des mathématiques ou tu fais des calculs comme ça, pour le plaisir ?
- Non, jamais.
- Mais tu aimais faire des mathématiques ?
- Oui, j'aimais beaucoup. Mais tu sais, quand j'étais à l'école, j'étais douée en maths. Mais j'étais aussi douée pour les activités physiques et les sports en général. À un moment donné, j'ai choisi les mathématiques. Mais, ensuite cette rupture dans ma vie a été si dure et si violente d'une certaine manière... Et le fait que toute ma famille soit complètement tournée vers des activités non intellectuelles, ça a joué aussi. Tout cela a fait que j'ai fini par penser que cette incursion dans le monde académique était un non-sens, que ce n'était pas pour moi. Vous savez, dans la communauté du yoga, et dans la communauté du développement personnel en général, c'est plus facile d'avoir l'impression qu'on est au bon endroit, là où on doit être.. Difficile à expliquer. Je sais que c'est un mode de vie très différent. Mais j'aime vraiment ce que je fais. Je m'occupe des gens. Je fais principalement de l'accompagnement individuel. J'ai une relation étroite avec les gens que j'encadre. J'ai l'impression de les aider. Je pense que je me sens plus utile que lorsque je faisais des maths. C'est peut-être un point de vue simpliste, à court terme. Mais c'est ce que je pense. Et puis... en maths, il y avait un fort esprit de compétition. Alors, ça va bien quand tu te sens fort, capable d'y arriver, mais dès que tu te sens blessé c'est plus compliqué. Et c'est ce qui m'arrivait en fait.

À un moment donné, Paul, qui semblait auparavant très enthousiaste dans la discussion, est devenu soudain très silencieux. Je l'ai remarqué et j'ai essayé de l'encourager à parler. Je savais qu'il avait des choses intéressantes à dire.
- Et toi, Paul, comment tu décrirais ton rapport aux mathématiques ? Tu as eu par ailleurs une vie des plus passionnantes, non ?
- Eh bien, passionnante... Oui, peut-être...
- Pour Alice, qui ne te connaît peut-être pas bien : tu étais mathématicien, tu as eu une carrière universitaire assez brillante dans divers départements de mathématiques, dans différents pays même. Ensuite, si je me souviens bien, tu as été président

d'université. Et plus récemment, conseiller du Premier Ministre sur les questions de recherche. C'est exact ?
- Oui, c'est exact.
- Est-ce que tu te considères toujours comme un mathématicien ?
- Certainement. Je fais des mathématiques autant que mon emploi du temps serré me le permet. Je pense que je n'ai jamais cessé de faire des mathématiques. Je ne suis peut être pas tout à fait honnête avec moi-même en disant cela. La façon dont je fais des mathématiques aujourd'hui n'est pas comparable avec ce que fait un chercheur professionnel. Je compte beaucoup sur mes collaborateurs. Ils font évidemment une bonne part du travail difficile que je n'ai pas le temps de faire. Mais je suis fier de dire que je n'ai jamais cessé d'écrire des mathématiques. Les mathématiques sont essentielles pour moi. Quand j'étais jeune, j'étais un type un peu en retard dans le développement de la personnalité. J'ai rencontré des gens, je peux le dire, avec des esprits précoces. Ce n'était pas mon cas. Mon père était également un type lent. Je viens d'une famille modeste. On peut commencer sa vie en connaissant les codes, les comportements sociaux. Ce n'était pas mon cas, et aujourd'hui encore, j'en ressens souvent les conséquences. Une sorte de lenteur, parfois même des dissociations cognitives. Un type qui n'est pas toujours bien connecté.
- Tu te juges peut-être un peu sévèrement, non ?
- Bien sûr. Je ne suis pas que ça, j'ai aussi une bonne capacité d'adaptation sociale. Mais ce que je voulais souligner, c'est le fait que les études supérieures et les mathématiques en général ont été pour moi, et sont toujours, une sorte de protection contre ces comportements étranges que je suis parfois tenté d'avoir. En tout cas, quand j'étais jeune, j'ai étudié les mathématiques, puis je suis devenu chercheur, et j'étais complètement investi là-dedans, passionné. D'ailleurs le Wolfgang Dahmen dont vous avez parlé tous les deux, je n'ai pas lu ses travaux, mais je l'ai rencontré au TTI, à Chicago, en 2007 ou 2008, je ne me souviens plus. Un type sympathique, on a discuté politique.

Au départ, j'étais complètement dans les maths, je n'avais aucune forme d'engagement actif dans des partis politiques ou des syndicats, même lorsque j'étais étudiant.

Je n'ai commencé à faire de la politique que lorsque j'ai obtenu mon premier poste de professeur. Mes enfants étaient nés, ma femme avait un poste dans un ministère en vue, on discutait politique à la maison... J'étais dans une université que je ne connaissais pas, avec une équipe plutôt amicale. Ils étaient tous très engagés politiquement. J'étais intéressé, ils m'ont en quelque sorte poussé. Puis, comme tu l'as décrit, Jane, je suis devenu président de l'université, et de fil en aiguille, je suis entré dans la sphère politique. Ça passe par des réunions. Beaucoup de réunions... Au début, je trouvais toutes ces réunions assez ennuyeuses. Puis on s'y habitue. Je pense qu'il existe de nombreuses différences entre le travail des hommes politiques et celui des mathématiciens. C'est compliqué de comparer les deux milieux. Dans les deux cas, il y a une grosse implication des gens, une certaine compétition aussi comme tu le racontes, Alice, mais des façons différentes de travailler. Par exemple, les mathématiciens, lorsqu'ils sont sur un problème, travaillent extrêmement dur, de manière très concentrée, jusqu'à ce qu'ils trouvent quelque chose de significatif.

Arrivés là, ils publient, et c'est comme s'ils ajoutaient leur petite pierre à un énorme édifice. En politique, je veux dire pour ce que je connais, les conseillers, on travaille aussi dur. Les types se couchent rarement avant minuit, et ne rentrent quasiment jamais diner chez eux. Le travail consiste surtout à organiser des réunions, préparer des actions officielles, aider à rédiger des discours, écrire des notes, beaucoup de notes, qui aboutissent quelque part, mais on ne sait pas forcément où, ni si elles seront lues ou pas. Il y a aussi ces réunions, toutes ces réunions, où il y a évidemment une énorme concentration d'intelligence. En général, les gens qui sont là sont quasiment tous extraordinaires, très impressionnants. Mais à la fin de la réunion, on a souvent l'impression que rien n'a été décidé, que les choses se sont comme effilochées... Mais, et toi, Jane, tu as bien des choses à nous raconter sur ce sujet là.

- Moi, je ne suis là que pour servir le thé !

-...

- Je plaisantais. Je dirais que je suis une version intermédiaire. Mais ce que vous avez raconté me parle beaucoup. En ce qui me concerne, j'ai arrêté de faire des mathématiques professionnellement il y a quelques années. Je fais toujours des maths mais uniquement pour le plaisir. Cela signifie que je continue à lire des papiers qui paraissent, que je fais mes petits calculs dans mon coin, mais je n'écris pas d'articles. Aujourd'hui, je concentre mon travail sur l'écriture de romans. En fait, je rejoins un peu Alim, sauf que moi, ce n'est pas un burn-out, c'est plus probablement du à l'age, ou au fait justement que je n'avais plus du tout l'esprit de compétition. En tout cas, à un moment, j'ai aussi senti comme une sorte de paresse, comme si des grosses cellules graisseuses venaient ralentir mon cerveau. En fait en ce qui me concerne, ce n'était pas le cerveau que je ressentais comme ralenti mais plutôt le désir, l'envie, comme si je devais me forcer... C'est compliqué à expliquer.

J'avais cette activité d'écrire des romans depuis très longtemps, vraiment très très longtemps, mais je n'avais pas trop de temps à y consacrer. Et tout d'un coup je me suis aperçue qu'écrire devenait beaucoup plus fluide. Comme un peu un principe de vases communicants : plus les maths devenaient difficiles, plus les romans devenaient accessibles. Alors, j'ai suivi la pente la plus facile... C'est amusant comme au fond, ce n'est pas si différent d'écrire des maths ou de la littérature. En tout cas, beaucoup moins différent qu'on pourrait le croire. Les ingrédients sont un peu les mêmes : l'imagination, la rigueur...

Alim m'avait interrompue, presque violemment :

- Mais, si on suit ton raisonnement, tu veux dire que les mathématiciens sont essentiellement motivés par la compétition ?

- Non, non, pas nécessairement. Je ne suis pas quelqu'un qui aime la compétition, donc je ne vais pas en faire une apologie. Ce que je dis c'est que en faisant des maths, on demande un effort assez intense au cerveau. Et je pense qu'il faut, en regard, lui proposer une gratification. Je pense que la compétitivité fournit cette gratification. Mais je ne dis pas qu'il n'y a pas d'autres motivations possibles. En politique, je pense que c'est aussi un peu le principe, mais là, la compétition est plus bien plus évidente.

C'est à ce moment là que la compagnie aérienne de Paul a téléphoné. Le blocage aérien était terminé. Cela m'a peut-être évité d'aller trop loin dans une analyse pour le moins incertaine.

English version

[1] There were four of us, in my house, drinking tea, maybe a few beers, and talking about our younger years.

Alim, Paul and Alice had arrived there, independently, but all blocked by air traffic. For a few hours, for a while, we didn't know. They'd preferred to come to my place rather than go to the hotel. And I have to say, that made me happy.

To be perfectly honest, they all liked coming to see me, but they perhaps liked my very cosy house even more - my grandmother's house in fact - and the dog Charly who knew how to put everyone at ease.

The atmosphere was very relaxed, even though they didn't really know each other.

At one point, the discussion turned to a particular question: when you're a mathematician, can you stop doing mathematics, or will you remain a mathematician for the rest of your life?

I looked at them. I wanted to know what each of them had to say on the subject. For the moment, they were watching each other. Alice seemed a little closed off. She was obviously paying close attention, but didn't want to take part in the discussion. I thought that, faced with this subject, Paul would probably be the one most keen to explain his life and his choices. In fact, it was Alim who initiated the discussion as if it were a major issue for him.

- You know, at the age I am, you start to draw conclusions and I think I've missed a lot - really a lot - of important opportunities. Not to become famous or anything like that, but to go much further in what I was doing.

In fact, in Turkey, with the education I received in Istanbul and then in Paris - my family was rich and cultivated - I was perfectly prepared to write a great many things. I'm talking about mathematics, of course, but also literature, poetry, novels... In fact, I'm a novelist who hasn't written a single line. I never had any ambitions, I just really liked understanding mathematics and communicating my way of understanding it. And I was very gifted. I can say that without immodesty. A thing of the past. Completely.

At the age of 26, I had a huge culture in mathematics, I was really brilliant, with great abilities. But I had a kind of accident.

I had just finished my studies, which I did very quickly, too quickly - it was a mistake, but I made a lot of mistakes... - and I started doing research. My thesis supervisor - a highly respected man - had given me an article to read during the holidays. It was a very rich, profound and interesting article on trigonometric approximation with exponential error rates, written by a very young researcher, Wolgang Dahmen.

[1] English version translated from French with the help of DeepL.

He must have been the same age as me or not much older. I learned in two months what very few people can assimilate in such a short time.

I then wrote a report and gave it to my thesis supervisor. In fact, he was very impressed. He obviously hadn't expected anything so well structured and written.

He was so enthusiastic that I got a job almost immediately. In Paderborn. I even had a choice. Paderborn or Paris, I chose Paderborn, because my German was better than my French. Even though I had studied at the French lycée in Istanbul. A family tradition, no doubt.

And three months later, I literally exploded. It's been a long time now, but I still don't know what really happened. The most likely explanation is burn-out. It's the most logical thing: I'd been working so hard on my studies over the last few months that my brain short-circuited and refused to continue working at that rate.

I thought it could be poisoning, I thought about a lot of things. In my family there have been many mental illnesses: my uncle died very young and is thought to have committed suicide, my youngest brother became schizophrenic...

It took me a long time to realise that I had some kind of illness. When I lost the ability to do maths, I didn't know whether it was my brain or the way I lived. I thought things would improve with a more balanced life, a wife and children. Unfortunately, that wasn't the case.

It took me a long time to understand even the fact that my brain didn't store things the way it used to. When I was young, I could read quantities and it would stay in my head forever. Today, if I read ten pages of maths, I forget them in a week.

And it never came back, I never did mathematics again with the total availability I had at 26. At some point, I had to face the facts. I said to myself: only do the maths you love, don't make your brain suffer.

And that's what I did. I continued to do my job, teaching mathematics to students. I liked the students, especially those in the first years. That's so important.

I also wrote books. I've always written a lot: even when I was a very young student, I was already writing notes to explain things to my friends.

My family had a huge library in Istanbul. It was my uncle who bought most of the books. Probably for us, because he died childless. He smoked a lot. A lot of people in my family smoked, bad tobacco.

Today, I have a large library in my own house, much bigger than the one we have in Istanbul. I'm passionate about books. I spend all my time reading, writing, trying to find rare mathematical works and looking after my publishing house. In fact, I set up two publishing houses, one in Germany and another in Istanbul. Unfortunately, the one in Istanbul didn't work because I wasn't there. I love books.

I don't think what I've done is dishonourable. I've 'done the job' as they say, I've even been useful in some ways... But certainly not as much as I could have been without this accident.

Alice had listened to Alim with obvious empathy. She hesitated before asking him:

- Have you ever tried seeing a psychotherapist? Maybe this disease in your brain could be cured or treated.

- Anyway, it's too late now, but of course I've seen many therapists. When my younger brother was diagnosed with schizophrenia, the whole family was treated. I'm not sure why, it's as if the doctors were carrying out some sort of deratting... I remember the psychiatrist saying to me: "I can't do anything for you because you're completely immature. You're a child. I took that as a compliment.

Alim laughed.

- A lot of people tell me that a lot of things have gone wrong in my life because of this schizophrenic brother I looked after: my career, my marriage which ended in divorce... But I don't think that's true. I consider him to be a blessing in my life. I love him very much.

But we were talking about mathematicians who have completely left mathematics to do something else. I'm not exactly in that category. I've just been operating on a sub-regime. But you, Alice, you are, it seems to me. Would you care to explain?

Alice sank back into her chair. I got the impression that she wasn't comfortable talking. Nevertheless, she began, bravely.

- Well, Alim, my story is actually a bit different. As you may know, I was a mathematician a long time ago. I mean, I probably wasn't as brilliant as you describe, but I did "reasonable" mathematics, I'd say. It wasn't a passion for me and there was no family tradition, not even an intellectual environment like you describe. I was the first person in my family to do mathematics. Worse still, I think they were completely agnostic about what I was doing. They're all very physical people, drawn to all sorts of sports. But they simply don't integrate intellectual activities into their lives. They're not stupid either, but they don't understand what mathematical research can mean, let alone make a living out of it.

In fact, I was very happy doing mathematics. Proud. I was quite respected in the mathematics department in South Carolina, where I was. I liked working, reading papers, going to meetings, visiting countries, having lots of mathematical discussions. It all fascinated me. By the way, it's funny, you mentioned Wolfgang Dahmen. I remember I started working on his work. He made me fantasise, he was super brilliant and also a Taekwondo champion... I never met him. But he came to South Carolina from time to time. Too bad, it might have changed my life! In any case, I married a university colleague, a physicist. Maybe not the best of ideas, but that's life... We were at the same university and, after a few years, something not very unusual happened. He left me for another woman. Another colleague, younger than me and brighter, or at least more ambitious. Perhaps I'm not being entirely fair to her. It's a long time ago now...

She was smiling but had stopped talking for a moment, taking a sip of tea as if to give herself courage:

- At that point, I felt pretty bad. I didn't know what to do, and maybe what I did wasn't the smartest thing. It's hard to say...

Anyway, I left, resigned from the department and moved to the town where my sister was. Further north, not far from DC. I took my son with me. He was 4 years old.

Paul reacted almost immediately:

- But didn't anyone tell you that you had rights, that you shouldn't just walk away, run away, as if you were the one at fault?

- Yes, yes, lots of people told me that, especially the head of the maths department. He really encouraged me to stay. No. It was my own decision. I didn't feel at home there any more. I felt that if I stayed, I wouldn't be able to shake off this sort of humiliating feeling that I had. I had to leave. As I said, it may not have been the wisest decision, but in fact I've never regretted it.

To make ends meet, I had to find a way of earning money. My husband gave me a little for his son's education, but he was never very generous. His contribution wasn't nearly enough. I didn't have any money to spare, so I had to find something as quickly as possible. My sister was great. She helped us out: we had a place to stay and lots of moral support. She's a very active, energetic woman. She immediately encouraged me to forget all that, to turn the page.

A bit like Alim, I asked myself what I could do to comfort and please myself. The answer came quite quickly with yoga. I was physically very flexible and had been doing yoga since I was a teenager. I felt I could be a good teacher. There was a yoga studio not far from my sister's house. We had a chat and they agreed to give me a month's trial. I still work at the same studio. I'm now one of the owners. Financially, everything's going well.

- Financially it's probably better than staying in the maths department. Paul quipped.
- Yes, much better, I must say.

Alim was impatient. He wanted to talk. He almost had trouble letting Alice finish her sentence.

- But really, you gave up maths and turned to physical activity? just like that, in an instant? without remorse?
- I wouldn't say it was in an instant. It took me a while to realise that the situation between my husband and me was irreversible. It took me another few weeks to decide that the best solution was to move out. Then a few more weeks to find a job...
- But do you read mathematics or do you just do calculations for fun?
- No, I don't.
- But you liked doing maths?
- Yes, I liked it a lot. But you know, when I was at school, I was good at maths. But I was also good at physical activities and sports in general. At a certain point, I chose mathematics. But then this break in my life was so hard and so violent in a way... And the fact that my whole family was completely focused on non-intellectual activities also played a part. All this meant that I ended up thinking that this foray into the academic world was nonsense, that it wasn't for me. You know, in the yoga community, and in the personal development community in general, it's easier to have the impression that you're in the right place, where you belong... It's hard to explain. I know it's a very different way of life. But I really love what I do. I look after people. I mainly do one-to-one support. I have a close relationship with the people I look after. I feel like I'm helping them. I think I feel more useful than when I was doing maths. That may be a simplistic, short-term point of view. But that's what I think. And also... in maths, there was a strong competitive spirit. So, it's fine when you feel strong, capable of succeeding, but as soon as you feel hurt it's more complicated. And that's what was happening to me.

At one point, Paul, who had previously seemed very enthusiastic in the discussion, suddenly became very quiet. I noticed this and tried to encourage him to talk. I knew he had some interesting things to say.
- And you, Paul, how would you describe your relationship with mathematics? You've had a very exciting life, haven't you?
- Well, exciting... Yes, perhaps...
- For Alice, who may not know you well: you were a mathematician, you had a fairly brilliant university career in various mathematics departments, in different countries even. Then, if I remember correctly, you were university president. And more recently, advisor to the Prime Minister on research issues. Is that correct?
- Yes, that's right.
- Do you still consider yourself a mathematician?
- I certainly do. I do as much mathematics as my tight schedule allows. I don't think I've ever stopped doing mathematics. Perhaps I'm not being entirely honest with myself when I say that. The way I do mathematics today is not comparable with what a professional researcher does. I rely heavily on my colleagues. They obviously do a lot of the difficult work that I don't have time to do. But I am proud to say that I have never stopped writing mathematics. Mathematics is essential to me. When I was young, I was a bit of a late bloomer in terms of personality development. I've met people, I can say, with precocious minds. I didn't.

My father was also a slow guy. I come from a modest family. You can start life knowing the codes and social behaviour. That wasn't my case, and even today I often feel the consequences. A sort of slowness, sometimes even cognitive dissociations. A guy who's not always well connected.
- Maybe you're judging yourself a bit harshly, aren't you?
- Of course, I do. I'm not just that, I also have a good capacity for social adaptation. But what I wanted to stress is the fact that higher education and mathematics in general have been, and still are, a kind of protection against the strange behaviour I'm sometimes tempted to display. In any case, when I was young, I studied mathematics, then I became a researcher, and I was completely invested in it, passionate about it. By the way, the Wolfgang Dahmen you both mentioned, I haven't read his work, but I met him at the TTI in Chicago in 2007 or 2008, I can't remember. He was a very nice guy and we talked politics.

Initially, I was completely into maths, I wasn't actively involved in any political parties or trade unions, even when I was a student.

I didn't start getting involved in politics until I got my first professorship position. My children were born, my wife had a job in a high-profile ministry, we were discussing politics at home... I was at a university I didn't know, with a fairly friendly team. They were all very politically committed. I was interested and they sort of pushed me. Then, as you described, Jane, I became president of the university, and one thing led to another and I entered the political sphere. That involves meetings. Lots of meetings... At first, I found all these meetings rather boring. But then you get used to them. I think there are a lot of differences between the work of politicians and that of mathematicians. It's complicated to compare the two. In both cases, there's a great deal of involvement on the part of people, and a certain amount of competition

too, as you say, Alice, but there are different ways of working. For example, when mathematicians are working on a problem, they work extremely hard, in a very concentrated way, until they find something significant. When they do, they publish, and it's like adding their little stone to an enormous edifice. In politics, as far as I know, councillors work just as hard. The guys rarely go to bed before midnight, and hardly ever go home for dinner. The action consists mainly of organising meetings, preparing official events, helping to draft speeches, writing notes, lots of notes, which end up somewhere, but we don't necessarily know where, or whether they will be read or not. There are also these meetings, all these meetings, where there is obviously a huge concentration of intelligence. In general, the people who are there are almost all extraordinary, very impressive. But at the end of the meeting, you often get the impression that nothing has been decided, that things have sort of unravelled...
But, and you, Jane, you have a lot to tell us about that.
- I'm just here to serve tea.
-...
- I was only joking. I'd say I'm an intermediate version. But what you said really spoke to me. As far as I'm concerned, I stopped doing maths professionally a few years ago. I still do maths, but only for pleasure. This means that I continue to read papers that come out, that I do my little calculations in my corner, but I don't write articles. These days, I concentrate on writing novels. In fact, I'm a bit like Alim, except that for me it wasn't a burn-out, it's more probably due to age, or to the fact that I just didn't have the competitive spirit any more. In any case, at one point I also felt a kind of laziness, as if big fat cells were slowing down my brain. In fact, as far as I was concerned, it wasn't the brain that I felt was slowing down but rather the desire, the urge, as if I had to force myself... It's complicated to explain.

I'd been writing novels for a very long time, a very, very long time, but I didn't have much time to devote to it. All of a sudden I realised that writing was becoming much more fluid. It's a bit like the principle of communicating vessels: the more difficult maths became, the more accessible novels became. So I took the easier route... It's funny how, in the end, it's not so different to write maths or literature. In any case, it's a lot less different than you might think. The ingredients are much the same: imagination, rigour...

Alim interrupted me, almost violently:
- But if we follow your reasoning, you mean that mathematicians are essentially motivated by competition?
- No, no, not necessarily. I'm not someone who likes competition, so I'm not going to make an apology for it. What I'm saying is that doing maths requires a fairly intense effort from the brain. And I think that, in return, you have to offer it some kind of reward. I think that competitiveness provides this gratification. But I'm not saying that there aren't other possible motivations. In politics, I think that's also the principle, but here the competition is much more obvious.

That's when Paul's airline phoned. The air blockade was over. This may have saved me from going too far in an analysis that was uncertain, to say the least.

Some Thoughts on Compositional Tensor Networks

Reinhold Schneider and
Mathias Oster

Abstract
In these notes we present some first ideas on the composition of tensor trains for the use in scientific computing. We discuss the relation to deep neural networks and the potential role compositional tensor trains can have in efficiently representing the solutions to PDEs. We illustrate the potential role compositional tensor trains might have for efficiently representing the solutions to high-dimensional PDEs circumventing the curse of dimensionality. Lastly, we embed the task of function regression on the set of compositional tensor trains into the context of semiglobal optimal control and mean field games.

1 Introduction

Recent developments in machine learning, complexity theory of numerical solutions to partial differential equations (PDEs) as well as infinite dimensional optimal control have shown a promising mathematical theory to overcome the computational challenges of contemporary high dimensional scientific computing tasks.

These notes shall provide a collection of thoughts interconnecting deep neural networks, tensor trains, compositional sparsity, high dimensional PDEs and mean field games in order to discover new relations within these fields, potentially providing tools to tackle problems deemed to be computationally infeasible at the moment. Since we cover a wide variety of seemingly unrelated fields, these notes will only provide a sketch on how to connect these different ideas.

Mathias Oster
RWTH Aachen, Templergraben 55, 52062 Aachen, Germany, e-mail: oster@igpm.rwth-aachen.de

Reinhold Schneider
Department of Mathematics, TU Berlin, Str. des 17. Junis 135, 10623 Berlin, Germany, e-mail: schneidr@math.tu-berlin.de

R. DeVore, A. Kunoth (eds.), *Multiscale, Nonlinear and Adaptive Approximation II*,
https://doi.org/10.1007/978-3-031-75802-7_19

As the dimension of the state space increases, traditional methods to represent a prescribed function, at least approximately, for example by Galerkin approximation with splines, finite elements or multi-polynomials, suffer from the so-called curse of dimensionality. This means that their computational complexity to approximate solutions to PDEs increases exponentially with the dimension of the state space.

It is well known from approximation theory that the approximation of functions in Sobolev classes (with fixed smoothness) is always exposed to the curse of dimensionality. There are well established lower bounds for the *Kolmogorov n-width* that show the limits of linear approximation techniques, see [12, 13, 51] among many others. However, there is also the observation that many functions we meet in practise, in particular solutions of some well known and important partial differential equations (PDEs), either have sufficient smoothness or lie in an unknown, non-linear class of functions, informally called manifolds, of much lower intrinsic dimension.

A wide variety of non-linear approximation techniques as, for example, *best n-term approximation*, which is based on the sparsity of the coefficient vector with respect to a given basis, tensor decomposition techniques and neural networks have been developed. In the following, we will focus primarily on tensor methods.

Rooted in quantum physics under the name *matrix product states*, c.f. [44], the so called *tensor trains* have been introduced to the mathematical community in [42] to tackle the curse of dimensionality. Tensor trains are a special case of the more general framework of hierarchical tensor networks or tree based tensor networks [16]; the correct name in the sense of Hackbusch would be *Hierarchical Tucker Tensor Format*. These networks have been developed in [26] where a well-founded mathematical framework is introduced, see also [6, 25, 5, 27, 50].

Subspaces of function spaces with tensorial structure have been introduced by Hackbusch, who employs the Tucker format to identify the best univariate subspaces which reduce the coefficients tensor, we refer also to [18]. Although a data sparse representation of functions was always part of the focus for the first mathematical developments, the now widely used name *functional tensor trains* and its present form has been coined in [17].

During the endless discussions concerning the superiority (or the lack thereof) of the tensor network approach in the approximation theory community, deep neural networks (DNN) have revolutionized machine learning technologies rapidly. After their success in practise and the establishment of powerful computational tools, they have attracted many mathematical researchers. For example, Nadav et al, [9], discovered that deep convolutional networks, developed in image and signal analysis, have a very similar structure as Hierarchical Tensor networks. They demonstrated that deep neural networks have an exponentially larger power of expressivity. Afterwards [55] and [7, 22, 23, 30, 45, 49, 56], as well as [39, 40, 46] among others have established convergence rates in many instances. Moreover, [21, 28] have shown the tractability for the solution of high-dimensional semi-linear parabolic PDEs. In [37] parametric PDEs have been investigated with these new tools. This is but an incomplete list of references.

An interesting recent result [4] states that for functions which are compositions of bi-variate functions, tensor network can provide the same convergence rates as

DNNs, (but possibly with a worse constant). Indeed the papers [39, 40, 46] and [55], including all papers mentioned above demonstrate that the use of compositions of functions provides a powerful tool for approximation. To our understanding the use of compositions of non-linear functions provides a revolutionary breakthrough.

Hierarchical tensor networks on the other hand are based only on the composition of bi-linear functions ($\sigma(x, y) := x \cdot y$) which is the simplest activation function. Recently, Dahmen, [15], has introduced the concept of *compositional sparsity* in order to identify function classes that allow for efficient compositional representations.

In these notes we apply this concept to compositional functional tensor trains and focus on polynomial (spectral) uni-variate basis functions. Generalizations to other basis functions and hierarchical tensor networks, [27], are straightforward.

The use of compositions, and therefore the possibility to use coordinate transforms, also alleviates the major disadvantages of tensor product approximation, namely its preference of coordinate parallel representations. This problem is inherent in the tensor product construction. However, with compositional structures we can apply multi-linear or even non-linear transformations to deform the problem into a coordinate parallel representation. Similarly to DNNs, when applying multiple compositions we are loosing the possibility of direct analytic integration that is possible in standard tensor networks. Additionally, the benefit of Hierachical SVD (HSVD) [26], [27], [42] is rather restricted after composition. We may call a single tensor network (HT Tensors) *shallow* and compositional networks as *deep*. Since we have to sacrifices benefits of shallow tensor representation in the compositional framework, it will depend on the actual application which concepts should be preferable (shallow versus deep).

To facilitate the presentation in this paper, we consider mainly morphism $\mathbb{R}^d \to \mathbb{R}^d$, like transport maps $\Psi : \Omega \subset \mathbb{R}^d \to \mathbb{R}^d$. A particular example is the flow induced by a dynamical system (ODEs). Together with the development of neural ODEs, normalizing flows and their intimate relation to recurrent neural networks we focus on flow driven compositions and/or networks, compare [36]. Here, the number of layers is the number of time steps or iteration steps, by which we can control the accuracy. We formulate and highlight further on a duality between local or microscopic, i.e, point-wise treatment, and a global representation given by multiple composition of tensor train functional mappings. In the microscopic or local perspective, we fix a variable $x \in \Omega$ and are interested in the function value $f(x)$. In the macroscopic or global view we are interested in the mapping $f : \Omega \to \mathbb{R}^d, x \mapsto f(x)$. Basic examples for this duality is the method of characteristics, the change between Eulerian and Lagrangian coordinates, and the treatment of parametric PDE's. It can be extended to a very large class of time dependent high-dimensional PDE's [14, 20, 63].

Beside its impressive power of expressivity there is another puzzling experience with DNN. Typically, they have an surprisingly small generalization error and are not prone to oversampling. Even if there are many more weights or parameters (DOFs) to be computed than samples given DNNs usually provide reasonable results. And this is not only due to redundancy. Although the full optimization problem is underdetermined the stochastic optimization seems to work reasonably well. To give some, not necessarily formalizable intuition, consider the composition of only two layers

$\Phi_2 \circ \Phi_1$ and some samples $x^i, i = 1, \ldots, N$. If we assume now, that we know the first layer, or have a reasonable guess for Φ_1 we can "create" new samples $y^i = \Phi_1(x^i)$ to optimize the second layer. In this sense the ratio of samples per parameters on a single layer enters. Although these arguments are for from rigorous, they might provide some insight into the role of composition.

However, the parametrization even of a single layer NN might be highly unstable or badly conditioned. For example in one dimension the ReLU activation function $\sigma(x) = \max\{0, x\}$ evaluated for some weights is a basis for 1D linear splines and known to be ill-conditioned. This has led to the development of B-Splines providing stable uni-variate basis functions. It has been experienced in early numerical mathematics that the use of unstable basis function can lead to disastrous results. That this problem can completely be cured in high-dimensions by a stochastic gradient descent method, as claimed in computer science cannot be completely true. Maybe the achieved accuracy is sufficient in statistic, but for high-fidelity solutions of PDE's and operator learning this stability and the well conditioning seem to play an essential role. Therefore, we are interested in more stable and efficient optimization steps. Experiences in optimal control, by Walter and Kunisch, [35, 33], using sparse polynomials instead of neural networks for semi-global feedback synthesis, demonstrate that solving well conditioned optimization problems is multiple magnitudes better than using neural networks. Also tensor methods have proven their value in various numerical studies in optimal control and for control related PDEs like Hamilton-Jacobi-Bellmann equations, see [57, 58, 59, 60, 61, 62].

Outline:

In the next chapter we will recall the notion of functional tensor trains and use them to introduce the concept of compositional tensor trains thereafter. In the fourth part we elaborate on a particular subfamily of compositional tensor trains, i.e. Flow Induced Tensor Trains.

Afterwards, we will give a brief tractability discussion of compositional tensor trains and demonstrate their application for (parameterized) PDEs and how the tensor cores can be optimized.

Lastly, we demonstrate that this optimization is closely related to the numerical solution of the HJB equation and variational mean field games. We include a chapter about some basics of closed and open loop loop optimal control, and present a local (or microscopic) solution of optimal control problem related to Pontryagin's Maximum Principle. A global, more averaged, perspective of the optimal control problem leads to some mean field games like equations. For sake of completeness we include also the hydro-dynamic formulation of the optimal transport problem, mainly to demonstrate how closely all these issues are related, and to underpin the fundamental character of these problems.

2 Functional Tensor Trains

Any quadratic form q on $\mathbb{R}^n$ can be identified with a symmetric $n \times n$ matrix A such that $q(x) = x^T Ax$. Similarly, we can use higher order tensors to represent polynomials of higher degree. However, for an order d tensor with input size n we need storage in $O(n^d)$, which might be not feasible. The same holds for tensor product spline and tensor product wavelets. The scaling $O(n^d)$ is often called the curse of dimensionality. It is clear that with this scaling the treatment of high-dimensional problems becomes intractable for higher dimensions $d \gg 1$. In order to circumvent this scaling further concepts are required. For example it may happen that only few basis functions contribute essentially to an approximation. This leads to a non-linear approximation selecting the best n-term approximation. Another method is focused in a best basis selection and low rank approximation. In this case the non-linearity is multi-linear.

In order to get some notion of compression, we introduce the TT-rank $r \in \mathbb{N}^{d-1}$ of the tensor $A \in \mathbb{R}^{(n_1,\ldots,n_d)}$ as element-wise smallest tuple such that

$$A(i_1, \ldots, i_d) = \sum_{k_1,\ldots,k_{d-1}=1}^{r_1,\ldots,r_{d-1}} \hat{U}_1(i_1, k_1) \cdot \hat{U}_2(k_1, i_2, k_2) \cdots \hat{U}_d(k_{d-1}, i_d)$$

holds for some $\hat{U}_i \in \mathbb{R}^{r_{i-1},n_i,r_i}$ for $i = 0, \ldots, d$ where we set $r_0 = r_d = 1$. The TT rank is well defined, and the tensors of fixed TT rank form a smooth manifold $\mathcal{T}_r$ of dimension in $O(dnr^2)$ [31]. In this paper we use this approach only as a first instance to tackle the curse of dimensionality.

We can use these TT tensors to represent polynomials as follows. Consider

$$P_i : \mathbb{R} \to \mathbb{R}^{n_i} \text{ with } P_i(x) = \begin{bmatrix} 1 \\ x \\ x^2 \\ \vdots \\ x^{n_i-1} \end{bmatrix}.$$

Then

$$p(x_1, \ldots, x_d) = \sum_{i_1,\ldots,i_d}^{n_1,\ldots,n_d} \sum_{k_1,\ldots,k_{d-1}}^{r_1,\ldots,r_{d-1}} \hat{U}_1(i_1, k_1) \cdots \hat{U}_d(k_{d-1}, i_d) \big(P_1(x_1)\big)_{i_1} \cdots \big(P_d(x_d)\big)_{i_d}$$

is a multivariate polynomial of degree $(\sum_i n_i)$. For advanced treatment we refer to [64]. Using the graphical tensor network representation, this polynomial can be interpreted as in figure 1.

Let us highlight that we can use other single variable functions or better (orthogonal) basis functions as well, like trigonometric polynomials, piecewise polynomial splines, quantized tensor trains (QTTs) etc.

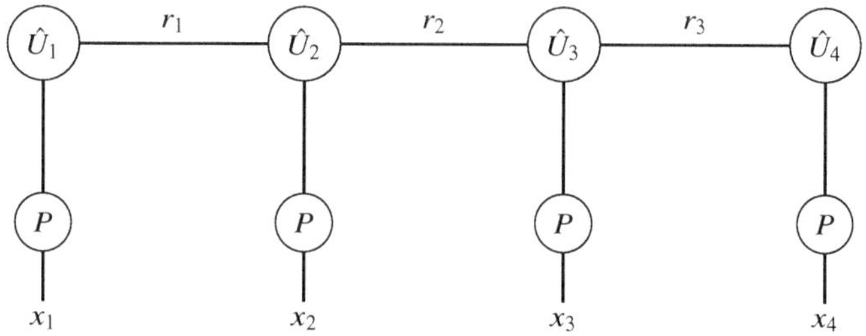

Fig. 1: Graphical representation of TT tensor train induced polynomial in four variables

In general, let us choose a basis $\{\phi_j\}_{j\in J}$

$$\mathcal{P}_{\leqslant n_i} := \operatorname{span}\{\phi_j : \mathbb{R} \to \mathbb{R} : j = 1, \dots, n_i\}$$

and define matrix valued functions

$$U_i(k_{i-1}, x_i, k_i) = \sum_{j=1}^{n_i} \hat{U}_i(k_{i-1}, j, k_i)\phi_j(x_i)$$

as well as as the representation

$$f(x) = \hat{f}(\hat{U}; x) = \sum_{k_i \leqslant r_i, i=1,\dots,d-1} U_1(x_1)U_2(k_1, x_2, k_2)\cdots U_d(k_{d-1}, x_d).$$

Here the cores or components $x_j \mapsto U_j(k_{j-1}, x, k_j)$ are matrix valued uni-variate functions.

This provides a representation of a global function $f : \mathbb{R}^d \to \mathbb{R}$ and for this reason it will be called a *Functional Tensor Train*. Other tree based tensor formats like HT can be used as well and Tensor Trains are only simple examples. It is important to point out that $\hat{f}$ is multi-linear in $\hat{U}$ and non-linear in the variables x.

Of course, this representation can also be used to express vector valued d-variate functions, by coordinate-wise using Functional Tensor Trains, i.e. given a function $x = (x_1, \dots, x_d) \mapsto f(x) : \Omega \subset \mathbb{R}^d \to \mathbb{R}^n$, we represent it coordinatewise for $i = 1, \dots, n$ by

$$f_i(x) = \hat{f}_i(\hat{U}; x) = \sum_{\substack{k_j \leqslant r_j \\ j=1,\dots,d-1}} U_1(x_1)U_2(k_1, x_2, k_2)\cdots U_d(k_{d-1}, x_d).$$

For given parameters $\hat{U}_j$ we have

$$x_j \mapsto U_j(k_{j-1}, x_j, k_j) \in \mathcal{P}_{\leqslant n_j}^{r_{j-1}, r_j}.$$

Observe that $f = (f_i)_{i=1,\ldots,d}$ can be represented in a single TT with an additional index

$$\begin{aligned} f_i(x) &= \hat{f}_i(\hat{U};x) \\ &= \sum_{\substack{k_j \leqslant r_j \\ j=1,\ldots,d-1}} U_1(x_1)U_2(k_1,x_2,k_2)\cdots U_d(k_{d-1},x_d,i),\ i=1,\ldots,n. \end{aligned}$$

Given a sufficiently rich basis, the universality of this concept, i.e. that any continuous functions can be represented in this form up to any desired accuracy $\epsilon > 0$, seems to be obvious. However, the functional analytic treatment of tensor products of Banach spaces require additional care about its topology, e.g. inductive and projective limits. To avoid these issues let us consider Hilbert spaces. Let $\mathcal{V}$ be a Hilbert space of uni-variate functions, e.g. $L_2(\mathbb{R})$. Moreover we assume that we have a complete Riesz basis $\{\phi_i : i \in \mathbb{N}\}$. It is clear that if there is a basis $\{\phi_i : i \in \mathbb{N}\}$ which is complete in the one-variate function space $\mathcal{V}$ the set of algebraic (finite) tensor products of this kind is dense in the tensor product space $\otimes\mathcal{V}_i$ if we allow arbitrary high ranks. More details can be found in the book of Hackbusch, who introduced and explained the fundamental concept of *optimal subspace approximation*, [24]. First approximation rates has been obtained by [48]. We like to mention recent related articles by Nouy and co-workers, [1, 2] and Griebel and co-workers, [19].

3 Compositional Tensor Networks

The Roman Empire has not been conquered in a single day!

Although standard tensor train representations might provide a data-sparse format of the vector valued function $f : \mathbb{R}^d \to \mathbb{R}^d$ if the ranks are bounded, this is not always the case and other hidden structures have to be exploited.

We will focus on compositional structures and will be inspired by the concept of *deep neural networks (DNN)*, where we emphasize the phrase "deep".

Such a *deep neural network* with L layers is of the form

$$f = \Phi^L \circ \Phi^{L-1} \circ \cdots \circ \Phi^1\ ,\ \Phi^j : \mathbb{R}^{d_j} \to \mathbb{R}^{d_{j+1}}\ ,\ d_1, d_{L+1} := d$$

where

$$\Phi^i(x) = \Phi^i(a_i, A_i, b_i; x) = \sigma(\mathbf{A}_i x + b_i)\ ,\ x \in \mathbb{R}^{d_i}\ ,\ \mathbf{A}_i \in \mathbb{R}^{d_{i+1}\times d_i},\ b_i \in \mathbb{R}^{d_{i+1}}$$

is a shallow neural network. Here σ denotes a uni-variate activation function, e.g. ReLU $\sigma(x) = \max\{0, x\}$.

Inspired by deep neural networks and their concept of the composition of simple functions, we can obtain a *deep tensor network* or particularly a *deep tensor train*, by replacing the shallow network in the above composition with a functional tensor network or tensor train. From the perspective of data complexity, the major point is that Φ^j has a data sparse representation e.g. shallow neural nets, tensor trains or sparse polynomials.

Remark 3.1 For neural networks with a fixed number of layers L polynomial activation functions do not admit universal approximation theorems. Indeed, a single layer NN with polynomial activation of degree p can only create polynomials of degree p. But if we consider the behaviour for an unlimited number of layers say $L \to \infty$ then any continuous function can be approximated e.g. with a quadratic activation function $\sigma(x) = x^2$. This has been used e.g. in [45, 37]. Indeed using the parallelogram law

$$xy = \frac{1}{4}((x+y)^2 - (x-y)^2)$$

one can deduce that arbitrary polynomials can be created by a DNN with a quadratic activation function, see e.g. Proposition 3.1 below.

Let us first consider a composition of two functional tensor trains: let $u = \hat{u}(\hat{U}, x), v = \hat{v}(\hat{V}, x) \in \mathcal{T}^d_{\leqslant r}$

$$x \mapsto u \circ v(x) = u(v(x)) = \hat{u}(\hat{U}, \cdot) \circ \hat{v}(\hat{V}, \cdot)(x).$$

If the two tensors are representations of polynomials the composition provides a new tensor. However, notice that if both tensors have ranks $\leqslant r$ the resulting tensor has a much higher rank in general. For example, if $d = 1$ and u, v are polynomials of degree ℓ the $u \circ v$ is in general of degree ℓ^2. Therefore, we do not intend to represent the resulting tensor with new cores but we keep the representation above as a data-sparse composite of two functions. Notice that the data complexity is only $\sharp u + \sharp v$ and scales linearly with the number of compositions or so to say the number of levels L i.e. the depth of the network. We observe that for given $x \in \mathbb{R}^d$ the value $u \circ v(x)$ is computable within reasonable complexity. Moreover, the derivatives $\partial_{x_i}, \partial_{U_i}$ of $u \circ v(x)$ at x are computable as well, as long as the uni-variate functions used are differentiable. However, we lose some advantages of tensor representation, e.g. integrals cannot be computed directly, but we have to apply quadrature methods e.g. Monte Carlo or Quasi-Monte Carlo integration.

We aim to circumvent one of the major disadvantages of tensor product approximation, namely the occurrence of singularities or cusps which are not axi-parallel. This can be remedied by a unitary transformation, and even curved shaped singularities could be resolved by non-linear coordinate transform. Furthermore, the ordering of variables plays a crucial role and can be improved by composition with a unitary linear coordinate transform and even an optimal permutation can be computed. Therefore, the approximation class should allow for (linear) coordinate transforms, i.e. let $v(x) := Ax$ then

$$u \circ v(x) = u(Ax).$$

Example 3.1 Let $f(x) = Ax$ for some $A \in \mathbb{R}^{d,n}$ and $\phi_1(x) = 1, \phi_2(x) = x$. Define

$$\hat{U}_1(:,:) = \begin{pmatrix} 1 & 0 \\ 0 & [A_{i1}]_i \end{pmatrix},$$

$$\hat{U}_j(:,1,:) = \begin{pmatrix} 1 & 0 \\ 0 & \mathbb{I}_n \end{pmatrix} \text{ and } \hat{U}_j(:,2,:) = \begin{pmatrix} 0 & [A_{ij}]_i \\ 0 & 0 \end{pmatrix}$$

and

$$\hat{U}_n(:,1,:) = \begin{pmatrix} 0 \\ \mathbb{I}_n \end{pmatrix} \text{ and } \hat{U}_n(:,2,:) = \begin{pmatrix} [A_{in}]_i \\ 0 \end{pmatrix}.$$

Then we have

$$U_1(:) = \begin{pmatrix} 1 & [A_{i1}x_1] \end{pmatrix},\ U_j(:,:) = \begin{pmatrix} 1 & [A_{ij}x_j]_i \\ 0 & \mathbb{I}_n \end{pmatrix} \text{ and } U_n(:,:) = \begin{pmatrix} [A_{in}x_n]_i \\ \mathbb{I}_n \end{pmatrix}$$

and, thus,

$$U_{n-1}U_n = \begin{pmatrix} 1 & [A_{i(n-1)}x_{n-1}]_i \\ 0 & \mathbb{I}_n \end{pmatrix} \begin{pmatrix} [A_{in}x_n]_i \\ \mathbb{I}_n \end{pmatrix} = \begin{pmatrix} [A_{in}x_n + A_{i(n-1)}x_{n-1}]_i \\ \mathbb{I}_n \end{pmatrix}.$$

By induction we have

$$U_2U_3\cdots U_n = \begin{pmatrix} [\sum_{k=2}^n A_{ik}x_k]_i \\ \mathbb{I}_n \end{pmatrix}$$

and thus

$$\prod_{i=1}^n U_i = \begin{pmatrix} 1 & [A_{i1}x_1]_i \end{pmatrix} \begin{pmatrix} [\sum_{k=2}^n A_{ik}x_k]_i \\ \mathbb{I}_n \end{pmatrix} = \left([\sum_{k=1}^n A_{ik}x_k + A_{i1}x_1]_i\right)$$
$$= \left([\sum_{k=1}^n A_{ik}x_k]_i\right).$$

Therefore,

$$f(x_1,\ldots,x_d) = U_1(x_1,:)\cdot U_2(:,x_2,:)\cdots U_d(:,x_d,:).$$

Hence, we can store it as TT vector with rank $r = n+1$. Let us also remark, that a scalar valued linear function $f : \mathbb{R}^n \to \mathbb{R}$, i.e. $f(x) = \sum_{i=1}^n a_i x_i$ can be expressed by TT with rank $r = 2$.

This observation gives rise to the following assumptions.

Assumptions 3.1. *We assume that the basis $\{\phi_j\}_{j\in J}$ is chosen such that*

- *linear maps $A : \mathbb{R}^d \to \mathbb{R}^n$, $x \to Ax$ can be represented,*
- *$x_k \to \phi_j(x_k)$ are locally Lipschitz.*

The next question is how we can approximate a continuous function by compositional tensor trains with an ultimately small uni-variate basis $\phi_1(x) := 1$, $\phi_2(x) := x$.

Proposition 3.1 *Let $\phi_1 := 1$, $\phi_2(x) := x$ and $-\infty < a < b < \infty$. Then for any continuous function $f : [a,b] \to \mathbb{R}$ and $\varepsilon > 0$ there exists a compositional tensor* *train network $\hat{f}(\hat{U}, \cdot)$ such that it holds*

$$\max_{x \in [a,b]} |f(x) - \hat{f}(\hat{U}, x)| \leqslant \varepsilon.$$

Proof. According to Weierstraß it is sufficient to represent arbitrary polynomials by compositional tensor trains. In principle this is easy, because we need to realize only $x \mapsto x^2$ as all other monomials can be build recursively by exploiting the multi-linear structure. To show that we can obtain this map, we will first represent a 'copy' layer that maps

$$(x, y) \mapsto (x, x)$$

and then we will use a bilinear layer to map

$$(x, x) \mapsto (x^2, x).$$

Let $\hat{U}_1^1 = \begin{pmatrix} 1 & 0 \\ 0 & 1 \end{pmatrix}$ and $\hat{U}_2^1(:, i, :) = \begin{pmatrix} 0 & 0 \\ 1 & 0 \end{pmatrix}$ for $i = 1, 2$. Then $U_1^1(x, :) = \begin{pmatrix} 1 & x \end{pmatrix}$ and $U_2^1(:, y, :) = \begin{pmatrix} 0 & 0 \\ 1 & 1 \end{pmatrix}$ and, therefore,

$$p_1(x, y) = U_1^1(x, :) \cdot U_2^1(:, y, :) = \begin{pmatrix} 1 & x \end{pmatrix} \cdot \begin{pmatrix} 0 & 0 \\ 1 & 1 \end{pmatrix} = \begin{pmatrix} x & x \end{pmatrix}.$$

Now consider $\hat{U}_1^2 = \begin{pmatrix} 0 \\ 1 \end{pmatrix}$ and $\hat{U}_2^2 = \begin{pmatrix} 0 & 1 \\ 1 & 0 \end{pmatrix}$. Then $U_1^2(x, :) = \begin{pmatrix} x \end{pmatrix}$ and $U_2^2(:, y, :) = \begin{pmatrix} y \\ 1 \end{pmatrix}$ and thus

$$p_2(x, y) = U_1^2(x, :) \cdot U_2^2(:, y, :) = \begin{pmatrix} yx \\ x \end{pmatrix}.$$

Hence $p_2 \circ p_1(x, y) = (x^2, x)$.

As any higher order monomial and, therefore, polynomial can generated by similar compositions the claim follows. We observe that the representation is not unique and depends on the architecture of the network. □

Remark 3.2 To demonstrate the expressive power of composition let us consider functions which are compositions of bi-variate functions like

$$f(x) = f_{\{1,2,3,4,5\}}(f_{\{1,2,3\}}(x1, f_{\{2,3\}}(x2, x3)), f_{\{4,5\}}(x4, x5)).$$

see e.g. [4] the dimension in each layer is at most two instead of n (here $n = 5$), and there is no curse of dimensionality. Corresponding function classes have been introduced by [40, 39] for an approximation theory of deep neural networks, at least for convolutional DNN. It is remarkable that this class is already rich enough such that most of multivariate function we "write on paper" are of this form. Clearly tensor train representations are contained in this class, [4].

Example 3.2 Let $p(x_1, \ldots, x_d) = (\sum_{i=1}^d x_i^2)^2$. Let $\phi_1 = 1$, $\phi_2(x) = x$ and $\phi_3(x) = x^2$. Then we can represent p as CTT with 2 layers of rank 3 and 1 respectively. This can be seen as follows. Let

$$\hat{U}_i^1(:,:,k) = \begin{pmatrix} 0 & 0 & 1 \\ 1 & 0 & 0 \end{pmatrix} \text{ for } k = 1, 2 \text{ and } i = 1, \ldots, d.$$

since then $U_i^1(:, x_i, :) = \begin{pmatrix} x_i^2 & x_i^2 \\ 1 & 1 \end{pmatrix}$ as well as

$$U_1^1(x_1, :) \cdot U_2^2(:, x_2, :) = \begin{pmatrix} 1 & x_1^2 \end{pmatrix} \cdot \begin{pmatrix} x_2^2 & x_2^2 \\ 1 & 1 \end{pmatrix} = \begin{pmatrix} x_1^2 + x_2^2 & x_1^2 + x_2^2 \end{pmatrix}$$

and, thus,

$$p_1(x_1, \ldots, x_d) = U_1^1(x_1, :) \cdot U_2^1(:, x_2, :) \cdots U_d^1(:, x_d, :) = \begin{pmatrix} \sum_{i=1}^d x_i^2 & \sum_{i=1}^d x_i^2 \end{pmatrix}.$$

Taking p_2 similar as in the proof of Proposition 3.1 with cores

$$\hat{U}_1^2 = \begin{pmatrix} 0 \\ 1 \end{pmatrix} \text{ and } \hat{U}_2^2 = \begin{pmatrix} 0 \\ 1 \end{pmatrix}$$

we get $p_2(x, y) = U_1^2(x, :) \cdot U_2^2(:, y, :) = xy$ and thus

$$p_2 \circ p_1(x_1, \cdots, x_d) = (\sum_{i=1}^d x_i^2)^2.$$

Remark 3.3 (Potential fields) In many cases, we are interested in the case where $u \in \mathcal{T}_r^d$ is a gradient field for a scalar valued Lipschitz-continuous $\varphi : \mathbb{R}^d \to \mathbb{R}$. Then $\varphi \in \mathcal{T}_r$ is called a *potential* of

$$u = \nabla_x \varphi.$$

Relation to Deep Neural Networks

Let σ be an activation function for a single layer neural network

$$f_i(x) = \sigma(\sum_{j=1}^n A_{i,j} x_j + b_i).$$

Since the linear operator $x \mapsto \sum_{j=1}^n A_{i,j} x_j$ is Laplace-like, it can be rewritten by a tensor train of multi-rank $n + 1$. If we set

$$y_i = \sum_{j=1}^n A_{i,j} x_j + b_i$$

and $v_i(y) := \sigma(y)$ then $f = v \circ u$ is a compositional tensor train with two layer. Accordingly a deep neural network with L layers can be written as a compositional tensor train with $2L$ layers. Hence, a deep neural network is a special compositional tensor train.

We can also extend the concept of DNN, by replacing the affine transformations

$$y_i = \sum_{j=1}^{n} A_{i,j} x_j + b_i$$

by polynomial maps $y_i := \Phi(\hat{U}_i, x_i)$. We may call this form *Deep Neural Tensor Trains* or *Deep Neural Tensor Networks*.

Since this transformation is in non-linear the use of activation functions is no longer essential and a default choice can be the trivial activation function $\sigma(x) = x$.

As a consequence the fundamental back-propagation algorithm can be applied also to compositional tensor networks with straight forward modifications.

Deep versus Shallow

Let us remark that the use of compositions can be a crucial tool in order to reduce the complexity in various circumstances. For example, if we want to apply inequality constraints e.g. $f(x) \geqslant 0$, for all $x \in \mathbb{R}^d$ and f has a data sparse representation, then the application $\tilde{f}(x) = (f(x))_+ = \sigma_{ReLU} \circ \tilde{f}(x)$ provides the projection onto the non-negative real cone and might be hard to approximate by shallow decomposition unless the ranks are chosen sufficiently high. However, f has a lower data complexity than $\tilde{f}$. Also other functions, quite regular functions that come to mind, for example the exponential function, do suffer from high TT ranks if approximated directly. On the other hand, the exponential in particular comes with an inbuilt compositional structure induced by its defining ODE. These situations appears frequently in applications. In this respect the *Deep Neural Tensor Networks* provide an interesting framework.

Observe that the power of expressivity of compositional tensor network is always higher compared to a shallow tensor network and also the deep neural networks.

However, we will not examine in detail computational aspects like the stability of optimization procedures. Since the expressivity of single tensor networks is impressively high and we are sacrifying benefits of single (shallow) tensor networks when emplying deeper structures. Hence, the decision shallow versus deep may depend on actual applications.

4 Flow Induced Tensor Trains - Flow of ODEs

In a very philosophical sense, all real variable computations on a physical machine are reduced to the basic algebraic operation as addition and multiplication and, thus, they are already compositional tensor networks. This implies that the present framework is in a certain extent too general. Therefore, we focus on particular network architectures, i.e. flow driven networks, which are considered in neural ODE techniques. This will allow the use of well-known tools from numerical analysis to understand the complexity and expressibility of these networks.

We will draw some inspiration from the neural ODE concept popularized in [8] in order to understand certain residual compositional structures from a continuous perspective.

4.1 Dynamical Systems and Flow

In order to define a parameterized transformation of coordinates, let us consider the initial value problem (IVP) for an ODE with Lipschitz vector field $u : \mathbb{R}^d \to \mathbb{R}^d$ and defining equations

$$\begin{aligned} \dot{x}(t) &= u(t, x(t)) \\ x(0) &= x \ , \ x \in \mathbb{R}^d . \end{aligned}$$

Let us define the *flow* Φ at time $t \in [0, T]$ as a map

$$x \mapsto \Phi_t(x) := x(t) \quad , \quad \Phi_0(x) = x \ ,$$

i.e. $\Phi_t \in C(\mathbb{R}^d, \mathbb{R}^d)$ or in other words $\Phi \in C^0([0, T], C(\mathbb{R}^d, \mathbb{R}^d))$. In particular, the flow map or transport map can be seen as a nonlinear coordinate transform.

Remark 4.1 Under Lipschitz regularity conditions on u this map is invertible (locally in time) and the the inverse Φ^{-1} is given by the backward flow. Furthermore, observe that the flow defines a (semi-) group

$$\Phi_{t_1+t_2} = \Phi_{t_2} \circ \Phi_{t_1} .$$

Remark 4.2 Not all transformations $\Psi : \mathbb{R}^d \to \mathbb{R}^d$ can be defined as a flow of a dynamical system.

Now let us parameterize the vector field u by tensor train fields, i.e. we represent a (velocity) field $u : [0, T] \times \mathbb{R}^d \to \mathbb{R}^d$, $u(t, \cdot) \in \mathcal{T}_r^d$ by

$$u_i(t, x) = \sum_{k_j \leqslant r_j, j=1,\dots,d-1} U_1(t, k_1, x_1) U_2(t, k_1, x_2, k_2) \cdots U_d(t, k_{d-1}, x_d, i).$$

Assumptions 4.1. *Let $\mathcal{P}_{\leqslant n_i} := span\{\phi_j : j = 1, \dots, n_i\}$ consists of (locally) Lipschitz functions and $U_i(t, k_{i-1}, x_i, k_i) = \sum_{j=1}^{n_i} \hat{U}_i(t, k_{i-1}, j, k_i)\phi_j(x_i)$. We assume for every $x_i \in \mathbb{R}$*

$$t \mapsto U_i(t, \cdot, x_i, \cdot) \in [C^2([0, T])]^{r_{i-1}, r_i}$$

and for all $t \in [0, T]$ we assume

$$x_i \longrightarrow U_i(t,\cdot,x_i,\cdot) \in \mathcal{P}^{r_{i-1},r_i}_{\leqslant n_i}$$

Remark 4.3 In order to ensure well-definedness one needs to either enforce global Lipschitz continuity within the choice of basis ϕ_j or one tries to expand the maximal time intervall of well-definedness by restricting the size of the cores e.g. $U_i(t,:,x_i,:)$ and thus minimizing the local Lipschitz constants.

Time Discretization

If one now discretizes the ODE in time an important modelling choice has to be made, as one either can use a new core $U_0(t,:)$ for a given basis set in time $\{\{t \mapsto \phi_{k_t}(t) : k_t = 1,\ldots,n_t\}\}$ or one can choose the cores to be time dependent, i.e. $U_i = U_i(t,k_{i-1},x_i,k_i)$. The last ansatz can for example lead to a sequence of cores if one assumes piecewise constant time dependence of the cores.

With this choices at hand the numerical approximation of the IVP

$$\dot{x} = u, \qquad x(0) := x$$

e.g. by *forward explicit Euler scheme*

$$x^{k+1} = x^k + \tau u_k(t_k, x^k) =: \tilde{\Phi}_k(x^k)$$

leads naturally to a CTT representation

$$\Phi_T(x) = \tilde{\Phi}_k \circ \cdots \circ \tilde{\Phi}_0(x).$$

Remark 4.4 Notice that as $x + \tau u(t_k, x)$ can be represented as TT vector with rank $d + 1 + \mathrm{rank}(u)$.

This is a special case architecture of compositional network knows as ResNets in DNN terminology. For $t \mapsto \hat{u}(t,\cdot)$ continuouly differentiable, and given tolerance $\epsilon > 0$ one needs $L \sim T\tau^{-1}$ layers when one choses $\tau \sim \epsilon$ to achieve the target accuracy. Notice that that the compositional network provides a global representation for all $x \in \Omega$, within a complexity $\sharp\Phi_T = \sum_{j=1,\ldots,L} \sharp\Phi_j \leqslant L \max_{j=1,\ldots,L} \sharp\Phi_j$ which scales linearly w.r.t. L and ϵ^{-1} independent of the spatial dimension of the state space $\mathbb{R}^d$. In particular, also the Picard-iteration and even the Newton iteration can be brought into this form, which then in turn are used for theoretical results, cf. compare [11].

Remark 4.5 This procedure can be used for a tractability analysis of of solving high-dimensional PDEs by compositional networks and may be applied to theoretical analysis of operator learning. It is based on the duality of local and global perspectives. Indeed, the tractability for computing the *global function* $x \mapsto \Phi_T(x)$

or $x \mapsto f(x)$, is shown by considering the *local* or microscopic problem: for given $x \in \Omega$ compute the vector $\Phi_T(x)$ or the value $f(x)$.

5 Relation to Parametric Partial Differential Equations (PDEs)

In the following we provide some thoughts on why compositional models, in particular ODE based models, can be beneficial to a vast variety of scientific computation problems. Foremost, will consider the case of elliptic and transport dominated PDEs.

Let us keep also in mind that many basic iteration scheme have inherently compositional structure and take the form (in the microscopic perspective, i.e. for a single x_k):

$$\begin{aligned} x^{k+1} &= C_k x^k + \tau u_k(t_k, x^k) =: \tilde{\Phi}_k(x^k) \\ &= \tilde{\Phi}_k(x_k) \end{aligned}$$

The contraction principle provides precise convergence estimates for all $x \in \Omega$. The resulting mapping $x \mapsto \Phi_N(x)$ provides a global (macroscopic) of the solution.

5.1 Elliptic equations with Parameterized Diffusion Field

Let us consider the discretization of a parametric diffusion equation well known in uncertainty quantification. Let

$$\kappa(x, y) := \kappa_0 + \sum_{k=1}^{d} \psi_k(x) y_k \ , \ y_k \in [a, b]$$

be uniformly elliptic, i.e.

$$0 < \gamma \leqslant \kappa(x, y) \leqslant \Gamma \ \text{ for almost all } x \in D \subset \mathbb{R}^3, y \in [a, b]^d$$

and let $f \in L_2(D)$. Define the operator

$$(A(y)u)(v) := -\langle \kappa(x, y) \nabla u, \nabla_x v \rangle$$

and consider the problem

$$A(y)u = f, \quad u = 0 \text{ on the boundary } \partial D,$$

with Galerkin (stiffness) matrix

$$\mathbf{A}_{i,j} := \langle \nabla_x \varphi_i, \kappa(x, y) \nabla \varphi_j \rangle$$

for an appropriate basis $\{\varphi_i : 1, \ldots, n\}$. Then the problem reduces to

$$\mathbf{A}(y)\mathbf{u}(y) = \mathbf{f}$$

which can be solved iteratively by

$$\mathbf{u}^{n+1}(y) = \mathbf{A}_0^{-1}(\mathbf{A}(y)\mathbf{u}^{(n)}(y) - \mathbf{f}) \tag{1}$$

with an appropriate preconditioner $\mathbf{A}_0^{-1}$ not depending on y and the maximal number of iterations $N = N(\epsilon)$ is chosen to guarantee an accuracy $\epsilon > 0$. Obviously the iteration scheme (1) is realized by a relatively simple compositional tensor train, where the complexity $\sharp\Phi^n \lesssim dn$ or $\lesssim dn^2$ depends linearly on d. Notice that for fixed parameter $y \in \Omega$ this is nothing but solving an elliptic PDE with fixed discretization, where me may assume an accuracy ϵ. Note that the convergence of the iteration scheme is linear which means the $\epsilon \leqslant e^{-cN}$ decays exponentially w.r.t. to the number of iterations. However, this rate is not required to obtain complexity estimates without a curse of dimensions. We obtain a final scaling of required data at most $\leqslant O(n^2(\epsilon)d \log \epsilon)$, if the stiffness matrices $\mathbf{A}(y)$ are considered as fully populated. Observe that in case of reduced basis functions n is relatively small. However, for full FEM systems where we have sparse matrices we obtain $\leqslant O(n(\epsilon)d \log \epsilon)$ with a much larger n. Let us keep in mind that the present compositional network is given a priori. Since it reproduces the results for computing $u(y)$ for a single parameter y with no benefit it is used only for theoretical purpose providing a rigorous upper bound for the complexity demonstrating the duality of local (microscopic) versus global, or view. This type of analysis has been used for DNN in [29, 37].

5.2 Transport Equation and Method of Characteristics

Let us now consider pure transport equations with regular vector fields but potentially irregular initial data or source terms. Then we can use the methods of characteristics to obtain solutions. To be precise, let $\Psi_t(x) := x(t)$, defined by

$$\dot{x} = u \ , \ x(0) = x$$

be the flow of the dynamical system, where $u(t, x) = \hat{u}(\hat{U}, t, x)$.

Then, given an initial distribution f_0, the function

$$f(t, x) := f_0(x(t)) = f_0 \circ \Psi_{\hat{U},t}(x) : \mathbb{R}^d \to \mathbb{R}$$

satisfies the *transport equation*

$$\partial_t f + \langle u, \nabla_x f \rangle = 0 \ , \ f(0, x) = f_0(x).$$

In a similar fashion we can solve solve the inhomogeneous transport equation

$$\partial_t f + \langle u, \nabla_x f \rangle = h(t,x) \ , \quad f(0,x) = f_0(x)$$

by defining

$$f(t, \Psi_t(x)) = f_0(x) + \int_0^t h(s, \Psi_s(x))ds.$$

Although the procedure above is nothing but the well known integration w.r.t. to characteristic curves, one can use this philosophy to derive the compositional network, e.g., by simple explicit Euler discretization with an appropriately small time stepping $\tau \to 0$ or Picard iterations to achieve any desired accuracy ϵ, [11]. One could say that each characteristic provides a local (microscopic) perspective by fixing an initial state $x \in \Omega$ while the flow Ψ and the induced composition operator $\mathcal{K}_t v(x) = v \circ \Psi_t(x) = v(x(t))$, $0 \leqslant t \leqslant T$, together with a simple explicit ODE discretization (e.g. forward Euler) , provide a more global perspective.

Remark 5.1 The method of characteristics is well known for systems of hyperbolic partial differential equations, where one often meets the difficult case of crossing characteristics. For the sake of simplicity let us assume that this does not happen in the present situation.

Lagrangian perspective

The notion of a flow of an ODE allows us to introduce and compare the concept of Eulerian and Lagrangian coordinates. Let us assume that the flow is invertible, e.g. if the characteristics are not crossing. Then the flow provides a coordinate transformation. Let $e_1, \ldots, e_d$ be a orthonormal basis in $\mathbb{R}^d$ and define a time-dependent basis $\xi_1(t), \ldots, \xi_d(t)$ via $\xi_i(t) = \Phi_t(e_i)$.

Let $x \in \Omega \subset \mathbb{R}^d$ be an initial point, often called state or abstract particle. Let x_i be its coordinates w.r.t e_i, i.e. $x = \sum_{i=1}^d x_i e_i$. If the particle is moved according to the ODE inducing the flow Φ we have that its coordinates w.r.t e_i, called Eulerian, are $x_i(t)$, while its coordinates w.r.t $\xi_i(t)$, called Lagrangian or material, are constant, i.e. $x(t) = \sum_{i=1}^d x_i \xi_i(t)$.

The change of variables w.r.t. the flow Φ_t induces the (linear) operator

$$\mathcal{K}_t f := f \circ \Phi_t$$

which is called *Koopman* or composition operator. The differential operator

$$f \mapsto \partial_t f + \langle \nabla f, u \rangle$$

is the corresponding infinitesimal generator of the semi-group $\mathcal{K}_t$. If we can move back and forth changing between Eulerian and Lagrangian coordinates, both approaches are completely equivalent. Indeed the Lagrangian perspective fits perfectly with the present concept of compositional tensor networks.

We would like to mention that the method of characteristics, and thus the Lagrangian approach, can also be used to solve (some) non-linear hyperbolic equations by fix point iteration.

Stochastic Processes

For some optimal control settings, the above strategy works also to some extent if we replace the deterministic dynamical system by a simple stochastic one, e.g.

$$dx_t = u(x_t)dt + \sigma dW_t, \quad \text{with some } \sigma > 0, \tag{2}$$

where dW_t is a Brownian motion.

The non-linear Feynman Kac theorem (cf. [7]) implies that some PDEs of the form

$$\partial_t f + \langle v, \nabla_x f \rangle + \sigma \Delta f = H(t, x, f, \nabla_x f)\ , \quad f(T, x) = f_T(x)$$

can be solved by forward-backwards SDEs related to (2) for which the convergence of the Picard iteration has been proved, [14]. This, in turn, provides upper bounds for the solution complexity by deep neural networks. These results apply in an straightforward fashion also for the present compositional tensor networks.

Although we confine ourselves to deterministic problems, the treatment of corresponding stochastic problems might be advantageous as the additional Laplacian in the PDE implies some additional regularity, which results for many non-linear PDEs in existence and approximability proofs.

Let us highlight that the previous chapter about tractability is only for theoretical purpose, since the tensor network is given a priori. It provides a global solution in a relatively large domain $\Omega \subset \mathbb{R}^d$, where all the issues discussed above are well defined. It is questionable whether this complexity can be reduced further in this uniform global setting. However, in practice one is often interested in smaller domain, or different subdomains are of different importance. This can be handled by introducing a weight function or probability density.

6 Computing flow driven tensor models

The next questions is how to efficiently recover a compositional tensor network from data, when the underlying model is unknown but suspected to exhibit a computational feasible composition structure. In practice the underlying probability distribution is unknown and one needs to resort to empirical density approximations as common in machine learning.

Let us highlight that optimization or training procedure is a typical *off-line* task. Here computational resources, e.g. hardware and computing time, can be extremely large and the focus is more on accuracy and reliability. However, after learning a surrogate Φ the evaluation given x and compute $\Phi(x)$ in the on-line phase needs to be done in resource efficient ways.

To understand the optimization of the compositional tensor networks, we will formulate it as a regression task and its empirical analog and point to the connection to semi-global optimal control problems that have recently gained substantial attraction.

6.1 Regression Problem for an Unknown Flow

Let us consider the task of approximating a mapping

$$\Phi : \mathbb{R}^d \to \mathbb{R}^d, x \mapsto \Phi(x).$$

Ideally, we would like to minimize

$$j_T(u, x) = \|\Phi(x) - \Psi_T(x)\|^2$$

for all $x \in \Omega \subset \mathbb{R}^d$ for some flow Ψ parameterized by a vector field u.

However, pointwise minimizations are not feasible in high-dimensions and we will relax the problem to minimizing an averaged cost functional

$$J_T(u, \rho) = \int_\Omega j_T(u, x) d\rho(x) = \int_\Omega \|\Phi(x) - \Psi_T(x)\|^2 d\rho(x) \tag{3}$$

for some measure ρ.

Notice that the above integration, possibly in high-dimensions, poses a significant numerical obstruction. Thus, we have to replace the above functional by an empirical functional

$$\mathbf{J}_T(u) := \frac{1}{N}\sum_{i=1}^{N} \|\Phi(x^i) - \Psi_T(x^i)\|^2 = \frac{1}{N}\sum_{i=1}^{N} \|y^i - \Psi_T(x^i)\|^2$$

where x^i are initial data of the ODE distributed according to ρ and $y^i = \Phi(x^i)$ are the pointwise evaluations of the target function. Notice that in many applications ρ is unknown. In order to obtain a minimizer we restrict ourselves to finite velocities $\|u\| \leqslant \Gamma < \infty$ by introducing a fixed regularization parameter $\lambda > 0$ and formulate the Banach space valued optimal control problem

$$\min_{\hat{U}_t \in C^1([0,T], \mathcal{T}^d_{\leqslant \hat{r}})} \mathcal{J}(U_t) = \frac{\lambda}{2} \int_0^T \int_\Omega \int_0^T \|u(\hat{U}_t, \Psi_t(x))\|^2 \, \mathrm{d}t \; d\rho(x) + J_T(u, \rho)$$

subject to

$$\frac{d}{dt}\Psi_t(x) = u(\Psi_t(x)) \,, \;\; \Psi_0(x) := x \,, \;\; u = \hat{u}(\hat{U}_t, \cdot).$$

Remark 6.1 In optimal transport the map Φ is called a transport map and is not known a priorily. When developing a machinery with flow driven networks we would like to incorporate the task of optimal transport into the present framework. Indeed it may provide a versatile tool for generative modeling, i.e. to create samples out of a probability density only known from given sampels.

6.2 Variational Monte Carlo Method

As we have observed earlier we need to replace the global functional (3) by an empirical functional, thus, let us have a closer look to so called *Variational Monte Carlo* methods.

In a fairly abstract setting, we want to minimize the global cost functional

$$\mathcal{J}(w) = \int j(x, w(x))dx,$$

i.e. we are looking for

$$u = \operatorname{argmin}\{\mathcal{J}(w) : w \in \mathcal{V}\}$$

on an appropriate Banach space $\mathcal{V}$.

In order to minimize this target functional $\mathcal{J} : \mathcal{V} \to \mathbb{R}$ numerically, we consider a (family of) model space

$$\mathcal{M} \subset \mathcal{V} \cap C_b^0(\mathbb{R}^M) \ , \ \mathcal{M} \subset\subset \mathcal{V} \text{ (compact)}$$

here,

$$\mathcal{M} := \{\psi \in \mathcal{V} : \psi = \psi(U, \cdot) \ , \ U \in \Pi \subset \mathbb{R}^p \text{ (compact) }\}$$

We call Π a finite real parameter set. This ansatz is valid for linear and non-linear approaches. For example, taking a basis set ϕ_k we can define

$$\mathcal{M} = \{\psi(U, x) = \sum_{k=1}^{p} U_k \varphi_k(x) \ , \ a \leqslant U_k \leqslant b\}$$

$(a, b$ sufficiently large). Here,

$$(U, x) \mapsto \psi(U, x)$$

is non-linear in x, but linear in U. However, imposing non-linear relations on the parameters U_k can identify non-linear subsets of $\mathcal{M}$. Hence, in contrast to classical Ritz Galerkin methods our model space is not a linear space, it is rather a compact metric space imbedded in some $\mathbb{R}^p$ where p is at least the number of parameters of our model.

As indicated before many target functionals are integrals and might be difficult to evaluate. Hence, we replace the functional by a computable *surrogate functional*

$$u_N := \operatorname{argmin} \{\mathcal{J}_{emp}(w) : w \in \mathcal{M}\} \ .$$

where the functional is approximated by numerical quadrature, e.g. simply standard Monte Carlo quadrature

$$\mathcal{J}_{emp}(w) := \frac{1}{N}\sum_{i=1}^{N} J(x^i, w(x^i)) \text{ where } x^i \text{ are sampled i.i.d. from } \rho_0.$$

Quasi-Monte Carlo provides an alternative possible, but it is not considered yet.

Remark 6.2 For TT methods, due to the tensor product structure, the Cross TT approximation can be used for numerical quadrature. This seems to improves the quality of the quadrature significantly. These methods are available yet only for functional tensor trains (TT). We keep cross TT approximation and related methods in mind, due to their potential for further improvements.

Remark 6.3 Since we are trying to minimize a surrogate functional instead of the exact functional we are committing variational crimes. The error can be decomposed into an approximation error originated by our model space and an additional consistency error. In Finite Element analysis this consistency is estimated by well known Strang-Fix lemmas.

In statistical learning theory there exists the theory of *empirical risk minimization* (our present functional is not a risk), here the consistency error is called *generalization error* and only estimates with high probability are provided. First results have been established by Chernovenkis and Vapnik [53, 54], a good reference to closer link to approximation and complexity theory is [10]. For tensor product approximation the theory is still in its infancy, starting with [15] progress has been obtained by A. Nouy et al, [52].

This generalization error plays a crucial role in machine learning where the above results are known to be notoriously pessimistic. Surprisingly, the experience using deep neural networks completely contradict the above theories. Compared to a relatively small amount of data the observed generalization error was spectacularly small, if stochastic gradient optimization is applied. Classical overfitting phenomena do not necessarily appear. This miracle has triggered strong theoretical efforts mostly in connection with an analysis of stochastic gradient method.

We believe that the accuracy of stochastic gradient is limited, and it does not provide an accuracy we need for *operator learning* in engineering applications.

6.3 Variational Monte Carlo (VMC) in the Optimal Control Setting

Now let us elude how the VMC concepts can be gainfully applied to our optimal control setting that mimics the regression with a compositional approximation structure.

Let N samples x^i be sampled from a measure ρ, then the empirical cost functional

$$\mathbb{J}_T(U_t) = \sum_{i=1}^{N} \frac{1}{N} \left(\int_0^T \|\hat{u}(U_t, \Psi_t(x^i))\|^2 dt + \|y^i - \Psi_T(x^i)\| \right)$$

$$s.t. \frac{d}{dt}\Psi_t(x^i) = \hat{u}(U_t, \Psi_t(x^i), \quad \Psi_0(x^i) = x^i$$

constitutes a classical optimal control problem, [32, 33, 34, 35, 43].

The Pontryagin Maximum Principle (PMP) gives as necessary conditions for a minimal control

$$\frac{d}{dt}\Psi_t(x^i) = \hat{u}(\hat{U}_t, \Psi_t(x^i), \qquad \Psi_0(x^i) = x^i$$

$$\frac{d}{dt}\lambda_i(t) = -\partial_y H(t, \Psi_t(x^i), \hat{U}_t, \lambda_i(t)) \qquad \lambda_i(T) = 2(y^i - \Psi_T(x^i))$$

$$\hat{U}_t(t) \in \arg\min_{\tilde{U} \in \mathcal{T}^d_{\leqslant r}} \frac{1}{N} \sum_{i=1}^{N} H(t, \Psi_t(x^i), \tilde{U}, \lambda_i)$$

where $H(t, \Psi_t(x^i), \hat{U}, \lambda_i) = \|\hat{U}\|^2 + \lambda_i^T \hat{u}(\hat{U}, \Psi_t(x^i)$, For the exact equation we refer to [32, 33, 34, 35, 34].

The optimality system is solved by gradient like methods commonly known in optimal control as open loop methods based on the Pontryagin Maximum Principle. For example, we can use the following iteration procedure. Assume we are having a guess $\hat{U}_t = \hat{U}_t^{(n)}$ for the (velocity) field, or control.

- First, solve for each sample x^i the state equation

$$\frac{d}{dt}\Psi_t^{n+1}(x^i) = \hat{u}(\hat{U}_t, \Psi_t^{n+1}(x^i)), \quad \Psi_0^{n+1}(x^i) = x^i.$$

- Then solve the adjoint problem with the new flow $\Psi_t^{(n+1)}$ by forward Euler scheme backwards in time. Let us note that this is a linear terminal condition problem.
- Update the new control $\hat{U}^{(n+1)}$ according the anti-gradient, which is given by the defect of the optimality condition numerically.

Let us remark that the control $u = \hat{u}(\hat{U}_t, x)$ is a globally defined function, however, in the first two steps we handle only finitely many function values, e.g $\Psi_t^{(n+1)}(x^i)$.

Semi-Global Optimal Control as Lagrangian Perspective on the Hamilton Jacobi Bellmann equation

One can observe that the control problem introduced here is a special case of optimal feedback synthesis and can be viewed as a Lagrangian treatment as opposed to the more classical Hamilton-Jacobi-Bellman approach that is closer to the Eulerian perspective.

Classically, in order to obtain an optimal feedback law for an optimal control problem with cost functional

$$J(t_0, x, u) := \int_0^T \ell(t, \Psi_t(x)) + |u(t)|^2 \,\mathrm{d}t + j_T(\Psi_T(x)). \tag{4}$$

restricted to

$$\frac{d}{dt}\Psi_t(x) = f(t, \Psi_t(x)) + g(t, \Psi(x))u(t), \quad \Psi_0(x) = x \tag{5}$$

one can exploit that the optimal feedback control is given by

$$\alpha^*(t, x) = -\frac{1}{2}g(t, x)^T \nabla v^*(t, x), \tag{6}$$

where the function $v(t, x) := \min_{u \in L^\infty([0,T],U)} J(t, x, u)$ obeys the inhomogeneous, non linear transport equation

$$\frac{\partial}{\partial t}v(t, x) + \min_{u \in U}\Big[- \nabla v(t, x) \cdot (f(t, x) + g(t, x)u) - c(t, x, u)\Big] = 0. \tag{7}$$

The Semi-Global ansatz is closer to the Lagrangian setting as its trying to find an optimally moving basis controlled by a family of vector fields. If the family of vector fields in the Semi Global approach is general enough the Lagrangian and Eulerian perspective are equivalent again. Observe, that the Lagrangian ansatz allows the use of particle methods which play an important role in very high-dimensional optimization problems.

7 Relation of the Regression Problem to Mean Field Games and Optimal Transport

Lastly, we want to embed the framework of semi-global optimal control and the regression tasks of ODE based models into the notion of mean field games and optimal transport, as the regression will be a subclass of these problem formulations.

7.1 Relation to Mean Field Games

The Semi Global Optimal Control problem is a special case of Mean Field games, [3]. This can be seen by considering the global functional

$$\mathcal{J}(\rho, u) = = \int_{\mathbb{R}^d} \left[\int_0^T (\ell(t, x) + \lambda|u(t, x)|^2)\rho(t, x)dt + j_T(x)\rho(T, x) \right] dx$$

with an unknown probability density ρ constraint to the continuity equation

$$\partial_t \rho + \mathrm{div}([f + gu]\rho) = 0 \ \text{ in } (0,1) \times \mathbb{R}^d$$

with initial conditions

$$\rho(0,x) = \rho_0(x) \ \geqslant 0.$$

Semi-global solutions can be obtained by restricting the support of $\rho_0 = \rho(0,\cdot)$ to some $\Omega \subset \mathbb{R}^d$ and using a change of variable formula as well as the method of characteristics. In the present setting there are no further boundary conditions required.

According to [38], this can be generalized to the following problem. For sake of simplicity let us assume $f = 0$, $g = I$ and introduce new functions $(t,x,\rho) \mapsto \mathcal{A}(t,x,\rho)$ and $(x,\rho) \mapsto \mathcal{G}_T(x,\rho)$. We want to optimize the functional

$$\mathcal{J}(\rho,u) = \int_{\mathbb{R}^n} \int_0^T \left(\lambda |u(t,x)|^2)\rho(t,x) + \mathcal{A}(t,x,\rho(t,x))\right) dxdt + \\ + \int_{\mathbb{R}^n} \mathcal{G}_T(x,\rho(T,x))dx$$

w.r.t. the continuity equation for the density $\rho(t,x)$

$$\partial_t \rho(t,x) + \ \mathrm{div}(u\rho)(t,x) = 0.$$

$$\rho(0,x) = \rho_0(x)$$

Setting up the formal Lagrangian, we obtain by differential calculus the first order conditions, ie. the *mean field equations* with $A := -\partial_\rho \mathcal{A}(\rho),\ G := -\partial_\rho \mathcal{G}(\rho)$

$$\begin{aligned} \partial_t \phi(t,x) - H(\nabla\phi) &= A(\rho(t,x)), \\ \phi(T,x) &= G(\rho(T,x)), \\ \partial_t \rho(t,x) + \ \mathrm{div}\ \rho \nabla H(\nabla\Phi)(t,x) &= 0\,, \\ \rho(0,x) &= \rho_0(x) \end{aligned}$$

where in this case the *Hamilton function* $p \mapsto H(p)$ is given by

$$\mathbb{R}^m \to \mathbb{R}\,,\ p \mapsto H(p) := \frac{1}{2\lambda}|p|^2\,.$$

For a stochastic counterpart the continuity equation has to be replaced by the Fokker Planck equation, and the deterministic HJB by the stochastic variant by adding the viscosity term.

Remark 7.1 Remark: For the optimal control problem, the continuity equation is almost completely decoupled from the HJB. There we choose any (appropriate) initial density ρ_0, the solution of the HJB does not depend on the density, whenever $\rho(t,x) > 0$. The above mean field equation (for deterministic optimal control) are more or less equivalent to the original optimal control problem. Perhaps it enlightens the nature of this problem from a more global perspective and also from a more abstract and mathematical point of view.

7.2 Relation to Optimal transport

The mean field formulation provides also an alternative formulation of the optimal transport problem. This formulation is often called the hydro-dynamic view of optimal transport or Bennamou-Brennier formulation. The only difference to standard MFGs is that the target density $\rho_1(x) = \rho(T, x)$ appear at a second boundary condition for the continuity equation

$$\mathcal{J}(\rho_1, \rho_2, \rho, u) := \int_{\mathbb{R}^d} [\int_0^T |u(t,x)|^2 \rho(t,x))dt]dx$$

restricted to

$$\partial_t \rho + \text{div } u\rho = 0, \quad \rho(0,\cdot) = \rho_0, \quad \rho(1,\cdot) = \rho_2.$$

It turns out, that

$$\min_{\rho,u} \mathcal{J}(\rho_1, \rho_2, \rho, u) = W_2^2(\rho_1, \rho_2),$$

i.e. exactly the Wasserstein distance. Taking the quadratic Hamiltonian $H(p) = \frac{1}{2}\|p\|^2$, we obtain the following optimality system

$$\begin{aligned} -\partial_t \phi + H(\nabla_x \phi) &= 0, \\ \partial_t \rho + div(\rho \nabla H(\nabla \phi)) &= 0, \\ \rho(0,\cdot) = \rho_0 \quad \rho(T,\cdot) &= \rho_T. \end{aligned}$$

The first two equations are the equations of Mean Field Games with vanishing source terms. The difference between semi global optimal control and transport is in the terminal conditions.

Solving the continuity equation with a Lagrangian perspective

In order to compute the density we will take a more Lagrangian perspective again as we can compute the pointwise time evolution of the density via the *Jacobi identity*:

For ρ_0 regular enough we have

$$\rho(y(t,x),t)\det\nabla_x\Psi_t(x) = \rho_0(x).$$

Furthermore, by denoting $s(t,x) = \log(\det \nabla_x \Psi_t(x))$ we have

$$\frac{d}{dt}s(t,x) = \operatorname{div}(u(\Psi_t(x),t))$$

and thus we can recover ρ by means of methods of characteristics, c.f. [41, 47].

8 Concluding Remarks

We have considered vector valued (non-linear) *functional tensor networks*, e.g. *functional tensor trains*. We have combined them with the compositions of non-linear functions in a similar way as in Deep Neural Networks, and will call them *Compositional Tensor Networks* or *Deep Tensor Networks*. The class of functions which admit a data-sparse representation is much larger than the class of single layer tensor networks. It contains also all deep neural networks. Furthermore, the standard optimization tools as back propagation can be used in this more general class as well.

We have considered the compositions of functions as an additional and powerful tool for approximation, and refer to recent development about *compositional sparsity*. We do not know how large the present classes are, but we have provided examples, mostly high dimensional PDEs where the curse of dimensionality can be avoided. We focused on flow driven models, e.g. normalizing flow, and provide a control based method and further methods for computing corresponding tensor trains. We have established a relationship between deterministic HJB equation and (variational) Mean Field Games including (a hydro-dynamic view of) optimal transport.

We have to admit that in contrast to a simple stochastic setting, the tractibility theory of HJB and MFG remains still open. The stochastic HJB with Brownian motion might be tractable, which has been shown by [20]. However, in real world applications, it is not the primary interest to solve the mean field equation with high ultimately accuracy. The primary goal is to treat the optimization problem appropriately. For examples, highly complex feed back laws (control) can be impractical for fast online computations. Although the approximation of viscosity solutions is still a challenging and highly interesting problem, one has to keep the requirements of real world applications in mind.

Acknowledgements

Mathias Oster acknowledges funding by the Deutsche Forschungsgemeinschaft (DFG, German Research Foundation) - Project number 442047500 through the Collaborative Research Center "Sparsity and Singular Structures" (SFB 1481 -B01).

References

1. M. Ali and A. Nouy. Approximation theory of tree tensor networks: Tensorized multivariate functions, 2023.
2. M. Ali and A. Nouy. Approximation theory of tree tensor networks: Tensorized univariate functions. *Constructive Approximation*, 58(2):463–544, 2023.
3. M. Annunziato, A. Borzì, F. Nobile, R.Tempone. On the connection between the Hamilton-Jacobi-Bellman and the fokker-planck control frameworks. *Applied Mathematics*, 5(16):2476, 2014.
4. M. Bachmayr, A. Nouy, and R. Schneider. Approximation by tree tensor networks in high dimensions: Sobolev and compositional functions. *Pure and Applied Functional Analysis*, 8(2):405–428, 2023.
5. M. Bachmayr. Low-rank tensor methods for partial differential equations. *Acta Numerica*, 32:1-121, 2023.
6. M. Bachmayr, R. Schneider, and A. Uschmajew. Tensor networks and hierarchical tensors for the solution of high-dimensional partial differential equations. *Found. Comput. Math.*, 16(6):1423–1472, December 2016.
7. C. Beck, S. Becker, P. Grohs, N. Jaafari, and A. Jentzen. Solving the Kolmogorov PDE by Means of Deep Learning, *J Sci Comput*, 88(3):73, 2018.
8. R. T. Q. Chen, Y. Rubanova, J. Bettencourt, and D. Duvenaud. Neural ordinary differential equations. In *Advances in Neural Information Processing Systems 31 (NeurIPS 2018)*, NIPS'18, page 6572–6583, Red Hook, NY, USA, 2018. Curran Associates Inc.
9. N. Cohen, O. Sharir, and A. Shashua. On the expressive power of deep learning: A tensor analysis. In *Annual Conference Computational Learning Theory*, 2015.
10. F. Cucker and S. Smale. On the mathematical foundations of learning. *Bulletin of the American mathematical society*, 39(1):1–49, 2002.
11. W. Dahmen. Compositional sparsity, approximation classes, and parametric transport equations, 2022.
12. R. A. DeVore and V. N. Temlyakov. Some remarks on greedy algorithms. *Advances in Computational Mathematics*, 5(1):173–187, 1996.
13. R. A. DeVore. Nonlinear approximation. *Acta Numerica*, 7:51–150, 1998.
14. W. E, J. Han, and A. Jentzen. Deep learning-based numerical methods for high-dimensional parabolic partial differential equations and backward stochastic differential equations. *Communications in Mathematics and Statistics*, 5(4):349–380, 2017.
15. M. Eigel, R. Schneider, and P. Trunschke. Convergence bounds for empirical nonlinear least-squares. *ESAIM: Mathematical Modelling and Numerical Analysis*, 56(1):79–104, 2022.
16. A. Falcó, W. Hackbusch, and A. Nouy. Geometry of tree-based tensor formats in tensor banach spaces. *Annali di Matematica Pura ed Applicata (1923 -)*, 202(5):2127–2144, 2023.
17. A. Gorodetsky, S. Karaman, and Y. Marzouk. High-dimensional stochastic optimal control using continuous tensor decompositions. *The International Journal of Robotics Research*, 37(2-3):340–377, 2018.
18. M. Griebel, H. Harbrecht, and R. Schneider. Low-rank approximation of continuous functions in Sobolev spaces with dominating mixed smoothness. *Mathematics of Computation*, 92(342):1729–1746, 2023. Also available as INS Preprint No. 2203.
19. M. Griebel, H. Harbrecht, and R. Schneider. Low-rank approximation of continuous functions in sobolev spaces with dominating mixed smoothness. *Mathematics of Computation*, 92(342):1729–1746, 2023.
20. P. Grohs and L. Herrmann. Deep neural network approximation for high-dimensional elliptic PDEs with boundary conditions. *IMA Journal of Numerical Analysis*, 42(3):2055–2082, 05 2021.
21. P. Grohs, F. Hornung, A. Jentzen, and R. Von Wurstemberger. *A proof that artificial neural networks overcome the curse of dimensionality in the numerical approximation of Black–Scholes partial differential equations*, volume 284. American Mathematical Society, 2023.

22. P. Grohs and F. Voigtlaender. Sobolev-type embeddings for neural network approximation spaces. *Constructive Approximation*, 57(2):579–599, 2023.
23. I. Gühring, G. Kutyniok, and P. Petersen. Error bounds for approximations with deep relu neural networks in ws,p norms. *Analysis and Applications*, 18(05):803–859, 2020.
24. W. Hackbusch. *Tensor Spaces and Numerical Tensor Calculus*, volume 42. Springer, 01 2012.
25. W. Hackbusch. Numerical tensor calculus. *Acta numerica*, 23:651–742, 2014.
26. W. Hackbusch and S. Kühn. A new scheme for the tensor representation. *Journal of Fourier Analysis and Applications*, 15(5):706–722, 2009.
27. W. Hackbusch and R. Schneider. *Tensor Spaces and Hierarchical Tensor Representations*. Springer International Publishing, Cham, 2014.
28. J. Han, A. Jentzen, and W. E. Solving high-dimensional partial differential equations using deep learning. *Proceedings of the National Academy of Sciences*, 115(34):8505–8510, 2018.
29. C. Heiß, I. Gühring, and M. Eigel. Multilevel CNNs for parametric PDEs. *Journal of Machine Learning Research*, 24(373):1–42, 2023.
30. L. Herrmann, C. Schwab, and J. Zech. Neural and spectral operator surrogates: unified construction and expression rate bounds. *arXiv preprint*, 2022.
31. S. Holtz, T. Rohwedder, and R. Schneider. On manifolds of tensors of fixed tt-rank. *Numerische Mathematik*, 120(4):701–731, 2012.
32. K. Kunisch, S. S. Rodrigues, and D. Walter. Learning an optimal feedback operator semiglobally stabilizing semilinear parabolic equations. *Applied Mathematics & Optimization*, 84(1):277–318, 12 2021.
33. K. Kunisch, D. Vásquez-Varas, and D. Walter. Learning optimal feedback operators and their sparse polynomial approximations *{JMLR}, 24(301):1–38, 2023.*
34. *K. Kunisch and D. Walter. Semiglobal optimal feedback stabilization of autonomous systems via deep neural network approximation.* ESAIM: COCV, *27:16, 2021.*
35. *K. Kunisch and D. Walter. Optimal feedback control of dynamical systems via value-function approximation,* Comptes Rendus. Mécanique, *351(S1):535–571, 2023.*
36. *A. Kunoth, M. Oster, and R. Schneider. Towards a continuous mathematical model for the analysis of classes of deep neural networks, 2024.*
37. *G. Kutyniok, P. Petersen, M. Raslan, and R. Schneider. A theoretical analysis of deep neural networks and parametric pdes.* Constructive Approximation, *55(1):73–125, 2022.*
38. *J.-M. Lasry and P.-L. Lions. Jeux à champ moyen. ii – horizon fini et contrôle optimal.* Comptes Rendus Mathematique, *343(10):679–684, 2006.*
39. *H. N. Mhaskar and T. Poggio. Deep vs. shallow networks: An approximation theory perspective.* Analysis and Applications, *14(06):829–848, 2016.*
40. *H. Mhaskar, Q. Liao, and T. Poggio. Learning functions: When is deep better than shallow, 2016.*
41. *D. Onken, L. Nurbekyan, X. Li, S. W. Fung, S. Osher, and L. Ruthotto. A neural network approach applied to multi-agent optimal control. In* 2021 European Control Conference (ECC), *pages 1036–1041, 2021.*
42. *I. V Oseledets. Tensor-train decomposition.* SIAM Journal on Scientific Computing, *33(5):2295–2317, 2011.*
43. *M. Oster.* On semi-global optimal control problems and their applications in machine learning. *PhD thesis, Fakulty II - Mathematics and Natural Sciences, TU Berlin, 2023.*
44. *D. Perez-Garcia, F. Verstraete, M. M. Wolf, and J. I. Cirac. Matrix product state representations.* Quantum Info. Comput., *7(5):401–430, 7 2007.*
45. *P. Petersen and F. Voigtlaender. Optimal approximation of piecewise smooth functions using deep relu neural networks.* Neural Networks, *108:296–330, 2018.*
46. *T. Poggio, H. Mhaskar, L. Rosasco, D. Miranda, and Q. Liao. Why and when can deep-but not shallow-networks avoid the curse of dimensionality: A review.* International Journal of Automation and Computing, *14(5):503–519, 10 2017.*
47. *L. Ruthotto, S. J. Osher, W. Li, L. Nurbekyan, and S. W. Fung. A machine learning framework for solving high-dimensional mean field game and mean field control problems.* Proceedings of the National Academy of Sciences, *117(17):9183–9193, 2020.*

48. *R. Schneider and A. Uschmajew. Approximation rates for the hierarchical tensor format in periodic sobolev spaces.* Journal of Complexity, *30(2):56–71, 2014.*
49. *C. Schwab, A. Stein, and J. Zech. Deep operator network approximation rates for lipschitz operators, 2023.*
50. *S. Szalay, M. Pfeffer, V. Murg, G. Barcza, F. Verstraete, R. Schneider, and Ö. Legeza. Tensor product methods and entanglement optimization for ab initio quantum chemistry.* International j. of quantum chemistry, *115(19):1342–1391, 2015.*
51. *J. F. Traub, G. W. Wasilkowski, and H. Woźniakowski.* Information, uncertainty, complexity. *Addison-Wesley Reading, MA, 1983.*
52. *P. Trunschke, A. Nouy, and M. Eigel. Weighted sparsity and sparse tensor networks for least squares approximation, 2023.*
53. *V. Vapnik.* The Nature of Statistical Learning Theory. *Information Science and Statistics. Springer New York, 1999.*
54. *V. Vapnik. Principles of risk minimization for learning theory. In* Advances in neural information processing systems, *pages 831–838, 1992.*
55. *D. Yarotsky. Error bounds for approximations with deep relu networks.* Neural Networks, *94:103–114, 2017.*
56. *J. Zech and Y. Marzouk. Sparse approximation of triangular transports, Part II: The infinite-dimensional case.* Constr. Approx., *55(3):987–1036, 2022.*
57. *L. Richter, L. Sallandt and N. Nüsken Solving high-dimensional parabolic PDEs using the tensor train format.* International Conference on Machine Learning *47: 8998-9009, 2021.*
58. *M. Oster, L. Sallandt and R. Schneider. Approximating optimal feedback controllers of finite horizon control problems using hierarchical tensor formats* SIAM Journal on Scientific Computing, *44(3): B746-B770, 2022.*
59. *C Bayer, M Eigel, L Sallandt and P Trunschke. Pricing high-dimensional Bermudan options with hierarchical tensor formats* SIAM Journal on Financial Mathematics, *14(2): 383-406, 2023.*
60. *K Fackeldey, M Oster, L Sallandt, R Schneider. Approximative policy iteration for exit time feedback control problems driven by stochastic differential equations using tensor train format* SIAM Multiscale Modeling & Simulation, *20(1): 379-403, 2022.*
61. *M. Eigel, R. Schneider, D. Sommer. Dynamical low-rank approximations of solutions to the Hamilton-Jacobi-Bellman equation.* Numer. Linear Algebra Appl. *30(3):e2463, 2023.*
62. *S. Dolgov, D. Kalise and K. Kunisch Tensor decomposition methods for high-dimensional Hamilton–Jacobi–Bellman equations* SIAM Journal on Scientific Computing, *43(3),:A1625-A1650, 2021.*
63. *P. Grohs and L. Herrmann. Deep neural network approximation for high-dimensional parabolic Hamilton-Jacobi-Bellman equations. arXiv preprint arXiv:2103.05744, 2021.*
64. *M. Götte, R. Schneider and P. Trunschke A Block-Sparse Tensor Train Format for Sample-Efficient High-Dimensional Polynomial Regression.* Frontiers in Applied Mathematics and Statistics, *7, 2021.*

Efficient least squares discretizations for Unique Continuation and Cauchy problems

Rob Stevenson

Abstract We consider least squares discretizations of Unique Continuation and Cauchy problems for the Poisson equation based on ultra-weak variational formulations. The dual norm that is present in the (regularized) least squares functional cannot be evaluated exactly, and so has to be discretized which leads to a saddle-point formulation. For uniformly stable pairs of 'trial' and 'test' finite element spaces, approximations are obtained that are quasi-best in view of the available conditional stability estimates.

Compared to standard variational formulations, conditional stability estimates that corresponds to ultra-weak formulations result in better convergence rates with the same error-norm. Globally C^1 finite element test spaces to accommodate the ultra-weak formulation will be avoided by the application of nonconforming test spaces. Thanks to the ultra-weak formulation, both Neumann and Dirichlet boundary conditions are natural ones, which in particular enables a convenient discretization of the Cauchy problem.

1 Introduction

This work is a continuation of the research from the joint papers [DSW22, DMS23] with Wolfgang Dahmen on the numerical solution of ill-posed boundary value problems $Au = f$ that are *conditionally stable*.

The latter means that for some Hilbert spaces X, V, and H, where $A \in \mathcal{L}(X, V)$, and some $L \in \mathcal{L}(X, H)$, for all $u \in X$ that satisfy an a priori bound $\|Lu\|_H \leqslant C$, there is a continuous dependency of the solution on the data in the following sense. For some $j: X \to \mathbb{R}_+$ and $\eta = \eta_C : \mathbb{R}^+ \to \mathbb{R}^+$ with $\lim_{t\downarrow 0} \eta(t) = 0$, it holds that $j(u) \leqslant \eta(\|Au\|_V)$. In applications $j(\cdot)$ is a (semi-) norm that is weaker than the

Korteweg-de Vries (KdV) Institute for Mathematics, University of Amsterdam, P.O. Box 94248, 1090 GE Amsterdam, The Netherlands. e-mail: rob.p.stevenson@gmail.com

R. DeVore, A. Kunoth (eds.), *Multiscale, Nonlinear and Adaptive Approximation II*,
https://doi.org/10.1007/978-3-031-75802-7_20

norm on X, and, e.g., $\eta(t) = t^s$ or even only $\eta(t) = (-1/\log t)^s$ for some $s \in (0, 1]$. Examples of conditionally stable problems will be given in Sec. 2

For a family of finite dimensional subspaces $(X^\delta)_{\delta \in I} \subset X$ ('δ' refers to 'discrete'), the approach that was developed in [DMS23] is to approximate u by the minimizer u_ε^δ over X^δ of the regularized least-squares functional $z \mapsto \sqrt{\|Az - f\|_V^2 + \varepsilon^2 \|Lz\|_H^2}$. It was shown that for a suitable selection of ε, both $\|L(u - u_\varepsilon^\delta)\|_H$ is uniformly bounded, so that $j(u - u_\varepsilon^\delta) \leqslant \eta\big(\|f - Au_\varepsilon^\delta\|_V\big)$, and $\|f - Au_\varepsilon^\delta\|_V$ is bounded by an absolute multiple of $\|f - Au\|_V + \min_{z \in X^\delta} \|u - z\|_X$, being the sum of the *consistency error* and the *error of best approximation*. Consequently, the bound on the error quantity $j(u - u_\varepsilon^\delta)$ that is realized is, *in view of the available conditional stability estimate*, qualitatively the best possible one for an approximation from X^δ.

In applications, V is a Cartesian product of the form $Y' \times W$. The norm on the dual space Y' cannot be evaluated and has to be replaced by the dual norm on a finite dimensional subspace $Y^\delta \subset Y$. By realizing uniform inf-sup conditions for the pair (X^δ, Y^δ) of trial and test spaces, this can be done whilst retaining above optimality results.

The conditional stability estimates that were applied in [DMS23] are the common ones derived from standard variational formulations of the PDEs. In this work, we will apply ultra-weak variational formulations, which will result in qualitatively the same conditional stability estimates in terms of a *weaker* $\|\cdot\|_X$-norm. Consequently, the resulting least-squares approximations are shown to converge faster as function of a decreasing mesh-size. A price to be paid for an ultra-weak formulation, however, is that Y will be a subspace of $H^2(\Omega)$, with $\Omega \subset \mathbb{R}^n$ being the domain on which the PDE is posed. In order not being forced to apply C^1 finite element spaces as test spaces, we will extend the analysis of the least-squares method to the use of *nonconforming* test spaces.

2 Examples of conditionally stable, ill-posed PDEs

In this work, we consider the following two problems: The Unique Continuation (UC) problem of finding $u\colon \mathbb{R}^n \supset \Omega \to \mathbb{R}$ that for some $\omega \Subset \Omega$ satisfies

$$\begin{cases} -\triangle u = \ell \text{ on } \Omega, \\ u|_\omega = q \text{ on } \omega, \end{cases}$$

and the Cauchy problem of finding $u\colon \Omega \to \mathbb{R}$ that for some $\Sigma \subset \partial\Omega$ satisfies

$$\begin{cases} -\triangle u = \ell \text{ on } \Omega, \\ u = g \text{ on } \Sigma, \\ \frac{\partial u}{\partial n} = \psi \text{ on } \Sigma. \end{cases}$$

For the moment we assume that the data, (ℓ, q) or (ℓ, g, ψ), is consistent so that solutions exist. Because with our numerical approximation all equations will be imposed in a least-squares sense, this will automatically accommodate the relevant case of having *inexact data*.

Besides for the Poisson problem, UC and Cauchy problems can also be formulated for the heat and wave equations.

A key ingredient in deriving conditional stability results is the following "Propagation of smallness result" [ARRV09, Thm. 5.1] for harmonic functions, which is a corollary of the "three balls estimate", e.g., see [ARRV09, Thm. 2.1].

Theorem 2.1 (Propagation of smallness) *For a bounded domain $\Omega \subset \mathbb{R}^n$, and subdomains $\omega \Subset G \Subset \Omega$, there exists an $\alpha \in (0,1)$ such that for all harmonic $w \in L_2(\Omega)$,*

$$\|w\|_{L_2(G)} \lesssim \|w\|_{L_2(\omega)}^{\alpha} \|w\|_{L_2(\Omega)}^{1-\alpha}. \tag{1}$$

The common variational formulation of the *UC problem* reads as seeking $u \in H^1(\Omega)$ that satisfies

$$\begin{cases} \int_\Omega \nabla u \cdot \nabla v \, dx &= \ell(v) \quad (v \in H_0^1(\Omega)), \\ u|_\omega &= q \qquad \text{on } \omega. \end{cases} \tag{2}$$

Following arguments as e.g. given in [BNO23, Sect. 2.2], we write $u = u_0 + \widetilde{u}$ where $u_0 \in H_0^1(\Omega)$ solves

$$\int_\Omega \nabla u_0 \cdot \nabla v \, dx = \ell(v) \quad (v \in H_0^1(\Omega)),$$

so that

$$\begin{cases} -\Delta \widetilde{u} &= 0 \qquad\qquad \text{on } \Omega, \\ \widetilde{u}|_\omega &= q - u_0|_\omega \ \text{on } \omega. \end{cases}$$

Repeated applications of the triangle-inequality, and an application of Theorem 2.1 to $\widetilde{u}$ shows that

$$\begin{aligned} \|u\|_{L_2(G)} &\leqslant \|u_0\|_{L_2(G)} + \|\widetilde{u}\|_{L_2(G)} \lesssim |u_0|_{H^1(\Omega)} + \|\widetilde{u}\|_{L_2(G)} \\ &\lesssim \|\ell\|_{H^{-1}(\Omega)} + \|\widetilde{u}\|_{L_2(\omega)}^{\alpha} \|\widetilde{u}\|_{L_2(\Omega)}^{1-\alpha} \\ &\lesssim \|\ell\|_{H^{-1}(\Omega)} + \big(\|u\|_{L_2(\omega)} + \|u_0\|_{L_2(\omega)}\big)^{\alpha} \big(\|u\|_{L_2(\Omega)} + \|u_0\|_{L_2(\Omega)}\big)^{1-\alpha} \\ &\lesssim \|\ell\|_{H^{-1}(\Omega)} + \big(\|q\|_{L_2(\omega)} + |u_0|_{H^1(\Omega)}\big)^{\alpha} \big(\|u\|_{L_2(\Omega)} + |u_0|_{H^1(\Omega)}\big)^{1-\alpha} \\ &\lesssim \big(\|q\|_{L_2(\omega)} + \|\ell\|_{H^{-1}(\Omega)}\big)^{\alpha} \big(\|u\|_{L_2(\Omega)} + \|\ell\|_{H^{-1}(\Omega)} + \|q\|_{L_2(\omega)}\big)^{1-\alpha}. \end{aligned} \tag{3}$$

Above estimate can be formulated as follows.

Theorem 2.2 (Conditional stability UC) *With $A \in \mathcal{L}\big(H^1(\Omega), H^{-1}(\Omega) \times L_2(\omega)\big)$ defined by*

$$Az = \Big(v \mapsto \int_\Omega \nabla z \cdot \nabla v \, dx, z|_\omega\Big),$$

for subdomains $\omega \Subset G \Subset \Omega$ and $\alpha \in (0,1)$ from Theorem 2.1, it holds that

$$\|z\|_{L_2(G)} \lesssim \|Az\|^{\alpha}_{H^{-1}(\Omega)\times L_2(\omega)} \big(\|z\|_{L_2(\Omega)} + \|Az\|_{H^{-1}(\Omega)\times L_2(\omega)}\big)^{1-\alpha} \quad (z \in H^1(\Omega)).$$

The estimate from this theorem is a conditional (Hölder) stability result. Indeed, provided that $\|z\|_{L_2(\Omega)} \lesssim 1$, it shows that

$$\|z\|_{L_2(G)} \lesssim \|Az\|^{\alpha}_{H^{-1}(\Omega)\times L_2(\omega)} \quad (\|Az\|_{H^{-1}(\Omega)\times L_2(\omega)} \to 0).$$

Notice that on several places in above derivation, we applied Friedrich's inequality $\|\cdot\|_{L_2(\omega)} \lesssim |\cdot|_{H^1(\Omega)}$ on $H^1_0(\Omega)$. For the application to estimate the error in our least squares approximations, this has the consequence that we need to estimate approximation errors in $H^1(\Omega)$, although in the end we arrive at an error estimate in $L_2(G)$-norm. To avoid this discrepancy, we consider the following ultra-weak formulation of the UC problem to seek $u \in L_2(\Omega)$ such that

$$\boxed{\begin{cases} -\int_\Omega u \Delta v \, dx &= \ell(v) \quad (v \in H^2_0(\Omega)), \\ u|_\omega &= q \qquad \text{on } \omega, \end{cases}} \tag{4}$$

and redefine A correspondingly.

Theorem 2.3 (Conditional stability UC ultra-weak) *With*

$$A \in \mathcal{L}\big(L_2(\Omega), H^{-2}(\Omega) \times L_2(\omega)\big)$$

defined by

$$Az = \Big(v \mapsto -\int_\Omega z \cdot \Delta v \, dx, z|_\omega\Big),$$

for subdomains $\omega \Subset G \Subset \Omega$ and $\alpha \in (0,1)$ from Theorem 2.1, it holds that

$$\|z\|_{L_2(G)} \lesssim \|Az\|^{\alpha}_{H^{-2}(\Omega)\times L_2(\omega)} \big(\|z\|_{L_2(\Omega)} + \|Az\|_{H^{-2}(\Omega)\times L_2(\omega)}\big)^{1-\alpha} \quad (z \in L_2(\Omega)).$$

Proof. It holds that

$$\|\cdot\|_{H^2(\Omega)} \eqsim \|\Delta \cdot\|_{L_2(\Omega)} \quad \text{on } H^2_0(\Omega). \tag{5}$$

Indeed, consider a ball $Z \supset \Omega$. From the H^2-regularity of the Poisson problem on Z with homogeneous Dirichlet boundary conditions, for $\varphi \in \mathcal{D}(\Omega)$ we have

$$\|\varphi\|_{H^2(\Omega)} = \|\varphi\|_{H^2(Z)} \lesssim \|\Delta\varphi\|_{L_2(Z)} = \|\Delta\varphi\|_{L_2(\Omega)}.$$

Let $\mathcal{H}(\Omega) := \operatorname{ran} \Delta|_{H^2_0(\Omega)} \subset L_2(\Omega)$. As a consequence of (5), we have that $\mathcal{H}(\Omega)$ is closed, and therefore is a Hilbert space. From $H^2_0(\Omega)$ being a proper subspace of $H^1_{\Delta,0}(\Omega) := \{z \in H^1_0(\Omega) \colon \Delta z \in L_2(\Omega)\}$ equipped with graph norm, and

$\triangle \in \mathcal{L}(H^1_{\triangle,0}(\Omega), L_2(\Omega))$ being boundedly invertible, we infer that $\mathcal{H}(\Omega)$ is a proper subspace of $L_2(\Omega)$.

The bilinear form $(z, v) \mapsto -\int_\Omega z\triangle v\,dx$ on $\mathcal{H}(\Omega) \times H_0^2(\Omega)$ is bounded,

$$\sup_{0\neq v\in H_0^2(\Omega)} \frac{\int_\Omega z\triangle v\,dx}{\|v\|_{H^2(\Omega)}} \eqsim \sup_{0\neq v\in H_0^2(\Omega)} \frac{\int_\Omega z\triangle v\,dx}{\|\triangle v\|_{H^2(\Omega)}} = \|z\|_{L_2(\Omega)} \quad (z \in \mathcal{H}(\Omega)),$$

and

$$\sup_{0\neq z\in\mathcal{H}(\Omega)} \frac{\int_\Omega z\triangle v\,dx}{\|z\|_{L_2(\Omega)}} = \|\triangle v\|_{H^2(\Omega)} \eqsim \|v\|_{H^2(\Omega)} \qquad (v \in H_0^2(\Omega)).$$

We conclude that for any $\ell \in H^{-2}(\Omega)$, there exists a unique $\widehat{u} \in \mathcal{H}(\Omega)$ with

$$-\int_\Omega \widehat{u}\triangle v\,dx = \ell(v) \quad (v \in H_0^2(\Omega)),$$

and $\|\widehat{u}\|_{L_2(\Omega)} \lesssim \|\ell\|_{H^{-2}(\Omega)}$.

We set $\widetilde{u} := u - \widehat{u}$. Then $\int_\Omega \widetilde{u}\triangle\varphi\,dx = 0$ ($\varphi \in \mathcal{D}(\Omega)$), and according to Weyl's lemma ([Wey40]) this means that $\widetilde{u}$ is harmonic.

As before, repeated applications of the triangle-inequality, and an application of Theorem 2.1 to $\widetilde{u}$ show that

$$\begin{aligned}\|u\|_{L_2(G)} &\leqslant \|\widehat{u}\|_{L_2(G)} + \|\widetilde{u}\|_{L_2(G)} \lesssim \|\ell\|_{H^{-2}(\Omega)} + \|\widetilde{u}\|^{\alpha}_{L_2(\omega)}\|\widetilde{u}\|^{1-\alpha}_{L_2(L_2(\Omega))}\\ &\lesssim \|\ell\|_{H^{-2}(\Omega)} + (\|u\|_{L_2(\omega)} + \|\widehat{u}\|_{L_2(\omega)})^{\alpha}(\|u\|_{L_2(\Omega)} + \|\widehat{u}\|_{L_2(\Omega)})^{1-\alpha}\\ &\lesssim (\|q\|_{L_2(\omega)} + \|\ell\|_{H^{-2}(\Omega)})^{\alpha}(\|u\|_{L_2(\Omega)} + \|\ell\|_{H^{-2}(\Omega)} + \|q\|_{L_2(\omega)})^{1-\alpha},\end{aligned}$$

which is the estimate to be proven. □

The common variational formulation of the *Cauchy problem* reads as seeking $u \in \{w \in H^1(\Omega) : w|_\Sigma = g \text{ in } H^{\frac{1}{2}}(\Sigma)\}$ such that

$$\int_\Omega \nabla u \cdot \nabla v\,dx = \ell(v) + \int_\Sigma \psi v\,ds \quad (v \in H^1_{0,\Sigma^c}(\Omega)), \tag{6}$$

where $\Sigma^c := \partial\Omega \setminus \overline{\Sigma}$. The following conditional stability result was shown in [ARRV09, Theorem 1.7].

Theorem 2.4 (Conditional stability Cauchy) *With* $A \in \mathcal{L}\big(H^1(\Omega), H^1_{0,\Sigma^c}(\Omega)' \times H^{\frac{1}{2}}(\Sigma)\big)$ *defined by*

$$Az = \Big(v \mapsto \int_\Omega \nabla z \cdot \nabla v\,dx, z|_\Sigma\Big),$$

for a subdomain $G \subset \Omega$ *that has positive distance to* Σ^c *there exists an* $\alpha \in (0, 1)$ *such that*

$$\|z\|_{L_2(G)} \lesssim \|Az\|^{\alpha}_{H^1_{0,\Sigma^c}(\Omega)'\times H^{\frac{1}{2}}(\Sigma)} \big(\|z\|_{L_2(\Omega)}+\|Az\|_{H^1_{0,\Sigma^c}(\Omega)'\times H^{\frac{1}{2}}(\Sigma)}\big)^{1-\alpha} \quad (z \in H^1(\Omega)).$$

So as with the standard variational formulation (2) of the UC problem, there is a mismatch between the $H^1(\Omega)$-space in which the solution is sought, and the $L_2(G)$-norm in which the error in the approximation for u will be measured.

By integrating the left hand side of (6) once more for smooth v with $\operatorname{supp} v \subset \Omega\cup\Sigma$, we obtain the problem to find $u \in L_2(\Omega)$ such that

$$\boxed{-\int_{\Omega} u\Delta v\, dx = f(v) := \ell(v) + \int_{\Sigma} \psi v - g\tfrac{\partial v}{\partial n}\, ds \quad (v \in H^2_{0,\Sigma^c}(\Omega)).} \tag{7}$$

This formulation has the *additional advantage* that also the Dirichlet boundary condition is natural, so that its imposition will be effortless.

A proof of the following conditional stability result for this ultra-weak formulation is given in [MS24].

Theorem 2.5 (Conditional stability Cauchy ultra-weak) *With the operator* $A \in \mathcal{L}(L_2(\Omega), (H^2_{0,\Sigma^c}(\Omega))')$ *defined by* $(Az)(v) = -\int_\Omega z\Delta v\, dx$, *for a subdomain* $G \subset \Omega$ *that has positive distance to* Σ^c *there exists an* $\alpha \in (0,1)$ *such that*

$$\|z\|_{L_2(G)} \lesssim \|Az\|^{\alpha}_{(H^2_{0,\Sigma^c}(\Omega))'}\big(\|z\|_{L_2(\Omega)} + \|Az\|_{(H^2_{0,\Sigma^c}(\Omega))'}\big)^{1-\alpha} \quad (z \in L_2(\Omega)).$$

3 Least squares approximations

We discuss the least squares method from [DMS23] for approximating the solutions of the ultra-weak formulations (4) and (7) for the UC and Cauchy problem, respectively. Let $(X^\delta)_{\delta\in I}$ be some (infinite) family of finite dimensional subspaces of $L_2(\Omega)$. Writing A from Theorem 2.3 as $A = (B, C)$, where $(Bz)(v) := -\int_\Omega z\Delta v\, dx$, and $Cz := z|_\omega$, the idea of this least squares method is to approximate u by the minimizer over X^δ of the regularized least squares functional

$$z \mapsto \|Bz - \ell\|^2_{H^{-2}(\Omega)} + \|Cz - q\|^2_{L_2(\omega)} + \varepsilon^2\|z\|^2_{L_2(\Omega)}, \tag{8}$$

or

$$z \mapsto \|Bz - f\|^2_{(H^2_{0,\Sigma^c}(\Omega))'} + \varepsilon^2\|z\|^2_{L_2(\Omega)},$$

respectively.

The choice of the $L_2(\Omega)$-norm in the regularizing terms is determined by its appearance in the conditional stability estimates.

The dual norms in both least squares functionals cannot be evaluated, and have to be replaced by discretized dual norms. *From here onwards we will focus on the functional* (8) *for the UC problem.* Obvious modifications (the replacement of $H^2_0(\Omega)$

by $H^2_{0,\Sigma^c}(\Omega)$, and the removal of $\|Cz-q\|^2_{L_2(\omega)}$) will give the results for the Cauchy problem.

Let $(Y^\delta)_{\delta\in I}$ be a family of finite dimensional (or closed) subspaces of $H^2_0(\Omega)$, which will be referred to as *test spaces*, such that

$$\inf_{\delta\in I}\inf_{\{z\in X^\delta:\,Bz\neq 0\}}\sup_{0\neq v\in Y^\delta}\frac{|(Bz)(v)|}{\|Bz\|_{H^{-2}(\Omega)}\|v\|_{H^2(\Omega)}}>0. \tag{9}$$

Then by replacing $\|\cdot\|_{H^{-2}(\Omega)}$ by $\sup_{0\neq v\in Y^\delta}\frac{|\cdot(v)|}{\|v\|_{H^2(\Omega)}}$ in (8), an approximation to u is obtained that is qualitatively equally good. Specifically, we have the following result.

Theorem 3.1 ([DMS23, Thm. 4.1]) *Assume* (9), *and let* $(\ell,q)\in H^{-2}(\Omega)\times L_2(\omega)$. *Then for* $\varepsilon>0$,

$$u^\delta_\varepsilon:=\underset{z\in X^\delta}{\arg\min}\Big\{\sup_{0\neq v\in Y^\delta}\frac{|(Bz-\ell)(v)|^2}{\|v\|^2_{H^2(\Omega)}}+\|Cz-q\|^2_{L_2(\omega)}+\varepsilon^2\|z\|^2_{L_2(\Omega)}\Big\} \tag{10}$$

satisfies, for $u\in L_2(\Omega)$,

$$\|A(u-u^\delta_\varepsilon)\|_{H^{-2}(\Omega)\times L_2(\omega)}+\varepsilon\|u-u^\delta_\varepsilon\|_{L_2(\Omega)}\lesssim \mathcal{E}_{\rm data}+\mathcal{E}_{\rm appr}(\delta)+\varepsilon\|u\|_{L_2(\Omega)}, \tag{11}$$

where

$$\mathcal{E}_{\rm data}:=\|Au-(\ell,q)\|_{H^{-2}(\Omega)\times L_2(\omega)},\qquad \mathcal{E}_{\rm appr}(\delta):=\min_{z\in X^\delta}\|u-z\|_{L_2(\Omega)}.$$

Corollary 3.1 *In the setting of Theorem 3.1, let* $\varepsilon>0$ *be such that*

$$\varepsilon\|u\|_{L_2(\Omega)}\eqsim \mathcal{E}_{\rm data}+\mathcal{E}_{\rm appr}(\delta). \tag{12}$$

Then with $\alpha\in(0,1)$ *from Theorem 2.1 it holds that*

$$\|u-u^\delta_\varepsilon\|_{L_2(G)}\lesssim\big(\mathcal{E}_{\rm data}+\mathcal{E}_{\rm appr}(\delta)\big)^\alpha\max\big(\mathcal{E}_{\rm data}+\mathcal{E}_{\rm appr}(\delta),\|u\|_{L_2(\Omega)}\big)^{1-\alpha}. \tag{13}$$

Proof. Substituting (12) in (11) shows that

$$\begin{cases}\|u-u^\delta_\varepsilon\|_{L_2(\Omega)}\lesssim\|u\|_{L_2(\Omega)},\\ \|A(u-u^\delta_\varepsilon)\|_{H^{-2}(\Omega)\times L_2(\omega)}\lesssim\mathcal{E}_{\rm data}+\mathcal{E}_{\rm appr}(\delta),\end{cases}$$

so that an application of the conditional stability result from Theorem 2.3 gives the result. □

The approximation u^δ_ε defined in (10) is *computed* as the first component of the pair $(u^\delta_\varepsilon,v^\delta_\varepsilon)\in X^\delta\times Y^\delta$ that solves the *saddle point problem*

$$\begin{cases}\langle\Delta v^\delta_\varepsilon,\Delta\tilde v\rangle_{L_2(\Omega)} \quad +(Bu^\delta_\varepsilon)(\tilde v) = \ell(\tilde v) & (\tilde v\in Y^\delta),\\ (B\tilde z)(v^\delta_\varepsilon)-\langle Cu^\delta_\varepsilon,C\tilde z\rangle_{L_2(\omega)}-\varepsilon^2\langle u^\delta_\varepsilon,\tilde z\rangle_{L_2(\Omega)}=-\langle q,C\tilde z\rangle_{L_2(\omega)} & (\tilde z\in X^\delta),\end{cases} \tag{14}$$

(see e.g. [DMS23]).

Remark 3.1 (Optimality of α in (13)) For $(X^\delta)_{\delta\in I}$ with $\overline{\cup_{\delta\in I}X^\delta} = L_2(\Omega)$, an estimate as (13) with an exponent $\widetilde{\alpha} > \widehat{\alpha} := \sup\{\alpha\colon (1) \text{ is valid for } \alpha\}$ cannot be valid for *any* scheme $(\ell, q) \mapsto (u^\delta)_{\delta\in I} \subset \prod_{\delta\in I} X^\delta$ that maps zero data $\ell = 0 = q$ onto $u^\delta \equiv 0$. Indeed, suppose such a scheme does exist, and consider $\ell = 0 = q$ and harmonic $u \in L_2(\Omega)$. Then

$$\|u\|_{L_2(G)} \lesssim \inf_{\delta\in I}(\|u\|_{L_2(\omega)} + \min_{z\in X^\delta} \|u - z\|_{L_2(\Omega)})^{\tilde{\alpha}} \|u\|_{L_2(\Omega)}^{1-\tilde{\alpha}} = \|u\|_{L_2(\omega)}^{\tilde{\alpha}} \|u\|_{L_2(\Omega)}^{1-\tilde{\alpha}},$$

in contradiction to the definition of $\widehat{\alpha}$.

Remark 3.2 (Discussion) To compare with results from the literature, let us choose the regularization parameter ε such that $\varepsilon\|u\|_{L_2(\Omega)} \eqsim \mathcal{E}_{\text{appr}}(\delta)$ assuming $\mathcal{E}_{\text{appr}}(\delta) \neq 0$, which corresponds to the choice made in Corollary 3.1 as long as $\mathcal{E}_{\text{appr}}(\delta) \gtrsim \mathcal{E}_{\text{data}}$. With this choice of ε, (11) shows

$$\begin{cases} \|u - u_\varepsilon^\delta\|_{L_2(\Omega)} \lesssim \left(1 + \frac{\mathcal{E}_{\text{data}}}{\mathcal{E}_{\text{appr}}(\delta)}\right)\|u\|_{L_2(\Omega)}, \\ \|A(u - u_\varepsilon^\delta)\|_{H^{-2}(\Omega)\times L_2(\omega)} \lesssim \mathcal{E}_{\text{data}} + \mathcal{E}_{\text{appr}}(\delta). \end{cases}$$

Assuming that $\mathcal{E}_{\text{data}} + \mathcal{E}_{\text{appr}}(\delta) \lesssim \|u\|_{L_2(\Omega)}$, substituting these bounds in the conditional stability result from Theorem 2.3 gives

$$\begin{aligned} \|u - u_\varepsilon^\delta\|_{L_2(G)} &\lesssim \big(\mathcal{E}_{\text{data}} + \mathcal{E}_{\text{appr}}(\delta)\big)^\alpha \\ &\qquad \cdot \left(\left(1 + \frac{\mathcal{E}_{\text{data}}}{\mathcal{E}_{\text{appr}}(\delta)}\right)\|u\|_{L_2(\Omega)} + \mathcal{E}_{\text{data}} + \mathcal{E}_{\text{appr}}(\delta)\right)^{1-\alpha} \\ &\eqsim \big(\mathcal{E}_{\text{data}} + \mathcal{E}_{\text{appr}}(\delta)\big)^\alpha \left(1 + \frac{\mathcal{E}_{\text{data}}}{\mathcal{E}_{\text{appr}}(\delta)}\right)^{1-\alpha} \|u\|_{L_2(\Omega)}^{1-\alpha} \\ &\eqsim \left(\mathcal{E}_{\text{appr}}(\delta)^\alpha + \frac{\mathcal{E}_{\text{data}}}{\mathcal{E}_{\text{appr}}(\delta)^{1-\alpha}}\right)\|u\|_{L_2(\Omega)}^{1-\alpha}. \end{aligned} \tag{15}$$

This upper bound is proportional to $\mathcal{E}_{\text{appr}}(\delta)^\alpha \|u\|_{L_2(\Omega)}^{1-\alpha}$ as long as $\mathcal{E}_{\text{appr}}(\delta) \gtrsim \mathcal{E}_{\text{data}}$, and thus is proportional to $(\mathcal{E}_{\text{data}})^\alpha \|u\|^{1-\alpha}$ when $\mathcal{E}_{\text{appr}}(\delta) \eqsim \mathcal{E}_{\text{data}}$, and it increases unboundedly for $\mathcal{E}_{\text{appr}}(\delta) \downarrow 0$ when $\mathcal{E}_{\text{appr}}(\delta) \lesssim \mathcal{E}_{\text{data}}$. [1]

Still for the purpose of comparison, let us consider X^δ to be a finite element space of degree k w.r.t. a quasi-uniform partition of Ω with mesh-size h_δ into uniformly shape regular n-simplices.

For the primal-dual weakly consistent regularized finite element method studied in [BNO23], where the UC problem is considered with forcing term $\ell = 0$, it was shown that

$$\|u - u^\delta\|_{L_2(G)} \lesssim h_\delta^{\alpha k} \|u\|_{H^{k+1}(\Omega)} + h_\delta^{(\alpha-1)k} \|u|_\omega - q\|_{L_2(\omega)}, \tag{16}$$

[1] Our choice in Corollary 3.1 of $\varepsilon\|u\|_{L_2(\Omega)} \eqsim \mathcal{E}_{\text{appr}}(\delta) + \mathcal{E}_{\text{data}}$ prevents this latter increase, but it results in a (qualitatively) equal minimal upper bound for $\|u - u_\varepsilon^\delta\|_{L_2(G)}$.

and so

$$\|u - u^\delta\|_{L_2(G)} \lesssim \|u|_\omega - q\|^\alpha_{L_2(\omega)} \|u\|^{1-\alpha}_{H^{k+1}(\Omega)}$$

for the minimum

$$h_\delta \eqsim \Big(\frac{\|u|_\omega - q\|_{L_2(\omega)}}{\|u\|_{H^{k+1}(\Omega)}}\Big)^{\frac{1}{k}}.$$

It was shown that the value of α in the exponents in (16) is optimal, but below we see that the value of k in these exponents can be improved.

Indeed, using that $\mathcal{E}_{\text{appr}}(\delta) \lesssim h_\delta^{k+1}|u|_{H^{k+1}(\Omega)}$ when $u \in H^{k+1}(\Omega)$, and by substituting this upper bound into (15) shows

$$\|u - u^\delta_\varepsilon\|_{L_2(G)} \lesssim \left(\left(h_\delta^{k+1}|u|_{H^{k+1}(\Omega)}\right)^\alpha + \frac{\mathcal{E}_{\text{data}}}{\left(h_\delta^{k+1}|u|_{H^{k+1}(\Omega)}\right)^{1-\alpha}}\right)\|u\|^{1-\alpha}_{L_2(\Omega)}.$$

Solving $h_\delta^{k+1}|u|_{H^{k+1}(\Omega)} \eqsim \mathcal{E}_{\text{data}}$ gives

$$\|u - u^\delta_\varepsilon\|_{L_2(G)} \lesssim (\mathcal{E}_{\text{data}})^\alpha \|u\|^{1-\alpha}_{L_2(\Omega)} \tag{17}$$

for

$$h_\delta \eqsim \Big(\frac{\mathcal{E}_{\text{data}}}{|u|_{H^{k+1}(\Omega)}}\Big)^{\frac{1}{k+1}}.$$

So thanks to the use of an ultra-weak formulation, our least squares method gives a qualitatively similar bound, but for a larger mesh-size, and so at lower computational cost (cf. [BNO23, Rem. 3]). Alternatively, by taking finite elements of one lower degree $k-1$, we find (17) for $h_\delta \eqsim \big(\frac{\mathcal{E}_{\text{data}}}{|u|_{H^k(\Omega)}}\big)^{\frac{1}{k}}$, so with a mesh-size as in [BNO23], but under a relaxed regularity condition.

Finally, notice that our least squares method is not restricted to quasi-uniform meshes, meaning that with appropriately locally refined meshes regularity conditions can be relaxed even further. On the other hand, our least squares method requires the construction of a family of pairs of trial and test spaces that satisfies the uniform inf-sup condition (9).

An example of a uniform inf-sup stable family of pairs of trial and test spaces is given by the following. Let $\mathcal{T}^\delta$ be a conforming (uniformly) shape regular triangulation of a *polygon* $\Omega \subset \mathbb{R}^2$. Let X^δ be the space of *piecewise constants* w.r.t. $\mathcal{T}^\delta$. For Y^δ we take the *Hsieh–Clough–Tocher* (HCT) macro-element finite element space w.r.t. $\mathcal{T}^\delta$, i.e.,

$$Y^\delta = \big\{v \in H_0^2(\Omega) \colon v|_K \in \{z \in C^1(K) \colon z|_{K_i} \in P_3(K_i),\ 1 \leqslant i \leqslant 3\}(K \in \mathcal{T}^\delta)\big\}, \tag{18}$$

where $\{K_i \colon 1 \leqslant i \leqslant 3\}$ is the subdivision of K obtained by connecting the centroid of K with the midpoints of its edges. A proof of the following result is given in [MS24].

Theorem 3.2 *The pair* $(X^\delta, Y^\delta)_{\delta \in I}$ *satisfies the uniform inf-sup condition* (9).

Also for higher order finite element trial spaces appropriate corresponding test finite element spaces in $H_0^2(\Omega)$ can be found. It is however fair to say that C^1-finite element spaces are not easy to implement. Therefore in the following section we investigate whether nonconforming test spaces can be applied instead.

4 Least squares with nonconforming test spaces

As before, let $(X^\delta)_{\delta\in I}$ be some (infinite) family of finite dimensional subspaces of $L_2(\Omega)$. Let $(Y^\delta)_{\delta\in I}$ be a family of (finite dimensional) Hilbert spaces, *not* necessarily included in $H_0^2(\Omega)$, with Hilbertian norms $\|\cdot\|_{Y^\delta}$. Let $B^\delta \in \mathcal{L}\big(L_2(\Omega), (H_0^2(\Omega)\times Y^\delta)'\big)$ be such that

$$(B^\delta\cdot)(v) = (B\cdot)(v) \quad (v \in H_0^2(\Omega)). \tag{19}$$

and

$$\inf_{\delta\in I} \inf_{\{z\in X^\delta:\, Bz\neq 0\}} \sup_{0\neq v\in Y^\delta} \frac{|(B^\delta z)(v)|}{\|Bz\|_{H^{-2}(\Omega)}\|v\|_{Y^\delta}} > 0. \tag{20}$$

For $(\ell^\delta, q) \in (Y^\delta)' \times L_2(\omega)$, where ℓ^δ approximates $\ell \in H^{-2}(\Omega)$, for $\varepsilon > 0$ we define

$$u_\varepsilon^\delta := \operatorname*{arg\,min}_{z\in X^\delta} \Big\{ \sup_{0\neq v\in Y^\delta} \frac{|(B^\delta z - \ell^\delta)(v)|^2}{\|v\|_{Y^\delta}^2} + \|Cz-q\|_{L_2(\omega)}^2 + \varepsilon^2\|z\|_{L_2(\Omega)}^2 \Big\}. \tag{21}$$

This u_ε^δ is the first component of the pair $(u_\varepsilon^\delta, v_\varepsilon^\delta) \in X^\delta \times Y^\delta$ that solves the *saddle point problem*

$$\begin{cases} \langle v_\varepsilon^\delta, \tilde v\rangle_{Y^\delta} + (B^\delta u_\varepsilon^\delta)(\tilde v) = \ell^\delta(\tilde v) & (\tilde v \in Y^\delta), \\ (B^\delta \tilde z)(v_\varepsilon^\delta) - \langle Cu_\varepsilon^\delta, C\tilde z\rangle_{L_2(\omega)} - \varepsilon^2\langle u_\varepsilon^\delta, \tilde z\rangle_{L_2(\Omega)} = -\langle q, C\tilde z\rangle_{L_2(\omega)} & (\tilde z \in X^\delta). \end{cases} \tag{22}$$

The following theorem generalizes Theorem 3.1 to possible nonconforming test spaces. A proof can be found in [MS24].

Theorem 4.1 *Assume* (19) *and* (20). *Let* $(\ell^\delta, q) \in (Y^\delta)' \times L_2(\omega)$ *and* $u \in L_2(\Omega)$. *Let* $E^\delta \in \mathcal{L}(Y^\delta, H_0^2(\Omega))$ *be such that*

$$\sup_{\delta\in I} \|E^\delta\|_{\mathcal{L}(Y^\delta, H_0^2(\Omega))} < \infty. \tag{23}$$

Then for $\varepsilon > 0$ *the solution* $u_\varepsilon^\delta \in X^\delta$ *of* (21) *satisfies*

$$\begin{aligned} &\|A(u-u_\varepsilon^\delta)\|_{H^{-2}(\Omega)\times L_2(\omega)} + \varepsilon\|u-u_\varepsilon^\delta\|_{L_2(\Omega)} \lesssim \mathcal{E}_{\text{data}} + \varepsilon\|u\|_{L_2(\Omega)} + \\ &\quad \min_{z\in X^\delta} \Big\{ \|u-z\|_{L_2(\Omega)} + \sup_{0\neq v\in Y^\delta} \frac{|(B^\delta z)(v - E^\delta v) + \ell(E^\delta v) - \ell^\delta(v)|}{\|v\|_{Y^\delta}} \Big\}. \end{aligned} \tag{24}$$

Consequently, if, in addition,

$$\sup_{0\neq v\in Y^\delta} \frac{|(B^\delta z)(v - E^\delta v) + \ell(E^\delta v) - \ell^\delta(v)|}{\|v\|_{Y^\delta}} \lesssim \|\ell - Bu\|_{H^{-2}(\Omega)} + \|u - z\|_{L_2(\Omega)} + \operatorname{osc}(\delta, \ell), \tag{25}$$

where $\operatorname{osc}(\delta,\ell)$ *is some* 'data-oscillation' *term, then*

$$\|A(u - u_\varepsilon^\delta)\|_{H^{-2}(\Omega)\times L_2(\omega)} + \varepsilon\|u - u_\varepsilon^\delta\|_{L_2(\Omega)} \lesssim \mathcal{E}_{\mathrm{data}} + \mathcal{E}_{\mathrm{appr}}(\delta) + \operatorname{osc}(\delta,\ell) + \varepsilon\|u\|_{L_2(\Omega)}. \tag{26}$$

For completeness we state the following analogue of Corollary 3.1.

Corollary 4.1 *In the setting of Theorem 4.1, let* $\varepsilon > 0$ *be such that*

$$\varepsilon\|u\|_{L_2(\Omega)} \eqsim \mathcal{E}_{\mathrm{data}} + \mathcal{E}_{\mathrm{appr}}(\delta) + \operatorname{osc}(\delta,\ell).$$

Then with $\alpha \in (0,1)$ *from Theorem 2.1 it holds that*

$$\|u - u_\varepsilon^\delta\|_{L_2(G)} \lesssim \big(\mathcal{E}_{\mathrm{data}} + \mathcal{E}_{\mathrm{appr}}(\delta) + \operatorname{osc}(\delta,\ell)\big)^\alpha \max\big(\mathcal{E}_{\mathrm{data}} + \mathcal{E}_{\mathrm{appr}}(\delta) + \operatorname{osc}(\delta,\ell), \|u\|_{L_2(\Omega)}\big)^{1-\alpha}.$$

As is shown in [MS24], an example of a suitable family of pairs of trial and nonconforming test spaces is given by the following. Let $\mathcal{T}^\delta$ be a conforming (uniformly) shape regular triangulation of a *polygon* $\Omega \subset \mathbb{R}^2$. Let X^δ be the space of *piecewise constants* w.r.t. $\mathcal{T}^\delta$. Let Y^δ be the *Morley finite element space* w.r.t. $\mathcal{T}^\delta$, being the space of piecewise quadratic polynomials that are continuous at the internal vertices of $\mathcal{T}^\delta$, and vanish at the boundary vertices of $\mathcal{T}^\delta$, and whose normal derivatives are continuous at the midpoints of the internal edges of $\mathcal{T}^\delta$, and vanish at the midpoints of boundary edges of $\mathcal{T}^\delta$. We equip Y^δ with the the Hilbertian norm $\|\cdot\|_{Y^\delta} := \sqrt{\sum_{K\in\mathcal{T}^\delta} |\cdot|^2_{H^2(T)}}$. With $(B^\delta z)(v) := -\sum_{K\in\mathcal{T}^\delta}\int_K z\Delta v\,dx$, the uniform inf-sup condition (20) is satisfied, and uniformly bounded $E^\delta \in \mathcal{L}(Y^\delta, H_0^2(\Omega))$ can be constructed such that for $\ell^\delta := \ell \in L_2(\Omega)$, for arbitrary but fixed $r \in \mathbb{N}$, (25) is valid for

$$\operatorname{osc}(\delta,\ell) := \sqrt{\sum_{T\in\mathcal{T}^\delta} \min_{\bar\ell\in\mathcal{P}_r(K)} h_K^4 \|\ell - \bar\ell\|^2_{L_2(K)}}.$$

Since $\operatorname{osc}(\delta,\ell)$ is of higher order than $\mathcal{E}_{\mathrm{appr}}(\delta)$, we conclude that under the condition that $\ell \in L_2(\Omega)$, the error bound with the nonconforming Morley test spaces is qualitatively the same as with the conforming HCT test spaces.

It can be envisaged that these results extend to n-dimensions for $n \geqslant 3$, cf. [WX06].

References

[ARRV09] G. Alessandrini, L. Rondi, E. Rosset, and S. Vessella. The stability for the Cauchy problem for elliptic equations. *Inverse Problems*, 25(12):123004, 47, 2009.

[BNO23] E. Burman, M. Nechita, and L. Oksanen. Optimal finite element approximation of unique continuation, 2023, 2311.07440.

[DMS23] W. Dahmen, H. Monsuur, and R.P. Stevenson. Least squares solvers for ill-posed PDEs that are conditionally stable. *ESAIM Math. Model. Numer. Anal.*, 57(4):2227–2255, 2023.

[DSW22] W. Dahmen, R. Stevenson, and J. Westerdiep. Accuracy controlled data assimilation for parabolic problems. *Math. Comp.*, 91(334):557–595, 2022.

[MS24] H. Monsuur and R.P. Stevenson. Ultra-weak least squares discretizations for Unique Continuation and Cauchy problems, 2024. In preparation.

[Wey40] H. Weyl. The method of orthogonal projection in potential theory. *Duke Math. J.*, 7:411–444, 1940.

[WX06] M. Wang and J. Xu. The Morley element for fourth order elliptic equations in any dimensions. *Numer. Math.*, 103(1):155–169, 2006.

GPSR Compliance

The European Union's (EU) General Product Safety Regulation (GPSR) is a set of rules that requires consumer products to be safe and our obligations to ensure this.

If you have any concerns about our products, you can contact us on ProductSafety@springernature.com

In case Publisher is established outside the EU, the EU authorized representative is:

Springer Nature Customer Service Center GmbH
Europaplatz 3
69115 Heidelberg, Germany

Batch number: 10732076

Printed by Printforce, the Netherlands